Vernon Bailey

North American fauna

No. 17, revision of American voles of the genus microtus by Vernon Bailey

Vernon Bailey

North American fauna
No. 17, revision of American voles of the genus microtus by Vernon Bailey

ISBN/EAN: 9783743323605

Manufactured in Europe, USA, Canada, Australia, Japa

Cover: Foto ©berggeist007 / pixelio.de

Manufactured and distributed by brebook publishing software
(www.brebook.com)

Vernon Bailey

North American fauna

COMMON VOLE OR MEADOW MOUSE (*Microtus pennsylvanicus*).

U. S. DEPARTMENT OF AGRICULTURE
DIVISION OF BIOLOGICAL SURVEY

NORTH AMERICAN FAUNA

No. 17

[Actual date of publication, June 6, 1900]

REVISION OF AMERICAN VOLES OF THE GENUS MICROTUS

BY

VERNON BAILEY
CHIEF FIELD NATURALIST

Prepared under the direction of

Dr. C. HART MERRIAM
CHIEF OF DIVISION OF BIOLOGICAL SURVEY

WASHINGTON
GOVERNMENT PRINTING OFFICE
1900

LETTER OF TRANSMITTAL.

U. S. DEPARTMENT OF AGRICULTURE,
Washington, D. C., March 10, 1900.

SIR: I have the honor to transmit for publication, as No. 17 of North American Fauna, 'A Revision of the American Voles of the Genus *Microtus*,' by Vernon Bailey, Chief Field Naturalist of the Biological Survey.

Respectfully,

C. HART MERRIAM,
Chief, Biological Survey.

Hon. JAMES WILSON,
Secretary of Agriculture.

CONTENTS.

ILLUSTRATIONS.

PLATES.

TEXT FIGURES.

No. 17. NORTH AMERICAN FAUNA. June, 1900.

REVISION OF AMERICAN VOLES OF THE GENUS MICROTUS.

By VERNON BAILEY.

INTRODUCTION.

The following synopsis of American voles is based on a study of between 5,000 and 6,000 specimens from more than 800 localities, including types or topotypes of every recognized species with a known type locality, and also types or topotypes of most of the species placed in synonymy. Voles, or meadow mice, occur throughout the greater part of the northern hemisphere north of the Tropics. In North America both species and individuals reach their maximum abundance in the Canadian and Transition zones, and from this broad belt the number of species decreases on both sides. On the north a few species occur in the Hudsonian and Arctic zones, and individuals are abundant even in the barren grounds, or 'tundras,' north to the arctic coast. South of the Transition zone the decrease in species and individuals is rapid. In the Upper Austral zone they are scarce; in the Lower Austral rare and exceedingly local; while in the Tropical only a single species, of very limited distribution, is known. To the south, as individuals decrease in abundance and species become restricted to distinct areas, the degree of specific and superspecific differentiation becomes more and more marked. Of the nine American subgenera, one (*Neofiber*) is confined to Florida, and two (*Orthriomys* and *Herpetomys*) are restricted to two isolated mountains in southern Mexico. Another (*Pitymys*) is mainly Austral, and is confined to the southeastern United States and a small area in southeastern Mexico. Three others (*Pedomys*, *Lagurus*, and *Chilotus*) are found mainly in the Transition zone, and reach but little north of the United States. The subgenus *Arvicola* belongs to mountains in the Hudsonian and Canadian zones; and the polymorphous subgenus *Microtus* is the only one that enters the arctic regions.

Voles adapt themselves to the most diverse conditions of environment. Many of the species inhabit moist or wet ground and several are mainly aquatic; others inhabit areas of excessive humidity, while a few live in dry and even arid regions. Some live in the perpetual shade of dense forests, others are exposed to the full effects of light on the open plains. Some of the most striking peculiarities of the different species result from these different conditions of environment. The development of oil and musk glands is most pronounced in the aquatic species of the subgenera *Neofiber* and *Arvicola* and least in the subgenera *Lagurus* and *Pedomys* of the dry regions. The color is palest in species most exposed to light and dryness, as in *curtatus* and *pallidus*, and darkest and richest in species from shaded and humid areas, as in *quasiater* and *umbrosus*.

The ranges of most of the species and subspecies conform to the limits of well-defined life zones, except in the subaquatic species, which follow water courses and often have the appearance of being out of their proper zones.

HABITS.

Certain peculiarities of habits are common to nearly all of the species. None are known to hibernate, but in the North they have snug winter homes under the snow, where they move about freely in numberless tunnels. They burrow in the ground, and are famous for their little roads or smooth trails which run through the grass from burrow to burrow or away to their feeding grounds. Bulky nests of grass and soft plant fibers are placed in underground cavities, or on the surface of the ground under cover of snow, logs, or dense vegetation. The nest is depressed globular in form, with an open chamber in the center, which contains a soft bed, and has one or two round entrances at the sides. These nests are the sleeping places of the old and the nurseries of the young. They are kept surprisingly clean and fresh, and new ones are frequently made to take the place of those that are old or imperfect.

Breeding.—Voles seem to have no definite breeding season. Four to eight young are usually produced at a birth, and as far north as Minnesota I have found them in the nests at all seasons of the year. Their increase is accordingly very rapid, and is only partially counterbalanced by the host of enemies that prey upon them. They form the principal food of nearly all owls and some hawks, while weasels, minks, foxes, coyotes, cats, badgers, skunks, and many other animals, as well as certain snakes, feed extensively on them. But in spite of their enemies they seem to hold their own, and tend to increase faster as the country becomes more thickly settled and the larger mammals and birds are destroyed.

Food.—Meadow mice choose a somewhat varied diet, but their food consists mainly of green vegetation, roots, and bark. Grass, especially the tender base of grass stems, forms the bulk of their food, but almost every plant with which they come in contact is eaten to some extent. Bark, both from roots and trunks of trees and shrubs, is a favorite winter food. Seeds and grain are eaten when found, but are not especially sought; flesh in any form is never refused. As the animals are active all winter and food is always abundant, they do not ordinarily lay up stores, although Mr. E. W. Nelson found *M. operarius*, of Alaska, storing roots.[1]

ECONOMIC STATUS.

Injury to trees and crops.—Though small enough to be commonly called mice (meadow mice, upland mice, field mice, pine mice, ground mice, bear mice, etc.), they make up in numbers what they lack in size, and over the whole breadth of the continent lay a heavy tribute on many products of the farm. Too small and too numerous to be successfully destroyed by traps, guns, or poison, they prove one of the most difficult enemies with which the farmer has to contend. If they would confine themselves to meadows, their mischief would be limited to the destruction of a comparatively small amount of grass; but they prefer growing grain to grass, and by running long tunnels under ground, or making little paths under cover of the vegetation, gain easy and safe access to the fields. With a stroke of their chisel-like teeth they fell the stalks of wheat and oats and eat the tender parts, together with some of the grain. It is so easy to cut down the stalks that they destroy many times as much as they need for food. The work of a few animals is insignificant, but the work of millions makes heavy inroads on growing crops. Later in the season, when the grain is cut and left standing in shocks or stacks, the field mice take possession, building their nests and establishing their homes under its cover. In shocks of corn and wheat left for a long time the grain is often completely devoured, and that remaining all winter in stacks suffers in proportion to the number of the little animals that make their homes in it. Even stacks of hay are often found in spring with the lower parts cut to chaff and filled with the nests of meadow mice.

When the snow comes these little rodents can safely leave their cover of weeds, grass, or bushes and plow their way under the snow on long exploring expeditions. The tunnels thus formed remain as open passages until the snow melts in spring, giving the animals free and safe conduct from the meadows to the uplands, into fields, orchards, gardens, and nurseries. There is no sign from above of what is happening below the surface; but later on, in spring, when the snow disappears, trees and shrubs are found stripped of their bark for a wide space near the ground. The marks of tiny teeth remain in the hard wood, and little piles of dry outer bark, mixed with characteristic pellets of excreta, show what animal has been at work. The uncovered roads may be seen leading from tree to tree, to winter nests on the surface of the ground, and back to the cover of brush or meadows. Shrubs and small trees are often stripped of their bark and killed, and sometimes even well-grown apple trees, 10 inches or a foot in diameter, are completely girdled. Usually, however, large trees are

[1] Proc. Biol. Soc. Wash., VIII, 140, 1893.

gnawed on only one side. In this case, although they are not killed at once, the wood, thus exposed usually decays in a few years, the trees become hollow at the base, their productiveness is impaired, and they die prematurely.

Protection of trees from voles.—Various means have been resorted to for protecting fruit trees and shrubs from these ravages, but with only partial success. Wire netting and tin cylinders placed around the bases of the trees in autumn are generally considered the surest protection, but in most cases this is too expensive to be practicable. Wrapping the trunks with burlap or twisted ropes of straw, or coating them with whitewash, tar, or other unpalatable substances, are common methods of protection used with varying degrees of success. But as some species of voles eat the bark from the roots below the surface of the ground, none of these resources insure perfect protection.

Destruction of voles.—The importance of placing every possible check on the increase of these animals and of reducing their numbers when they become too numerous is obvious. No direct method of accomplishing these ends has as yet been devised, but the desired result can be attained indirectly by avoiding or preventing the useless destruction of their natural enemies. Owls and some species of hawks live almost exclusively upon them, watching for them night and day in the grass, and are always ready to pounce on any that appear above the snow. Weasels run through their burrows and trails, and not only kill enough for food, but destroy great numbers for the mere pleasure of killing. In spite of these well-known and often reiterated facts, bounties are still paid for the destruction of hawks and owls in counties where the annual loss in fruit trees and grain from the ravages of field mice if computed would amount to a startling sum. In the spring of 1895 I examined a small apple orchard in Washtenaw County, Mich., in which several choice trees had been killed and many others injured during the preceding winter by the common vole (*Microtus pennsylvanicus*). The owner of the orchard considered $50 a low estimate of the damage done. At the same time the county of Washtenaw was taxing the farmers to pay a bounty of 25 cents each on all hawks and owls, while the several gun clubs of the county gave these birds a high count in their competitive hunting matches. Many similar instances could be cited. Who was ever known to miss an opportunity to destroy a hawk or weasel? The diminution of foxes, minks, coyotes, and such predatory mammals may be necessary, but if so, the protection and encouragement of other less harmful species becomes doubly important, and in fact imperative, if we are to escape such devastating hordes of voles as have occasionally swept over certain parts of Europe, particularly in Scotland,[1] Germany,[2] Italy,[2] Russia,[2] and Thessaly.[1]

[1] Parliamentary Report of Plague of Field Voles in Scotland, London, 1893.
[2] U. S. Consular Reports, L, No. 187, 539–543, 1896.

It is not many years since certain prominent writers treated as mere varieties, or subspecies, animals that belong to widely different subgenera, while others described and named with full specific rank every different condition of pelage in a single species. In some cases the original type was not preserved, or no type was designated by the describer, or still worse, the type locality was not given, so that subsequent writers renamed these same species or confounded them with others. The resulting confusion can now be cleared up by means of series of specimens collected within the past ten years at most of the known type localities, and in the general region of those not definitely known. The series of specimens available, and the number of localities represented, make it possible to define almost every North American species from typical specimens, and in most cases to give the various changes of pelage due to season and age. When possible, the original types have been compared with the new series of specimens from the type localities, and in this way the names *californicus*, *trowbridgi*, *edax*, *occidentalis*, *townsendi*, *longirostris*, and *modestus* have been sifted out with the following result: *californicus* stands for a widely distributed western species with *trowbridgi* as a synonym; *edax* as a well-marked species, but one in which the name has been persistently misapplied; *occidentalis* as a synonym of *townsendi*; *longirostris* as a synonym of *montanus*; and *modestus* as a western form of *pennsylvanicus*. The type of *montanus* is lost, but a series of 57 specimens from the type locality agrees with Peale's description of the species. The types of *modestus* and *edax* are immature specimens made up with the skulls inside the skins. It was only by the removal of the skulls that even the group to which the species belonged could be determined.[1]

MATERIAL EXAMINED.

The following synopsis of the genus *Microtus* is based mainly on a study of specimens in the collection of the Biological Survey and that of Dr. C. Hart Merriam, both of which are in the United States National Museum. For the use of much additional material, including types and topotypes, my thanks are heartily extended to Dr. F. W. True, executive curator, and Mr. Gerrit S. Miller, jr., assistant curator of mammals, United States National Museum; to Dr. J. A. Allen, curator of mammals and birds, and Mr. Frank M. Chapman, assistant curator, American Museum of Natural History; to Mr. D. G. Elliot, curator of the Department of Zoology, Field Columbian Museum; and to Mr. Outram Bangs. Most of all, I am indebted to Dr. C. Hart Merriam,

[1] Through the kindness of Dr. True and Mr. Miller, skulls have also been removed from a large number of specimens from Alaska and Arctic America, so that it has been possible for the first time to identify the species and make use of the localities in determining their ranges.

who, after doing much work on the genus, has placed his manuscript, drawings, and large private collection of specimens at my disposal, besides giving me constant criticism and advice. Among others who have contributed material or notes my thanks are especially due to Mr. E. W. Nelson, who has collected all the known Mexican species of *Microtus* and has contributed the notes on their zonal distribution.

Seventy species and subspecies are here recognized. Of these 54 actual types and series of topotypes of 13 additional forms have been examined, while of the three remaining forms, for which no type exists or is accessible and no definite type locality is known, specimens have been examined from the type region, or as near to it as can be determined. Three forms, *Microtus californicus constrictus*, *M. ludovicianus* and *M. scirpensis* are described as new. Except for a relatively small number of alcoholics and a few skeletons, the specimens are mostly well-prepared skins with cleaned skulls and are accompanied by collectors' measurements.

All measurements are in millimeters, and external measurements, unless otherwise stated, are taken in the flesh by collectors. Skull measurements are my own, made from perfect skulls unless otherwise stated. The skull drawings are by Dr. James C. McConnell. Most of the drawings of teeth have been used in previous publications of the Biological Survey.

Subfamily MICROTINÆ Cope.[1]

The subfamily *Microtinæ* includes the Voles of the genera *Microtus*, *Evotomys*, and *Phenacomys;* the Lemmings of the genera *Lemmus*, *Dicrostonyx*, and *Synaptomys;* and the Muskrats of the genus *Fiber*. As the genera and subgenera of the family have been recently treated in detail by Mr. Gerrit S. Miller, jr.,[2] it is only necessary to give briefly the characters distinguishing the genus *Microtus*.

Genus MICROTUS Schrank.

Generic characters.—Lower incisors with roots extending far behind and on outer side of molar series; upper incisors not grooved; molars rootless, with outer and inner reentrant angles approximately equal. Palate with median ridge, distinct lateral pits, complete lateral bridges[3] (not terminating in posterior shelf in any American species). Tail as long as or longer than hind foot, terete; claw of thumb pointed, not strap-shaped.

SUBGENERA.

Nine subgenera are here recognized among the living species of North America.[4] Five of these (*Chilotus*, *Pedomys*, *Herpetomys*, *Orthriomys*, and *Neofiber*) are found only in North America. The remaining

[1] *Microtidæ* Cope, Syllabus Lectures Geol. and Paleont., p. 90, 1891. *Microtinæ* Rhoads, Am. Nat., XXIX, 940, Oct., 1895.

[2] North American Fauna No. 12, Genera and Subgenera of Voles and Lemmings, 1896.

[3] Usually incomplete in *Neofiber*.

[4] The extinct species of *Microtus* are not included in the present paper.

four (*Microtus*, *Pitymys*, *Arvicola*, and *Lagurus*) include also Old World species. All of the nine subgenera, save *Microtus*, are sharply defined and easily distinguished by either cranial or external characters. The subgenus *Microtus* contains many more species than all of the other subgenera together, and species differing so widely that only the most general characterization can be applied to it. It is a composite group containing all forms that do not fit into the other more restricted subgenera and yet are not sufficiently differentiated to merit subgeneric rank.

KEY TO SUBGENERA OF MICROTUS.

m3 with 3 transverse loops and no closed triangles. Plantar tubercles 5 or 6.
  m3 with 3 closed triangles,[1] mammæ 8.[2]
    Plantar tubercles 6, side glands on hips in adult males (on flanks in *xanthognathus*) .... *Microtus*
    Plantar tubercles 5, side glands on flanks or else inconspicuous.
      Side glands conspicuous on flanks of adult males, size large .... *Arvicola*
      Side glands obscure or wanting, size small .... *Chilotus*
  m3 with 2 closed triangles, mammæ 4 or 6.
    Skull wide and flat, tail very short, fur short and dense, mammæ 4 .. *Pitymys*
    Skull high and narrow, tail medium, fur coarse, mammæ 6 .... *Pedomys*
m3 with 2 transverse loops and 2 median triangles, plantar tubercles 5.
  m1 with 5 closed triangles.
    Side glands conspicuous in both sexes, mammæ 6, size very large, tail long .... *Neofiber*
    Side glands obscure, mammæ 8, size small, tail very short .... *Lagurus*
  m1 with 3 closed triangles.
    m3 with 3 closed triangles, mammæ 6, tail short .... *Herpetomys*
    m3 with 2 closed triangles, mammæ 4, tail long .... *Orthriomys*

LIST OF SPECIES AND SUBSPECIES, WITH TYPE LOCALITIES.

Microtus abbreviatus Miller. Hall Island, Bering Sea, Alaska.
  acadicus Bangs. Digby, Nova Scotia.
  alleni (True). Georgiana, Brevard County, Florida.
  alticolus (Merriam). Little Spring, San Francisco Mountain, Arizona, 8,200 feet.
  angusticeps Bailey. Crescent City, California.
  arizonensis Bailey. Springerville, Arizona.
  arvicoloides (Rhoads). Lake Keechelus, Washington, 8,000 feet.
  auricularis Bailey. Washington, Mississippi.
  austerus (Le Conte). Racine, Wisconsin.
  aztecus (Allen). Aztec, Rio Arriba County, New Mexico, 5,900 feet.
  bairdi Merriam. Crater Lake (Glacier Peak), Oregon, 7,800 feet.
  breweri (Baird). Muskeget Island, Massachusetts.
  californicus (Peale). San Francisco Bay, California.
  canescens Bailey. Conconully, Washington.
  canicaudus Miller. McCoy, Oregon.
  chrotorrhinus (Miller). Mount Washington, head of Tuckerman Ravine, New Hampshire, 5,300 feet.

---

[1] Except *Microtus breweri*, in which 2 are usually confluent, and *chrotorrhinus*, which has 5 closed triangles.

[2] Except in the *Microtus mexicanus* group, in which the number is 4.

Microtus constrictus Bailey. Cape Mendocino, California.
curtatus (Cope). Pigeon Spring, Mount Magruder, Nevada.
drummondi (Aud. & Bach.). Rocky Mountains, vicinity of Jasper House, Alberta, Canada.
dutcheri Bailey. Big Cottonwood Meadows, near Mount Whitney, California, 10,000 feet.
edax (Le Conte). California (south of San Francisco).
enixus Bangs. Hamilton Inlet, Labrador.
fisheri Merriam. St. Matthew Island, Bering Sea, Alaska.
fontigenus Bangs. Lake Edward, Quebec.
fulviventer Merriam. Cerro San Felipe, Oaxaca, Mexico.
guatemalensis Merriam. Todos Santos, Huehuechenango, Guatemala, 10,000 feet.
haydeni (Baird). Fort Pierre, South Dakota.
innuitus Merriam. St. Lawrence Island, Bering Sea, Alaska.
kadiacensis Merriam. Kadiak Island, Alaska.
labradorius Bailey. Fort Chimo, Ungava, Labrador.
leucophæus (Allen). Graham Mountains, Arizona.
longicaudus (Merriam). Custer, South Dakota.
ludovicianus Bailey. Iowa, Calcasieu Parish. Louisiana.
macfarlani Merriam. Fort Anderson (north of Great Bear Lake), Arctic America.
macropus (Merriam). Pahsimeroi Mountains, Idaho, 9,700 feet.
macrurus Merriam. Lake Cushman, Olympic Mountains, Washington.
mexicanus (De Saussure). Mount Orizaba, Mexico.
minor (Merriam). Bottineau, North Dakota.
modestus (Baird). Sawatch Pass (Cochetopa Pass). Colorado.
mogollonensis (Mearns). Baker Butte, Mogollon Mountains, Arizona.
montanus (Peale). Headwaters of Sacramento River, near Mount Shasta, California.
mordax (Merriam). Sawtooth (or Alturas) Lake, Idaho, 7,200 feet.
nanus (Merriam). Pahsimeroi Mountains, Idaho.
nemoralis Bailey. Stillwell (Boston Mountains), Indian Territory.
nesophilus Bailey. Great Gull Island, New York.
nevadensis Bailey. Ash Meadows, Nye County, Nevada.
nigrans Rhoads. Currituck, North Carolina.
operarius (Nelson). St. Michael, Alaska.
oregoni (Bachman). Astoria, Oregon.
pallidus (Merriam). Fort Buford, North Dakota.
pauperrimus (Cooper). Plains of Columbia, near Snake River, Washington.
pennsylvanicus (Ord). Pennsylvania (near Philadelphia).
phæus Merriam. North slope Sierra Nevada de Colima, Jalisco, Mexico, 10,000 feet.
pinetorum (Le Conte). Pine forests of Georgia (probably near the old Le Conte plantation at Riceboro, Georgia.)
popofensis Merriam. Popof Island, Shumagin Islands, Alaska.
quasiater (Coues). Jalapa, Vera Cruz, Mexico.
rivularis Bailey. St. George, Utah.
richardsoni (De Kay). Near foot of Rocky Mountains, vicinity of Jasper House, Alberta, Canada.
scalopsoides (Aud. & Bach.). Long Island, New York.
scirpensis Bailey. Amargosa River, California, near California-Nevada line.
serpens Merriam. Agassiz, British Columbia.
sitkensis Merriam. Sitka, Alaska.

Microtus terrænovæ (Bangs). Codroy, Newfoundland.
tetramerus (Rhoads). Beacon Hill Park, Victoria, British Columbia.
townsendi (Bachman). On or near Wappatoo (Sauvie) Island, Willamette River, Oregon.
umbrosus Merriam. Mount Zempoaltepec, Oaxaca, Mexico, 8,200 feet.
unalascensis Merriam. Unalaska Island, Alaska.
vallicola Bailey. Lone Pine, Inyo County, California.
yakutatensis Merriam. Yakutat Bay, Alaska.
xanthognathus (Leach). Hudson Bay.

Subgenus MICROTUS Schrank.

Type.—*Microtus terrestris* Schrank (=*Mus arvalis* Pallas).
Microtus Schrank, Fauna Boica, I, 1ste Abth., 72, 1798.
Microtus Miller, N. Am. Fauna No. 12, 63, July 23, 1896 (subgenus).

Geographic distribution (in North America).—From the Arctic Ocean southward to southern Mexico, and across the continent, mainly in Boreal, Transition, and Upper Austral zones.

Subgeneric characters.—Plantar tubercles 6; lateral glands on hips in adult males;[1] mammæ normally 8, 4 inguinal and 4 pectoral;[2] ears usually overtopping fur; m1 normally with 5 closed triangles;[3] m3 with 3 transverse loops and no triangles; m2 with 4 closed sections, and in most eastern species an additional posterior inner loop; m3 with 3 closed triangles (except in *chrotorrhinus* and *abbreviatus* groups).

GROUPS IN THE SUBGENUS MICROTUS.

The subgenus *Microtus* is readily divided into 10 fairly well-marked groups of slightly superspecific rank that may be conveniently designated by the name of their best-known or most characteristic species. These groups are not of great importance or of equal rank, but for showing the relationship of species and for convenience in arrangement they serve a useful purpose.

1. **Pennsylvanicus Group,** characterized by a posterior fifth loop to middle upper molar, includes *pennsylvanicus, nigrans, acadicus, modestus, fontigenus, labradorius, enixus, aztecus, drummondi, terrænovæ, nesophilus,* and *breweri.*
2. **Montanus Group,** characterized by moderately short tail and constricted incisive foramina, includes *montanus, arizonensis, nanus, canescens, canicaudus, nevadensis, rivularis,* and *dutcheri.*
3. **Townsendi Group,** characterized by large size, long tail, and dark-brown color, includes *townsendi* and *tetramerus.*
4. **Californicus Group,** characterized by large size and wide-open incisive foramina, includes *californicus, constrictus, vallicola, edax,* and *scirpensis.*
5. **Longicaudus Group,** characterized by long tail and gray color, includes *longicaudus, mordax, macrurus, angusticeps, alticolus,* and *leucophæus.*
6. **Mexicanus Group,** characterized by short tail, brown color, and only 4 mammæ, includes *mexicanus, phæus, fulviventer,* and *mogollonensis.*

---

[1] In front of hips in *xanthognathus* and probably in *chrotorrhinus.*
[2] Four in the *mexicanus* group, a pair of inguinal and a pair of pectoral.
[3] With only four closed triangles in most of the Alaska species.

7. **Operarius Group**, characterized by short tail and only 4 closed triangles in anterior lower molar, includes *operarius, macfarlani, kadiacensis, unalascensis, sitkensis, yakutatensis, popofensis*, and *innuitus*.
8. **Abbreviatus Group**, characterized by robust form, very short tail, 5 closed triangles in anterior lower molar, and two closed and one open in posterior upper, includes *abbreviatus* and *fisheri*.
9. **Chrotorrhinus Group**, characterized by yellow nose and five closed triangles in posterior upper molar, includes *chrotorrhinus* and *rarus*.
10. **Xanthognathus Group**, characterized by yellow nose, large size, glands on flanks, and 3 closed triangles in posterior upper molar, includes one species, *xanthognathus*.

In using the following key it will be necessary to have both skins and skulls in hand, and even then it will be impossible to identify some of the forms without actual comparison with their nearest allies. Whenever possible, several specimens should be examined, to avoid the danger of being led astray by abnormal molar patterns, for even the widest ranges of subgeneric differences are sometimes covered by individual variation or abnormal tooth pattern.

KEY TO SPECIES AND SUBSPECIES OF THE SUBGENUS MICROTUS.

m2 with 4 closed angular sections and a rounded posterior loop.
 m3 with two of the 3 triangles usually confluent.
  Interparietal about as wide as long, colors pale........................*breweri*
  Interparietal much wider than long, colors dark........................*nesophilus*
 m3 with 3 closed triangles.
  m1 with usually a sharp point or spur at base of posterior triangle; belly white with a median dusky line........................*terraenovæ*
  m1 with normal truncate posterior triangle; belly without median dusky line.
   Interparietal more than half as long as wide, belly white.
    Skull long and narrow, braincase long; feet and tail stout........................*aztecus*
   Interparietal about half as long as wide, belly usually dull colored.
    Skull wide, braincase short, molars small........................*enixus*
    Skull not wide, braincase medium, molars medium.
     Colors dusky or blackish.
      Size large, hind foot 23........................*nigrans*
      Size small, hind foot 21........................*fontigenus*
     Colors brownish or dark grayish.
      Size medium.
       Belly white or whitish........................*acadicus*
       Belly dull.
        Colors bright or dark brownish........................*pennsylvanicus*
        Colors paler, size less........................*modestus*
      Size small, feet and tail very slender.
       Skull low, incisors projecting, bullæ not large........................*labradorius*
       Skull high, incisors decurved, bullæ large........................*drummondi*
m2 with 4 closed sections and no posterior loop (except irregularly in Californicus group).
 Mammæ 4, inguinal, 1-1; pectoral, 1-1. Skull short and wide. Incisive foramina not constricted.
  Colors bright rich brown above and below........................*fulviventer*
  Colors dull brownish above and below.
   Belly but little lighter than back.
    Size medium........................*mexicanus*
    Size slightly larger........................*phæus*
   Belly much lighter than back; size small........................*mogollonensis*

Mammæ 8, inguinal, 2-2; pectoral, 2-2.
m1 with normally 4 closed triangles (sometimes 5 in *sitkensis*) and rounded anterior loop.
Bullæ very small and narrow, molars very light.
Skull narrow and slender .... *operarius*
Skull wider and heavier .... *kadiacensis*
Bullæ medium, molars moderately heavy.
Incisors strongly projecting.
Size large, hind foot 23 .... *innuitus*
Size small, hind foot 19 .... *macfarlani*
Incisors not strongly projecting, size medium.
Frontals heavily ridged in adult males.
Prezygomatic notch deep, color dusky gray or ochraceous.
Color dark ochraceous, belly dusky .... *sitkensis*
Color dusky gray, belly buffy gray .... *yakutatensis*
Prezygomatic notch shallow, color ochraceous .... *unalascensis*
Frontals not ridged in adults, color ochraceous .... *popofensis*
m1 with 5 or 6 closed triangles.
A pair of glands on flanks of males, nose yellowish.
Size large, side glands conspicuous in adult males, m3 with 3 closed triangles .... *xanthognathus*
Size smaller, glands obscure or wanting, m3 with 5 closed triangles.
Color bister .... *chrotorrhinus*
Color grayish .... *ravus*
A pair of glands on hips of males, nose not yellow.
Incisive foramina not constricted posteriorly, m2 with or without posterior loop.
Size large, colors dark, young blackish.
Nasals emarginate posteriorly .... *edax*
Nasals truncate posteriorly .... *scirpensis*
Size smaller, colors grayish, young dusky .... *californicus*
Colors clearer gray, bullæ smaller.
Skull wide .... *vallicola*
Skull narrow .... *constrictus*
Incisive foramina constricted posteriorly, m2 normally without posterior loop.
Tail very short, size medium.
Belly dusky, lips and tip of nose white .... *dutcheri*
Belly, lips, and nose buffy.
Rostrum and nasals slender .... *abbreviatus*
Rostrum and nasals heavy .... *fisheri*
Tail medium, size large or small.
m1 with 6 closed triangles and deep-lobed trefoil.
Size large, hind foot 24 .... *nevadensis*
m1 with 5 closed triangles and anterior trefoil.
Size large, hind foot 23 .... *rivularis*
Size medium or small, hind foot 18-22, belly gray or whitish, ears large.
Hind foot 20 or more.
Color dark gray above .... *montanus*
Color rusty gray above .... *arizonensis*
Hind foot 20 or less.
Lateral pits of palate deep, tail bicolor.
Color grizzled gray .... *nanus*
Color ashy gray .... *canescens*
Lateral pits of palate shallow, tail mostly gray .... *canicaudus*

KEY TO SPECIES AND SUBSPECIES OF THE SUBGENUS MICROTUS—Cont'd.

Tail long, about ⅓ of total length.
  Hip glands conspicuous in males, colors dark brown.
    Hind foot averaging 25.4 .... *townsendi*
    Hind foot averaging 22 .... *tetramerus*
  Hip glands not conspicuous, colors grayish, belly whitish.
    Size large, hind foot 24 .... *macrurus*
    Size medium, hind foot 22.
      Anterior arm of frontal acuminate.
        Skull narrow, bullæ small .... *angusticeps*
        Skull normal, bullæ large.
          Sides much grayer than back .... *mordax*
          Sides scarcely grayer than back .... *longicaudus*
      Anterior arm of frontal obliquely truncate.
        Size small, foot 20 .... *alticolus*
        Size larger, foot 22 .... *leucophæus*

MICROTUS PENNSYLVANICUS (Ord). Meadow Vole.

Mus pennsylvanica Ord, Guthrie's Geography, 2d American edition, II, 292, 1815. (Rhoads' reprint.) Based on Wilson's description of the meadow mouse from meadows below Philadelphia and along the seashore.

Mynomes pratensis Rafinesque, Am. Monthly Mag., II, 45, 1817. Based on Wilson's description of meadow mouse.

Lemmus noveboracensis Rafinesque, Annals of Nature, 3, 1820. (New York and New Jersey.)

Arvicola riparius Ord, Journ. Acad. Nat. Sci. Philadelphia, IV, Pt. II, 305–306, 1825. (Type locality not given.)

Arvicola palustris Harlan, Fauna Americana, 136–138, 1825. (Swamp along the shores of the Delaware.)

Arvicola hirsutus Emmons, Rept. Quad. Mass., 60, 1840.

Arvicola alborufescens Emmons, Rept. Quad. Mass., 60–61, 1840. (Williamstown, Mass.)

Arvicola fulva Aud. and Bach., Proc. Acad. Nat. Sci. Phila., I, 96, 1841. ("One of the Western States; we believe Illinois.")

Arvicola nasuta Aud. and Bach., Proc. Acad. Nat. Sci. Phila., I, 96–97, 1841. (Near Boston, Mass.)

Arvicola rufescens DeKay, Zool. N. Y., Mammals, I, 85, 1842. (Oneida Lake, N. Y.)

Arvicola oneida DeKay, Zool. N. Y., Mammals, I, 88–89, 1842. (Oneida Lake, N. Y.)

Arvicola dekayi Aud. and Bach., Quad. N. Am., III, 287–288, 1854. (New York or Illinois.)

Arvicola riparia var. *longipilis* Baird, Mammals N. Am., 524, 1857. (West Northfield, Ill., and Racine, Wis.)

Arvicola rufidorsum Baird, Mammals N. Am., 526, 1857.[1] (Holmes Hole, Marthas Vineyard, Mass.)

Type locality.—Pennsylvania (meadows below Philadelphia).

Geographic distribution.—Eastern United States and westward as far as Dakota and Nebraska, shading into *modestus* of the western plains and Rocky Mountains. In a general way it occupies the Transition zone from the Atlantic coast to the edge of the Great Plains.

Habitat.—Meadows, fields, and especially grassy places near water.

[1] Not having seen the type of *rufidorsum* or any specimen from Marthas Vineyard, I hesitate to place this name in synonomy.

General characters.—Size medium; tail at least twice as long as hind foot; fur long, overlaid with coarse hairs; ears moderate, conspicuous above fur in summer, almost concealed in winter pelage; colors dusky gray or brownish; skull long, well arched, and rather smooth; middle upper molar with four triangles and a posterior loop.

Color.—*Summer pelage:* Upperparts dull chestnut brown, varying to bright yellowish chestnut, darkened along the back with coarse black hairs; belly dusky gray or tinged with cinnamon; feet brownish; tail dusky above, slightly paler below. *Winter pelage:* Duller and grayer throughout; tail indistinctly bicolor. *Young:* Blackish.

Cranial characters.—Skull long, usually not angular or much ridged; incisors projecting well in front of nasals; incisive foramina long, occupying two-thirds of the space between molars and incisors; bullæ moderately large and well rounded; molar series long; m2 with 4 closed triangles and a posterior loop; m3 with an anterior crescent, 3 closed triangles, and a posterior loop with two inner lobes; m1 with 5 closed triangles, anterior trefoil, 4 outer and 5 inner salient angles; m3 with 3 long inner and 3 short outer salient angles.

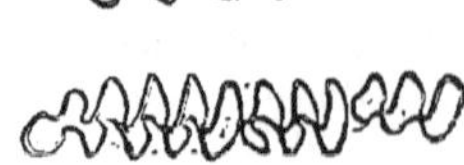

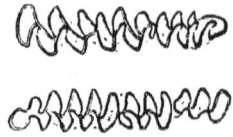

FIG. 1.—Molar enamel pattern of *Microtus pennsylvanicus* (× 5).

Measurements.—Average of 5 adults from Washington, D. C.: Total length, 171; tail vertebræ, 46; hind foot, 21.2. *Skull* (No. 30321, ♀ ad., from Washington, D. C.): Basal length, 27.4; nasals, 8.3; zygomatic breadth, 17.2; mastoid breadth, 12.7; alveolar length of upper molar series, 7.2.

Remarks.—The above description is based on a good series of specimens from the District of Columbia, showing the seasonal changes of pelage and agreeing perfectly with the Pennsylvania animal. From Pennsylvania south along the Atlantic coast, specimens show a noticeable increase in size and intensity of coloration, which reaches its maximum in the subspecies *nigrans* of North Carolina; while to the north they show a corresponding decrease in size and intensity of coloration, which reaches its extreme in the subspecies *acadicus* of Nova Scotia. To the westward *pennsylvanicus* is fairly typical as far as southern Michigan and Iowa, but on the plains of Nebraska and South Dakota it grows paler as it grades into *modestus*. Northward in Minnesota it becomes smaller until scarcely distinguishable from and perhaps grading into *drummondi* of northwest Canada. Three skulls in the U. S. National Museum, that seem to be typical *pennsylvanicus*, are labeled as coming from Prairie Mer Rouge, La., but I am inclined to question the authenticity of the labeling.

Specimens examined.—Total number, 716, from the following localities:

Pennsylvania: Philadelphia, 1; Chester County, 1; Carlisle, 6; Columbia, 2; Drurys Run (near Renovo), 6; Foxbury, 2; Pine Glen, 1; Leasuresville, 2; Bear Lake (Warren County), 2.

New York: Owego, 2; Nichols, 24; Lake George, 20; Alder Creek, 2; Locust Grove, 55; Peterboro, 9; Troy, 5; Geneva, 5; Brandon, 4; Catskill Mountains, 3; Highland Falls, 12; Mott Haven, 1; Oyster Bay, 2; Lake Grove, 2; Montauk Point, 45; Shelter Island, 6; Plum Island, 14; Roslyn, 4.
Connecticut: East Hartford, 2.
Massachusetts: Wilmington, 13; Middleboro, 19; Newtonville, 4; Holmes Hole, 1; Woods Holl, 1.
Vermont: Burlington, 4.
New Hampshire: Ossipee, 15.
Maine: Addison, 1; Calais, 1.
New Jersey: Tuckerton, 4; Mays Landing, 1; Sea Island City, 1.
Maryland: Laurel, 23; Hyattsville, 7; Bladensburg, 1; Mountain Lake Park, 2; Finzel, 1; Grantsville, 1.
District of Columbia: Washington, 64.
Virginia: Falls Church, 2; Dunn Loring, 2; Arlington, 2; Fort Myer, 1; Bristoe, 1.
West Virginia: Travellers Repose, 2; White Sulphur Springs, 3.
North Carolina: Roan Mountain, 45; Old Richmond, 3; Raleigh, 20.
Ohio: Garrettsville, 10; Salem, 1.
Michigan: Detroit River, 1; Manchester, 3; Ann Arbor, 2.
Illinois: West Northfield, 6.
Wisconsin: Racine, 14; Busseyville, 1; Milwaukee, 4; Saxeville, 1; Fisher Lake (Iron County), 14.
Iowa: Knoxville, 2.
Missouri: St. Louis, 5.
Louisiana: Prairie Mer Rouge, 3.
Nebraska: Blair, 3; Valentine, 2.
Minnesota: Elk River, 112; Fort Snelling, 3; Heron Lake, 1; Ortonville, 1; Tower, 6.
South Dakota: Vermilion, 2; Pierre, 2; Travere, 1; Flandreau, 4; Fort Sisseton, 18; Fort Wadsworth, 2.
Ontario: Toronto, 1; Lorne Park, 6.

MICROTUS PENNSYLVANICUS NIGRANS Rhoads. Albemarle Meadow Vole.

Microtus pennsylvanicus nigrans Rhoads, Proc. Acad. Nat. Sci. Phila., 1897, 307–308.

Type locality.—Currituck, N. C.

Geographic distribution.—(Typical form.) Coast region of northern North Carolina and southern Virginia, in the Austroriparian zone.

Habitat.—Marshes and localities close to water.

General characters.—Slightly larger than *pennsylvanicus*, with noticeably larger hind feet and darker coloration.

Color.—*Summer pelage:* Upperparts dull bister, much obscured by black hairs; belly smoky gray to dull cinnamon; tail black above, sooty below; feet blackish. *Winter pelage* (partly retained in April specimens): Darker, with dorsal area almost black. *Young* (to nearly half grown): Sooty black all over.

Cranial characters.—Skull averaging slightly larger than in typical *pennsylvanicus;* rostrum slightly heavier, incisive foramina wider; dentition the same.

Measurements.—Average of three not fully adult males from type locality: Total length, 165; tail vertebræ, 48; hind foot, 23. *Skull* (No. 72374, ♂ ad., from Eastville, Va.): Basal length, 29; nasals, 8.5; zygomatic breadth, 17.5; mastoid breadth. 13.4: alveolar length of upper molar series, 7.3.

Remarks.—Specimens of *Microtus* from a chain of localities along the Atlantic coast, from North Carolina to Nova Scotia, show a decrease in size and intensity of coloration from the south northward. Unfortunately the type of *pennsylvanicus* was taken from an intermediate locality, and it becomes necessary to recognize the extremes—*acadicus* and *nigrans*—as slightly differentiated forms.

Specimens examined.—Total number of typical specimens, 16, from the following localities:

North Carolina: Currituck, 6.
Virginia: Wallaceton, 7; Eastville, 1; Smiths Island, 2.

MICROTUS PENNSYLVANICUS ACADICUS Bangs. Acadian Vole.

Microtus pennsylvanicus acadicus Bangs, Am. Nat., XXXI, 239-240, March, 1897.

Type locality.—Digby, Nova Scotia.

Geographic distribution.—Nova Scotia and Prince Edward Island.

Habitat.—Fields and fresh-water marshes.

General characters.—Slightly smaller than *M. pennsylvanicus;* color slightly paler, both in summer and winter.

Color.—*Summer pelage* (July to October): Upperparts yellowish bister, slightly lined with black hairs; belly washed with white or smoky gray; tail indistinctly bicolor, brownish black above, slightly paler below; feet dusky plumbeous. *Winter pelage:* Back buffy gray; sides paler; ears nearly concealed under bright ochraceous patch; belly washed with pure white; tail sharply bicolor, blackish above, white below: feet plumbeous. *Young:* Not so dark as those of *pennsylvanicus.*

Cranial characters.—Skull usually distinguishable from that of *pennsylvanicus* by projecting posterior point of palate; posterior tip of nasals slightly emarginate or truncate, never rounded. Dentition as in *pennsylvanicus.*

Measurements.—Type: Total length, 172; tail vertebræ, 49; hind foot, 20. Average of 5 topotypes: 170; 47; 21. *Skull* (No. 2145, ♂—not fully adult): Basal length, 25.5; nasals, 7.7; zygomatic breadth, 14.8; mastoid breadth, 12; alveolar length of upper molar series, 6.5.

Remarks.—None of the 19 topotypes before me are old, and the majority are not fully adult, but in a series of 40 specimens from Prince Edward Island, including fully adult specimens in both summer and winter pelages, the principal characters of *M. acadicus* are accentuated. The winter pelage is rather more distinctive than the summer.

Specimens examined.—Total number, 67, from the following localities:

Nova Scotia: Digby, 19; Halifax, 1.
Prince Edward Island: 47.

MICROTUS PENNSYLVANICUS MODESTUS (Baird). Sawatch Vole.

Arvicola modesta Baird, Mamm. N. Am., 535–536, 1857.
Arvicola insperatus Allen, Bul. Am. Mus. Nat. Hist., 347, 1894 (Custer, S. Dak.).

Type locality.—"Sawatch Pass, Rocky Mountains" [same as Cochetopa Pass], Colorado.

Geographic distribution.—Rocky Mountains and western Plains from New Mexico to British Columbia, and from the Black Hills of South Dakota to central Idaho, and beyond, with slight variation, to the plains of the Columbia, mainly in Transition zone.

Habitat.—Marshes and damp grassy places.

General characters.—Size of *M. pennsylvanicus*, tail slightly shorter, color paler, more yellowish, never chestnut in summer pelage; skull heavier, becoming more ridged and angular with age.

Color.—*Summer pelage:* Upperparts dull ochraceous, darkened with black-tipped hairs; belly washed with soiled whitish, smoky gray or pale cinnamon; feet plumbeous; tail indistinctly bicolor, blackish above, dull grayish below. *Winter pelage:* Much darkened above by long black hairs, especially early in the season, later becoming paler than in summer as the under-fur grows longer; belly heavily washed with creamy white; feet paler; tail more sharply bicolor. *Young:* Slightly less blackish than in *pennsylvanicus.*

Cranial characters.—Skull not positively distinguishable from that of *pennsylvanicus*, but in adults averaging heavier and more ridged.

Measurements.—Average of 10 adults from Cochetopa Pass, Colorado: Total length, 176; tail vertebræ, 44; hind foot, 20.6. *Skull* (No. 48053, ♀ ad.): Basal length, 27; nasals, 7.6; zygomatic breadth, 16; mastoid breadth, 12.4; alveolar length of upper molar series, 6.7.

Remarks.—Baird's type of *modestus* was collected at Sawatch or Cochetopa Pass in the Cochetopa Mountains. The type specimen in the United States National Museum is a half-grown young in the black pelage, and agrees perfectly in both external and cranial characters with specimens of the same age since collected at the type locality. The other specimen from Sawatch Pass (No. 593), which Professor Baird examined and believed to be distinct from *modestus*, but refrained from describing from a single immature specimen,[1] is also in the United States National Museum, and proves to be *Microtus nanus*, a good series of which has since been collected at a point 3 miles east of Cochetopa Pass. *Microtus mordax* is the only other species known to occur in this part of Colorado. These three widely different species are readily distinguishable at any age.

Microtus modestus decreases in size to the northward until, in northwestern Montana, it seems to merge into the little *drummondi* of the region farther north. Westward it becomes darker, specimens from Salt Lake Valley, Utah, being practically indistinguishable from typical *pennsylvanicus*, while those from Cœur d'Alene, Idaho, and the

[1] Mamm. N. Am., p. 535, 1857.

plains of the Columbia in eastern Washington are too small and dark to be typical *modestus*.

Specimens examined.—Total number, 259, from the following localities:

Colorado: Cochetopa Pass, 89; Fort Garland, 15; Loveland, 7; Twin Lakes, 1.

Wyoming: Newcastle, 1; Bear Lodge Mountains, 2; Sundance, 1; Lower Geyser Basin, Yellowstone Park, 1.

South Dakota: Custer, 2; Hill City, 1.

North Dakota: Fort Buford, 1.

Montana: Little Bighorn River, 2; Fort Custer, 3; Bozeman, 2; Fort Ellis, 1; Big Snowy Mountains, 13; Philbrook, 1; Stanford, 1; Choteau, 1; Robare, 1; Blackfoot, 1; Fort Assinniboine, 1; Tobacco Plains, 3; Stillwater Lake, 8; Flathead Lake, 9; Little Bitterroot Creek, 2; Hot Springs Creek, 1; Horse Plains (8 miles east), 1.

Idaho: Lemhi, 1; Salmon River, 3; Challis, 3; Birch Creek, 24; Cœur d'Alene, 3; Fort Sherman, 1.

Washington: Marshall, 15; Coulee City, 4; Conconully, 4; Colville, 20; Marcus, 1.

Utah: Ogden, 7; Salt Lake, 1.

MICROTUS PENNSYLVANICUS FONTIGENUS (Bangs). Forest Vole.

Microtus fontigenus Bangs, Proc. Biol. Soc. Wash., X, 48–49, March 9, 1896.
Microtus pennsylvanicus fontigenus Miller, Proc. Boston Soc. Nat. Hist., XXVIII, 14, April, 1897.

Type locality.—Lake Edward, Quebec.

Geographic distribution. Eastern Canada, in the Hudsonian zone.

Habitat.—Marshes, fields, dry banks, and deep woods.

General characters.—Smaller than *pennsylvanicus*, with short wide skull, large round bullæ, and short incisive foramina.

Color.—*Autumn pelage* (September specimens in long fur): Upperparts dark bister mixed with black, slightly paler on sides and cheeks; belly washed with whitish or smoky gray; tail bicolor, blackish above, grayish below; feet plumbeous.

Cranial characters.—Skull light and smooth, not ridged or angular; rostrum and incisive foramina short; braincase wide; audital bullæ large and smoothly rounded; interpterygoid space narrow, ending squarely at palate; dentition as in *pennsylvanicus*.

Measurements.—Type, ♀ ad.: Total length, 151; tail vertebræ, 41; hind foot, 21. Topotype, ♂ ad.: 150; 45; 21. *Skull* (No. 3839, ♂): Basal length, 23, nasals, 6.2; zygomatic breadth, 14.3; mastoid breadth, 11.5; alveolar length of upper molar series, 6.

Remarks.—The short rostrum, short, wide braincase, and short incisive foramina distinguish *fontigenus* from both *pennsylvanicus* and *drummondi*, with both of which species it seems to intergrade.

In size it is intermediate, and in external characters not very different from either. It is recorded by Mr. Miller from Nepigon and Peninsula Harbor, Ontario, but he considers the specimens obtained at those places not quite typical.[1] A series of 10 specimens from Godbout, Quebec, are rather nearer *fontigenus* than *acadicus*.

Specimens examined.—Total number, 6, topotypes (from the Bangs Coll.).

---

[1] Proc. Boston Soc. Nat. Hist., XXVIII, 14, April, 1897.

MICROTUS PENNSYLVANICUS LABRADORIUS Bailey. Little Labrador Vole.

Microtus pennsylvanicus labradorius Bailey, Proc. Biol. Soc. Wash., XII, 88, April 30, 1898.

Type locality.—Fort Chimo, Ungava, Labrador.

Geographic distribution.—Known only from the type locality.

General characters.—Size of *Microtus drummondi* and of approximately the same proportions. Skull flatter, with smaller audital bullæ and more protruding upper incisors.

Color.—(Much changed by alcohol.) Upperparts dark brownish; belly whitish; tail bicolor; feet pale.

Cranial characters.—Skull low, not much ridged or angled; postorbital ridge prominent; nasals short, cuneate and scarcely reaching base of incisors; audital bullæ small; incisive foramina short; first upper molar usually with an inner posterior point, molar pattern otherwise as in *pennsylvanicus.* The skull is readily distinguishable from that of either *drummondi* or *fontigenus* by the protruding incisors and small audital bullæ.

Measurements.—Type, ♀ ad. (in alcohol), measured by Dr. C. Hart Merriam: Total length, 139; tail vertebræ, 39; hind foot, 20. Average of 7 alcoholic specimens from type locality: 137; 37; 19. *Skull* (of type): Basal length, 24.3; nasals, 6.7; zygomatic breadth, 14.4; mastoid breadth, 11; alveolar length of upper molar series, 6.2.

Remarks.—*M. labradorius* shows closer affinity with *drummondi* than with *fontigenus*, though no doubt meeting and grading into the latter. It is widely different from *enixus*, and the two occur together at Fort Chimo.

Specimens examined.—Total number, 9, from the type locality.

MICROTUS DRUMMONDI (Aud. & Bach.). Drummond Vole.

Arvicola drummondi Aud. and Bach., Quad. N. Am., III, 166-167, 1854.

Arvicola (Mynomes) microcephalus Rhoads, Proc. Acad. Nat. Sci. Phila., 1894, 286-287 (Lac La Hache, B. C.).

Microtus stonei Allen, Bul. Am. Mus. Nat. Hist., XII, 4, March, 1899 (Liard River, N. W. T.).

Type locality.—Rocky Mountains, vicinity of Jasper House, Alberta.

Geographic distribution.—From Hudson Bay to the west slope of the Rocky Mountains and Alaska, and from the northern edge of the United States north to Fort Anderson, N. W. T., in Canadian and Hudsonian zones.

Habitat.—Both marshes and dry upland.

General characters.—Similar to *Microtus pennsylvanicus*, but much smaller, with slenderer feet and tail, and paler color.

Color.—*Summer pelage:* Upperparts yellowish bister with numerous dark-brown- or black-tipped hairs, sides of nose and hairs in front of ears more decidedly yellowish; belly white or rarely tinged with buffy, sometimes dusky during the molt; feet silvery gray; tail bicolor, blackish above, whitish below. *Winter Pelage:* Paler than in summer; yellow on ears and nose more conspicuous; *Young:* Paler and not so sooty as young *pennsylvanicus.*

Cranial characters.—Skull not much arched and rather flat topped, slender but sharply ridged in adults; audital bullæ large and smoothly rounded; palate flattened in immature specimens, becoming higher with deep lateral pits in adults. Dentition as in *pennsylvanicus.* Except for the larger bullæ and a few characters of minor weight, the skull of *drummondi* is a miniature of the skull of *pennsylvanicus.*

Measurements.—Average of 6 adult males and females from Muskeg Creek, Alberta: Total length, 145; tail vertebræ, 39; hind foot, 17.8. Largest specimen from Muskeg Creek: 160; 41; 18. *Skull* (No. 81487, ♀ ad., same locality): Basal length, 23; nasals, 6; zygomatic breadth, 14; mastoid breadth, 11; alveolar length of upper molar series, 6.

Remarks.—The characters separating *drummondi* from *pennsylvanicus* and *modestus* are relative. There is no sharp distinction and the forms either merge into each other, or after approaching each other in size overlap in range and occur together at the same localities. Specimens from Blackfoot, Montana, are nearer *modestus*, while those from Summit and St. Marys Lake are almost typical *drummondi.* At Elk River, Minnesota, about half of the specimens are almost typical *drummondi*, while the others are a small form of *pennsylvanicus;* but, as others fall between and cannot positively be placed with either form, it is impossible to decide whether the difference is due to individual variation in an intermediate form or whether two species meet and interbreed.

To the eastward *drummondi* merges into *fontigenus*, from which it differs in such slight degree that the two are not easily distinguishable. On the west slope in British Columbia *drummondi* becomes slightly darker, and in that respect less unlike *fontigenus* in appearance, but retains its cranial characters, Alaska and northwestern specimens (from Fort Wrangel, Nulato, and Fort Simpson) are larger and in general appearance less different from *pennsylvanicus*, but detailed cranial characters show them to be but a robust form of *drummondi.* Specimens from Liard River, including the type of *M. stonei*, are indistinguishable from typical *drummondi.*

Specimens examined.—Total number, 355, from the following localities:

Alberta: Muskeg Creek (15 miles south of Smoky River), 13; Smoky Valley (50 miles north of Jasper House), 5; Fishing Lake (90 miles north of Jasper house), 2; Henry House (15 miles south), 4; South Edmonton, 110; St. Alberts, 26; Canmore, 2; Banff, 1; Red Deer, 1.

Athabasca: Lesser Slave Lake, 1.

Assiniboia: Indian Head, 30; Medicine Hat, 1.

British Columbia: Shuswap, 13; Sicamous, 1; Cariboo Lake, 2; Okanagan, 11; Ducks, 2; Stuart Lake, 1.

Saskatchewan: Cumberland House, 2; St. Louis, 1.

Manitoba: Carberry, 2.

Ontario: Rat Portage, 2; Coney Island (Lake of the Woods), 2.

Northwest Territory: Fort Churchill, 2; Fort Simpson, 2; Fort Rae, 3; Fort Reliance, 1; Big Island (Great Slave Lake), 1; Fort Good Hope, 1; Fort MacPherson, 1; Fort Anderson, 2; head of Liard River, 1; Chandindu River, 1; Dawson, 1; Sixty-Mile Creek, 1; Fort Selkirk, 3; 50 miles below Fort Selkirk, 1; Rink Rapids, 6; Thirty-Mile River, 3; Lake Lebarge, 4; Fifty-Mile River, 6; Lake Marsh, 22; Caribou Crossing, 16.

Alaska: Fort Wrangell, 1; Nulato, 1; Charlie Creek (Yukon River), 2; Canadian Boundary and Yukon River, 4.
North Dakota: Portland, 4; Lisbon, 1; Bottineau, 1.
Montana: St. Marys Lake, 6; Summit, 8.
Idaho: Priest Lake, 5.
Washington: Loon Lake, 9.

MICROTUS AZTECUS (Allen). Aztec Vole.

Arvicola (Mynomes) aztecus Allen, Bul. Am. Mus. Nat. Hist., V, 73–74, April 28, 1893.

Type locality.—Aztec, New Mexico (altitude, 5,900 feet).

Geographic distribution.—Valley of the San Juan River in northwestern New Mexico, in Transition zone.

Habitat.—Grassy places along the river banks and near irrigation ditches.

General characters.—In size similar to *M. pennsylvanicus*, but with shorter tail and larger hind foot; skull long; braincase narrow; interparietal long.

Color.—*Winter pelage:* Upperparts dull buffy, heavily mixed with black hairs; belly washed with creamy or pale buff; feet plumbeous; tail sharply bicolor, black above, soiled whitish below. *Young* (half-grown specimens in Dec.): Scarcely different in color from adults. (*Summer pelage* not examined.)

Cranial characters.—Skull long; braincase high and narrow; interparietal more than half as long as wide, extending back to plane of foramen magnum; audital bullæ large; mandible short and heavy; angular process of mandible wide; dentition as in *pennsylvanicus*.

Measurements.—Average of 7 adults from type locality: Total length, 171; tail vertebræ, 43; hind foot, 22. *Skull* (No. 57432, ♂ ad.): Basal length, 28.8; nasals, 8.4; zygomatic breadth, 17; mastoid breadth, 12.5; alveolar length of upper molar series, 7.2.

Remarks.—*Microtus aztecus* belongs to the *pennsylvanicus* group. Externally it is not very different from *modestus*, but none of the specimens show any signs of intergradation; and the skull characters are so well marked that there seems no doubt of its full specific rank.

Specimens examined.—Total number, 45, from the following localities:

New Mexico: Aztec, 42; La Plata, 3.

MICROTUS ENIXUS Bangs. Large Labrador Vole.

Microtus enixus Bangs, Am. Nat., XXX, 1051–1052, Dec. 5, 1896.

Type locality.—Hamilton Inlet (north shore), Labrador.

Geographic distribution.—Eastern coast of Labrador from Hamilton Inlet to Ungava Bay, in Hudsonian zone.

General characters.—Size slightly larger than *M. pennsylvanicus*, with relatively longer tail and larger ears; coloration duller and darker; skull smaller, shorter, and wider, with lighter molars. Hip glands present in adult males.

Color.—Summer pelage (July and Aug. specimens): Upperparts dark yellowish bister mixed with blackish; belly smoky gray or soiled whitish, sometimes tinged with buffy; feet dusky or blackish; tail black above, grayish brown below.

Cranial characters.—Outline of skull shorter, wider, and less arched than in *pennsylvanicus*; prezygomatic notch deep; coronoid notch of mandible wide and rounded; molar series small and slender; m2 with posterior loop completely closed and circular in form; m3 with terminal loop shortened; dentition otherwise as in *pennsylvanicus*.

Measurements.[1]—Type, ♀ ad.: Total length, 210; tail vertebræ, 67; hind foot, 22.5. Average of 10 adult topotypes: 189.4; 60.4; 22.4. *Skull* (No. 4018, ♀ ad.): Basal length, 27.3; nasals, 8.6; zygomatic breadth, 16.7; mastoid breadth, 12.3; alveolar length of upper molar series, 6.5.

Remarks.—Microtus enixus appears to be a distinct and well marked species of the *pennsylvanicus* group, the long tail and small molars being the most convenient characters for recognition. From its geographically nearest neighbors, *Microtus p. fontigenus*, of Lake Edward, Quebec, and *Microtus p. labradorius*, of Ungava, it shows a wider difference than from typical *pennsylvanicus*.

Specimens examined.—Total number, 16, from the following localities:

Labrador: Hamilton Inlet, 13; Fort Chimo, Ungava, 3.

MICROTUS TERRÆNOVÆ (Bangs). Newfoundland Vole.

Arvicola terrænovæ Bangs, Proc. Biol. Soc. Wash., IX, 129-132, July 27, 1894.

Type locality.—Codroy, Newfoundland.

Geographic distribution.—Newfoundland and Penguin Island.

General characters.—Slightly larger than *pennsylvanicus*, with decidedly larger hind foot, and more yellowish colors; belly with dusky median line; nose patch buffy; skull wide and angular.

Color.—Summer pelage (in July and August specimens): Upperparts dark russet, darkened by brown-tipped hairs, becoming paler on sides and across face; nose patch dark buffy or dull russet; belly whitish or smoke gray with a median streak of dusky cinnamon; tail distinctly bicolor, blackish above, soiled whitish below; feet grayish brown. *Winter pelage* (retained in April specimens): Slightly paler russet above, whiter below, with sharper markings throughout. *Young* (nearly half-grown): Similar to adults.

Cranial characters.—Skull short with wide-spreading zygomata; prezygomatic notch deep; nasals terminating even with arms of premaxillæ; lateral pits of palate deep and wide; interpterygoid space narrow with median constriction; m1 with an anterior spur or loop; m3 with posterior loop short, irregularly rounded, triangular or trifoliate; $\overline{\text{m2}}$ and m3 normally with anterior point or spur.

---

[1] From original description.

Measurements.—Average of 10 adults from type locality: Total length, 182; tail vertebræ, 52; hind foot, 23.4. *Skull* (No. 74029, ♂ ad.): Basal length, 28.5; nasals, 8.4; zygomatic breadth, 17; mastoid breadth, 13; alveolar length of upper molar series, 7.

Remarks.—*Microtus terrænovæ* shows very distinctive characters, and no close affinity with any neighboring species. Specimens showing very young and full winter pelage are still needed for a comprehensive description

Specimens examined.—Total number, 43; from the following localities:

Newfoundland: Codroy, 35; Penguin Island, 8.

MICROTUS BREWERI (Baird). Beach Vole.

Arvicola breweri Baird, Mamm. N. Am., 525-526, 1857.

Type locality.—Muskeget Island, Massachusetts.

Geographic distribution.—Muskeget Island.

Habitat.—Beach plum thickets on the sandy island.

General characters.—Size a little larger than *pennsylvanicus;* colors pale grayish; pelage long and coarse; skull heavy with wide nasals and quadrate interparietal.

Color.—*Summer pelage* (July 18): Upperparts buffy gray with scattered brown- and black-tipped hairs, paler on sides; belly tinged with sulphur yellow; feet silvery gray; tail bicolor, rusty brown or blackish above, soiled whitish below. *Young:* Rather paler and duller than adults.

Cranial characters.—Skull massive; nasals wide anteriorly; interparietal more than half as long as wide; inner edges of zygomata sharply notched close to lachrymals; rostrum heavy; upper incisors bent abruptly downward; molar pattern not very different from that of *pennsylvanicus.*

Measurements.—Average of 10 adults from Muskeget Island: Total length, 182; tail vertebræ, 54; hind foot, 22.3. *Skull* (No. 73141, ♂ ad.): Basal length, 28.7; nasals, 8.3; zygomatic breadth, 17.3; mastoid breadth, 13; alveolar length of upper molar series, 7.2.

Remarks.—Nine of the 26 specimens have a small white spot on the forehead. This may be accidental or an only partially acquired character.

Specimens examined.—Total number, 26; all from the type locality.

MICROTUS NESOPHILUS Bailey. Gull Island Vole.

Microtus insularis Bailey, Proc. Biol. Soc. Wash., XII, 86, April 30, 1898. Name preoccupied by *Lemmus insularis*, Nillson (= *Microtus agrestis* L.).

Microtus nesophilus Bailey, Science, N. S., VIII, 782, Dec. 2, 1898.

Type locality.—Great Gull Island, New York.

Geographic distribution.—Great Gull Island (at entrance to Long Island Sound).

General characters.—Size of *pennsylvanicus;* colors darker; skull shorter and wider with more spreading zygomata and deeper prezygomatic notches.

Color.—*Summer pelage:* Upperparts, dark yellowish bister heavily mixed with black hairs, darkest on nose and face; belly dusky, washed with cinnamon; feet blackish; tail blackish above, dark brown below.

Cranial characters.—The skull differs from that of *pennsylvanicus* in shorter, wider braincase, wider and more abruptly spreading zygomatic arches, more expanded jugal, and smaller audital bullæ; palate short, with a median point or spur and deep lateral pits; m3 normally with anterior inner and outer triangles approximately opposite and confluent; dentition otherwise similar to that of *pennsylvanicus.*

Measurements.—Type (measured in dry skin): Tail, 29; hind foot, 20. No. 1943, Am. Mus., ♂ ad., 185: 41 : 21. *Skull* (No. 53969): Basal length, 26; zygomatic breadth, 16.2; mastoid breadth, 12.3; alveolar length of upper molar series, 6.8. (No. 1943) 26.6; 8.5; 16.2; 13; 6.6.

Remarks.—*Microtus nesophilus* needs no comparison with *breweri* or *terrænovæ*, the other two insular forms from the Atlantic coast. In general appearance it more nearly resembles *pennsylvanicus*, but in cranial characters it is as distinctly different as either of the other island species.

During the month of August, 1898, Mr. A. H. Howell visited Great Gull Island for the purpose of getting specimens of *Microtus*, but he found their old haunts covered by the earth moved in grading the island for fortifications, while no trace of the animals remained. He thinks they are completely exterminated.

Specimens examined.—Total number, 15; all from the type locality.

MICROTUS MONTANUS (Peale). Peale Vole.

Arvicola montanus Peale, U. S. Exploring Exp'd., Mammalogy, 44, 1848.
Arvicola longirostris Baird, Mamm. N. Am., 530–531, 1857. (From upper Pitt River, California.)

Type locality.—Headwaters of Sacramento River, near Mount Shasta, California.

Geographic distribution.—Northeastern California, eastern Oregon, northern Utah and Nevada, in the Upper Sonoran and Transition zones.

Habitat.—Marshes, meadows, and tule swamps.

General characters.—Size medium (about as in *pennsylvanicus*); tail about twice as long as hind foot; colors dark; hip glands conspicuous in adult males; incisors projecting well in front of nasals; incisive foramina narrow and constricted posteriorly.

Color.—*Summer pelage:* Upperparts bister or ashy mixed with blackish; belly washed with soiled whitish, giving a smoky gray or dusky color; feet plumbeous; tail indistinctly bicolor, blackish above, plumbeous below; lips usually showing a trace of whitish.

Cranial characters.—Skull generally slender and smooth, becoming angular and ridged in only a few very old individuals; nasals narrow and short; interparietal wide and normally strap-shaped; incisive foramina narrow and constricted posteriorly; bullæ medium and well rounded; dentition rather light; m2 with 4 closed sections; m3 with 3 closed triangles; m1 normally with 5 closed triangles.

FIG. 2.—Molar enamel pattern of *Microtus montanus* (×5).

Measurements.—Average of 10 specimens from Sisson, Cal.: Total length, 175; tail vertebræ, 52; hind foot, 21.5. Extremely large specimens from Sisson run as high as 192; 54; 23. *Skull* (No. 98689, ♂ ad., from Sisson): Basal length, 28; nasals, 8; zygomatic breadth, 17; mastoid breadth, 13.5; alveolar length of upper molar series, 7.3.

General remarks.—The original description of *M. montanus*, though meager, agrees in all particulars with the animal from Sisson, at the west base of Mount Shasta. The measurements (total length, 6⅛ inches; tail, 1½ inches=156 mm. and 38 mm.) give it too short a tail, which only serves to restrict it more closely to this form in distinction from either of the longer-tailed species (*mordax* or *californicus*) that occur at or near the type locality. Three mounted specimens in the United States National Museum, which Baird referred to *montanus*,[1] and which came from Upper Klamath Lake and the Upper Des Chutes, are identical with those of the present series from Sisson, Fort Klamath, and Fort Crook. Specimens from the south end of Goose Lake, which is the source of Pitt River, are the same as those from Sisson and from Fort Crook, lower down the river, and also the same as Baird's type of *M. longirostris* from 'Upper Pitt River.'

M. montanus has a somewhat scattered and interrupted distribution and shows considerable geographic variation in widely separated localities. The extreme development of large size, large feet, and heavy angular skull is found in the big marshes of the Carson Sink, Nevada, while specimens from higher levels in the Transition zone are smaller, with slenderer feet and grayer coloration. The variation is mainly, but not entirely, zonal. To separate either extreme would tend to confusion rather than convenience, as the extremes point to *nevadensis* on the one hand and to *nanus* on the other.

Specimens examined.—Total number, 263, from the following localities:

California: Sisson, 57; Fort Crook, 23; Hayden Hill, 2; Fall Lake, 2; Cassel, 1; Tule Lake, 2; Goose Lake, 8; Greenville (8 miles NW.), 3; Bucks Ranch (Plumas Co.), 1; Quincy, 3; Summit, 1.

Nevada: Washoe, 1; Deep Hole (south end of Granite Range), 1; Pine Forest Range, 1; Mountain City, 4; Wells, 13; Austin, 7; Carson, 16; Stillwater, 10; Newark Valley, 5; Monitor Valley, 5; Ruby Lake, 5; Ruby Mountains, 3.

[1] Mamm. N. Am., 530, 1857.

Oregon: Klamath Basin (Lost River), 8; Klamath Falls, 1; Fort Klamath, 5; Swan Lake Valley, 5; Plush (west side Warner Lake), 9; Shirk, 2; Steen Mountains (east slope), 1; Summit NE. of Steen Mountains), 2; Burns, 2; Wapinitia, 4.
Utah: Ogden, 16; Salt Lake City, 3; Provo, 1; Fairfield, 10; Manti, 20.

MICROTUS MONTANUS ARIZONENSIS Bailey. Arizona Vole.

Microtus montanus arizonensis Bailey, Proc. Biol. Soc. Wash., XII, 88, April 30, 1898.

Type locality.—Springerville, Ariz.

Geographic distribution.—Plateau country of eastern Arizona, at head of Little Colorado, in the Transition Zone.

Habitat.—Creek banks and damp meadows.

General characters.—Similar to *M. montanus*, but brighter and more ferruginous in color; lateral pits of palate shallower.

Color.—*Early winter pelage* (October and November specimens): Upperparts yellowish or rusty brown; belly washed with white; feet dark grayish; tail bicolor, blackish above, grayish below; lips whitish. Slightly immature specimens are a little duller colored than adults.

Cranial characters.—Skull very similar to that of *montanus*, but easily distinguished by the flatter palate with shallower lateral pits and by thicker pterygoids; condyloid process of mandible slightly shorter. Dentition not different.

Measurements.—Type: Total length, 184; tail vertebræ, 55; hind foot, 20. Average of 7 specimens from type locality: 158; 41; 20.6. *Skull* (of type): Basal length, 27.3; nasals, 8; zygomatic breadth, 16; mastoid breadth, 12.2; alveolar length of upper molar series, 6.5.

General remarks.—Although widely separated geographically from *M. montanus* by desert country through which continuity of range is improbable, this form is so closely related to that species that its position is best indicated by subspecific rank.

Specimens examined.—Total number, 12, from the following localities:

Arizona: Springerville, 11.
New Mexico: Nutria, 1.

MICROTUS MONTANUS RIVULARIS Bailey. Utah Vole.

Microtus nevadensis rivularis Bailey, Proc. Biol. Soc. Wash., XII, 87, April 30, 1898.

Type locality.—St. George, Utah.

Geographic distribution.—Known only from type locality, probably restricted to Lower Sonoran zone.

Habitat.—Tule marshes along the banks of the Virgin River. The runways were always found in wet places among sedges and rushes.

General characters.—Larger and lighter colored than typical *montanus;* skull more arched; nasals narrower.

Color.—*Winter pelage:* Upperparts dull bister, darkened with blackish-tipped hairs; sides scarcely paler; belly washed with white; feet dull grayish; tail bicolor, blackish above, grayish below. *Young:* Darker than adult, but not black backed as in *nevadensis.*

Cranial characters.—Skull well arched, not much ridged; nasals conspicuously narrower than in *montanus;* frontals narrower posteriorly; basioccipital more constricted anteriorly; dentition essentially the same.

Measurements.—Type: Total length, 179; tail vertebræ, 48; hind foot, 23. A nearly adult female topotype: 163; 43; 21. *Skull* (of type): Basal length, 28.2; nasals, 8.3; zygomatic breadth, 17; mastoid breadth, 13.3; alveolar length of upper molar series, 7.3.

General remarks.—Since *rivularis* was described in 1898 as a subspecies of *nevadensis*, a series of 50 specimens has been collected at the type locality of *montanus*, showing for the first time the real characters and the range of variation in that species, and, moreover, as Dr. Merriam had previously suggested to me, that *rivularis* comes nearer to typical *montanus* than to *nevadensis*. So far as at present known it has an isolated range in a Lower Sonoran valley, but it may readily extend northward to meet and merge into *montanus* in central Utah.

Specimens examined.—Total number, 4, from the type locality.

MICROTUS NANUS (Merriam). Dwarf Vole.

Arvicola nanus Merriam, North American Fauna No. 5, 62-63, pl. II, figs. 5 and 6, July 30, 1891.

Type locality.—Pahsimeroi Mountains, Idaho (altitude 9,350 feet).

Geographic distribution.—Rocky Mountains and outlying ranges, from central Idaho southward to central Nevada and southern Colorado, in Canadian zone.

Habitat.—Dry, grassy parks on mountain slopes.

General characters.—Size small; tail short; ears short and rounded; color dark grayish; skull slender.

Color.—*Summer pelage:* Upperparts uniformly grizzled gray mixed with sepia and blackish hairs; belly washed with white; feet grayish or plumbeous; tail bicolor, dusky gray above, whitish below. (*Winter pelage* unknown.) *Young:* Similar to adult, but slightly duller throughout.

Cranial characters.—Skull small, slender and well arched, with slender zygomata and large well-rounded bullæ; superciliary ridges prominent, sometimes confluent in old age; incisors projecting well beyond nasals; molars light, with short, wide triangles; enamel pattern scarcely distinguishable from that of *mordax* and *montanus*.

Measurements.—Type, ♂ ad.: Total length, 151; tail vertebræ, 41; hind foot, 18. Average of five adults from type locality: 143; 37; 18.4. *Skull* (of type): Basal length, 23.7; nasals, 6.7; zygomatic breadth, 14; mastoid breadth, 10.8; alveolar length of upper molar series, 5.8.

Remarks.—*Microtus nanus* belongs to the *montanus* group but occupies a higher zone and has more of the habits of *Pedomys* or *Lagurus*. It is rarely found in wet places or near water. Specimens from certain isolated localities are not entirely typical, but do not differ enough to warrant separation.

Specimens examined.—Total number, 114, from the following localities:

Idaho: Pahsimeroi Mountains, 13; Lost River Mountains, 1; Challis, 7; Sawtooth Lake, 5; Three Creek, 3; Montpelier Creek, 3; Seven Devils Mountains, 4.

Utah: Uinta Mountains, head of Smith Fork, 1.

Wyoming: Fort Bridger, 9; Kinney Ranch, 6; Beaver, 1 (no skull); La Barge Creek (near head), 1; Cheyenne, 3; Sherman, 2; Laramie, 2; South Pass City, 20; Bighorn Mountains, head of Powder River, 9.

Montana: Beartooth Mountains, 3; Big Snowy Mountains, 1.

Colorado: Estes Park, 1; Cochetopa Pass, 17; Twin River, 1; Twin Lakes, 1.

MICROTUS NANUS CANESCENS Bailey. Gray Vole.

Microtus nanus canescens Bailey, Proc. Biol. Soc. Wash., XII, 87, April 30, 1898.

Type locality.—Conconully, Washington.

Geographic distribution.—Northern Washington and southern British Columbia, east of the Cascades. Apparently confined to the Transition zone.

Habitat.—Dry grassy ground.

General characters.—Like *nanus* but lighter, clearer gray; skull with larger bullæ and greater mastoid breadth; zygomatic arches less widely spreading; upper incisors bent more abruptly downward. Hip glands conspicuous in adult males.

Color.—*Summer pelage:* Upperparts clear, dark grayish, formed by pale buffy and black-tipped hairs; sides shading to lighter gray and belly to white; feet dark gray; tail bicolor, blackish above, grayish below. (*Young* and *winter pelage* not shown in present material.)

Cranial characters.—Skull slightly narrower and more elongate than in *nanus;* interparietal averaging longer; bullæ decidedly larger and fuller; mastoid breadth relatively greater; incisors scarcely reaching beyond nasals; molar pattern as in *nanus.*

Measurements.—Type: Total length, 149; tail vertebræ, 42; hind foot, 20. *Skull* (of type): Occipital condyle to anterior base of molars, 17.4; posterior tip of nasals to foramen magnum, 19.2; zygomatic breadth, 15; mastoid breadth, 12.3; alveolar length of upper molar series, 6.3.

General remarks.—In its extreme development this northern form is readily distinguishable from typical *nanus.* From intermediate localities, Flathead Lake and the Plains of the Columbia, specimens are not typical of either but show slight peculiarities of local development interesting in themselves but not sufficiently marked for even subspecific distinction. To a certain extent they are intermediate between *nanus* and *canescens.*

Specimens examined.—Total number, 47, from the following localities:

British Columbia: Okanagan, 11; Ducks, 2; Vernon, 7.

Washington: Conconully, 1; Wenatchee, 1; Fort Walla Walla, 1; Oakesdale, 2; Wawawai, 4; Cheney, 1.

Oregon: Elgin, 2; Wallowa Mountains (near Joseph), 6.

Montana: West arm of Flathead Lake, 5; Hot Spring Creek (a branch of the Little Bitterroot), 4.

MICROTUS CANICAUDUS Miller. Gray-tailed Vole.

Microtus canicaudus Miller, Proc. Biol. Soc. Wash., XI, 67–68, April 21, 1897.

Type locality.—McCoy, Oregon.

Geographic distribution.—Willamette Valley, Oregon, and the east base of the Cascades in southern Washington, in Transition zone.

General characters.—Size and proportions about as in *nanus;* ears larger, skull heavier, more arched, with fuller, rounder bullæ, and shallower lateral pits of palate, coloration more yellowish, tail grayer.

Color.—*Winter pelage:* Upperparts bright yellowish bister, darkened with blackish-tipped hairs, slightly paler on sides; belly and whole lower parts whitish-gray; feet grayish or pale plumbeous; tail in winter adults uniformly grayish, with a half-concealed dusky dorsal line. In the only *summer* specimen (from North Yakima, Wash., and perhaps not typical) the tail is sharply bicolor with a blackish dorsal line. *Young* (half-grown November specimens): Sooty gray above and scarcely lighter below; feet dusky; tail gray, with a blackish dorsal stripe.

Cranial characters.—Skull high, smooth, and well arched, with scarcely a trace of superciliary ridges; interparietal lozenge-shaped; bullæ larger and more rounded than in *nanus;* interpterygoid fossa narrower and more acuminate; lateral pits of palate shallower; incisors less protruding; enamel pattern of molars the same as in *nanus.*

Measurements.—Type: Total length, 135; tail vertebræ, 33; hind foot, 20. Average of 8 adults from type locality: 141; 35.7; 20. *Skull* (of type): Basal length, 24.2; nasals, 7.3; zygomatic breadth, 15.3; mastoid breadth, 12.8; alveolar length of upper molar series, 6.

Remarks.—A single specimen with a badly broken skull from North Yakima, Wash., seems to be true *canicaudus* in summer pelage, and indicates that the range of the species is much more extensive than is at present known.

Specimens examined.—Total number, 14, from the following localities:

Oregon: McCoy, 9; Beaverton, 2; Sheridan, 2 (im).
Washington: North Yakima, 1.

MICROTUS DUTCHERI Bailey, Dutcher Vole.

Microtus dutcheri Bailey, Proc. Biol. Soc. Wash., XII, 85, April 30, 1898.

Type locality.—Big Cottonwood Meadows, near Mount Whitney, California (altitude, 10,000 feet).

Geographic distribution.—Hudsonian zone of the southern Sierra Nevada.

Habitat.—Wet alpine meadows.

General characters.—Size rather small; tail short; ears small, nearly concealed by fur; colors dark above and below; lips and usually nose white; hip glands present in adult males.

Color.—*Summer pelage:* Upperparts dark bister with brown tips to the long hairs; below, dull cinnamon or buffy-brown; feet whitish or plumbeous-gray; tail bicolor, brown or blackish above, whitish below; lips and usually tip of nose white. (*Winter pelage* unknown.) *Young:* Dull brown above and scarcely lighter below; feet and tail blackish; lips and nose usually white.

Cranial characters.—Skull similar to that of *montanus* but differing in many details; rostrum slightly longer; bullæ smaller and less globular; lateral pits of palate shallower; dentition the same.

Measurements.—Type, ♂ ad.: Total length, 167; tail vertebræ, 35; hind foot, 20. Average of 10 adults, 5 males and 5 females, from type locality: 163; 37; 20.6. *Skull* (of type): Basal length, 27.4; nasals, 8; zygomatic breadth, 16.7; mastoid breadth, 12.2; alveolar length of upper molar series, 6.5.

General remarks.—The nearest relative of *M. dutcheri* is *montanus*, but the two species occupy widely separated zones, and show no evidence of intergradation.

Specimens examined.—Total number, 65, from the following localities in the Sierra Nevada:

California: Big Cottonwood Meadows, 28; Whitney Meadows, 11; Menache Meadows, 2; Olancha Peak, 3; Head of Kern River, 1; Mammoth, 12; Pine City, 3; Head of San Joaquin River, 5.

MICROTUS NEVADENSIS Bailey. Nevada Vole.

Microtus nevadensis Bailey, Proc. Biol. Soc. Wash., XII, 86, April 30, 1898.

Type locality.—Ash Meadows,[1] Nye County, Nevada.

Geographic distribution.—Known only from the type locality and Pahranagat Valley, about 100 miles to the northeast. Both localities are in the Lower Sonoran zone.

Habitat.—Salt grass and tule marshes in alkaline valleys. Runways always found in wet, muddy places, and often extending through shallow water.

General characters.—Size large; ears small; tail rather short; fur coarse and lax; colors dark; hip glands conspicuous in adult males. Skull massive and angular; incisive foramina narrow and closing to a point posteriorly.

Color.—*Winter pelage* (March specimens): Upperparts dark sepia or bister, much obscured by blackish hairs; sides lighter; belly smoky gray; feet dark gray; tail indistinctly bicolor, blackish above, gray or brownish below; lips usually white; tip of nose usually whitish. *Young:* With a blackish dorsal stripe and dusky feet and tail.

Cranial characters.—Skull heavy, angular, and much ridged; frontals high; rostrum bent downward; nasals truncate or rounded posteriorly, terminating even with arms of premaxillæ; incisive foramina short, rather narrow and constricted to a point posteriorly; dentition heavy; upper incisors curved abruptly downward; molar pattern vari-

[1] Ash Meadows is on the Nevada side of the Nevada-California line near where the Amargosa River crosses the boundary. The exact locality is a big salt marsh below Watkins ranch.

18392—No. 17——3

able; m2 with 4 closed sections in 8 out of 16 specimens, in the other 8, with a slight inner lobe or loop at base of posterior triangle; m3 with anterior crescent, three closed triangles, and a posterior loop with two inner lobes; m1 usually with 6 closed triangles.

Measurements.—Type: Total length, 210; tail vertebræ, 55; hind foot, 25.5. Average of 8 specimens from type locality: 176; 47; 23. *Skull* (of type): Basal length, 32; nasals, 10.2; zygomatic breadth, 19.3; mastoid breadth, 14.3; alveolar length of upper molar series, 8.

Remarks.—Three specimens taken May 26 in Pahranagat Valley differ slightly from the type series, but the cranial differences are slight and the darker color may be only seasonal. The species inhabits marshes and wet places, which are so rare and isolated in the desert region that it can have no extensive continuous range. It is one of the few forms of *Microtus* inhabiting a part of the Lower Sonoran zone.

Specimens examined.—Total number, 19, from the following localities:

Nevada: Ash Meadows, 16; Pahranagat Valley, 3.

MICROTUS CALIFORNICUS (Peale.) California Vole.

Arvicola californica Peale, U. S. Expl. Expd., Mammalogy, 46, 1848.
Arvicola trowbridgi Baird, Mamm. N. Am., 529, 1857. (Monterey, California.)

Type locality.—San Francisco Bay, California.

Habitat.—Dry meadows and grassy uplands, Upper Sonoran zone.

Geographic distribution.—California, west of the Colorado Desert and the Sierra Nevada, and from Santa Ysabel, San Diego County, Calif., north to the Rogue River and Umpqua valleys, Oregon.

General characters.—Size rather large, ears conspicuous above fur; pelage coarse and harsh, color similar to that of the house mouse; skull of adult heavy and angular, incisive foramina wide and open, usually widest posteriorly; a trace of hip glands in adult males.

Color.—*Summer pelage:* Upperparts dull buffy or clay-colored, slightly lined with blackish-tipped hairs; sides paler; belly light buffy or soiled whitish; tail bicolor, brownish above, buffy below; feet clear gray. *Winter pelage:* Much darker than the summer, with an excess of black-tipped hairs over the back; tail more sharply bicolor, blackish above. *Young:* Fur woolly and soft, duller and darker throughout than in the adult; belly dusky or plumbeous; feet and tail dusky.

Cranial characters.—Skull of adult heavy, angular, and ridged; nasals long, bent well down, widening abruptly in front, narrow and notched at posterior end, not reaching tip of ascending arm of premaxillæ; prezygomatic notch deep; postorbital processes prominent; frontals concave posteriorly; incisive foramina open, rounded at both ends and usually widest posterior to middle. Incisors heavy, the upper bent abruptly downward, not extending beyond tip of nasals; molars large and irregular, posterior triangle of m1 normally with an inner point or angle; posterior triangle of m2 with an inner point or angle or loop,[1]

[1] In 100 specimens, 77 have 4 closed triangles in m2, 20 have an open posterior loop, and 3 have a closed posterior loop as in *pennsylvanicus*.

suggesting the posterior loop in *pennsylvanicus;* m3 with three closed triangles, 3 outer and 4 inner salient angles; m1 with 5 closed triangles and 9 or 11 salient angles.

Measurements.—Average of 4 adults, 2 males and 2 females from Walnut Creek, Calif.: Total length, 171; tail vertebræ, 49; hind foot, 21.1. Of 10 adults from Monterey: 172; 52; 22.3. *Skull* (♂ ad., No. 44678, from Walnut Creek, Calif.): Basal length, 27.5; nasals, 8.5; zygomatic breadth, 16.6; mastoid breadth, 13.6; alveolar length of upper molar series, 6.8.

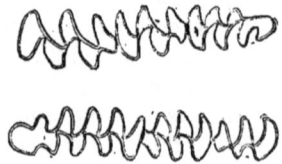

FIG. 3.—Molar enamel pattern of *Microtus californicus* (×5).

Remarks.—But slight variation is shown throughout the range of the species. Specimens from Santa Ysabel and the base of San Bernardino Mountains are indistinguishable from those of the shores of San Francisco Bay or from the Rogue River Valley, Oregon. A slight brightening in color is noticeable in specimens from Auburn.[1] The species is conspicuously absent from the bottom, or Lower Sonoran area, of the San Joaquin and Sacramento valleys, its place being taken in the tule marshes of these valleys by the larger, darker, longer tailed species, *M. edax*.

The skin of the type of *M. californicus* in the United States National Museum agrees perfectly with series of specimens from Berkeley, Walnut Creek, and other localities around San Francisco Bay, but the slightly immature skull is either abnormal or else never came from the same animal as did the skin. By some error it was given the same catalogue number as the type skull of *M. occidentalis*, a synonym of *M. townsendi*, but it is not the skull of that species.

Specimens examined.—Total number, 338, from the following localities:

California: Walnut Creek, 6; Oakland, 1; Berkeley, 44; San Lorenzo, 3; San Mateo, 14; Novato, 7; Glen Ellen, 9; Nicasio, 16; Point Reyes, 34; Cape Mendocino, 2; Mill Valley, Marin County, 3; Olema, 5; Cloverdale, 14; Ukiah, 2; Laytonville, 2; Round Valley, 1; Upper Lake, 1; Leesville, 10; Bartlett Mountain, 1; Rio Dell, 1; Hornbrook, 3; Little Shasta, 4; Mayten, 2; Cassel, 6; Red Bluff, 1; near Edgewood, 2; Grindstone Creek, Tehama County, 2; Auburn, 5; Jackson, 7; Chinese Camp, 2; Boulder Creek, 2; Monterey, 22; Jamesburg, 12; Jolon, 4; San Simeon, 1; Paso Robles, 1; Morro, 4; Pozo, 1; Gaviota Pass, Santa Barbara County, 4; Santa Barbara, 2; Ventura River, 3; San Emigdio Canyon, Kern County, 7; Mount McGill, Ventura County, 2; Calabasas, 3; San Fernando, 4; San Bernardino, 2; San Bernardino Mountains, 9; San Diego, 1; Riverside, 5; Techelote Canyon, Riverside County, 1; Little Bear Valley (San Bernardino Mountains), 1; Las Virginias Creek, 1; Radec, Riverside County, 1; Santa Ysabel, 10; near Tejon Pass, 1; Fort Tejon, 1; South Fork Kern River, 6.

Oregon: Rogue River Valley (near Grants Pass), 7; Siskiyou, 4; Drain, 6.

---

[1] The effect of red soil is noticeable in these as in some other mammals from the vicinity of Auburn.

MICROTUS CALIFORNICUS VALLICOLA Bailey. Valley Vole.

Microtus californicus vallicola Bailey, Proc. Biol. Soc. Wash., XII, 89, 1898.

Type locality.—Lone Pine, Inyo County, California.

Geographic distribution.—Valleys east of the Sierra Nevada, California. Confined mainly to Upper Sonoran zone.

Habitat.—Dry, grassy banks, upland meadows, and old weedy fields.

General characters.—Similar to *californicus*, but averaging slightly larger and darker. Proportions the same.

Color.—*Summer pelage:* Upperparts dull sepia, darkened by black-tipped hairs, darker and with less buffy suffusion than in *californicus;* belly dull grayish or smoky plumbeous; feet dusky; tail bicolor, blackish above, grayish below. *Winter pelage:* Darker throughout, with the black hairs of the back longer and more conspicuous. *Young:* Sooty gray above, plumbeous below, not black backed; feet and tail dusky.

Cranial characters.—Skull like that of *californicus*, but usually with smaller audital bullæ, more abruptly truncated occiput, and nasals reaching nearer to tips of premaxillæ; lobe at base of 4th triangle of middle upper molar sometimes developed into a loop.

Measurements.—Type, ♀ ad.: Total length, 200; tail vertebræ, 57; hind foot, 23. Average of 7 specimens from type locality: 188; 56; 23. *Skull* (of type): Basal length, 29.4; nasals, 9.5; zygomatic breadth, 17.6; mastoid breadth, 13.4; alveolar length of upper molar series, 7.4.

Remarks.—The range of this form is not widely separated from that of *californicus* on the west slope of the mountains, and the two forms may meet by way of Walker Pass and the South Fork of Kern River. The difference is not sufficiently marked to warrant full specific separation, in view of the fact that their ranges are so nearly continuous that they occupy the same zone and have essentially the same habits.

Specimens examined.—Total number, 52, from the following localities:

California: Lone Pine, 26; Olancha, 3; Cartago (west side of Owens Lake), 3; Independence Creek, 1; Alvord, 8; Bishop Creek, 2; Panamint Mountains (head of Willow Creek at east end of Nelson Range), 9.

MICROTUS CALIFORNICUS CONSTRICTUS subsp. nov. Coast Vole.

Type from Cape Mendocino, California. No. 98347, U. S. Nat. Mus., Biological Survey Collection. Collected Sept. 6, 1899, by Vernon Bailey. Original number, 7174.

Geographic distribution.—Coast region near Cape Mendocino.

Habitat.—Open grassy hillsides and old fields and pastures.

General characters.—Smaller and grayer than *californicus*, with narrower skull, smaller audital bullæ, and narrower interpterygoid fossa.

Color.—*Summer pelage* (in September specimens): Buffy gray above, whitish below; tail almost concolor, dull grayish; feet gray.

Cranial characters.—Skull smaller and especially narrower than that of *californicus*, with slender nasals and rostrum; bullæ small and narrow; pterygoids close together; zygomatic arches not abruptly spreading and not notched at anterior junction with premaxillæ; dentition as in *californicus;* tooth rows noticeably closer together.

Measurements.—Average of 4 adults from type locality: Total length, 163; tail vertebræ, 55; hind foot, 21.5. *Skull* (of type): Basal length, 26; nasals, 8.9; zygomatic breadth, 15.6; mastoid breadth, 12; alveolar length of upper molar series, 6.7.

General remarks.—There is a striking similarity in the characters separating this narrow-skulled form from its widely distributed species *californicus*, and those separating *angusticeps* of the coast region a little farther north from the still more widely distributed *mordax*, that shows an interesting parallelism in geographic modifications. Both forms are from the wind-beaten coast strip where arboreal vegetation is scanty and dwarfed, and, like some of the trees, they apparently represent depauperate forms of widely distributed and more protected inland species. At Capetown, just back of Cape Mendocino, California, I found *constrictus* in great abundance on the open grassy hills. In some places the ground was perforated with their burrows, while on the surface their runways crossed in all directions.

Specimens examined.—Four from the type locality, besides a large number in the flesh.

MICROTUS EDAX (Le Conte). Tule Vole.

Arvicola edax Le Conte, Proc. Acad. Nat. Sci. Phila., VI, 405, 1853.

Type locality.—California [south of San Francisco].[1]

Geographic distribution.—Bottom of the San Joaquin and Sacramento valleys, in Lower Austral zone.

Habitat.—Tule swamps and wet places, under heavy grass, where the runways usually extend through mud and water and in places are flooded by the tide.

General characters.—Size large; feet large and stout; hair long and coarse; skull long, angular, and much ridged in adults; hip glands inconspicuous or rudimentary in adult males.

Color.—*Winter pelage:* Much blacker than in *californicus;* gray of upperparts more or less obscured by black; that in full, ripe pelage is glossed with iridescent purple;[2] sides more grayish; belly washed with whitish; feet dusky; tail bicolor, black above and gray below or dusky gray above and whitish below. *Summer pelage:* Upperparts less glossed with black. *Young:* With black back, dusky sides, and paler dusky belly; feet and tail dark.

Cranial characters.—Skull similar to that of *californicus*, but larger, more elongated, more heavily ridged in adults, with more expanded jugal and heavier dentition; molar pattern similar; m2 with usually an open posterior fifth loop.

Measurements.—No. 58128, ♂ ad., from near Tracy, Calif.: Total length, 217; tail vertebræ, 72; hind foot, 25. An immature ♀ from the same place: 167; 49; 23. An adult ♂, No. 70602, from near Marysville Buttes: 208; 67; 25. *Skull* (No. 70602, ♂ ad.): Basal length, 30; nasals, 9.2;

[1] Baird, Mamm. N. Am. 532, 1857.

[2] *M. californicus* and many other species show a purple gloss in high pelage, but less marked than in some specimens of *edax*.

zygomatic breadth, 17.7; mastoid breadth, 13.9; alveolar length of upper molar series, 7.8.

Remarks.—The type of *M. edax* in the United States National Museum, at Dr. Merriam's request, has been relaxed, the skull removed, and the skin made over and greatly improved for purposes of comparison. The base of the skull has been cut away, but enough remains to show that the specimen is immature and is the large swamp species, instead of *californicus*. The hind foot gives the only reliable measurement. In the dry skin it measures, flattened out, 23.5, and is proportionately stout. This is fully up to the flesh measurement of No. 57909 from Tracy, though the skull shows the latter to be slightly older.

M. californicus and *M. edax* differ widely in habits, their ranges conform to different zones, the distinctive characters are certainly strong enough for full specific recognition, and the present series shows no intergradation.

Specimens examined.—Total number, 50, from the following localities:

> **California**: Tracy, 2; Marysville, 1; near Marysville Buttes, 1; Union Island (San Joaquin River), 1; Suisun, 21; Tulare Lake, 2; Mendota, 19.

MICROTUS SCIRPENSIS sp. nov. Desert Vole.

Type from Amargosa River (near Nevada line), Inyo County, Calif. No. $\frac{33372}{45485}$, ♀ ad., U. S. Nat. Mus., Biological Survey Collection. Collected February 26, 1891, by Vernon Bailey. Original number, 2520.

Geographic distribution.—Known only from the type locality.

Habitat.—Wet ground under tall tules (*Scirpus olneyi*), where the runways extend through mud and water in a little marsh around a warm spring.

General characters.—Size and proportions about as in *edax*, colors not so dark, tail long, skull heavy and angular, middle upper molar with rounded open or closed posterior loop.

Colors.—*Winter pelage:* Upperparts dark buffy gray, slightly darker than in *californicus*, but not so black as in *edax;* belly smoky gray, tail indistinctly bicolor, brown above, grayish below; feet brownish gray, not dusky. *Young:* Upperparts black, belly grayish, a black dorsal stripe retained until the animals are half grown.

Cranial characters.—Skull of adult angular and heavily ridged; in general characters resembling that of *edax*, but with more truncate posterior tip of nasals, heavier dentition, and well-developed inner posterior loop of middle upper molar. The same characters and larger size distinguish it from those of *vallicola* and *californicus*, and the wide incisive foramina with many other characters distinguish it from that of its nearest neighbor—*nevadensis*.

Measurements.—Type: Total length, 210; tail vertebræ, 67; hind foot, 25. Average of 6 adults: 203; 65; 25.1. *Skull* (of type): Basal length, 31; nasals, 10; zygomatic breadth, 19; mastoid breadth, 13.6; alveolar length of upper molar series, 8.7.

General remarks.—*Microtus scirpensis* stands nearest to *M. edax*, and, except for the more completely developed posterior loop of middle upper molar, fits into the *californicus* group. Among 14 specimens the loop is closed in 7 and open in 7, while among 43 specimens of *edax* it is closed in 2, open in 32, and absent in 9, and among 100 specimens of *californicus* it is closed in 3, open in 20, and absent in 77. Although resembling *pennsylvanicus* in the fifth loop, in other characters it does not approach that group or any of its forms. In range it comes nearer to *vallicola* than to *edax* or *californicus*, but from *vallicola* it differs in the same way as from *californicus*.

Specimens examined.—Total number, 14, from the type locality.

MICROTUS OPERARIUS (Nelson). Tundra Vole.

Arvicola operarius Nelson, Proc. Biol. Soc. Wash., VIII, 139, Dec. 28, 1893.

Type locality.—St. Michael, Alaska.

Geographic distribution.—Barren grounds from Bristol Bay, St. Michael, and Kowak River, Alaska, east to Anderson River.

Habitat.—Mossy tundras.

General characters.—Size small; tail short, densely haired; ears small and wholly concealed in long winter fur; colors yellowish; skull slender and narrow; dentition light.

Color.—*Winter pelage:* Upperparts dark rich buff, slightly tinged along back with black-tipped hairs; sides paler; belly pale buffy or creamy white; tail soiled whitish below and on sides, a partly concealed blackish dorsal line; feet gray; heels tinged with dusky. *Summer pelage:* Darker yellowish above, more buffy below.

Cranial characters.—Skull rather slender and narrow, angular and well ridged in adults; nasals slender, ending even with arm of premaxillæ; bullæ small and narrow; palate low; incisive foramina short, constricted posteriorly; incisors projecting well in front of nasals; molars very light; m2 with 4 closed sections; m3 with 3 closed triangles, 3 outer and 3 inner salient angles and terminal loop; m1 with 4 closed triangles, 3 outer and 5 inner salient angles, and fifth triangle open and confluent with short terminal loop, as in *M. ratticeps* of Europe.

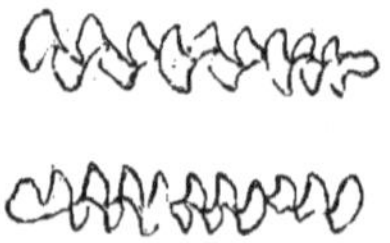

Fig. 4.—Molar enamel pattern of *Microtus operarius* (× 5).

Measurements.—Type (immature, measured in dry skin): Total length, 110; tail vertebræ, 28; hind foot, 18. Average of 10 adult topotypes, 168; 40; 19.7. *Skull* (of type): Basal length, 22.4; nasals, 5.8; zygomatic breadth, 12.4; mastoid breadth, 10.7; alveolar length of upper molar series, 5.5. Skull (of adult ♂, No. 9899): 27; 8; 16; 12.3; 6.3.

General remarks.—Mr. E. W. Nelson found these voles abundant along the coast tundras of Bering Sea from Cape Vancouver north to Bering

Strait, and on Nelson, St. Michael, and Stewart islands.[1] Mr. W. H. Osgood found them extending southward into the timbered region as far as the point where the Yukon crosses the Alaska boundary.

Specimens examined.—Total number, 81, from the following localities:

> **Alaska:** St. Michael, 65; Kowak River, 1; Kagiktowik, 1; Bristol Bay, 1; Fort Yukon, 8; Circle, 1; 40 miles above Circle, 2; International boundary on Yukon, 1; Yukon River (200 miles southwest of Fort Yukon), 1.

MICROTUS MACFARLANI Merriam. Macfarlane Vole.

Microtus macfarlani Merriam, Proc. Wash. Acad. Sci., II, 24, March 14, 1900.

Type locality.—Fort Anderson, Anderson River, Northwest Territory.

Geographic distribution.—Tundra region of Arctic America, east of the Mackenzie River.

General characters.—Similar to *operarius* in external characters, but with shorter tail, shorter, wider skull and more projecting incisors. Fur very long and soft in winter specimens.

Color.—*Winter pelage:* Upperparts light buffy gray; belly whitish; feet silvery gray; tail sharply bicolor, black above, white below. *Summer pelage:* Darker and brighter buff or ochraceous. *Young:* More grayish.

Cranial characters.—Compared with *operarius:* Skull short and wide; nasals shorter; incisors more projecting; bullæ wider; incisive foramina shorter; molars slightly heavier; enamel pattern the same. With *yakutatensis:* Size smaller; coloration brighter; skull flatter; nasals shorter; incisors more projecting; interparietal smaller.

Measurements.—Type (in dry skin): Tail vertebræ, 29; hind foot, 18.5; topotype (No. 9144): tail, 27; hind foot, 19. *Skull* (of type): Basal length, 26; nasals, 7; zygomatic breadth, 15.5; mastoid breadth, 12.5; alveolar length of upper molar series, 6.

General remarks.—Material is so scanty from the Arctic regions that little is known of the range of this form, whether it meets and grades into *operarius* or *yakutatensis*, or whether it has a restricted and isolated range.

Specimens examined.—Total number, 18, from the following localities:

> **Northwest Territory:** Fort Anderson, 4; Mackenzie River, 11; 'Arctic Coast,' 3.

MICROTUS YAKUTATENSIS Merriam. Yakutat Vole.

Microtus yakutatensis Merriam, Proc. Wash. Acad. Sci., II, 22. March 14, 1900.

Type locality.—Yakutat Bay (north shore), Alaska.

Geographic distribution.—Mainland of Alaska from Glacier Bay to Prince William Sound.

---

[1] Proc. Biol. Soc. Wash., VIII, 140, 1893.

General characters.—Size medium, about equal to *operarius*, less than that of any of the island forms of the group. Color dusky, as in *sitkensis*, but belly whitish. Skull of adult male heavily ridged; bullæ medium, rounded; interparietal large, shield-shaped.

Color.—*Summer pelage:* Upperparts dusky gray, with a trace of brownish, darker dorsally; belly washed with soiled white or pale buffy; tail sharply bicolor, sooty or black above, whitish below; feet silvery gray, soles black. *Young:* Quarter-grown specimen (June 19), darker gray than adult, with black nape, whitish belly, sharply bicolor, black and white tail.

Cranial characters.—Skull heavy, ridged and angular in adult male; interparietal large, shield-shaped; nasals long, with median constriction; dentition heavy. From the skull of *operarius* it differs in greater width, larger bullæ, heavier dentition; from that of *unalascensis* in smaller size, larger interparietal, slenderer nasals, smaller bullæ, shorter pterygoids. Molar pattern as in *operarius*. Skulls of adult females conspicuously smoother and less ridged than in males.

Measurements.—Average of 10 adults (5 males and 5 females) from type locality: Total length, 161; tail vertebræ, 37; hind foot, 20.6. *Skull* (of type, ♂ ad., No. 98005): Basal length, 28; nasals, 8; zygomatic breadth, 16; mastoid breadth, 13; alveolar length of upper molar series, 7.

General remarks.—This mainland form is readily distinguished from any of the island species of the group by either cranial or external characters, although it shows closer relationship with some of them than with the neighboring mainland species, *operarius*. If it has an uninterrupted range to the north it may grade into *operarius*, but at present there is no intermediate material to show whether it does or not.

Specimens examined.—Total number, 47, from the following localities:

Alaska: Yakutat, 29; Glacier Bay, 17; Prince William Sound, 1.

MICROTUS KADIACENSIS Merriam. Kadiak Vole.

Microtus kadiacensis Merriam, Proc. Biol. Soc. Wash., XI, 222, July 15, 1897.

Type locality.—Kadiak Island, Alaska.

Geographic distribution.—Known only from the type locality.

General characters.—Size about that of *sitkensis;* belly white; ears very small; bullæ small and narrow; basioccipital short and wide.

Color.—*Summer pelage:* Yellowish brown above, with scattered black hairs; sides paler; belly washed with pure white; feet silvery gray, heels dusky, soles blackish; tail not sharply bicolor, black above, whitish below. *Young* (in June): Dull buffy gray above, maltese below.

Cranial characters.—Skull flat, long, and narrow; audital bullæ small and laterally compressed; basioccipital short and wide; palate low, with sloping median ridge, lateral pits deeper, and incisive foramina wider than in *sitkensis;* nasals short, not reaching posterior tips of premaxillæ; incisors projecting; molars small; m3 with 3 closed triangles and elliptical terminal loop, making 3 inner and 3 outer salient angles, as in *operarius*.

Measurements.—Average of five adult males: Total length, 188; tail vertebræ, 50; hind foot, 21. *Skull* (of topotype, ♂ ad. No. 97969,): Basal length, 28; nasals, 8; zygomatic breadth, 16.5; mastoid breadth, 13.2; alveolar length of upper molar series, 6.4.

Remarks.—This insular species belongs to the *operarius* group, but differs from *operarius* in larger size, smaller bullæ, and transversely longer and narrower interparietal.

Specimens examined.—Total number, 12, from the type locality.

MICROTUS UNALASCENSIS Merriam. Unalaska Vole.

Microtus unalascensis Merriam, Proc. Biol. Soc. Wash., XI, 222, July 15, 1897.

Type locality.—Unalaska, Alaska.

Geographic distribution—Island of Unalaska.

General characters.—Larger and more robust than *operarius*; belly white; feet light gray; skull well arched and heavily ridged; bullæ large and well rounded.

Color.—Upperparts dull yellowish brown, darkest on head and rump; end of nose whitish; belly white or slightly soiled whitish; feet light gray with dusky soles; tail bicolor, a narrow line of blackish above, soiled white below. *Young:* Similar to adult.

Cranial characters.—Skull considerably arched, deep, heavy, and angular; frontals heavily ridged in old age; bullæ medium, much larger than in *operarius* or *kadiacensis*; basioccipital narrowly constricted between bullæ; dentition actually and relatively heavier than in *operarius*; molar pattern the same; m3 with 3 closed triangles and an inner salient angle confluent with rounded posterior loop; m1 with 4 closed triangles, the fifth triangle confluent with shortened anterior loop.

Measurements.—Type, ♀ im.: Total length, 122; tail vertebræ, 28; hind foot, 19. Adult ♂ topotype, No. 97963: 181; 38; 22. *Skull* (of topotype): Basal length, 30; nasals, 8; zygomatic breadth, 17.7; mastoid breadth, 14; alveolar length of upper molars, 7.

Specimens examined.—Total number, 9, from the type locality.

MICROTUS UNALASCENSIS POPOFENSIS Merriam. Popof Island Vole.

Microtus unalascensis popofensis Merriam, Proc. Wash. Acad. Sci., II, 22, March 14, 1900.

Type locality.—Popof Island, Shumagin group, Alaska.

Geographic distribution.—Known only from Popof Island.

General characters.—Similar to *kadiacensis*, but slightly larger, with larger feet, relatively shorter and sharply bicolor tail. Skull less ridged, with larger bullæ and heavier molars. In size and proportions more nearly agreeing with *unalascensis*, from which it differs in wholly dusky nose, less ridged skull, smaller audital bullæ and deeper prezygomatic notch.

Color.—General coloration not readily distinguishable from that of *kadiacensis;* upperparts dark yellowish brown; nose dusky to tip; belly soiled white or pale buffy; tail sharply bicolor, whitish below, dusky or black above; feet silvery gray, with black soles and dusky heels.

Cranial characters.—Skull rather long and narrow; frontals not ridged in adults; prezygomatic notch deep; audital bullæ medium, not narrowly constricted as in *kadiacensis*, nor large and rounded as in *unalascensis;* palate with posterior point projecting into pterygoid fossa; incisive foramina short and wide; molar pattern as in *operarius* except in m3, which has normally 4 inner and 3 or 4 outer salient angles; m1 has 4 closed and 1 open triangle as in *operarius*.

Measurements.—Average of 3 topotypes: Total length, 165; tail vertebrae, 38; hind foot, 22.4. An adult ♀, No. 97959: 188; 43; 22. *Skull* (of type, No. 97956, ♂ ad.): Basal length, 29.4; nasals, 8.5; zygomatic breadth, 17.5; mastoid breadth, 13.5; alveolar length of upper molar series, 7.2.

General remarks.—This species needs comparison only with *unalascensis* and *kadiacensis*, from both of which it differs in slight external and well-marked cranial characters. The three are evidently from the same original stock that from long insular separation has been modified by somewhat varied conditions.

Specimens examined.—Total number, 7, from the type locality.

MICROTUS SITKENSIS Merriam. Sitka Vole.

Microtus sitkensis Merriam, Proc. Biol. Soc. Wash., XI, 221, July 15, 1897.

Type locality.—Sitka, Alaska.

Geographic distribution.—Known only from Baranof Island, Alaska.

General characters.—Size medium, about that of *unalascensis;* color yellowish brown above and below; skull rather flat, wide interorbitally; interparietal triangular; molars small; m1 with 4 or 5 closed triangles.

Color.—*August pelage:* Upperparts rusty brown, brightest on rump and nose, besprinkled with blackish hairs; sides paler; belly washed with dark buff; nose blackish; feet silvery plumbeous; heels and soles black; tail sharply bicolor, black above, pale buff below.

Cranial characters.—Skull long and flat, with no trace of superciliary ridges, wide interorbitally; tip of nasals reaching back of premaxillæ; interparietal narrow, subtriangular; bullæ medium and globose; palate long and flattened, lateral bridges low, lateral pits shallow; incisive foramina short and narrow; incisors projecting well beyond nasals; dentition slightly more intricate than in *operarius;* m1 has 4 or 5 closed triangles and a rounded terminal loop with a sharp inner salient angle; m3 has 3 closed triangles and 4 inner and 4 outer salient angles.

Measurements.—Type, ♂ ad.: Total length, 155; tail vertebræ, 42; hind foot, 23. Adult ♀ topotype: 190; 45; 22. *Skull:* Basal length, 25.5; nasals, 7; zygomatic breadth, 14.2; mastoid breadth, 11.3; alveolar length of upper molar series, 6. *Skull* (of topotype, ♀ ad.): 30; 8; 17.7; 14; 7.

Remarks.—*Microtus sitkensis* belongs to a well-marked group of the subgenus *Microtus* with the molar pattern of *M. ratticeps* of Europe, although in *sitkensis* m1 is usually closed up, making 5 triangles instead of 4.

Specimens examined.—Total number, 2, from the type locality.

MICROTUS INNUITUS Merriam. Innuit Vole.

Microtus innuitus Merriam, Proc. Wash. Acad. Sci., II, 21, March 14, 1900.

Type locality.—St. Lawrence Island, Bering Sea, Alaska.

Geographic distribution.—Known only from St. Lawrence Island.

General characters.—Size large; tail of medium length, sharply bicolor; skull wide and low, with projecting incisors; dentition mainly as in the *operarius* group.

Cranial characters.—Skull ridged and angular, not much arched; braincase short and wide; nasals short and cuneate, falling considerably back of base of incisors; interparietal small, semicircular; bullæ large, somewhat flattened and angular; pterygoids short; interpterygoid fossa very narrow: dentition heavy, incisors conspicuously projecting; molars with sharply constricted enamel folds; m1 with only 4 closed triangles, m3 with three closed triangles, a short posterior loop, and long posterior inner salient angle.

Measurements.—Tail vertebræ, 44; hind foot, 23. *Skull* (of type): Basal length, 32.5; nasals, 9; zygomatic breadth, 19.5; mastoid breadth, 15.3; alveolar length of upper molar series, 7.2.

General remarks.—The specimens from St. Lawrence Island were taken from regurgitated pellets of owls and jaegers, and consist of skulls, feet, tails, and imperfect skeletons. The animals are abundant. Many were seen running in the grass by members of the Harriman party who landed for a short time on the island. The species of *Microtus* coming geographically nearest to St. Lawrence Island is *tshuktshorum* Miller, from Plover Bay, on the Siberian coast, a tiny species bearing little resemblance to the present one.

Specimens examined.—Ten more or less imperfect skulls, besides feet, tails, and parts of skeletons, from the type locality.

MICROTUS ABBREVIATUS Miller. Hall Island Vole.

Microtus abbreviatus Miller, Proc. Biol. Soc. Wash., XIII, 13, Jan. 31, 1899.

Type locality.—Hall Island, Bering Sea, Alaska.

Geographic distribution.—Known only from Hall Island.

General characters.—Size rather large; tail very short, and densely haired; ears concealed in long fur; feet large and stout, measuring about 23.

Color.—July pelage: Upperparts dark buff or yellowish brown, brightest over ears, face, and rump; belly creamy white or pale buff; tail sharply bicolor, a narrow line of dark brownish above, creamy below; feet soiled white. *Young* (half-grown specimens): Duller and darker.

Cranial characters.—Skull similar to that of *unalascensis* in size and general appearance, but more heavily ridged, with deeper prezygomatic notches, larger, more quadrangular interparietal, shallower lateral pits of palate; slightly smaller and especially narrower audital bullæ; rather lighter dentition, with different molar pattern. m1 and m2 with base of posterior triangles broadly open; m3 shortened, with but 2 closed triangles, 3 outer, and 4 inner salient angles, the third triangle opening into short posterior loop; $\overline{\text{m1}}$ with 5 completely closed triangles, and a well developed anterior trefoil, 4 outer, and 5 inner sharp, salient angles back of terminal loop. From the St. Matthew Island subspecies *fisheri*, it differs in molar pattern as from *unalascensis*, and also in slenderer skull and rostrum, slenderer zygomata, and wider interpterygoid fossa.

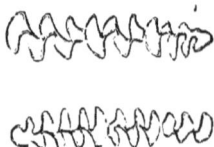

FIG. 5.—Molar enamel pattern of *Microtus abbreviatus* (5).

Measurements.—Average of 4 adult topotypes: Total length, 160; tail vertebræ, 25.7; hind foot, 23.3. *Skull* (of No. 97981, ♂ ad.): Basal length, 30.3; nasals, 8.8; zygomatic breadth, 18; mastoid breadth, 14; alveolar length of upper molar series, 7.

General remarks.—*Microtus abbreviatus* was described from an alcoholic specimen retaining none of the original colors. On the Harriman expedition 7 specimens were collected July 14 on Hall Island, and for the first time the natural appearance of the animal was made known. In external characters it strongly resembles a lemming on account of the short tail, long fur, and stout form, but the skull is that of a robust *Microtus*. It belongs to the subgenus *Microtus*, and in general character comes nearest to the *operarius* group, from which it is excluded, however, by its unique molar pattern—m3 having but 2 closed triangles, while m1 has 5.

Specimens examined.—Total number, 8, from the type locality.

MICROTUS ABBREVIATUS FISHERI Merriam. St. Matthew Island Vole.

Microtus abbreviatus fisheri Merriam, Proc. Wash. Acad. Sci., II, 23, March 14, 1900.

Type locality.—St. Matthew Island, Bering Sea, Alaska.

Geographic distribution.—Known only from St. Matthew Island.

General characters.—Similar to *abbreviatus*, but slightly larger and darker; rostrum longer and heavier; nasals anteriorly expanded and posteriorly notched.

Color.—Summer pelage (July specimens): Upperparts dark rich buff, brightest over ears, face, and rump, sprinkled with black hairs over back; belly strong clear buff; tail buff, with concealed dusky line above; feet pale buffy. *Young* (half-grown specimens): Duller and darker.

Cranial characters.—Skull larger than in *abbreviatus*, with relatively narrower braincase; rostrum longer and heavier on account of the longer, anteriorly spreading nasals; posterior tip of nasals distinctly notched; dentition slightly heavier than in *abbreviatus*, but with essentially the same molar enamel pattern.

Measurements.—Average of 7 adults, 1 male and 6 females, from St. Matthew Island: Total length, 166; tail vertebræ, 27; hind foot, 22.7. One adult ♂: 178; 32; 24. *Skull* (No. 97976, ♂ ad.): Basal length, 31.3; nasals, 9.8; zygomatic breadth, 19; mastoid breadth, 14; alveolar length of upper molar series, 7.3.

General remarks.—*Microtus a. fisheri* needs comparison only with *abbreviatus*, from which it differs in well-marked subspecific characters. The external differences are less marked than the cranial, as a natural result of the very similar conditions on the neighboring islands occupied by the two forms. Hall Island, the home of *abbreviatus*, is separated from St. Matthew Island, the home of *fisheri*, by about 4 miles of deep sea.

Specimens examined.—Total number, 8, from the type locality.

MICROTUS TOWNSENDI (Bachman). Townsend Vole.

Arvicola townsendi Bachman, Journ. Acad. Nat. Sci. Phila., VIII, 60, Pl. I, 1839.
Arvicola occidentalis Peale, U. S. Expl. Expd., Mammalogy, 45, 1848. (Puget Sound.)

Type locality.—Lower Columbia River, near mouth of Willamette, on or near Wappatoo (or Sauvie) Island.

Geographic distribution.—Low country west of the Cascades, from Port Moody, British Columbia, south to the Willamette Valley and to Yaquina Bay, Oregon, in Transition zone.

Habitat.—Open grass land, pastures, fields, and dry meadows.

General characters.—Size large; fur thin and harsh; ears conspicuous above fur; color dark brownish; a pair of conspicuous glands on hips in adult males.

Color.—*Summer pelage:* Back vandyke brown, much darkened with long black hairs; sides dark buffy gray; belly grayish or dusky; tail blackish, scarcely lighter below; feet plumbeous gray. *Winter pelage* (imperfect in late October and April specimens): Slightly grayer above and lighter below. *Young:* Darker than adult, with dusky belly and blackish feet and tail.

Cranial characters.—Skull long and not much arched; angular and heavily ridged in old age; superciliary ridges conspicuous; incisive foramina long and narrow, constricted posteriorly; lateral pits of palate deep; bullæ medium in size and well rounded; dentition heavy; m2 with four closed sections; m3 with 3 closed triangles, 4 inner and 3 outer salient angles; m1 with 5 closed triangles, 5 inner and 4 outer salient angles. The long, narrow incisive foramina distinguish the skull most readily from that of *californicus*.

Measurements.—Adult male from Oregon City, Oreg.: Total length, 226; tail vertebræ, 66; hind foot, 26. Average of 10 adults, 5 ♂ and 5 ♀, from Avon, Wash.: 193; 64; 25.4. *Skull* (No. 56907, from Oregon City, Oreg.): Basal length, 29.2; nasals, 8.4; zygomatic breadth, 17.1; mastoid breadth, 13.5; alveolar length of upper molar series, 7.5.

FIG. 6.—Molar enamel pattern of *Microtus townsendi* (× 5).

General remarks.—*Microtus townsendi* has no close affinities with any species except *tetramerus* of Vancouver Island. Except for the conspicuous hip glands, it comes nearest to the *longicaudus* and *mordax* group in combination of general characters. Almost no variation is shown throughout its rather limited range, and there are no characters by which to recognize *occidentalis* even as a subspecies.

Specimens examined.—Total number, 74, from the following localities:

Oregon: Oregon City, 1; Aumsville, 2; Yaquina Bay, 5; Newport, 6; Shelburn, 1 (im); Salem, 1.

Washington: Tenino, 2; Steilacoom, 3; Roy, 2; Kent, 1; Lake Washington, Seattle (south end), 3; Avon, 25; Mount Vernon, 4; Sauk, 1.

British Columbia: Port Moody, 16; Chilliwack, 1.

MICROTUS TETRAMERUS (Rhoads). Vancouver Vole.

Arvicola (*Tetramerodon*) *tetramerus* Rhoads, Proc. Acad. Nat. Sci. Phila., Oct., 1894, 283.

Type locality.—Beacon Hill Park, Victoria, British Columbia.

Geographic distribution.—Southern end of Vancouver Island.

General characters.—Like *townsendi*, but slightly smaller, with slenderer feet and tail and narrower, slenderer skull.

Color.—Indistinguishable from that of *townsendi*, in either winter or summer pelage.

Cranial characters.—Skull smaller, slenderer, and less arched than in *townsendi*, with superciliary ridges never quite meeting; nasals relatively shorter and more spreading anteriorly; incisors slenderer; molars smaller but with the same enamel pattern.

Measurements.—Average of 6 adult males from Goldstream (near Victoria), Vancouver Island: Total length, 177; tail vertebræ, 54.3; hind foot, 22. The largest of a series of 14: 190; 60; 23. *Skull* (No. 91901, ♂ ad.): Basal length, 27.3; nasals, 7.8; zygomatic breadth, 16; mastoid breadth, 12.4; alveolar length of upper molar series, 6.7.

General remarks.—The marked difference in size, together with slight cranial characters, separates this insular form as an easily recognizable species. Specimens of *townsendi* from the nearest localities on the mainland, Port Moody, British Columbia, and Avon, Wash., show no tendency toward *tetramerus*. The small series of specimens includes both summer and winter pelage.

Specimens examined.—Total number, 17, from near the type locality:

Vancouver Island, B. C.: Goldstream, 16; near Victoria, 1.

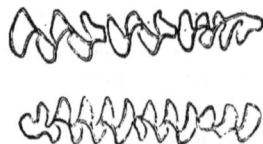

MICROTUS LONGICAUDUS (Merriam). Long-tailed Vole.

Arvicola (Mynomes) longicaudus Merriam, Am. Nat., XXII, 934–935, Oct., 1888.

Type locality.—Custer, S. Dak. (in the Black Hills at an altitude of about 5,500 feet).

Geographic distribution.—Boreal cap of the Black Hills and down some of the cold streams well into the Transition zone.

Habitat.—Banks of cold streams and in mountain meadows.

General characters.—Size of body about equal to that of *Microtus pennsylvanicus;* tail much longer; ears larger; colors grayer; skull flatter; braincase wider.

Color.—*Summer pelage:* Upperparts dull bister, darkened with numerous black-tipped hairs, becoming grayish on the sides and shading into dull, buffy gray on belly; feet plumbeous; tail dimly bicolor, blackish above, soiled whitish below. *Winter pelage* (old and faded in a June specimen from Sundance, Wyo.): Upperparts grayish bister, mixed with blackish-tipped hairs, shading gradually into slightly paler sides and dull whitish belly; tail distinctly bicolor; feet soiled whitish.

Cranial characters.—Skull long and not much arched; rostrum long; nasals reaching to anterior plane of incisors; bullæ large and rounded; molar pattern similar to that of *pennsylvanicus*, except for absence of posterior loop in middle upper molar; m3 with 3 closed triangles, 3 outer and 4 inner salient angles; $\overline{\text{m}}$ 1 with anterior loop, 5 closed triangles, 4 outer and 5 inner salient angles. From *mordax* it differs in slightly shorter, heavier rostrum and wider nasals; narrower interpterygoid fossa; wider expansion of jugal; shorter and wider condyloid ramus of mandible.

Measurements.—Type, ♀ ad.: Total length, 185; tail vertebræ, 65; hind foot, 21. Topotype, ♀ ad.: 184; 61; 22. *Skull* (of type): Basal length, 25; nasals, 7.8; zygomatic breadth, 15.2; mastoid breadth, 11.6; alveolar length of upper molar series, 6.3.

General remarks.—*Microtus longicaudus* stands as one of the few outlying and isolated forms, though the first-described species of its widely distributed group. Its nearest neighbor is *M. mordax* of the Bighorn Mountains, Wyoming, between which range and the Black Hills neither species is known to occur.

Specimens examined.—Total number, 6, from the following localities:

South Dakota: Custer, 2.
Wyoming: Sundance (in the western edge of the Black Hills), 4.

MICROTUS MORDAX (Merriam). Cantankerous Vole.

Arvicola (Mynomes) mordax Merriam, North American Fauna No. 5, 61, July 30, 1891.

Microtus vellerosus Allen,[1] Bul. Am. Mus. Nat. Hist., XII, 7, March, 1899. (Liard River, Northwest Territory.)

Microtus cautus Allen,[1] Bul. Am. Mus. Nat. Hist., XII, 7, March, 1899. (Hell Gate, Liard River, Northwest Territory.)

---

[1] The types of *Microtus vellerosus* and *M. cautus*, kindly loaned me for comparison by Dr. Allen, agree in every character with specimens in corresponding pelage of *M. mordax* from its type locality. The type of *vellerosus*, collected May 4, shows the dark brownish-gray back of imperfect summer pelage, while the type of *cautus*, collected November 29, shows the light gray pelage of early winter.

Type locality.—Sawtooth (or Alturas) Lake, east foot of the Sawtooth Mountains, Idaho.

Geographic distribution.—Rocky Mountains and outlying ranges from latitude 60° to northern New Mexico, and south in the Cascades and Sierra Nevada as far as Kaweah and Kern rivers, California. In the Cascades mainly confined to the east slope, but extending west to the Siskiyous, in southern Oregon, and Salmon and Trinity mountains, in northern California. Found in most of the isolated ranges of eastern Oregon and northern and central Nevada. Common in Canadian and Hudsonian zones.

Habitat.—Marshes and wet woods, but more especially the banks of cold mountain streams, down which it often extends into the Transition zone.

General characters.—Size medium; tail long; ears large; feet small; no conspicuous side or hip glands in males. Very similar to *longicaudus.*

Color.—*Summer pelage:* Back grayish bister; sides olive gray; belly washed with whitish; nose dusky; feet plumbeous; tail dimly bicolor, dusky above, soiled whitish below. *Winter pelage:* Lighter colored than in summer; dorsal stripe of yellowish bister more sharply contrasted with the deeper gray of sides and face; belly heavily washed with pure white; tail sharply bicolor; feet whitish. *Young:* Darker, less sharply marked than the adults; feet and tail dusky.

Cranial characters.—Skull light and slender, similar to that of *longicaudus*, but with slightly longer, slenderer rostrum and nasals; slenderer zygomata, and longer condylar ramus of mandible; dentition essentially the same; m2 with 4 closed sections, the posterior open; m3 with anterior crescent, 3 closed triangles, and posterior loop with 2 inner salient angles; m1 with 5 closed triangles, 5 inner and 4 outer salient angles back of anterior loop; second and third lower molars each with 3 outer and 3 inner salient angles.

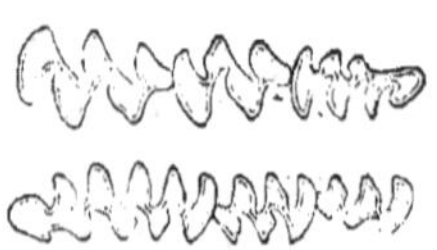

FIG. 7.—Molar enamel pattern of *Microtus mordax* (× 5).

Measurements.—Type: Total length, 200; tail vertebræ, 77; hind foot, 22. Average of five adults from type locality, 182; 66; 22. *Skull* (of type): Basal length, 26.5; nasals, 8.6; zygomatic breadth, 16.2; mastoid breadth, 12.8; alveolar length of upper molar series, 6.6.

General remarks.—The species has a wide and frequently interrupted range, but shows remarkably slight variation of characters. Even from the southern extremities of the Rocky Mountains and Sierra Nevada the variation is too slight for subspecific recognition. Specimens from isolated ranges in Nevada are practically typical.

Specimens examined.—Total number, 708, from the following localities:

Idaho: Sawtooth Lake, 35; Lemhi, 4; Lost River Mountains, 2; Salmon River Mountains, 1; Three Creek, 1; Preuss Mountains, 1; Montpelier Creek, 3; Kingston, 1; Osborn, 1; Mullan, 10; Cœur d'Alene, 7; Craig Mountains, 1; Seven Devils Mountains, 3; Priest Lake (east side), 2.

Utah: Laketown, 6; Park City, 1; Barclay, 3; near Barclay, 4.

Colorado: Estes Park, 2; Ward, 4; Gold Hill, 2; Longs Peak, 10; Canyon City, 1; Lake City, 2; Silverton, 5; Fort Garland, 12.

New Mexico: Chama, 1; Martinez, 1; Agua de Lobo, 1 (no skull).

Wyoming: Bridger Pass, 14; Bighorn Mountains, 1; Lake Fork, near Bull Lake, 4; Clark Fork, mouth of Crandle Creek, 3; Tower Falls, Yellowstone Park, 2.

Montana: Red Lodge, 2; Pryor Mountains, 8; Fort Custer, 2; Big Snowy Mountains, 2; Jefferson River, 1; Blackfoot, 1; St. Marys Lake, 3; Java, 4; Summit, 5; Flathead Lake, 8; Tobacco Plains, 2; Horse Plains, 3; Upper Stillwater Lake, 6; Prospect Creek, 11; Thompson Falls, 2; Silver, 4.

Nevada: Reese River, 18; Arc Dome, 15; Indian Creek, 1; Shoshone Mountains, north of Cloverdale, 5; Pine Forest Range, 3; Granite Creek, 8; Mountain City, 15; Bull Run Mountains, 1; White Mountains, 6; Ruby Mountains, 14; Monitor Mountains, 2.

California: Near Mount Whitney, 31; Olancha Peak, 1; Mulkey Meadows (near Olancha Peak), 1; Soda Springs (on North Fork of Kern River), 1; Mineral King (on East Fork Kaweah River), 19; Upper San Joaquin River, 2; Lone Pine, 2; Bishop Creek, 2; Queen Station, 1; Sequoia National Park, 24; Yosemite Valley, 3; Pine City, 1; Mammoth, 7; White Mountains, 4; Donner, 3; Sierra Valley, 3; Hope Valley, 2; Carberry Ranch, 10; Emerald Bay, 5; Goose Lake, 2; Warner Creek, 1; Lassen Creek, 2; Lassen Peak, 19; Etna, 1; Trinity Mountains, 4; Canyon Creek, 16; Plumas County (20 miles southwest of Quincy), 2; Mount Shasta, 59; Sisson, 15; Goose Nest Mountain, 1; Bear Creek (Shasta County), 1.

Oregon: Siskiyou, 6; Fort Klamath, 18; Crater Lake, 5; Diamond Lake, 5; Sink Creek (east of Mount Thielson), 1; Upper Des Chutes River, Little Meadows (near head of Des Chutes River), 2; Farewell Bend (Des Chutes River, 30 miles southwest Prineville), 2; Swan Lake Valley, 2; head of Drews Creek, 1; Warner Mountains, 3; Steen Mountains, 6; Summit (east of Malheur Lake), 1; 10 miles north of Harney, 9; Maury Mountains, 5; Wallowa Mountains, 3; Lone Rock, 1.

Washington: Cleveland, 2; Wenatchee, 5; head of Lake Chelan, 12; Easton, 2; Conconully, 3.

British Columbia: Mount Richter, 2; Nelson, 1; Sicamous, 5; Hope, 2; Shuswap, 1; Glacier, 3; Okanagan, 2; Bennett City, 6.

Alberta: Henry House, 19; 15 miles south of Henry House, 4; south of Smoky River, 3; Banff, 1.

Northwest Territory: Liard River, 2; Rink Rapids (upper Yukon), 1; Lake Lebarge, 2; Lake Marsh, 1.

Alaska: Charlie Creek (Upper Yukon), 4; Circle, 1; White Pass, 5; Skagway, 1.

MICROTUS MACRURUS Merriam. Olympic Vole.

Microtus macrurus Merriam, Proc. Acad. Nat. Sci. Phila., Aug., 1898, 353.

Type locality.—Lake Cushman, Olympic Mountains, Washington.

Geographic distribution.—(The typical form) Olympic Mountains. (With slight variation) along the coast strip of British Columbia and Alaska north to Yakutat.

Habitat.—Marshes and borders of cold streams.

General characters.—Like *mordax*, but considerably larger, with conspicuously larger hind foot and darker coloration.

Color.—*Summer pelage:* Upperparts dark bister, shaded with numerous black hairs, becoming sooty gray in some specimens; sides slightly paler; belly washed with dull buffy or whitish; feet plumbeous; tail distinctly bicolor, blackish or brownish above, soiled whitish below, usually white-tipped. (*Winter pelage* unknown.) *Young:* Darker than adult, with blackish feet and tail.

Cranial characters.—Skull averaging much larger than *mordax*, with wider interorbital region, heavier rostrum, smaller audital bullæ, and heavier dentition; molars especially wider; lower jaw conspicuously more massive, with wide, heavy molars.

Measurements.—Type: Total length, 220; tail vertebræ, 88; hind foot, 24. Average of five specimens from three localities in the Olympic Mountains: 204; 80; 24.3. *Skull* (of type, No. 66151, ♂ ad.): Basal length, 27.7; nasals, 8.2; zygomatic breadth, 16; mastoid breadth, 12.5; alveolar length of upper molar series, 7.

General remarks.—*Microtus macrurus* is the most conspicuously marked and easily recognizable form of the *longicaudus* group, though with less deeply seated characters than some forms that are externally scarcely distinguishable from each other. In the Olympic Mountains its range is completely isolated, being separated from that of *mordax* by the intervening low country, the habitat of the larger *townsendi*, and by the high Cascades, in which neither form occurs. To the northward it again occurs, in nearly typical form, on the coast at Lund, British Columbia (lat. 50°), and extends northward along the coast to Yakutat, Alaska, becoming slightly smaller and less markedly different from true *mordax*.

Specimens examined.—Total number, 84, from the following localities:

> **Washington**: Lake Cushman, 7; head of Skokomish River, 1; head of Soleduc River, 1; Quinault Lake, 4; Granville, 1.
>
> **British Columbia**: Lund, on Malaspina Inlet, 3; River Inlet (head), 14; Fort Simpson, 6.
>
> **Alaska**: Loring, 3; Wrangell, 4; Juneau, 12; Yakutat, 7; Yakutat Bay (north shore), 10; Glacier Bay, 11.

MICROTUS ANGUSTICEPS Bailey. Coast Vole.

Microtus angusticeps Bailey, Proc. Biol. Soc. Wash., XII, 86, April 30, 1899.

Type locality.—Crescent City, California.

Geographic distribution.—Coast region of northwestern California and southwestern Oregon.

Habitat.—Damp pastures in the Sitka spruce belt.

General characters.—Smaller and darker colored than typical *mordax*, with very narrow, slender skull and small audital bullæ.

Color.—*Summer pelage:* Upperparts dark bister, lined with black hairs, darkest on face and nose; sides paler; belly washed with creamy white; feet plumbeous gray; tail distinctly bicolor, blackish above, soiled white below.

Cranial characters.—Skull small and very narrow, distinctly ridged in adults; nasals projecting in front of incisors; incisive foramina short; audital bullæ very small and constricted; coronoid notch of mandible narrow; incisors slender; molars small, with narrow, sharp angles; enamel pattern as in *mordax*.

Measurements.—Type, ♂ ad.: Total length, 170; tail vertebræ, 56; hind foot, 22. An adult ♀ topotype: 170; 55; 22. *Skull* (of type): Basal length, 23.4; nasals, 7.6; zygomatic breadth, 13.5; mastoid breadth, 10.8; alveolar length of upper molar series, 6.

General remarks.—Externally this species is not very different from true *mordax*, but the skull shows such marked characters as to warrant full specific rank.

Specimens examined.—Total number, 45, from the following localities:

California: Crescent City, 31; Arcata (Humboldt Bay), 13.
Oregon: Gold Beach, 1.

MICROTUS ALTICOLUS (Merriam). Mountain Vole.

Arvicola (*Mynomes*) *alticolus* Merriam, North American Fauna No. 3, 67-69, Pl. V, figs. 1 and 2; Pl. VI, figs. 1, 2, 3, and 4; Sept. 11, 1890.

Type locality.—San Francisco Mountain, Arizona (Little Spring, on northwest side of mountain, altitude 8,200 feet).

Geographic range.—Boreal zone of San Francisco Mountain, from 8,200 feet altitude up to timberline at 11,000 feet.

Habitat.—Vicinity of springs and cold streams on the slopes of the mountain.

General characters.—Similar to *longicaudus*, but tail shorter, hind foot and ear smaller, and skull with truncate, instead of pointed, anterior end of frontal and deeper lateral pit of palate.

Color.—*Summer pelage:* Upperparts uniform sepia or dull bister, darkened with blackish-tipped hairs; sides scarcely lighter; belly pale buffy or whitish; feet dull grayish or dirty whitish; tail not sharply bicolor, blackish above, grayish below. *Young:* Similar to adults, but with woolly fur and long, scattered, black-tipped hairs. (*Winter pelage* unknown.)

Cranial characters.—Skull similar to that of *mordax*, but readily distinguished from it and those of all other forms of the group by truncate end of anterior arm of frontal. Other characters are, deeper lateral pits of the palate; wider interpterygoid fossa; slightly longer, more open, incisive foramina. Dentition similar to that of *longicaudus* and *mordax*.

Measurements.—Type, ♀ ad.: Total length, 170; tail vertebræ, 56; hind foot, 20. Average of 5 adults from type locality: 178; 56; 20. *Skull:* Basal length, 25; nasals, 7.5; zygomatic breadth, 14.8; mastoid breadth, 12.3; alveolar length of upper molar series, 6.5.

General remarks.—*Microtus alticolus*, with its subspecies *leucophæus*, is the most isolated form in the *longicaudus* group. Its geographically nearest neighbor and probably nearest relative is *mordax*, in the mountains of Colorado and northern New Mexico.

Specimens examined.—Total number, 13, from the type locality. One immature specimen from Springerville, in the White Mountains, may be either *alticolus* or *leucophæus*.

MICROTUS ALTICOLUS LEUCOPHÆUS (Allen). Graham Mountain Vole.

Arvicola leucophæus Allen, Bul. Am. Mus. Nat. Hist., VI, 320–321, Nov. 7, 1894.

Type locality.—Graham Mountains, Arizona.

Geographic distribution.—Known only from the type locality.

General characters.—Similar to *alticolus*, and of the same proportions, but slightly larger, color the same, skull wider interorbitally and with other slight differences.

Color.—*Summer pelage:* Upperparts sepia or dull bister, but little paler on sides and faintly lined with blackish hairs; belly washed with soiled whitish; feet dull grayish; tail distinctly bicolor, brown above, grayish below. (*Young* and *winter pelages* not represented.)

Cranial characters.—Skull wider interorbitally than in *mordax* or *alticolus;* anterior arm of frontal with triangular instead of truncate point; incisive foramina wider than in *alticolus*, slightly constricted posteriorly; lateral pits of palate wide and shallow; coronoid notch narrow and sharp. Dentition as in *alticolus* and *mordax*.

Measurements.—Type, ♀ ad.: Total length, 173; tail vertebræ, 50; hind foot, 22.5. Topotype, ♀ ad.: 183; 50; 23. *Skull* (of type): Basal length, 26.5; nasals, 8.3; zygomatic breadth, 15.2; mastoid breadth, 12.3; interorbital width, 4.2; alveolar length of upper molar series, 6.3.

General remarks.—*Microtus leucophæus* belongs to the *longicaudus* group. It is closely related to *alticolus*, from which size and slight cranial characters separate it as a fairly well-marked subspecies.

Through the kindness of Dr. J. A. Allen, of the American Museum of Natural History, I have the type and a topotype of *M. leucophæus* for comparison with the Biological Survey series of *alticolus*, *mordax*, and *longicaudus*.

Specimens examined.—Total number, 2, from the type locality.

MICROTUS MEXICANUS (De Saussure). Mexican Vole.

Arvicola (*Hemiotomys*) *mexicanus* De Saussure, Revue et Mag. de Zool., 2e sér., XIII, 3, Jan., 1861.

Type locality.—Mount Orizaba, Puebla, Mexico.

Geographic distribution.—Eastern Puebla and to the north and west, grading into its subspecies *phæus*.

Habitat.—Grassy places in open forests, in upper Austral and Transition zones.

General characters.—Size rather small; tail short; ears conspicuous; pelage coarse and lax; colors brownish; skull wide, with short, wide incisive foramina; m1 normally with 6 inner salient angles.

Color.—*Winter pelage:* Upperparts grizzled brown, from a mixture of dull cinnamon and black; sides paler; belly washed with cinnamon or buffy, or rarely with whitish; sides of nose and ear coverts usually a brighter shade of cinnamon; feet clear gray; tail dusky above, gray below. *Summer pelage* (imperfectly represented): Evidently darker and less ferruginous. *Young:* Duller and darker than adult.

Cranial characters.—Skull rather angular, with wide-spreading zygomatic arches, narrow interorbital constriction, and large, well-rounded audital bullæ; incisive foramina short and wide, truncate posteriorly; zygomata broadly flattened; palate with a median groove between the lateral pits where a spur or ridge appears in most species of *Microtus;* upper incisors abruptly decurved; molar pattern differing from that of *nanus* and *montanus*—mainly in extra angle of anterior trefoil of m1; most of the salient angles acute; m2 has 4 closed sections; m3 has 3 closed triangles, 3 outer and 4 inner salient angles; m1 has 5 closed triangles, 5 outer and 6 inner salient angles.[1]

Measurements.—Average of 10 adults from the type locality (5 ♂ 5 ♀): Total length, 138; tail vertebræ, 29; hind foot, 19.35; maximum: 148; 30; 20. *Skull* (of topotype, No. 54496, ♂ ad.): Basal length, 24.5; nasals, 7.4; zygomatic breadth, 15.3; mastoid breadth, 11.6; alveolar length of upper molar series, 6.6.

General remarks.—*Microtus mexicanus, phæus, fulviventer*, and *mogollonensis* form a well-marked and closely united group of small, short-tailed, brownish voles, distinguished by the arrangement of mammæ in two pairs, a pair of inguinal, and a pair of pectoral; by wide-spreading zygomatic arches and narrow interorbital constriction; by wide incisive foramina and grooved posterior ridge of palate; and by similar habits and habitat. They need comparison only among themselves. While *mogollonensis* and *fulviventer* are well-marked forms, occupying widely separated and probably disconnected areas, *mexicanus* and *phæus* merely show the extremes of differentiation found in one wide-ranging and somewhat variable form. As only those from the type localities are really typical, any line separating *mexicanus* and *phæus* is purely arbitrary.

Specimens examined.—Total number, 194, from the following localities in Mexico:

Puebla: Mount Orizaba, 27; Chalchicomula, 22.
Vera Cruz: Cofre de Perote, 29; Las Vegas, 11.
Tlaxcala: Mount Malinche, 1; Huamantla, 2.
Hidalgo: Sierra de Pachuca, 7; Tulancingo, 7; Real del Monte, 10.
Morelos: Huitzilac, 4.
Mexico: Ajusco, 6; Toluca Valley, 20; North slope Volcan de Toluca, 9; Mount Popocatepetl, 19; Amecameca, 1; Salazar, 19.

MICROTUS MEXICANUS PILEUS (Merriam). Colima Vole.

Microtus phæus Merriam, Proc. Biol. Soc. Wash., VII, 171-172, Sept. 29, 1892.

Type locality.—North slope of Sierra Nevada de Colima, Jalisco, Mexico (altitude 10,000 feet).

Geographic distribution.—Southern Jalisco and northward to northwestern Chihuahua (to the eastward, grading into *mexicanus*), occupying Boreal and Transition zones.

[1] This extra number of angles is a weak character depending on the slightly unusual development of the anterior trefoil of m1, so that a pair of short points or angles may in most cases be counted on its inner and outer sides.

Habitat.—Grassy parks in open timber.

General characters.—Similar to *mexicanus*, but slightly larger and a shade darker, and with slight cranial differences.

Color.—*Winter pelage:* Upperparts uniform dark cinnamon brown mixed with blackish; belly a lighter shade of cinnamon or buffy, or sometimes whitish; feet brownish gray; tail brownish gray, paler below. *Summer pelage:* Not shown in specimens from near the type locality, but June specimens from El Salto, Durango, are brighter and darker ferruginous than topotypes in winter pelage. *Young:* Dull brownish.

Cranial characters.—Skull similar to that of *mexicanus*, but with less constricted interorbital region, slightly shorter incisive foramina, and shallower prezygomatic notches. Dentition essentially the same.

Measurements.—Average of 10 adult topotypes (5 ♂ and 5 ♀): Total length, 151; tail vertebræ, 35; hind foot, 20.5. *Skull* (topotype, ♂ ad., No. 45645): Basal length, 25.2; nasals, 7.3; zygomatic breadth, 15.5; mastoid breadth, 12; alveolar length of upper molar series, 6.08.

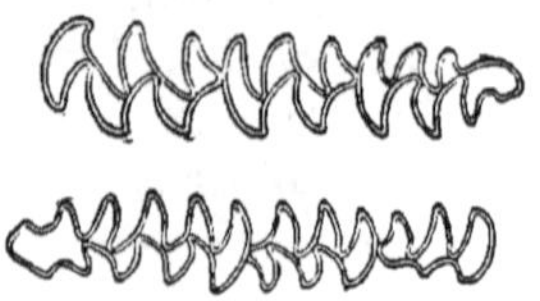

Fig. 8.—Molar enamel pattern of *Microtus phæus* (× 5).

General remarks.—*Microtus m. phæus* is not a strongly or sharply marked form of *mexicanus*, although apparently the more widespread form. Ajusco and Salazar specimens might as well be referred to it as to *mexicanus*. A large series from near Guadalupe, in southwestern Chihuahua, are indistinguishable from typical *phæus*, and those from Miquihuana, western Tamaulipas, are nearer to *phæus* than to *mexicanus*.

Specimens examined.—Total number, 136, from the following localities in Mexico:

Jalisco: Sierra Nevada de Colima, 17.
Michoacan: Nahuatzin, 23.
Queretaro: Pinal de Amoles, 9.
Tamaulipas: Miquihuana, 20.
Durango: El Salto, 25.
Chihuahua: Sierra Madre near Guadalupe y Calvo, 19; Colonia Garcia, 23.

MICROTUS FULVIVENTER Merriam. Oaxaca Vole.

Microtus fulviventer Merriam, Proc. Biol. Soc. Wash., XII, 106, April 30, 1898.

Type locality.—Cerro San Felipe, Oaxaca, Mexico.

Geographic distribution.—Central part of the State of Oaxaca.

Habitat.—Open grassy places and along edges of fields in the Boreal zone.

General characters.—Slightly larger than *mexicanus* and of nearly the same proportions; darker and richer in coloration; ears conspicuous above fur; tail a little more than one and a half times the length of hind foot.

Color.—*Summer pelage:* Upperparts umber brown, darkened by black hairs; under parts fulvous or dull chestnut brown; feet grayish brown;

tail dusky brown above, fulvous below, darker toward the end. *Winter pelage* (in October and March specimens): Less deeply colored. *Young:* Dull sooty, with scarcely a trace of fulvous.

Cranial characters.—Skull similar to that of *mexicanus*, but with smaller bullæ, longer incisive foramina; sharper posterior point of frontals; molars slightly heavier, enamel pattern almost the same: m1 has a more rounded anterior loop.

Measurements.—Average of 10 adult topotypes (5 ♂ and 5 ♀): Total length, 152; tail vertebræ, 35; hind foot, 20.5. Type (♂ ad.): 154; 38; 20. *Skull* (of type): Basal length, 25.4; nasals, 7.4; zygomatic breadth, 15.5; mastoid breadth, 12.4; alveolar length of upper molar series, 6.5.

General remarks.—*M. fulviventer* belongs to the *mexicanus* group, but is sharply separated in its distinguishing characters as well as in geographic range. But little variation is shown throughout its range and although specimens from the mountains near Ozolotepec show differences, these are slight and unimportant.

Specimens examined.—Total number, 126, from the following localities in Mexico:

> **Oaxaca**: Cerro San Felipe, 32; Reyes, 23; 15 miles west of Oaxaca, 20; Mount Zempoaltepec, 28; near Cajones, 5; Guajamaloya, 1 (im.); mountains near Ozolotepec, 9; Totontepec, 8.

MICROTUS MOGOLLONENSIS (Mearns). Mogollon Mountain Vole.

Arvicola mogollonensis Mearns, Bul. Am. Mus. Nat. Hist., II, No. 4, 283–284, Feb., 1890.

Type locality.—Baker Buttes, Mogollon Mountains, Arizona.

Geographic distribution.—Plateau country of central Arizona.

Habitat.—Dry grassy parks among the yellow pines of the Transition zone.

General characters.—Size small; tail and feet short; color dull rusty brown; fur long and soft; ears not concealed; skull short, wide, and angular; lateral pits of palate very deep; an inner projecting point at base of posterior triangle of middle upper molar.

Color.—Upperparts dull rusty brown, brightest on tips of ears; sides slightly paler; belly cinnamon or buffy gray; feet grayish brown; tail brownish gray above, grayish below.

Cranial characters.—Skull short and well arched, with wide-spreading zygomata and sharply constricted interorbital region; zygomatic shield broad and flat; interparietal small and narrow; nasals notched posteriorly, falling considerably short of terminus of premaxillæ; bullæ full and rounded; incisive foramina short, wide, and open; incisors wider than in *nanus* and bent more abruptly downward; molar pattern as in *mexicanus*, except in $\underline{\text{m2}}$, which has an inner point at base of posterior triangle, and in $\overline{\text{m1}}$, which has 5 closed triangles and only 5 inner and 4 outer salient angles and an abbreviated terminal loop.

Measurements.—Average of 10 adults from San Francisco Mountain, Arizona: Total length, 131; tail vertebræ, 28.5; hind foot, 18. *Skull* (adult ♂, No. 24563): Basal length, 23.6; nasals, 7; zyg matic breadth, 15.2; mastoid breadth, 12; alveolar length of upper molar series, 6.3.

Remarks.—*M. mogollonensis* is widely separated, both geographically and specifically, from the other members of its group. Its nearest ally is *phæus* from Mexico.

Specimens examined.—Total number, 51; from the following localities:

Arizona: San Francisco Mountain (Little Spring on north side of mountain), 15; Springerville, 35.

New Mexico: Fort Wingate, 1.

MICROTUS XANTHOGNATHUS (Leach). Yellow-cheeked Vole.

Arvicola xanthognatha Leach, Zool. Miscel., I, 60, 1814.

Type locality.—Hudson Bay.

Geographic distribution.—Northwestern Canada and Alaska, from central Alberta north to the Arctic coast and west to central Alaska.

General characters.—Size large, almost equaling that of *richardsoni*, but tail shorter and ears larger; colors dull; nose and ear patch yellow; skull heavy, ridged, and angular. Side glands as in *richardsoni*, or a little farther back on flanks.

Color (March and May specimens).—Upperparts dark sepia to bister, heavily lined with coarse black hairs over the back; sides of nose and ear patch bright rusty yellowish, a tinge of the same around eyes and on cheeks; belly dusky gray; breast sooty; tail indistinctly bicolor, blackish above, dusky gray below; feet sooty.

Cranial characters.—Skull smaller than that of *richardsoni* and relatively longer and narrower, with less projecting incisors; nasals long and narrow; bullæ large; incisive foramina long and narrow; dentition heavy; molar pattern scarcely different from that of *townsendi;* anterior loop of m.1 small and triangular; middle section of m.3 frequently divided into two nearly closed triangles.

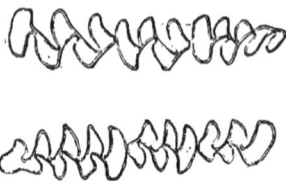

FIG. 9.—Molar enamel pattern of *Microtus xanthognathus* (× 5).

Measurements.—Two dry skins from Fort Resolution, Great Slave Lake, Canada, adult females, in U. S. Nat. Mus. No. 4504: Total length, 210; tail vertebræ, 50; hind foot, 27. No. 4502: 218; 45; 25. *Skull* (No. 4504): Basal length (approximately), 34.5; nasals, 10.3; zygomatic breadth, 20; mastoid breadth, 15.7; alveolar length of upper molar series, 8.5.

Remarks.—*Microtus xanthognathus* shows no close relationship to any other American species. In the position of side glands it resembles *richardsoni*, but in no other characters. It shows a strong superficial resemblance to *chrotorrhinus* in color, but the great difference in size prevents the possibility of one ever being mistaken for the other.

Specimens examined.—Total number, 44, from the following localities:

Canada: Nelson River, N. W. T., 1; Cache Apocotte (40 miles east of Henry House, Alberta), 1; Fort Resolution, 22; Great Slave Lake, 1; Fort Rae, 1; Liard River, 1; La Pierre House, 1; Anderson River, 1; Arctic coast (east of Fort Anderson), 2.

Alaska: Mouth of Porcupine River, 1; Yukon (200 miles southwest of mouth of Porcupine), 3; Charlie Creek (Upper Yukon), 9.

MICROTUS CHROTORRHINUS (Miller). Rock Vole.

Arvicola chrotorrhinus Miller, Proc. Boston Soc. Nat. Hist., XXVI, 189-193, pl. 3, 1894.

Type locality.—Mount Washington, New Hampshire, head of Tuckerman Ravine (altitude, 5,300 feet).

Geographic distribution.—Mount Washington, the Catskills, central Quebec, and northern New Brunswick, in the Hudsonian zone.

Habitat.—Rocky places near water on the mountains, and in deep spruce forests farther north.

General characters.—Size and proportions of *pennsylvanicus* except slightly smaller hind foot; ears larger: fur lax; conspicuously yellowish about nose, ears and rump; skull comparatively thin-walled and smooth; dentition unique.

Color.—*Summer pelage* (July 14): Upperparts bright glossy bister, lined with black hairs; nose to eyes dull orange rufus; hairs around ears and on rump yellowish; belly plumbeous; feet dark gray; tail grayish brown, slightly paler below. Worn, left-over *winter pelage:* Darker and more rusty above.

Cranial characters.—Skull light and smooth, somewhat flattened on top, superficially resembling that of *Evotomys;* bullæ large and rounded; incisive foramina short and wide; dentition rather light; incisors bent down at right angles to axis of skull, extending scarcely beyond tip of nasals; m2 with 4 closed sections; m3 normally with 5 closed triangles, 5 inner and 5 outer salient angles and a double-lobed posterior loop; m3 with outer salient angles prominent and reentrant angles deep.

Measurements.—Type: Total length, 165; tail vertebræ, 45; hind foot, 19.4. Average of 4 adult topotypes: 170; 48; 19.6. *Skull* (of ad. ♂, No. 2523, Bangs Coll.): Basal length, 25.4; nasals, 7.2; zygomatic breadth, 15; mastoid breadth, 12; alveolar length of upper molar series, 6.4.

Remarks.—*Microtus chrotorrhinus* shows a marked superficial resemblance to *M. xanthognathus*, but in cranial characters it differs widely from this and all other American species and is quite unique in the subgenus *Microtus*. In the specimens examined there is no trace of hip glands, but in two adult males (3845 and 3849) there appear to be rudiments of side glands on the flanks.

Specimens examined.[1]—Total number, 8, from the following localities:[2]

New Hampshire: Mount Washington, 3.

New York: Catskill Mountains, 1.

Quebec: Lake Edward, 4.

---

[1] Type in collection of G. S. Miller, jr.; other specimens in the collection of E. A. and O. Bangs.

[2] Mr. Miller records a specimen in the Am. Mus. Nat. Hist. from Trousers Lake, New Brunswick—Proc. Bost. Soc. Nat. Hist., XXVI, 193, 1894.

MICROTUS CHROTORRHINUS RAVUS Bangs. Gray Rock Vole.

Microtus chrotorrhinus ravus Bangs, Proc. Biol. Soc. Wash., XII, 187. Nov. 16, 1898.

Type locality.—Black Bay (north shore of Strait of Belle Isle), Labrador.

Geographic distribution.—Known only from the type locality.

General characters.—Similar to *chrotorrhinus*, but slightly grayer, and with noticeably more yellowish on nose and face. Skull slenderer, with lighter dentition.

Color.—*Summer pelage* (July specimens): Upperparts grayish bister, becoming yellowish on rump; whole face from behind ears suffused with yellowish, brighter on nose; belly thinly washed with white over the plumbeous; feet buffy gray; tail brownish above, slightly paler below.

Cranial characters.—*Skull*, compared with that of *chrotorrhinus*, slightly thinner, lighter, and slenderer throughout; interorbital constriction narrower; rostrum longer and narrower; incisive foramina longer; molar series shorter and narrower; tooth pattern as in *chrotorrhinus*.

Measurements.—Type,[1] ♂ ad.: Total length, 170; tail vertebræ, 50; hind foot, 22. Average of 4 adult topotypes: 159; 46; 21.25. *Skull* (No. 7952, ♀ ad.): Basal length, 25; nasals, 7; zygomatic breadth, 14.3; mastoid breadth, 11.3; alveolar length of upper molar series, 6.2.

General remarks.—July specimens from the type locality are in full, long pelage, with a freshness of appearance and brightness of color quite different from the type of *chrotorrhinus* of nearly the same date. There is a question as to whether the real summer pelage is shown.

Specimens examined.—Total number, 5, from the type locality.

Subgenus ARVICOLA[2] Lacépède.

Arvicola Lacépède, Nouv. Tableau Méthod. Mamm., in Mém. de l'Instit., Paris, III, 495, 1801. Type, *Mus terrestris* Linnæus (genus).

Arvicola Lataste, Le Naturaliste, II, 349, 1883 (subgenus).

Geographic distribution.—(In America) Boreal zone of the Cascades and Rocky Mountains of Canada and the northern United States.

Subgeneric characters.—(In American species) plantar tubercles, 5; side glands on flanks of males conspicuous; a musk-bearing anal gland; mammæ, 8; pectoral, 2–2; inguinal, 2–2; feet large; tail long; fur full and long; bullæ very small; incisors projecting far beyond premaxillæ; molars with constricted and tightly closed sections; m2 with 4 closed sections; $\overline{m3}$ with 3 closed triangles; m1 with 5 closed

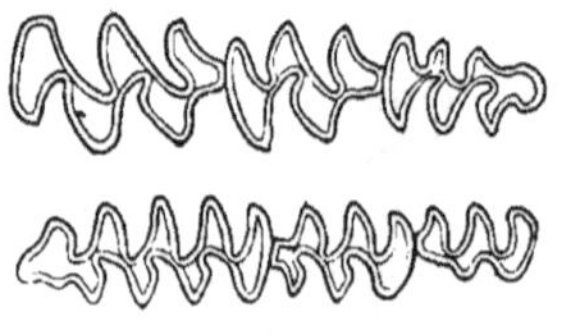

FIG. 10.—Molar enamel pattern of *Microtus* (*Arvicola*) *macropus* (× 5).

---

[1] The measurements of type and 4 topotypes are from original description. In 3 adult topotypes the hind foot measures uniformly 20 mm. in dry skins with toes straight.

[2] For full synonymy of the subgenus *Arvicola*, see Miller, N. Am. Fauna No. 12, 66, 1896.

triangles, 5 inner and 4 outer salient angles; m3 with 3 transverse loops. (In the European section of the subgenus m3 has but 2 closed triangles, and m1 but 3.)

MICROTUS RICHARDSONI (De Kay). Richardson Vole.

Arvicola richardsoni De Kay, Zoology of New York, Mammalia, 91, 1842.

Type locality.—"Near the foot of the Rocky Mountains." (Type collected by Drummond in the vicinity of Jasper House, Alberta.)

Geographic distribution.—The typical form is known only from the vicinity of Jasper House and Henry House, Alberta, Canada. To the south it apparently grades into *macropus* in the Rocky Mountains, and *arvicoloides* in the Cascades.

General characters.—Size very large (probably not exceeded in America except by *alleni*); tail long; feet large; fur long and heavy; ears mostly concealed; colors dull; skull large and angular, with protruding incisors.

Color.—Early winter pelage (October 14 to 18): Upperparts uniformly grayish sepia, darkened by black-tipped hairs, becoming paler on the sides; belly washed with white over the plumbeous under-fur; feet silvery gray; tail bicolor, dusky above, soiled whitish below.

Cranial characters.—Skull large, with wide-spreading zygomatic arches, long rostrum, very long and protruding incisors; nasals rather short and wide, rounded at both ends, not reaching to base of incisors; audital bullæ small for a *Microtus*, but slightly larger and more rounded than in either of the subspecies; incisive foramina longer and less constricted posteriorly; superoccipital smooth without median ridge; terminal loop of third upper molar normally not recurved.

Measurements.—Young adult, No. 81381, from 25 miles west of Henry House: Total length, 208; tail vertebræ, 61; hind foot, 28. *Skull* (of same): Basal length, 32.5 (over incisors); nasals, 8.3; zygomatic breadth, 19.9; mastoid breadth, 14.6; alveolar length of upper molar series, 7.5.

General remarks.—Richardson's specimens, which he referred to *Arvicola riparius*, and which were later re-described by De Kay, were collected by Thomas Drummond in the foothills of the Rocky Mountains, probably west of Jasper House. In October, 1896, J. Alden Loring collected a series of 8 specimens at points 15 and 25 miles west of Henry House (40 or 50 miles southwest of Jasper House), Alberta, on the very trail by which Drummond crossed from Jasper House to the Columbia in 1826. Unfortunately, none of this series are fully adult, and but one condition of pelage is shown. By comparing specimens of the same age it becomes evident that the species equals the larger of its two southern forms, *arvicoloides*, though in color it agrees more nearly with the Rocky Mountain form, *macropus*. A single specimen from Glacier, British Columbia, is fairly intermediate between *richardsoni* and its two southern subspecies.

Specimens examined.—Total number, 8, from west of Henry House, Alberta, Canada.

MICROTUS RICHARDSONI MACROPUS (Merriam). Big-footed Vole.

Arvicola macropus Merriam, North American Fauna No. 5, 59–60, 1891.

Type locality.—Pahsimeroi Mountains, Idaho (altitude about 9,700 feet).

Geographic distribution.—Boreal zone of the Rocky Mountains from the Wasatch north to Canada, of the Wind River Mountains of Wyoming, the Blue Mountains of Oregon, and most of the intermediate ranges.

General characters.—Similar to *richardsoni*, but evidently somewhat smaller; colors the same in October specimens; distinguished by less projecting incisors and other cranial characters.

Color.—*Summer pelage:* Upperparts dark sepia, lined with long, black hairs; slightly paler on sides; belly washed with silvery white; feet dusky gray; tail distinctly bicolor throughout its length, sooty above, whitish below. *Winter pelage* (imperfect in October and May specimens): Lighter, clearer gray above, black hairs less conspicuous, more heavily washed with white below. *Young:* Like adult or slightly darker, with long, woolly fur and dusky feet and tail; during one stage of pelage with entirely dusky belly.

Cranial characters.—Skull similar to that of *richardsoni*, from which it differs in less projecting upper incisors, longer nasals with narrower posterior tips; supraoccipital with a median ridge; bullæ smaller, more compressed and angular; incisive foramina sharply constricted posteriorly; third upper molar with terminal loop recurved in about half of the specimens.

Measurements.—Type, ♀ ad.: Total length, 220; tail vertebræ, 71; hind foot, 26. Male from type locality (not fully adult): 202; 68; 28. *Skull* (of type): Basal length, 31.5; nasals, 9; zygomatic breadth, 19.7; mastoid breadth, 14.5; alveolar length of upper molar series, 7.5.

General remarks.—In a large series of specimens from numerous ranges of the Rocky Mountains the subspecies remains rather uniform. There is a slight increase in size toward the north, specimens from northern Montana averaging larger than from the type locality. Those from Strawberry Butte and the Wallowa Mountains of eastern Oregon are indistinguishable from the type series. Two half-grown specimens from the top of the Wasatch Mountains, near Park City, Utah, are too young to indicate reliable specific variation.

Specimens examined.—Total number, 113, from the following localities:

Idaho: Pahsimeroi Mountains, 8; Salmon River Mountains, 11; Sawtooth (or Alturas) Lake, 18; head of Wood River, 2; Summit, Alturas County, 1; Seven Devils Mountains, 2; head of Crow Creek, Preuss Mountains, 4; Thompson Pass, 3; Priest Lake, 1.

Utah: Park City, 2.

Wyoming: La Barge Creek, 5; South Pass City, 1; Lake Fork, Wind River Mountains, 10.

Montana: Beartooth Mountains, 17; Summit, Teton County, 2; St. Marys Lake, Teton County, 4.

Oregon: Strawberry Butte, 2; Wallowa Mountains, 20.

MICROTUS RICHARDSONI ARVICOLOIDES (Rhoads). Cascade Water-Vole.

Aulacomys arvicoloides Rhoads, Am. Nat., XXVIII, 182–185, Feb. 11, 1894.
Microtus principalis Rhoads, Am. Nat., XXIX, 940, Oct., 1895. (Mount Baker Range, British Columbia.)

Type locality.—Lake Keechelus, near Snoqualmie Pass, Kittitas County, Washington (altitude 8,000 feet).

Geographic distribution.—Boreal zone of the Cascade Mountains, in Washington and Oregon.

General characters.—Apparently equal to *richardsoni* in size, larger than *macropus*, and slightly darker than either. In cranial characters nearer *macropus* than *richardsoni*.

Color.—*Summer pelage:* Upperparts dark sepia, considerably darkened with coarse black hairs; belly thinly washed with pearl gray or silvery whitish; feet dusky gray; tail bicolor, blackish above, soiled whitish below. *Winter pelage:* Darker than summer, with an excess of black hairs above; belly strongly washed with white; feet and tail as in summer. *Young* not different from young *macropus*.

Cranial characters.—Skull like that of *macropus*, but larger; rostrum and incisors slightly heavier; nasals more broadly spreading anteriorly, with a slight constriction near the middle, narrow, and pointed posteriorly; audital bullæ, incisive foramina, and arc of upper incisors as in *macropus*.

Measurements.—Average of 6 adults, 3 males and 3 females, from Easton, Wash. (near the type locality): Total length, 234; tail vertebræ, 81; hind foot, 29.3. Largest specimen of the series: 253; 89; 29. *Skull* (No. 41578, ♂ ad., from Easton): Basal length, 36; nasals, 10.8; zygomatic breadth, 23; mastoid breadth, 16.3; alveolar length of upper molar series, 8.3.

General remarks.—I have before me a series of 13 specimens, collected at Easton, on the outlet of Lake Keechelus, about 12 miles from the type locality, and a large number of specimens from the upper slopes of Mount Rainier, as well as farther north and south in the Cascades. It is safe to assume that these Easton specimens are typical, especially as there is little variation shown throughout the Cascades of Washington. As at present known, the ranges of *arvicoloides* and *macropus* are widely separated, but no doubt they meet and coalesce in *richardsoni* of the Canadian Rockies.

Specimens examined.—Total number, 101, from the following localities:

Washington: Easton, 13; head of Cascade River, 2; Mount Rainier and vicinity, 34; Mount St. Helens, 4; Wenatchee, 1.

Oregon: Mount Hood, 11; Marmot, 1; Mount Jefferson, 2; Detroit, 1; Crater Lake, 22; Anna Creek, base of Mount Mazama, 10.

Subgenus PITYMYS [1] McMurtrie.

Psammomys Le Conte, Ann. Lyceum Nat. Hist. N. Y., III, 132, 1830. Type *Psammomys pinctorum* Le Conte. (Not *Psammomys* Cretzschmar, 1828).
Pitymys McMurtrie, Cuvier's Animal Kingdom, Am. edition, I, 434, 1831. Type *Psammomys pinetorum* Le Conte. (New name for *Psammomys* Le Conte.)

[1] For full synonymy of the subgenus *Pitymys*, see Miller, N. Am. Fauna No. 12, 58, 1896.

Geographic distribution (in America).—Southeastern United States, mainly in Upper Austral zone, and a small area in the Tropical zone of southeastern Mexico.

Subgeneric characters.—Plantar tubercles, 5; mammæ, 4, two pairs of inguinal; lateral glands on hips in adult males; ears very small; tail short; fur short, dense, and glossy. *Skull* flat and wide, with quadrate braincase; bullæ small; molars narrow; m3 with 2 closed triangles; $\overline{m1}$ with 3 closed and 2 open triangles; m2 with anterior pair of triangles confluent; m3 with 3 transverse loops.

Fig. 11.—Molar enamel pattern of *Microtus* (*Pitymys*) *pinetorum* (×5).

MICROTUS PINETORUM (Le Conte). Pine Vole.

Psammomys pinetorum Le Conte, Ann. Lyc. Nat. Hist. N. Y., III, 133, Pl. II, 1830 (read Dec. 21, 1829).

Type locality.—Pine forests of Georgia. Probably the old Le Conte plantation at Riceboro.

Geographic distribution.—Georgia and the Carolinas.

Habitat.—Fields, open woods, and grassy uplands.

General characters.—Size small; ears very small; tail short; fur short and fine; colors bright.

Color.—Upperparts bright russet brown with a distinct gloss, becoming lighter on sides; belly dusky, lightly washed with color of back: tail brownish, darker above; feet grayish brown; ears concealed in the fur.

Cranial characters.—Skull short and wide with a quadrangular braincase and abruptly truncate occiput; interparietal wide and normally somewhat lozenge-shaped; mastoids and bullæ relatively small; interpterygoid fossa normally V-shaped. *Molar* series rather short and crowded; two middle triangles of m3 often confluent; first pair of reentrant angles in $\overline{m1}$ usually not meeting between anterior loop and first pair of salient angles.

Measurements.—Average of 2 adult females from Georgetown, S. C.: Total length, 113; tail vertebræ, 18.5; hind foot, 15.5. *Skull* (No. 1523, Merriam Coll., ♂ ad., from Frogmore, S. C.): Basal length, 22.3; nasals, 7.3; zygomatic breadth, 15; mastoid breadth, 12.5; alveolar length of upper molar series, 6.

General remarks.—No definite type locality was given in the original description of *pinetorum*, but the species was said to inhabit the sandy soil of the pine barrens of Georgia. Very probably the type came from the vicinity of the old Le Conte plantation, near Riceboro, Ga. Thirteen specimens in the Merriam collection, from Beaufort County, S. C., about 60 miles from Riceboro, are probably fairly typical, and are taken for the basis of the above description. They represent the extreme development of the bright cinnamon brown and small-eared form of the Atlantic coast region. Northward through the Atlantic States the specimens become darker and duller colored without other important modifications, except a slight average increase in size. The species described as *scalopsoides* from Long Island, and later as *apella* from Pennsylvania, includes this Northern form and seems worthy of recognition as a subspecies.

Specimens examined.—Total number, 32, from the following localities:

Georgia: Columbus, 2.
South Carolina: Beaufort County (Beaufort, Frogmore, and St. Helena Island), 13; Georgetown, 2; Society Hill, 1.
North Carolina: Old Richmond, 1; Raleigh, 11; Tarboro, 2.

MICROTUS PINETORUM SCALOPSOIDES (Aud. & Bach.). Mole like Vole.

Arvicola scalopsoides Aud. & Bach., Proc. Acad. Nat. Sci. Phila., I, 97, 1841.
Arvicola apella Le Conte, Proc. Acad. Nat. Sci. Phila., VI, 405, 1853. Type from Pennsylvania.
Arvicola kennicotti Baird, Mamm. N. Am., 517, 1857. Type from Illinois.

Type locality.—Long Island, New York.

Geographic distribution.—Southern New York and westward to Illinois, southward along the coast, blending into true *pinetorum.*

Habitat.—Open grassy country, meadows, pastures, and waste places.

General characters.—Larger, darker, and duller than true *pinetorum.*

Color.—Upperparts dull brownish chestnut, slightly darkened by dusky-tipped hairs; sides paler; belly lightly washed with dull buff over plumbeous under-fur; feet brownish gray; tail indistinctly bicolor, sooty above, grayish below.

Cranial characters.—Skull similar to that of *pinetorum*, but larger, with heavier molars.

Measurements.—Average of three adults from Lake Grove, Long Island: Total length, 125; tail vertebræ, 20; hind foot, 16.3. *Skull* (No. 88732, same locality): Basal length, 23.5; nasals, 7.4; zygomatic breadth, 16; mastoid breadth, 12.6; alveolar length of upper molar series, 6.6.

Specimens examined.—Total number, 83, from the following localities:

New York: Lake Grove, Long Island, 4; Millers Place, Long Island, 2; Oyster Bay, Long Island, 1; Sing Sing, 4; Lake George, 1; Locust Grove, 1.
Pennsylvania: Philadelphia, 1.
New Jersey: Tuckerton, 3.
Maryland: Laurel, 2; Kensington, 1; Bladensburg, 2.
District of Columbia: Washington, 27.
Virginia: Falls Church, 4; Dunn Loring, 1; Fort Myer, 1; Clark County, 2; Cape Charles, 4; Bellehaven, 1; Wallaceton (Dismal Swamp), 4.
North Carolina: Currituck, 2; Magnetic City, 1.
West Virginia: White Sulphur Springs, 6.
Ohio: A specimen in the U. S. Nat. Mus., collected by Kennicott, is labeled 'Ohio.'
Indiana: Brookville, 2 (approaching *auricularis*); Terre Haute, 1.
Illinois: West Northfield, 2; Warsaw, 2.

MICROTUS PINETORUM AURICULARIS Bailey. Bluegrass Vole.

Microtus pinetorum auricularis Bailey, Proc. Biol. Soc. Wash., XII, 90, April 30, 1898.

Type locality.—Washington, Mississippi.

Geographic distribution.—Northern Mississippi, Tennessee, Kentucky, and southern Indiana, or in a general way the region between the Allegheny Mountains and the Mississippi River, mainly in the Lower Austral zone.

General characters.—Size small, about equaling that of *pinetorum*; ears very large for a *Pitymys* and conspicuous above fur; colors dark and rich, not always darker than *scalopsoides* but richer and more intense; fur short and dense like that of *pinetorum*.

Color.—Upperparts dark rich chestnut darkened by dusky-tipped hairs; belly washed with paler chestnut over dark under-fur; projecting tip of ear with scattered dusky hairs; tail not bicolor, scarcely darker above, like the back or slightly darker; feet dull brownish.

Cranial characters.—Skull like that of *pinetorum* in form and general characters; interpterygoid fossa normally U-shaped instead of V-shaped.

Measurements.—Type: Total length, 120; tail vertebræ, 22; hind foot, 16. Average of six adult specimens from the type locality, measured in the flesh by collector: 119; 22; 17. *Skull* (of type): Basal length, 22.3; nasals, 7; zygomatic breadth, 15.2; mastoid breadth, 12.3; alveolar length of upper molar series, 6.

General remarks.—A series of 31 specimens in the Merriam collection from Eubank, Ky., average darker and richer in coloration than the type series and have equally large ears. Specimens from Brookville, Ind., are dark and dull colored and might pass for either this species or *scalopsoides*. A single specimen from Hickman, Ky., is immature but apparently typical. A flat skin with crushed skull from Barron Springs, near Fredericksburg, Tex., has the large ear and small foot of *auricularis* but the dull color of *nemoralis*.

Specimens examined.—Total number, 45, from the following localities:

Mississippi: Washington, 10.
Kentucky: Hickman, 1; Eubank, 31.
Indiana: Brookville, 1.
Tennessee: Rogersville, 1.
Alabama: Greensboro, 1.

MICROTUS NEMORALIS Bailey. Woodland Vole.

Microtus nemoralis Bailey, Proc. Biol. Soc. Wash., XII, 89, April 30, 1898.

Type locality.—Stilwell (Boston Mountains), Indian Territory.

Habitat.—Open woods and brushland.

Geographic distribution.—West of the Mississippi River from central Arkansas north to Council Bluffs, Iowa.

General characters.—Size, larger than any other species of *Pitymys* in the United States; ears, relatively large; fur, comparatively long and coarse; colors, duller than in *pinetorum*, not so dark as in *scalopsoides* or *auricularis*.

Color.—Upperparts dull chestnut, slightly lined with blackish-tipped hairs over the back and rump, becoming paler on the sides; belly washed with bright cinnamon; tail indistinctly bicolor; feet thinly clothed with pale buffy or sometimes dusky hairs. *Young:* Plumbeous or dark maltese, with a slight tinge of chestnut suffusing the back as maturity is approached.

Cranial characters.—Skull large and relatively elongated; supraoccipital sloping; interparietal short and wide; mastoids and audital bullæ large and projecting farther back than in *pinetorum;* palate often with a posterior point projecting into the U-shaped interpterygoid fossa.

Measurements.—Type: Total length, 130; tail vertebræ, 24; hind foot, 18. Average of five females and five males from the type locality: 135; 25; 18.1. *Skull* (of type): Basal length, 25.3; nasals, 7.7; zygomatic breadth, 16.5; mastoid breadth, 13.4; alveolar length of upper molar series, 7.

General remarks.—Specimens from London, Nebraska, and Council Bluffs, Iowa, are typical or slightly larger than those from the type locality. Those from central Arkansas and eastern Missouri are nearly or quite typical. So far as shown by the present series of specimens, the species stands distinct and apparently unconnected with the other forms of *Pitymys* east of the Mississippi River.

Specimens examined.—Total number, 46, from the following localities:

Indian Territory: Stilwell, Boston Mountains, 16.
Arkansas: Beebe, 5; Hardy, 1.
Missouri: Hunter, 3; Williamsville, 5; Kimswick, 5.
Iowa: Council Bluffs, 4.
Kansas: Neosho Falls, 1.
Nebraska: London, 6.

MICROTUS QUASIATER (Coues). Jalapa Vole.

Arvicola (*Pitymys*) *pinetorum quasiater* Coues. Proc. Acad. Nat. Sci. Phila., 1874, 191–192.

Type locality.—Jalapa, Vera Cruz, Mexico.

Geographic distribution.—Central Vera Cruz and eastern Puebla, on the east slope of the mountains (altitude 4,000 to 5,000 feet), in Humid Tropical and lower edge of Lower Austral zones.

Habitat.—Brushy woodland.

General characters.—Size of *pinetorum;* tail about as long as hind foot; ears large for a *Pitymys;* colors dark; fur glossy.

Color.—*Summer pelage:* Uniformly dark umber or seal brown, slightly paler on belly, feet, and tail; tail slightly paler below than above. *Winter pelage* (in January, specimens from Orizaba and Huauchinango): Darker, richer, and more glossy. *Young:* Darker and duller, inclining to sooty or plumbeous.

Cranial characters.—Skull similar to that of *pinetorum*, but with longer, more quadrate braincase, more prominent postorbital ridges, narrower interorbital space, and larger audital bullæ; dentition slightly heavier; molar pattern the same.

Measurements.—Average of 6 adult males and females from the type locality: Total length, 130; tail vertebræ, 23; hind foot, 17.7. *Skull* (No. 55048, ♀ adult): Basal length, 24; nasals, 7; zygomatic breadth, 12; alveolar length of upper molar series, 6.3.

General remarks.—*Microtus quasiater* is by far the most divergent form of *Pitymys* in America. Its range, so far as known, is restricted to a comparatively small area, 1,000 miles from that of its nearest relative, and reaches into a zone not known to be inhabited by any other species of *Microtus*.

Specimens examined.—Total number, 44, from the following localities:

Vera Cruz: Jalapa, 6; Jico, 8; Orizaba, 10.
Puebla: Huauchinango, 20.

Subgenus LAGURUS Gloger.

Lagurus Gloger, Gemeinnütz. Hand- u. Hilfsbuch d. Naturgesch., I, 98, 1841. Type *Lagurus migratorius* Gloger.
Lagurus Merriam, Am. Naturalist, XXIX, 758, Aug., 1895 (subgenus).

Geographic distribution (in America).—Transition zone of the semiarid parts of the northwestern United States, east of the Cascades and Sierra Nevada.

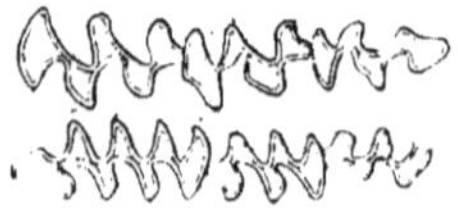

Subgeneric characters.—(In American species) plantar tubercles, 5; mammæ, 8, inguinal 2–2, pectoral, 2–2; lateral glands on flanks; tail little little longer than hind foot; colors pale; fur lax. *Skull* low and wide; bullæ very large; mastoids reaching plane of exoccipital condyles; molars slender, with wide reentrant angles; m3 with 2 closed triangles and narrow posterior loop; m1 with 5 closed triangles, 4 inner and 4 outer salient angles; m̄3 with two terminal transverse loops and a pair of median triangles.

FIG. 12.—Molar enamel pattern of *Microtus (Lagurus) pallidus* (× 5).

MICROTUS CURTATUS (Cope). Short-tailed Vole.

Arvicola curtata Cope, Proc. Acad. Nat. Sci. Phila., 1868, 2.
Arvicola decurtata Coues, Mon. N. Am. Rodentia, 215 (in text), 1877, nomen nudum.

Type locality.—Pigeon Spring, Mount Magruder, Nevada.

Geographic distribution.—Transition zone of the low mountain ranges in western Nevada and eastern California, east of the Sierra Nevada and north of Death Valley.

Habitat.—Dry, barren country, usually in sagebrush.

General characters.—Tail very short; feet hairy; fur long and lax; color pale buffy gray; skull wide and flat, with very large audital bullæ.

Color.—*Winter pelage:* Upperparts uniform pale buffy gray, or ashy gray becoming paler on the sides, and silvery white or soiled whitish below; tail like belly, except an indistinct dusky dorsal line; ears slightly buffy, more noticeably so in young than in adults; feet soiled silvery whitish. *Summer pelage:* Slightly darker. *Young:* Darkened above by long, dusky-tipped hairs; ears distinctly buff tipped.

Cranial characters.—Skull wide and flat, with short rostrum, spreading zygomatic arches, and great mastoid breadth; audital bullæ and mastoids much inflated, and with thick, spongy walls; mastoids projecting back to plane of exoccipital condyles. Molar series rather light, with narrow, tightly closed triangles and wide reentrant angles.

Measurements.—Average of five adults from the type locality: Total length, 141; tail vertebræ, 27; hind foot, 17.6. *Skull* (No. 41019, ♀ ad.): Basal length, 24; nasals, 6.6; zygomatic breadth, 15; mastoid breadth, 13; alveolar length of upper molar series, 6.

General remarks.—This is the largest of the three species of *Lagurus* at present known in North America, size alone being sufficient to distinguish it from either *pauperrimus* or *pallidus*. It shows but slight variation throughout its somewhat restricted and probably interrupted range. Specimens from the head waters of Reese River, which is separated from the type locality by Sonoran valleys, show slightly larger audital bullæ and mastoids, but no characters of specific or subspecific value.

Specimens examined.—Total number, 54, from the following localities:

Nevada: Mount Magruder (near Pigeon Spring), 14; Reese River Valley, 7; Indian Creek (near head of Reese River), 2; head of Reese River, 10.
California: Inyo Mountains, 17; White Mountains, 4.

NOTE.—*Microtus* (*Lagurus*) *pumilus* Elliot (Field Columbian Museum, Zool. Series, Vol. I, No. II, p. 226, Feb. 1, 1899) from the Olympic Mountains, Washington, proves to be a young *Phenacomys*, as I have ascertained by examination of the type specimen, kindly loaned me by Mr. D. G. Elliot, curator of mammals in the Field Columbian Museum.

MICROTUS PALLIDUS (Merriam.) Pallid Vole.

Arvicola pallidus Merriam, Am. Nat., XXII, 702–705, Aug., 1888.

Type locality.—Fort Buford, N. Dak. The type was taken on a northeast slope, near the top of a high, barren hill, 2 miles east of the fort.

Geographic distribution.—Transition prairies of western North Dakota, Montana, and as far north as Calgary, Alberta.

Habitat.—High, semi-arid prairies, usually on shady slopes.

General characters.—Slightly paler than *curtatus;* smaller; with relatively much smaller audital bullæ. The palest species of *Microtus* found in America, and probably the shortest tailed.

Color.—Upperparts uniform pale buffy gray with an extra tinge of buff about ears and nose; belly white or soiled whitish; tail silvery whitish below, dusky above; feet silvery whitish or pale gray. The type series was taken in September and shows what is probably the darkest phase of summer pelage.

Cranial characters.—Skull like that of *curtatus* in general, but averaging slightly smaller and with decidedly smaller audital bullæ and narrower mastoid breadth. Teeth relatively heavy; incisors fully equaling those of larger skulls of *curtatus;* molar series heavy and actually longer in the smallest adults than in much larger specimens of *curtatus;* enamel pattern essentially the same.

Measurements.—Type, ♀ ad.: Total length, 121; tail vertebræ, 20; hind foot, 18 (measured dry). *Skull* (of type): Basal length, 22.3; nasals, 6.5; zygomatic breadth, 14.6; mastoid breadth, 11.6; alveolar length of upper molar series, 6.3.

General remarks.—In size *pallidus* falls between *curtatus* and *pauperrimus*, but in relative size of teeth *curtatus* comes in the middle, while in geographic position *pauperrimus* separates the other two. There is nothing in the material before me to indicate any intergradation between the forms or any subspecific relationship.

Specimens examined.—Total number, 8, from the following localities:

North Dakota: Fort Buford, 4.
Montana: Philbrook, 1.
Canada: Calgary, Alberta, 3.

MICROTUS PAUPERRIMUS (Cooper). Pigmy Vole.

Arvicola pauperrima Cooper, Am. Nat., II, 535–536, Dec., 1868.

Type locality.—Plains of the Columbia, near Snake River, Washington.

Geographic distribution.—Eastern Washington and Oregon, central Idaho, and the north slope of the Uinta Mountains, Utah, in Transition zone.

Habitat.—Open grassy ridges or high prairie, except in the Uinta Mountains, where they were found in grassy parks near the lower edge of pine timber.

General characters.—The smallest species of the subgenus *Lagurus*, with colors a shade darker than in *curtatus* or *pallidus;* skull small and very flat-topped, often concave postorbitally.

Color.—*Summer pelage:* Upperparts uniform buffy gray, slightly darkened with dusky-tipped hairs; ears and nose strongly tinged with buff; belly pale buffy; tail darkened above by a dusky line, buffy below; feet like belly. *Young:* Less buffy and slightly more dusky than adult.

Cranial and dental characters.—Skull small, relatively smooth, not ridged or angled, flat or concave on top; audital bullæ relatively as well as actually smaller than in *curtatus;* hamular process of mandible short and slender, inclosing a wide two-angled or rounded notch; incisors slender; molars differing from those of *curtatus* only in smaller size; enamel pattern essentially the same.

Measurements.—Average of 3 adults from the vicinity of Antelope, Oreg.: Total length, 115; tail vertebræ, 20; hind foot, 16. *Skull* (of adult ♀, No. 78534, from Antelope, Oreg.): Basal length, 20; nasals, 5.5; zygomatic breadth, 13.4; mastoid breadth, 11.3; alveolar length of upper molar series, 5.2.

General remarks.—The above description is based mainly on a series of 6 specimens collected near Antelope, Oreg., on top of the high prairie ridge between the John Day and Des Chutes rivers, and not more than 150 miles from where Dr. J. G. Cooper collected his type of *pauperrimus* on the plains of the Columbia, near Snake River, October 9, 1860. Antelope is in reality on the south edge of the plains of the Columbia, and specimens from that point agree in every way with the somewhat mutilated type of *pauperrimus* still in the United States National Museum. Specimens from the Salmon River Mountains, Idaho, do not differ perceptibly from the type or from the Antelope series. A single specimen from the top of Steen Mountain is not typical, but the characters are not sufficient to warrant separating it on a single specimen, and they may prove only individual. Four specimens from the north slope of the Uinta Mountains, Utah, show but litttle deviation from the typical series.

Specimens examined.—Total number, 19, from the following localities:

Washington: Plains of the Columbia, 1 (the type).
Oregon: Antelope, 6; Bake Oven, 1 (im.); Steen Mountain, 1.
Idaho: Salmon River Mountains, 6.
Utah: Uinta Mountains, 4.

Subgenus CHILOTUS Baird.

Chilotus Baird, Mam. N. Am., 516, 1857. Type, *Arvicola oregoni* Bachman.

Geographic distribution.—Pacific slope from northern California to southern British Columbia.

Subgeneric characters.—Plantar tubercles, 5; mammæ, 8, inguinal, 2–2; pectoral, 2–2; side glands obscure or wanting;[1] ears rather small; fur dense, without stiff hairs. *Skull* short, low, and with elliptical braincase; molars small; m3 with 2 or 3 closed triangles; m1 with 5 closed triangles; m2 with anterior pair of triangles usually confluent; m3 with 3 transverse loops.

MICROTUS OREGONI (Bachman). Oregon Vole.

Arvicola oregoni Bachman, Journ. Acad. Nat. Sci. Phila., VIII, Pl. 1, 60–61, 1839.
Microtus morosus Elliot, Field Columbian Mus., Zool. Series, Vol. I, No. II, 227, Feb. 1, 1899. (Olympic Mountains, Washington.)

Type locality.—Astoria, Oregon.

Geographic distribution.—Pacific coast region from northern California to Puget Sound.

[1] In a large number of skins of adult males, about a dozen show what appear to be ill-defined glands on the sides, midway between hips and shoulders; but before stating postively the presence and position of these glands it will be necessary to examine specimens in the flesh.

Habitat.—Dry open ground, under cover of grass and low vegetation, and under logs in the open redwood forest of California.

General characters.—Size rather small; tail long; colors dark; fur short, without long hairs.

Color.—Upperparts mixed bister and blackish, with a pepper-and-salt appearance; belly dusky, lightly washed with dull buffy; feet dusky gray; tail blackish, slightly lighter below; ears blackish, scantily haired, protruding from the fur.

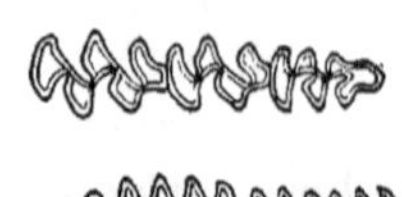

FIG. 13.—Molar enamel pattern of *Microtus* (*Chilotus*) *oregoni* (× 5).

Cranial characters.—Skull, compared with those of other species ot the subgenus, long and slender, with narrower braincase, longer rostrum, more arched and less abruptly spreading zygomata, more quadrangular interparietal; superciliary ridges in adults well marked, approaching or meeting interorbitally; audital bullæ small and globose; dentition not peculiar.

Measurements.—Adult ♂, No. $\frac{17330}{24235}$, from Astoria: Total length, 140; tail vertebræ, 42; hind foot, 17. *Skull* (of same): Basal length, 22.2; nasals, 17; zygomatic breadth, 14.8; mastoid breadth, 11.8; alveolar length of molar series, 5.5.

General remarks.—The species shows some slight variation throughout its range, but all of the specimens examined from the low country south of Puget Sound are clearly referable to *oregoni*, and those from timberline in the Olympics do not vary sufficiently for even subspecific recognition.

Specimens examined.—Total number, 103, from the following localities:

Oregon: Astoria, 6; Oregon City, 6; Yaquina Bay, 1; Aumsville, 1; Elk Head, 1.

California: Crescent City, 12; Humboldt Bay, 1; Hoopa Valley, 2; Dyerville, 3.

Washington: Tenino, 9; Roy, 1; Skamania County (45 miles southeast of Toledo), 1; Kent, 2; Steilacoom, 1; Aberdeen, 3; Granville, 6; Quinaault Lake, Chehalis County, 3; La Push, 3; Suez, 1; Neah Bay, 10; Olympic Mountains (near head of Soleduc River), 9; Lake Cushman, 21.

MICROTUS SERPENS Merriam. Creeping Vole.

Microtus serpens Merriam, Proc. Biol. Soc. Wash., XI, 75, 1897.

Type locality.—Agassiz, British Columbia.

Geographic distribution.—Low country of southern British Columbia and northern Washington between the Cascade Mountains and Puget Sound.

General characters.—Size, largest of the subgenus; tail short; colors dark.

Color.—*Winter pelage:* Upperparts uniformly sooty gray, becoming slightly lighter on sides; belly dusky, washed with dull buff; tail sooty above, silvery gray below; feet dusky gray; ears nearly naked, concealed in long fur. *Summer pelage:* Paler and more brownish above, belly lightly washed with buffy; ears projecting slightly from thinner, coarser fur.

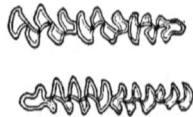

Cranial characters.—Skull rather wide and short; superciliary ridges not well defined; interorbital width greater than in *oregoni;* zygomata abruptly spreading anteriorly, interparietal lozenge-shaped; audital bullæ full and globose; incisors larger and stronger and molars slightly larger than in *oregoni.*

Measurements.—Type: Total length, 130; tail vertebræ, 31; hind foot, 18. Average of 7 adults from type locality: 129; 32; 17.5. *Skull* (of type): Basal length, 22.4; nasals, 6.6; zygomatic breadth, 14; mastoid breadth, 11.2; alveolar length of upper molar series, 5.9.

General remarks.—The range of this northern, more robust form of *Chilotus*, as at present known, is rather limited, but future collections may show it to be continuous with that of *oregoni.*

Specimens examined.—Total number, 56, from the following localities:

British Columbia: Agassiz, 7; Port Moody, 10; Langley, 3; Sumas, 10; Mount Baker Range, 4.
Washington: Mount Vernon, 19; Hamilton, 2; Sauk, 1.

MICROTUS BAIRDI Merriam. Baird Vole.

Microtus bairdi Merriam, Proc. Biol. Soc. Wash., XI, 74, 1897.

Type locality.—Glacier Peak, Crater Lake, Oregon (altitude, 7,800 feet).

Geographic distribution.—Known only from the type locality, but probably restricted to the Hudsonian zone of the higher Cascades.

Habitat.—Beds of *Phyllodoce* and *Lutkea* at timberline.

General characters.—Slightly smaller than *M. oregoni;* color yellowish brown; fur short and glossy; tail short; ears almost concealed in the fur.

Color.—Upperparts glossy yellowish bister; sides paler; belly washed with whitish; tail indistinctly bicolor, dusky above, dark gray below; feet dusky gray; nose dusky.

Cranial characters.—Skull relatively short, wide, and flat, with short rostrum; braincase subquadrate; interparietal narrow; audital bullæ large; ascending arm of premaxillæ not extending beyond nasals; incisive foramina short and wide; dentition not peculiar.

Measurements.—Type, No. 79906, ♀ ad.: Total length, 131; tail vertebræ, 33; hind foot, 17.5. A young adult ♂ from type locality: 130; 32; 17. *Skull* (of type): Basal length, 22; nasals, 6.6; zygomatic breadth, 14; mastoid breadth, 11.5; alveolar length of upper molar series, 5.5.

General remarks.—This species of *Chilotus* is as yet known only from 2 specimens from the type locality. No doubt it will eventually be found to extend along the crest of the Cascade Range in Oregon.

Specimens examined.—Total number, 2, from the type locality.

Subgenus PEDOMYS Baird.

Pedomys Baird, Mamm. N. Am., 517, 1857. Type, *Arvicola austerus* Le Conte.

Geographic distribution.—Middle United States from southern Louisiana to Plains of the Saskatchewan.

Subgeneric characters.—Plantar tubercles, 5; side glands obscure or wanting,[1] rarely discernible; mammæ, 6, inguinal, 2–2, pectoral, 1–1; ears medium; fur long and coarse. *Skull* high and narrow; molars with wide reentrant angles; m3 with 2 closed triangles; m1 with 3 closed and 2 open triangles; m2 with anterior pair of triangles confluent; m3 with 3 transverse loops, the middle loop sometimes constricted, or even divided into 2 triangles.

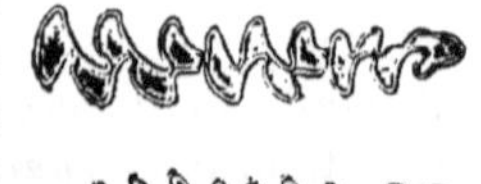

FIG. 14.—Molar enamel pattern of *Microtus (Pedomys) austerus* (×5).

MICROTUS AUSTERUS (Le Conte). Prairie Vole.

Arvicola austerus Le Conte, Proc. Acad. Nat. Sci. Phila., VI, 405–406, 1853.

Arvicola (Pedomys) cinnamonea Baird, Mamm. N. Am., 541, 1857. (Type from Pembina, N. Dak.)

Type locality.—Racine, Wisconsin.

Geographic distribution.—Central part of Mississippi Valley from southern Wisconsin to southern Missouri and Fort Reno, Oklahoma, and west into eastern Nebraska and Kansas.

Habitat.—Dry upland prairie under low grass, and in rose and hazel thickets.

General characters.—Size of *Microtus pennsylvanicus*, but with slightly shorter tail and apparently coarser pelage. Color, dark peppery gray above, dull fulvous below.

Color.—*Winter pelage:* Upperparts dark gray, with a peppery appearance from the mixture of black and pale fulvous tips of long hairs, black tips predominating; sides paler; belly washed with pale cinnamon or fulvous; tail sharply bicolor, feet dusky; a tuft of fulvous hairs in front of ear. *Summer pelage:* Darker throughout, with fewer light-tipped hairs and thinner fulvous wash below. *Young:* Slightly paler than adult.

Cranial characters.—Skull high, narrow, and well arched; interparietal small, lozenge shaped; premaxillæ extending well back of nasals; audital bullæ small and narrow; incisive foramina wide posteriorly; molar pattern, that of the subgenus.

Measurements.—No. 2928, ♂ ad., from Racine, Wis., (measured from alcohol by Baird)[2]: Total length, 127; tail vertebræ, 32; No. 2897: hind foot, 19. *Skull* (No. 1999, ad., from Racine): Basal length, 27; nasals, 7.9; zygomatic breadth, 15.5; mastoid breadth, 12.6; alveolar length of upper molar series, 6.8. *Skull* (No. 948—not fully adult—from Racine): 25; 7.9; 15.4; 11.8; 6.

[1] A large number of skins of males show no trace of side glands, but a few show what appear to be very small glandular areas on the middle of the sides. It will be necessary to examine fresh specimens of old males before the presence or position of the glands is fully determined.

[2] Mamm. N. Am., 541, 1857.

Considerable variation is shown throughout the range of the species. To the southwest, at Orlando and Fort Reno, Okla., the individuals show slightly deeper coloration and slight modifications of cranial characters. Except for a slightly abnormal tooth pattern Baird's type of *cinnamonea* is a large specimen of typical *austerus*. I cannot believe that it ever came from Pembina.

Specimens examined.—Total, 211, from the following localities:

Wisconsin: Racine, 4.
Illinois: West Northfield, 14; Warsaw, 1.
Indiana: Wheatland, 4.
Iowa: Fairfield, 1; Knoxville, 93.
Nebraska: Blair, 1 (im.) Columbus, 7; London, 13; Norfolk, 1.
Kansas: Cairo, 4; Onaga, 13; Burlington, 1; Doniphan County, 1; Fort Leavenworth, 21.
Missouri: Golden City, 2; Piedmont, 10; Bismark, 6; Kimswick, 6.
Oklahoma: Orlando, 1; Fort Reno, 7.

MICROTUS LUDOVICIANUS sp. nov. Louisiana Vole.

Type from Iowa, Calcasieu Parish, Louisiana. No. 96624, ♂ ad., U. S. Nat. Mus., Biological Survey Collection. Collected April 7, 1899, by Vernon Bailey. Collector's number, 6767.

Geographic distribution.—Coast prairie of southwestern Louisiana, in Lower Austral zone.

Habitat.—Dry grassy mounds on the flat, half-marshy coast prairie.

General characters.—Size and proportions about as in *M. austerus*, color similar; rostrum and nasals slenderer and audital bullæ larger.

Color.—*Winter pelage* (in April specimens): Upperparts dark gray, with a coarse, peppery appearance, produced by the mixture of black-, brown-, and whitish-tipped hairs, and varying in color as these different colored hairs predominate; below dull fulvous or dark buffy; tail indistinctly bicolor, dusky above, buffy below; feet dusky. *Young* (quarter to half grown): Darker, more dusky, and less brownish than adult.

Cranial characters.—Skull like that of *austerus* with larger, more rounded audital bullæ, larger molars, and slenderer nasals. Middle section of m3 often constricted or separated into two closed triangles. (This may occur in any species of *Pedomys*.)

Measurements.—Average of 10 adults (5 males and 5 females) from type locality: Total length, 164; tail vertebræ, 33; hind foot, 18.5. Average of hind foot of males, 19; of females, 18. Type: 146; 36; 19. *Skull* (of type): Basal length, 25.8; nasals, 8; zygomatic breadth, 15; mastoid breadth, 11.5; alveolar length of upper molar series, 6.3.

Remarks.—A single imperfect skull in the United States National Museum collection from Calcasieu Parish, La., showed such pronounced characters as to suggest the collection of the present series of specimens. Some of the characters in the old skull prove to be abnormal, and the actual differences between this southern form and true *austerus* are not strongly marked. There is no known and probably no actual intergradation or continuity of range between the two forms, and perhaps subspecific rank would show better the close relationship of *ludovicianus* to *austerus*.

Specimens examined.—Total number, 26, from Calcasieu Parish, La.

MICROTUS HAYDENI (Baird). Hayden Vole.

Arvicola haydeni Baird, Mamm. N. Am., 543–544, 1857.

Type locality.—Fort Pierre, South Dakota.

Geographic distribution.—Plains region of western South Dakota, Nebraska, and Kansas, eastern Colorado and Wyoming, and southern Montana, in Transition zone.

Habitat.—Dry prairies. At the type locality, in sagebrush on badland hills.

General characters.—Considerably larger and lighter colored than *M. austerus*, with little or none of the fulvous or cinnamon wash below; fur very long and lax in winter and spring pelages; skull heavy and angular.

Color.—(May specimens from Fort Pierre): Upperparts uniform light gray, the color formed by a combination of whitish- and blackish-tipped hairs, with the white-tipped predominating; belly washed with silvery white, or sometimes soiled white, over plumbeous under-fur; feet dusky gray; tail bicolor. *Summer pelage:* Somewhat darker, with sometimes a slight wash of buff below. *Young* (one-fourth to one-half grown): Very woolly and slightly darker than adult.

Cranial characters.—Skull larger, more angular, and more heavily ridged than in *austerus;* anterior part of zygomatic arches more abruptly spreading; prezygomatic notch deeper; interparietal larger; palate higher, with more prominent median ridge.

Measurements.—Adult ♀ from type locality (No. 4239, Merriam Coll.): Total length, 180; tail vertebræ, 47; hind foot, 22. *Skull* (No. 4971 from Fort Pierre): Basal length, 28; nasals, 8; zygomatic breadth, 17.6; mastoid breadth, 12.6; alveolar length of upper molar series, 7.4.

General remarks.—Probably *haydeni* intergrades with true *austerus*, and is merely a more robust and paler northwestern form. The ranges of the two almost meet, if they are not continuous.

Specimens examined.—Total number, 110, from the following localities:

South Dakota: Fort Pierre, 4; Pierre, 3; Buffalo Gap, 4; Rapid City, 4.
Nebraska: Valentine, 10; Kennedy, 11; Sidney, 2; Callaway, 4; Alliance, 2.
Kansas: Pendennis, 10; Banner, 11.
Colorado: Loveland, 1; Canyon City, 1.
Wyoming: Beaver, 1; Newcastle, 1; Sundance, 1; Dayton, 1; Pass, 4.
Montana: Little Bighorn Valley, 8; Fort Custer, 24; Custer Station, 1; Lake Basin, 2.

MICROTUS MINOR (Merriam). Least Upland Vole.

Arvicola austerus minor Merriam, Am. Nat., XXII, 598–601, July, 1888.

Type locality.—Bottineau, at base of Turtle Mountains, North Dakota.

Geographic distribution.—Northern border of the Great Plains from northeastern North Dakota to Edmonton, Alberta, and southeastward to Minneapolis, Minn.

Habitat.—Dry upland prairie.

General characters.—Size very small, scarcely as large as *Evotomys gapperi* and of about the same proportions; color peppery gray; pelage long, lax, and coarse; sixth tubercle on hind foot usually present, though small; skull small and slender.

Color.—*Winter pelage:* Upperparts uniform, clear peppery gray, from a combination of black- and whitish-tipped hairs; belly washed with soiled white or pale buffy; tail sharply bicolor, dusky above, buffy below; feet gray. *Summer pelage:* With a mixture of fulvous above; belly with thinner wash of light-tipped hairs over dark under-fur. *Young:* Slightly darker than adult with less peppery appearance of fur.

Cranial characters.—Skull very small, not much arched, slender and narrow, with relatively heavy rostrum, narrow strap-shaped interparietal and slender zygomata; audital bullæ small and laterally compressed; molars with enamel pattern of the subgenus.

Measurements.—Type: Total length, 133; tail vertebræ, 36; hind foot, 16.5. Average of four adults from Sherbrook, N. Dak.: 128; 30; 16.7. *Skull* (No. 49230, ♀ ad., from Sherbrook): Basal length, 22.3; nasals, 6.4; zygomatic breadth, 12.2; mastoid breadth, 10; alveolar length of upper molar series, 5.

General remarks.—A mere glance at the skulls shows *minor* to be widely separated from any other species of the subgenus, differing from *austerus* in much smaller size, narrower braincase, and relatively smaller and narrower audital bullæ. The species shows little variation throughout its range over the prairie region, but those occupying the half-timbered region of south-central Minnesota show a marked intensity of color.

Specimens examined.—Total number, 94, from the following localities:

North Dakota: Bottineau, 3; Sherbrook, 4; Devils Lake, 1.
South Dakota: Traverse, 7.
Minnesota: Ortonville, 6; Elk River, 40; Fort Snelling, 7; Hamlington, 1.
Canada: Carberry, Manitoba, 1; Indian Head, Assinaboia, 11; Wingard, Saskatchewan, 10; Red Deer, Alberta, 1; Edmonton, Alberta, 2.

Subgenus ORTHRIOMYS Merriam.

Orthriomys Merriam, Proc. Biol. Soc. Wash., XII, 107, April 30, 1898. Type, *Microtus umbrosus* Merriam.

Geographic distribution.—That of its only known species.

Subgeneric characters.—Plantar tubercles, 5 (a rudiment of 6th); side glands wanting or very rudimentary; mammæ, 4, pectoral, 2–2; ears large and almost naked; feet large; tail long and scantily haired. *Skull* long and narrow; bullæ very small; posterior median ridge of palate sloping and grooved; m3 with 2 closed rounded triangles, and a third open one; m1 with 3 closed triangles, 4 inner and 3 outer salient angles; m2 with the anterior pair of triangles confluent; m3 with 4 closed sections including 2 median triangles.

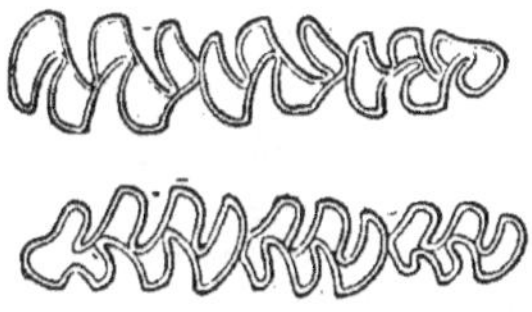

FIG. 15.—Molar enamel pattern of *Microtus (Orthriomys) umbrosus* (x 5).

MICROTUS UMBROSUS Merriam. Zempoaltepec Vole.

Microtus umbrosus Merriam, Proc. Biol. Soc. Wash., XII, 108, Apr. 30, 1898.

Type locality.—Mount Zempoaltepec, Oaxaca, Mexico (altitude 8,200 feet).

Geographic distribution.—Known only from the east slope of Mount Zempoaltepec, in the humid Upper Austral zone.

Habitat.—Dense oak forests, living in burrows and long underground tunnels.

General characters.—Size rather large; tail long; ears large; fur long and lax; colors dark; skull long and flat, with small bullæ and peculiar dentition.

Color.—Upperparts uniform dusky, with brown-tipped hairs; below dark plumbeous thinly washed with fulvous; feet and tail thinly haired, concolor, dark brown.

Cranial characters.—Skull long, narrow, and but little arched, with smooth outlines, and slender zygomatic arches; bullæ very small; palate low, with slender or incomplete lateral bridges, shallow lateral pits and grooved posterior ridge; interpterygoid fossa wide and quadrate; incisive foramina short and widest in the middle. Dentition heavy; incisors abruptly decurved; inner salient 'angles' of upper and posterior lower molars rounded instead of acute; m3 with a small outer and a large inner closed triangle and a posterior trefoil with large inner and small outer lobe; m3 with 2 median closed triangles, an outer and inner, and broad terminal loops.

Measurements.—Average of 7 specimens from type locality: Total length, 184; tail vertebræ, 65; hind foot, 23. Type: 177: 61; 23.5. *Skull* (of type): Basal length, 26.5; nasals, 7.3; zygomatic breadth, 16; mastoid breadth, 12; alveolar length of upper molar series, 7.

Specimens examined: Total number, 15; from the following localities in Mexico.

Oaxaca: Mount Zempoaltepec (above Totontepec), 8; Totontepec, 7.

Subgenus HERPETOMYS Merriam.

Herpetomys Merriam, Proc. Biol. Soc. Wash., XII, 107, April 30, 1898. Type, *Microtus guatemalensis* Merriam, from Todos Santos, Guatemala.

Geographic distribution.—That of the type species.

Subgeneric characters.—Plantar tubercles, 5; side glands[1] on flanks of males small and obscure or sometimes wanting; mammæ, 6, pectoral, 2–2, inguinal, 1–1 (the latter apparently rudimentary and functionless); ears large; pelage long and soft; colors dark brownish. *Skull* with smooth outlines and large globose audital bullæ; m3 with 3 closed triangles; m1 with 3 closed triangles and an interior confluent

[1] In some specimens no side glands can be discovered, and in others they are marked by a pencil of white hairs. There is some doubt as to whether the white hairs are a product of the glands or occur there accidentally or from injury, as they sometimes do over other parts of the body.

pair of triangles opening into terminal loop, and with 5 inner and 4 outer salient angles; m3 with 4 closed sections including a pair of subequal median triangles.

MICROTUS GUATEMALENSIS Merriam. Guatemalan Vole.

Microtus guatemalensis Merriam, Proc. Biol. Soc. Wash., XII, 107, April 30, 1898.

Type locality.—Todos Santos, Huehuetenango, Guatemala (altitude 10,000 feet).

Geographic distribution.—Known only from type locality.

Habitat.—Open ground on damp north slopes under rank growth of brush and weeds in the Boreal zone (altitude 9,800–11,000 feet).

General characters.—Size medium; tail less than twice the length of hind foot; ears large but nearly concealed in the long fur; colors dark.

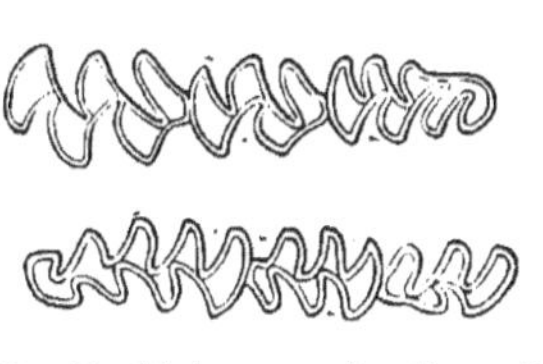

FIG. 16.—Molar enamel pattern of *Microtus (Herpetomys) guatemalensis* (× 5).

Color.—*Winter pelage:* Upperparts dark umber brown; nose blackish; lips white; belly clear plumbeous or lightly washed with dull ochraceous; feet and tail dusky brown, concolor, and thinly haired. (*Summer pelage* not seen.) *Young* (half grown individuals in December): Slightly duller than adult.

Cranial characters.—Skull rather long and but little arched, without conspicuous ridges and angles; interorbital space wide; braincase long; bullæ large and globose (larger than those of any other Mexican species); palate with steep and lightly grooved posterior median ridge; incisive foramina wide and short. *Dentition* heavy; most of the salient angles of molars acute; prisms deltoid; m3 with anterior crescent, 3 closed triangles, and a posterior crescent with two inner horns; m3 with posterior and anterior transverse crescents and a pair of subequal median triangles.

Measurements.—Average of 20 specimens from the type locality: Total length, 150; tail vertebræ, 37; hind foot, 21. Type (♂ ad.): 155; 40; 21. *Skull* (of type): Basal length, 25.6; nasals, 7.6; zygomatic breadth, 15; mastoid breadth, 12.3; alveolar length of upper molar series, 7.

General remarks.—So far as at present known, this is the southernmost species of *Microtus* in America. Its nearest relatives are *umbrosus* and *mexicanus*, with both of which it has some characters in common, but from which it differs so widely as to require subgeneric separation.

Specimens examined: Total number, 34, from the type locality.

Subgenus **NEOFIBER** True.

Neofiber True, Science, IV, 34, July 11, 1884 (genus). Type, *Neofiber alleni* True.

Neofiber Merriam, North American Fauna No. 5, 59, July, 1891 (subgenus).

Geographic distribution.—That of the type species.

Subgeneric characters.—Plantar tubercles, 5; side glands conspicuous in both sexes and in young, situated half-way between hips and shoulders, the glandular area marked by brownish base of fur and half-encircled above by a semilunar area of fur with white base; mammæ, 6, inguinal, 2–2, pectoral, 1–1; feet and fur modified for aquatic life; soles naked; a dorsal keel of long hair on rump. *Skull* massive; palate long with incomplete lateral bridges; pterygoids wing-like; m3 with 2 closed triangles; m1 with 5 closed triangles; m3 with 2 median triangles and 2 transverse terminal loops.

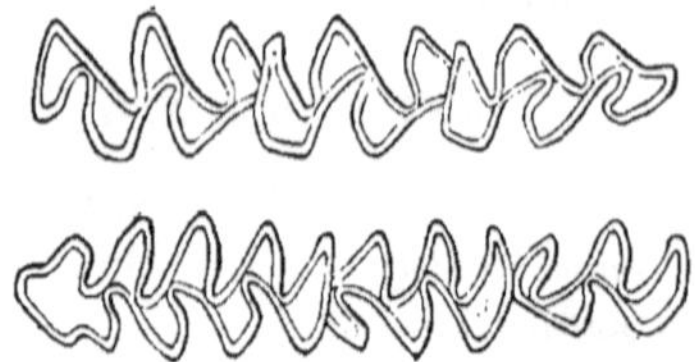

FIG. 17.—Molar enamel pattern of *Microtus* (*Neofiber*) *alleni* (× 5).

MICROTUS ALLENI (True). Florida Water-Rat.

Neofiber alleni True, Science, IV, 34, July 11, 1884.
Microtus (*Neofiber*) *alleni* Miller, North American Fauna No. 12, 70, July 23, 1896.

Type locality.—Georgiana, Brevard County, Florida.

Geographic distribution.—Eastern and central Florida.

Habitat.—Marshes, shallow lakes, and banks of streams.

General characters.—In appearance very similar to a small muskrat, but with a round tail, a tuft of long hair above the tail, hind feet less modified for aquatic life; fur dense, with color and texture of muskrat fur; skull resembling that of the muskrat, but with the rootless molars of *Microtus*.

Color.—Upperparts dark brown, darkened on head and along back by coarse blackish hairs; nose black; chin dusky; belly pale buff or soiled silvery whitish; tail dark brown or blackish, darker toward the tip; feet dark brown. *Young:* Dark maltese, with sooty backs.

Cranial characters.—Skull high and short, with heavy ridges and sharp angles; prezygomatic notches deep; postorbital shelf projecting; palate bone longer than in any other *Microtus*, shorter than in *Fiber;* lateral bridges of palate interrupted; pterygoids wing-like (as in *Fiber*); dentition heavy; upper incisors bent abruptly downward.

Measurements.—Average of 3 adult specimens from Canaveral, Fla.: Total length, 320; tail vertebræ, 121; hind foot, 44. Largest adult, ♂: 330; 130; 44. *Skull* (No. 23450, ♀ ad.): Basal length, 44.6; nasals, 12.5; zygomatic breadth, 26; mastoid breadth, 20.5; alveolar length of upper molar series, 12.

General remarks.—The striking resemblance between *M. alleni* of Florida and *M. amphibius* of England proves on comparison of cranial characters to be only superficial; the differences are subgeneric.

Specimens examined: Total number, 17, from the following localities:

Florida: Georgiana, 3; Titusville, 1; Eden, 3; Canaveral, 5; Geneva, 3; Lake Harney, 1; Oaklodge (on peninsula opposite Micco), 1.

PLATE II.

Skulls of 9 subgenera, upper view.

[Enlarged one and one-half times.]

FIG. 1. *Microtus* (*Microtus*) *pennsylvanicus*. Hyattsville, Md.
(No. 87163, ♀ ad., U. S. Nat. Mus.)

2. *Microtus* (*Arvicola*) *macropus*. Sawtooth Lake, Idaho.
(No. 31451, ♀ ad., U. S. Nat. Mus.)

3. *Microtus* (*Neofiber*) *alleni*. Eden, Fla.
(No. 24112, ♀ ad., U. S. Nat. Mus.)

4. *Microtus* (*Pedomys*) *austerus*. Racine, Wis.
(No. 92851, ♂ ad., U. S. Nat. Mus.)

5. *Microtus* (*Pitymys*) *pinetorum*. Frogmore, S. C.
(No. 1523, ♂ ad., Merriam collection.)

6. *Microtus* (*Lagurus*) *curtatus*. Mount Magruder, Nev.
(No. 41017, ♂ ad., U. S. Nat. Mus.)

7. *Microtus* (*Chilotus*) *oregoni*. Astoria, Oreg.
(No. 24255, ♂ ad., U. S. Nat. Mus.)

8. *Microtus* (*Orthriomys*) *umbrosus*. Mount Zempoaltepec, Oaxaca, Mexico.
(No. 68469, ♀ ad., U. S. Nat. Mus.)

9. *Microtus* (*Herpetomys*) *guatemalensis*. Todos Santos, Guatemala.
(No. 76776, ♂ ad., U. S. Nat. Mus.)

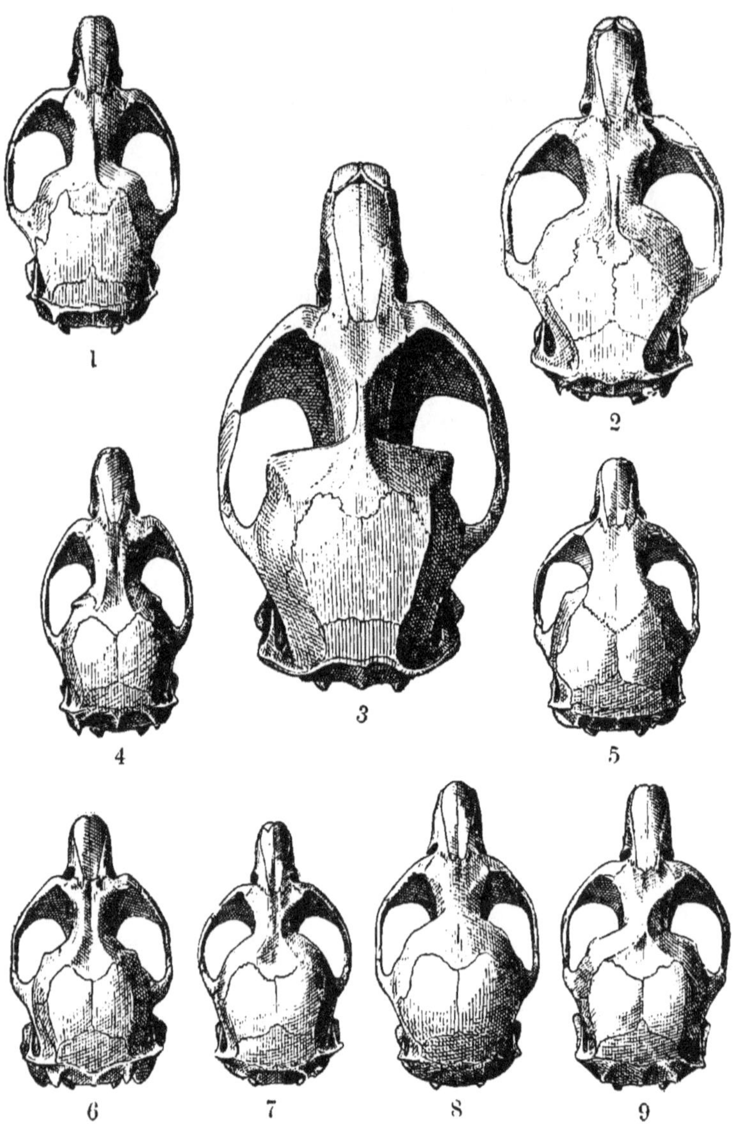

SKULLS OF REPRESENTATIVE SPECIES OF THE NINE SUBGENERA OF *Microtus* (top view).

1. *Microtus* (*Microtus*) *pennsylvanicus*.
2. *Microtus* (*Arvicola*) *macropus*.
3. *Microtus* (*Neofiber*) *alleni*.
4. *Microtus* (*Pedomys*) *austerus*.
5. *Microtus* (*Pitymys*) *pinetorum*.
6. *Microtus* (*Lagurus*) *curtatus*.
7. *Microtus* (*Chilotus*) *oregoni*.
8. *Microtus* (*Orthriomys*) *umbrosus*.
9. *Microtus* (*Herpetomys*) *guatemalensis*.

18392—No. 17——6

PLATE III.

Skulls of 9 subgenera, lower view.

[Enlarged one and one-half times.]

FIG. 1. *Microtus* (*Microtus*) *pennsylvanicus*. Hyattsville, Md.
(No. 87163, ♀ ad., U. S. Nat. Mus.)

2. *Microtus* (*Arvicola*) *macropus*. Sawtooth Lake, Idaho.
(No. 31451, ♀ ad., U. S. Nat. Mus.)

3. *Microtus* (*Neofiber*) *alleni*. Eden, Fla.
(No. 24112, ♀ ad., U. S. Nat. Mus.)

4. *Microtus* (*Pedomys*) *austerus*. Racine, Wis.
(No. 92851, ♂ ad., U. S. Nat. Mus.)

5. *Microtus* (*Pitymys*) *pinetorum*. Frogmore, S. C.
(No. 1523, ♂ ad., Merriam Collection.)

6. *Microtus* (*Lagurus*) *curtatus*. Mount Magruder, Nev.
(No. 41017, ♂ ad., U. S. Nat. Mus.)

7. *Microtus* (*Chilotus*) *oregoni*. Astoria, Oreg.
(No. 24255, ♂ ad., U. S. Nat. Mus.)

8. *Microtus* (*Orthriomys*) *umbrosus*. Mt. Zempoaltepec, Oaxaca, Mexico.
(No. 68469, ♀ ad., U. S. Nat. Mus.)

9. *Microtus* (*Herpetomys*) *guatemalensis*. Todos Santos, Guatemala.
(No. 76776, ♂ ad., U. S. Nat. Mus.)

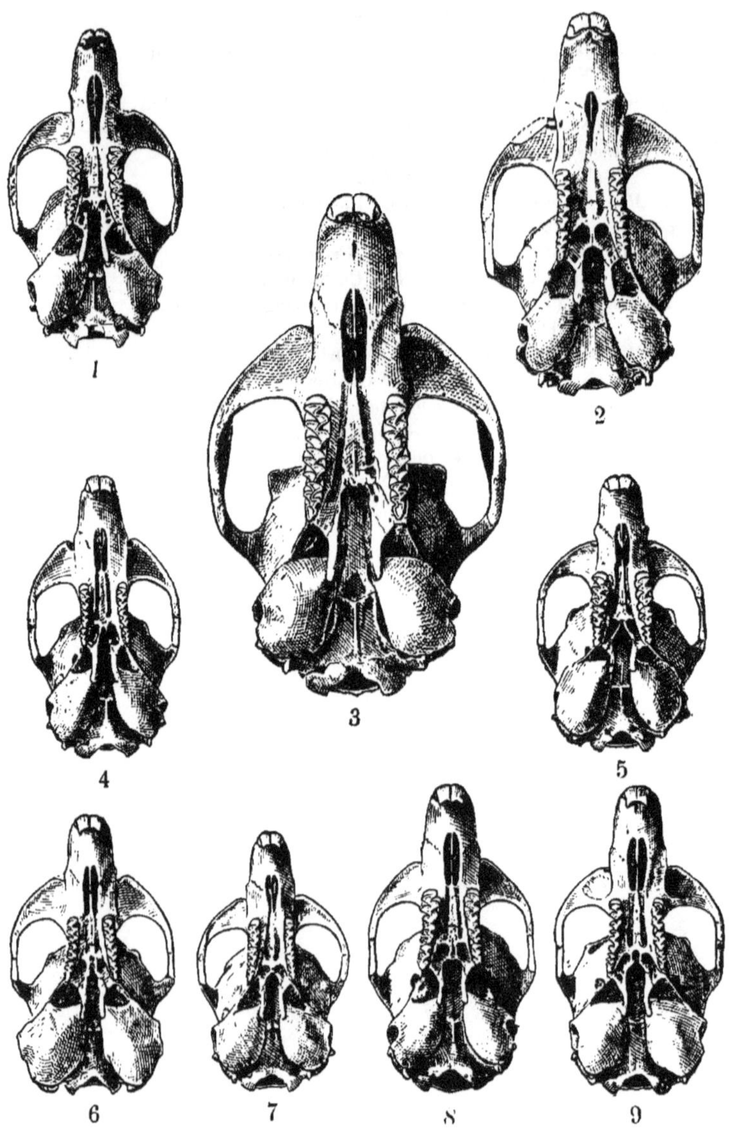

SKULLS OF REPRESENTATIVE SPECIES OF THE NINE SUBGENERA OF *Microtus* (bottom view).

1. *Microtus* (*Microtus*) *pennsylvanicus*.
2. *Microtus* (*Arvicola*) *macropus*.
3. *Microtus* (*Neofiber*) *alleni*.
4. *Microtus* (*Pedomys*) *austerus*.
5. *Microtus* (*Pitymys*) *pinetorum*.
6. *Microtus* (*Lagurus*) *curtatus*.
7. *Microtus* (*Chilotus*) *oregoni*.
8. *Microtus* (*Orthriomys*) *umbrosus*.
9. *Microtus* (*Herpetomys*) *guatemalensis*.

PLATE IV.

Skulls of 7 groups in subgenus *Microtus*, upper view.

[Enlarged one and one-half times.]

FIG. 1. ***Microtus mordax.*** **Red Lodge, Mont.**
(No. 67305, ♀ ad., U. S. Nat. Mus.)
2. ***Microtus nevadensis.*** **Ash Meadows, Nev.**
(No. 39663, ♂ ad., U. S. Nat. Mus.)
3. ***Microtus nanus.*** **Sawtooth Lake, Idaho.**
(No. 75181, ♂ ad., U. S. Nat. Mus.)
4. ***Microtus operarius.*** **St. Michael, Alaska.**
(No. 22214, ♂ ad., U. S. Nat. Mus.)
5. ***Microtus chrotorrhinus.*** **Mount Washington, N. H.**
(No. 1501, ♂ ad., Bangs Collection.)
6. ***Microtus townsendi.*** **Steilacoom, Wash.**
(No. 42921, ♂ ad., U. S. Nat. Mus.)
7. ***Microtus californicus.*** **Walnut Creek, Cal.**
(No. 44678, ♂ ad., U. S. Nat. Mus.)
8. ***Microtus mexicanus.*** **Orizaba, Puebla, Mexico.**
(No. 53406, ♀ ad., U. S. Nat. Mus.)

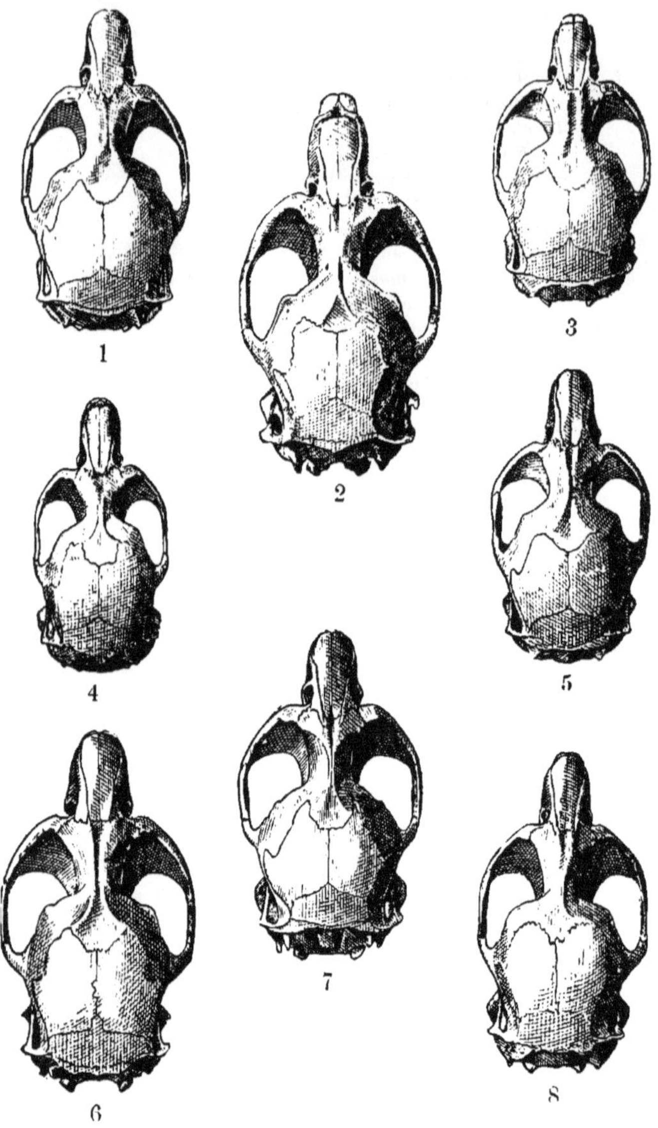

SKULLS OF REPRESENTATIVE SPECIES OF SEVEN OF THE GROUPS IN THE SUBGENUS *Microtus* (top view).

1. *Microtus mordax.*
2. *Microtus nevadensis.* }
3. *Microtus nanus.* }
4. *Microtus operarius.*
5. *Microtus chrotorrhinus.*
6. *Microtus townsendi.*
7. *Microtus californicus.*
8. *Microtus mexicanus.*

PLATE V.

Skulls of 7 groups in subgenus *Microtus*, lower view.

[Enlarged one and one-half times.]

FIG. 1. *Microtus mordax*. Red Lodge, Mont.
(No. 67305, ♀ ad., U. S. Nat. Mus.)
2. *Microtus nevadensis*. Ash Meadows, Nev.
(No. 39663, ♂ ad., U. S. Nat. Mus.)
3. *Microtus nanus*. Sawtooth Lake, Idaho.
(No. 75181, ♂ ad., U. S. Nat. Mus.)
4. *Microtus operarius*. St. Michael, Alaska.
(No. 22214, ♂ ad., U. S. Nat. Mus.)
5. *Microtus chrotorrhinus*. Mount Washington, N. H.
(No. 1501, ♂ ad., Bangs Collection.)
6. *Microtus townsendi*. Steilacoom, Wash.
(No. 42921, ♂ ad., U. S. Nat. Mus.)
7. *Microtus californicus*. Walnut Creek, Cal.
(No. 44678, ♂ ad., U. S. Nat. Mus.)
8. *Microtus mexicanus*. Orizaba, Puebla, Mexico.
(No. 53406, ♀ ad., U. S. Nat. Mus.)

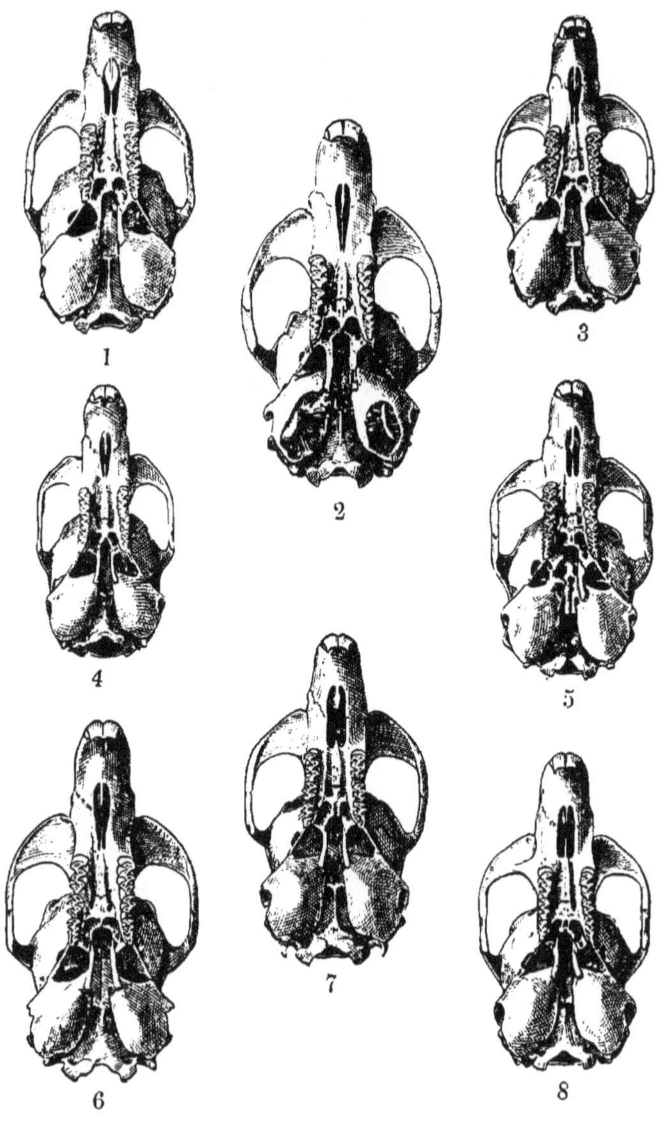

SKULLS OF REPRESENTATIVE SPECIES OF SEVEN OF THE GROUPS IN THE SUBGENUS *Microtus* (bottom view).

1. *Microtus mordax.*
2. *Microtus nevadensis.*
3. *Microtus nanus.*
4. *Microtus operarius.*
5. *Microtus chrotorrhinus.*
6. *Microtus townsendi.*
7. *Microtus californicus.*
8. *Microtus mexicanus.*

INDEX.

[Names of new species in black-face type, synonyms in *italics*.]

()

U. S. DEPARTMENT OF AGRICULTURE

DIVISION OF BIOLOGICAL SURVEY

NORTH AMERICAN FAUNA

No. 18

[Actual date of publication, September 20, 1900]

REVISION OF THE POCKET MICE OF THE GENUS PEROGNATHUS

BY

WILFRED H. OSGOOD

ASSISTANT BIOLOGIST, BIOLOGICAL SURVEY

Prepared under the direction of

Dr. C. HART MERRIAM

CHIEF OF DIVISION OF BIOLOGICAL SURVEY

WASHINGTON

GOVERNMENT PRINTING OFFICE

1900

STATISTICAL DEPARTMENT
OCT 2 1900
BOSTON PUBLIC LIBRARY

LETTER OF TRANSMITTAL.

U. S. DEPARTMENT OF AGRICULTURE,
DIVISION OF BIOLOGICAL SURVEY,
Washington, D. C., July 13, 1900.

SIR: I have the honor to transmit for publication, as No. 18 of North American Fauna, a 'Revision of the Pocket Mice of the Genus Perognathus,' by Wilfred H. Osgood, assistant in the Biological Survey. A preliminary revision of this group, based on the study of about 170 specimens, was published by me in 1889 as the first number of North American Fauna. In this early paper certain fundamental points in the history and synonym of the group were for the first time cleared up and the number of known forms was increased from 6 to 21. Five years later the rapid growth of the Biological Survey's collections enabled me to publish descriptions of a dozen additional species and to undertake a new revision of the group, which was brought down to date in 1896. The publication of this revision, with its accompanying illustrations, and colored maps showing the distribution of the various species, was deferred in order to obtain additional material still needed to settle a few remaining questions of distribution and relationship. This material was subsequently obtained, bringing the total number of specimens available up to 3,000; and my assistant, Mr. Osgood, to whom I had referred certain unsolved problems, undertook to bring the study of the whole group down to date. The result is here offered for publication.

Respectfully,

C. HART MERRIAM,
Chief, Biological Survey.

Hon. JAMES WILSON,
Secretary of Agriculture.

CONTENTS.

ILLUSTRATIONS.

PLATES.

TEXT FIGURES.

REVISION OF THE POCKET MICE OF THE GENUS PEROGNATHUS.

By Wilfred H. Osgood.
Assistant Biologist, Biological Survey.

INTRODUCTION.

HISTORY AND MATERIAL.

Pocket mice were first discovered by Maximilian, Prince of Wied, during his journey up the Missouri River. In 1839 he described and figured *Perognathus fasciatus* from specimens taken on the Missouri, near the mouth of the Yellowstone, in the present State of North Dakota. During the following half century several additional species were discovered; and in 1889 Dr. Merriam tentatively revised the group,[1] added many new species, and established the subgenus *Chætodipus* to include the large coarse-haired species. He also reviewed the history of the genus, so that it is now necessary to consider only events subsequent to 1889. Since then hardly a year has passed without the publication of additional species. In 1890 *fuliginosus* was proposed by Merriam;[2] in 1891 *femoralis*, by Allen;[3] in 1892 *merriami*, by Allen;[4] in 1893 *alticola* and *copei*, by Rhoads,[5] and *infraluteus*, by Thomas.[6] During 1894 two papers by Merriam greatly increased the knowledge of the group. The first[7] added *baileyi*, *canescens*, *columbianus*, *mexicanus*, *nelsoni*, *nevadensis*, *panamintinus*, and *stephensi*; the second,[8] *arenarius*, *bryanti*, *margaritæ*, and *peninsulæ*. In 1894 also, *conditi* and *pricei* were published by Allen[9] and

[1] North Am. Fauna, No. 1, 1889.
[2] Ibid., No. 3, 74, 1890.
[3] Bull. Am. Mus. Nat. Hist., N. Y., III, 281, 1891.
[4] Ibid., IV, 45, 1892.
[5] Proc. Acad. Nat. Sci. Phila., 1893, 404.
[6] Ann. and Mag. Nat. Hist., 6th ser., XI, 406, 1893.
[7] Proc. Acad. Nat. Sci. Phila., September, 1894, 262–268.
[8] Proc. Calif. Acad. Sci., 2d ser., IV, 460, 1894.
[9] Bull. Am. Mus. Nat. Hist., N. Y., VI, 318, 1894.

latirostris, by Rhoads;[1] in 1896 *mearnsi*, by Allen,[2] and in 1898 *pernix*, by Allen,[3] and *bangsi*, *eremicus*, and *pacificus*, by Mearns.[4]

The preliminary revision of the genus *Perognathus* by Dr. Merriam in 1889 was based on less than 200 specimens, practically all that were available in this country at that time. Nearly 3,000 specimens, all accumulated in the past decade, have been used in the present revision. This large collection, like those recently studied in other groups, proves the existence of many new forms,[5] shows the true status of doubtful ones, and clears up troublesome questions of relationship, nomenclature, and geographic distribution.

Most of the names of doubtful application in 1889 may now be disposed of definitely. In the case of *longimembris*, the name is found applicable to the species inhabiting the San Joaquin Valley, California, and a new name, *brevinasus*, is given to the San Bernardino form heretofore assumed to be true *longimembris*. The acquisition of topotypes settles previous questions regarding *flavus* and *mollipilosus*, and the possession of large series of specimens from Washington and Oregon makes it possible to fix the types of *lordi*, *parvus*, and *monticola*, though a slight uncertainty still attaches to the last. Abundance of material also makes available the name *hispidus*, under which *paradoxus* is placed as a subspecies, and of which *conditi* and *spilotus* become synonyms.

This material embraces all the specimens of *Perognathus* in the collections of the United States Biological Survey, the United States National Museum, the American Museum of Natural History, New York, and the private collections of Messrs. E. A. and O. Bangs and Dr. C. Hart Merriam. Besides these, various important specimens from other sources have been examined. All the types known to exist have been seen except those of *lordi*, *infraluteus*, and *pernix*, which are in the British Museum. In making acknowledgments, I wish first to express my obligations to Dr. C. Hart Merriam for the privilege of using his private collection and that of the Biological Survey, and also for much generous criticism and advice. For the privilege of using the collections in their charge, thanks are also due Dr. F. W. True, executive curator, and Mr. G. S. Miller, jr., assistant curator of mammals, in the United States National Museum; to Dr. J. A. Allen, curator of mammals and birds in the American Museum of Natural History; to Mr. Witmer Stone, curator of birds, Academy of Natural Sciences, Philadelphia; and to Mr. Outram Bangs, Mr. S. N. Rhoads, and Mr. W. W. Price.

---

[1] Am. Nat., XXVII, 185, 1894.

[2] Bull. Am. Mus. Nat. Hist., N. Y., VIII, 237, 1896.

[3] Ibid., X, 149, 1898.

[4] Ibid., X, 299, 1898.

[5] Five new species and eight new subspecies are described in the present revision.

The illustrations of skulls in Plates I and II and a few of the text figures were drawn by Dr. James E. McConnell; the outline figures of skulls in the text are republished from the plates in North American Fauna No. 1.

DISTRIBUTION.

The genus *Perognathus* is confined to North America and is restricted to the region west of the Mississippi River. Its northern limit is Ashcroft, British Columbia; its southern, Tlalpam, in the valley of Mexico. On the east its limits coincide approximately with those of the arid divisions of the Austral and Transition zones; on the west it extends to the Pacific coast. It may be said in a general way that the subgenus *Perognathus* inhabits the Sonoran and the lower part of the Transition zones (see Pl. III), while *Chætodipus* is seldom found outside of the Lower Sonoran zone except on the Great Plains (see Pl. IV). A curious exception to this distribution is found in central California, where *Perognathus* (*Chætodipus*) *californicus* is found in the Upper Sonoran zone and *Perognathus* (*Perognathus*) *longimembris* in the Lower Sonoran. Pocket mice usually choose plains and deserts for their habitat, and one or more species may be found in nearly all the desert and semi-desert country in the western part of the United States. As a rule, they are not found in mountainous regions, except where the aridity is considerable and the conditions are otherwise favorable. They abound in southern California, Lower California, and the Great Basin region, and in Mexico large areas are well populated with them.

COLOR AND PELAGES.

The general pattern of marking and coloration found in the genus is subject to little variation. The upperparts show varying shades of buff with greater or less admixture of black; the underparts are nearly always white. Most species have a distinct side stripe or lateral line and a minute white subauricular spot. Among the desert forms are numerous examples of protective coloration and adaptation to environment. A peculiar rump armature found in some species of the subgenus *Chætodipus* consists of grooved spiny bristles which extend beyond the rest of the pelage. What its function may be is little more than conjecture.

So far as known no species has more than one molt. This usually occurs in late summer after the breeding season, but is somewhat irregular, as specimens in entirely different pelages may be taken at the same place and date. The pelage acquired by the molt continues throughout the year, becoming more or less worn and patchy in early summer, just before its renewal:[1] hence in most species the seasonal

[1] Unless otherwise stated, the specimens described in this paper are in the new, unworn, or post-breeding pelage.

differences are not very great, the winter and spring pelage being simply paler than that of late summer and fall. The young invariably pass through a stage in which the pelage is soft and plumbeous.

HABITS.

The habits of pocket mice, as of most other small mammals, are not very well known. Most species are strictly nocturnal and very shy, and many of them are difficult to trap, as they do not readily take such bait as rolled oats or meat. They live in small burrows, from the entrances of which they throw out miniature mounds of earth like those of the pocket gopher. These burrows usually have two or more entrances, which often open under small bushes, and are closed with earth during the day, so that a casual observer might easily overlook them, particularly in the case of the smaller species. The food consists of seeds, which are carried in the cheek pouches and stored in chambers in the burrows. No species is known to hibernate, but it is possible that some of the more northern ones may do so.

CLASSIFICATION.

The genus *Perognathus* is a member of the family Heteromyidæ, one of the most peculiar groups of New World mammals. The other genera of this family are *Heteromys*, *Dipodomys*, *Perodipus*, and *Microdipodops*. Of these, *Heteromys* may be readily separated from the others by its very hispid pelage, which consists almost entirely of grooved spines, and by its rather murine skull, smooth upper incisors, and small mastoids and audital bullæ.

The genus *Perognathus* is commonly divided into two subgenera—*Perognathus* proper, including the small soft-haired species, and *Chætodipus*, containing the large coarse-haired and long-tailed forms. All the species except three fall naturally into one or the other of the two subgenera. One of these (*formosus*) is a *Perognathus* with strong inclination toward *Chætodipus;* another (*baileyi*) presents the reverse case; and the third (*hispidus*) must be classed as a *Chætodipus*, though it is aberrant in some ways. *P. femoralis* and *P. flavus* represent the extremes of the two subgenera and would certainly be placed in different genera if no other species were known, but between them may be found species showing almost every degree of differentiation. For convenience the genus has been divided into groups (see pp. 17–18) in order to show the affinities of the species and, to a certain extent, of the groups themselves.

In distinguishing species, dental peculiarities are of some service and cranial characters indispensable, showing relationship when external characters do not, and demonstrating intergradation to a degree of nicety otherwise almost unattainable. The best characters for comparison are the relative sizes of the mastoids and consequent dimensions of the interparietal. The shape of the interparietal varies somewhat, but its proportions and dimensions are generally reliable. The rostrum and interorbital space also furnish good characters. The hairiness of the feet is important, but of value only for separating species or groups in which other good differences are not apparent. The size and shape of the ears are also occasionally of use. In most species the males are slightly larger than the females, and in some the young adults are slightly different from fully mature individuals. Slight local variations are abundant; in some species it seems almost impossible to find two local series which are absolutely alike. But after making allowance for variation due to age and sex, individual variation will not be found very great; although so far as size is concerned it is greater in the *parvus* and *hispidus* groups than in the others.

NEW SPECIES.

Thirteen forms here characterized have not heretofore been described. These include five species and eight subspecies, as follows:

Perognathus merriami gilvus
apache melanotis
callistus
panamintinus brevinasus
amplus
parvus magruderensis
hispidus zacatecæ
Perognathus penicillatus angustirostris
pernix rostratus
goldmani
artus
anthonyi
californicus dispar

All measurements in the present paper are in millimeters.

Genus **PEROGNATHUS** Maximilian, 1839.

Perognathus Maximilian, Nova Acta Acad. Cæs. Leop.-Carol., XIX, I, 369–373, Pl. XXXIV, 1839; Reise Nord-Am., I, 449, 1839. Type, *Perognathus fasciatus* Maximilian, 1839, from the Upper Missouri River.
Cricetodipus Peale, U. S. Expl. Exp'd., VIII, 52–54, 1848.
Abromys Gray, Proc. Zool. Soc. London, 1868, 202.
Otognosis Coues, Proc. Acad. Nat. Sci. Phila., 1875, 305.
Chætodipus Merriam, N. Am. Fauna No. 1, 5, 1889.

Characters.—Size medium or small; form murine, rather slender; tail nearly as long as or longer than head and body; ears small; hind legs and feet rather long; external cheek pouches lined with hair. Skull rather small and light, flattened above; mastoids very large; audital bullæ inflated, more or less triangular in outline, anteriorly apposed to pterygoids; jugals light and thread-like; rostrum attenuate, nasals somewhat tubular anteriorly; infraorbital foramen reduced to a lateral opening in the maxillary. Teeth 20; molars rooted and tuberculate; upper incisors strongly sulcate.

Characters of subgenera.

I. PEROGNATHUS.

Size medium or small; pelage soft, no spines or bristles. Soles of hind feet more or less hairy (except in *formosus*).

Mastoids greatly developed, projecting beyond plane of occiput; mastoid side of parietal longest. Interparietal width less than interorbital width (rarely equal in *longimembris*).

Audital bullæ meeting or nearly meeting anteriorly. Supraoccipital without lateral indentations by mastoids (except in *formosus*); ascending branches of supraoccipital slender and thread-like.

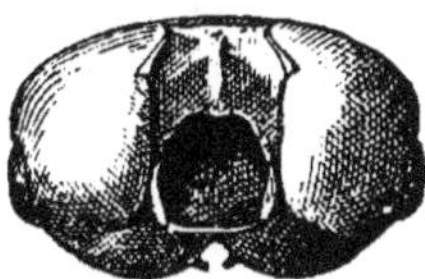

FIG. 1.—Posterior view of skull of *Perognathus* (*Perognathus*) *bimaculatus*.

II. CHÆTODIPUS.

Size medium or large; pelage harsh, often with spiny bristles on rump. Soles of hind feet naked.

Mastoids relatively small, not projecting beyond plane of occiput; mastoid side of parietal equal to or shorter than other sides. Interparietal width equal to or greater than interorbital width.

Audital bullæ separated by nearly full width of basisphenoid. Supraoccipital with deep lateral indentations by mastoids (except in *hispidus*); ascending branches of supraoccipital heavy and laminate.

FIG. 2.—Posterior view of skull of *Perognathus* (*Chætodipus*) *intermedius*.

Key to species and subspecies.

[Based on typical adults.]

I. Subgenus PEROGNATHUS.

Antitragus lobed; hind foot more than 20.
- Tail long and heavily crested; soles naked........................*formosus* (p. 40)
- Tail moderate; soles of hind feet somewhat hairy.
  - Interparietal narrow, ratio of its width to basilar length of Hensel[1] about 25; color grayish; size large.
    - Mastoids moderate ....*lordi* (p. 39)
    - Mastoids larger..*columbianus* (p. 40)
  - Interparietal wide, ratio of its width to basilar length of Hensel about 27.
    - Ears white; upper side of tail faintly dusky at tip..............*alticola* (p. 39)
    - Ears buff or dusky; upper side of tail dusky throughout entire length.
      - Size large, hind foot 23 to 26............................*magruderensis* (p. 38)
      - Size medium, hind foot 21 to 23.
        - Audital bullæ meeting anteriorly in a strong symphysis; color usually slaty buff ....*parvus* (p. 34)
        - Audital bullæ meeting anteriorly in a weak symphysis or not meeting; color cinnamon or ochraceous buff.
          - Premaxillæ exceeding nasals; color cinnamon buff; ears medium. *olivaceus* (p. 37)
          - Premaxillæ not exceeding nasals; color ochraceous buff; ears large. *mollipilosus* (p. 36)

---

[1] The basilar length of Hensel is measured from the anterior margin of the foramen magnum to the posterior rim of alveolus of the middle incisor.

Antitragus not lobed; hind foot 20 or less.
Tail longer than head and body.
Total length more than 150; mastoids very large................*amplus* (p. 32)
Total length less than 150; mastoids moderate.
Interorbital space narrow (less than 5); basilar length of Hensel about 17.......................................*longimembris* (p. 33)
Interorbital space wide (5 or more); basilar length of Hensel 15 or less.
Hairs of belly plumbeous at base.......................*nevadensis* (p. 31)
Hairs of belly white to roots.
Nasals short (about 7); tail 70 or less.................*brevinasus* (p. 30)
Nasals long (about 8); tail more than 70.
Color pale vinaceous buff...........................*bangsi* (p. 29)
Color grayish buff..........................*panamintinus* (p. 28)
Tail about equal to or shorter than head and body.
Size rather large; interparietal width 4 or more; hind foot 18 or more.
Inside of ears chiefly black ....*melanotis* (p. 27)
Inside of ears chiefly buff.
Color grayish olive buff...................................*callistus* (p. 28)
Color buff or ochraceous buff ....*apache* (p. 26)
Size medium or small; hind foot less than 18.
Tail about 60.
Color olivaceous.
Hairs of belly white to roots..........................*fasciatus* (p. 18)
Hairs of belly plumbeous at base.....................*infraluteus* (p. 19)
Color not olivaceous.
Total length about 130; lower premolar smaller than last molar. *flavescens* (p. 20)
Total length 120 or less; lower premolar about equal to last molar.
Rostrum heavy; mastoids small ....*merriami* (p. 21)
Rostrum light; mastoids larger ....*gilvus* (p. 23)
Tail about 50.
Lower premolar larger than last molar ....*pacificus* (p. 31)
Lower premolar smaller than last molar.
Hind foot about 15; color salmon buff.....................*flavus* (p. 23)
Hind foot about 17.
Upper parts sooty or black.........................*fuliginosus* (p. 25)
Upper parts salmon buff. (Northern Arizona)......*bimaculatus* (p. 24)
Upper parts buff, strongly mixed with black. (Central Mexico.) *mexicanus* (p. 26)

II. Subgenus Chætodipus.

Rump with more or less distinct spines or bristles.
Lateral line well marked; pelage not very hispid; bristles moderate, usually confined to rump.
Ears elongate (length 10 to 12); mastoids quite small; ratio of mastoid breadth to basilar length of Hensel about 70.
Size medium; total length less than 200; hind foot about 24.. *californicus* (p. 58)
Size very large; length more than 200; hind foot about 26.
Interorbital space moderate; mastoids relatively small ...... *dispar* (p. 58)
Interorbital space wider; mastoids larger .... *femoralis* (p. 57)
Ears rounded.
Ears large and orbicular (length about 10); color dark; rostrum heavy. (Mexico.)

Mastoids large .... *goldmani* (p. 54)
Mastoids small.. *artus* (p. 55)
Ears medium (length about 8); rostrum light.
Color of upperparts drab gray .... *canescens* (p. 54)
Color of upperparts not drab gray.
Pelage rather hispid; color dark; rostrum relatively heavy. (Central Mexico) .... *nelsoni* (p. 53)
Pelage softer; color lighter; rostrum slender.
Rump spines weak; interparietal strap-shaped; mastoids large.
intermedius (p. 52)
Rump spines stronger; interparietal somewhat produced anteriorly.
Mastoids large .... *fallax* (p. 55)
Mastoids smaller. (Cerros Island, Lower California) .*anthonyi* (p. 56)
Lateral line very faint or not evident; pelage very hispid; bristles strong, extending to sides.
Size large; tail 120 or more. (San Jose Island, Lower California)..*bryanti* (p. 61)
Size smaller; tail less than 110; hind foot about 24............*peninsulæ* (p. 60)
Size smaller; hind foot about 22.
Mastoids moderate. (Southern California)..................*spinatus* (p. 59)
Mastoids very small ....*margaritæ* (p. 61)
Rump without spines or bristles.
Tail not crested, shorter than head and body; skull in adults with a supraorbital bead.
Size very large; hind foot about 26; color pale ochraceous. (Kansas and Nebraska) ....*paradoxus* (p. 44)
Size smaller; hind foot about 24; color bright ochraceous. (Texas.)
hispidus (p. 42)
Size large; hind foot about 26; color olive brown. (Zacatecas, Mexico.)
zacatecæ (p. 45)
Tail crested, longer than head and body; skull without supraorbital bead.
Size very large; tail much longer than head and body; interparietal width about equal to interorbital width ....*baileyi* (p. 41)
Size medium or large; tail slightly longer than head and body; interparietal width exceeding interorbital width.
Interorbital width less than 6; color of upperparts hair-brown. (West coast of Mexico.)
Skull narrow and elongate; rostrum slender..................*pernix* (p. 50)
Skull short; rostrum heavy..................................*rostratus* (p. 51)
Interorbital width more than 6; color of upperparts pale vinaceous buff to broccoli brown.
Hind foot 20 to 21.
Size small; tail short, less than 80; hind foot 20. (Lower California).... .... *arenarius* (p. 50)
Size larger; tail 90 or more; hind foot 21................*stephensi* (p. 49)
Hind foot 22 to 26.
Hind foot 23 to 26; color vinaceous buff.
Large; rostrum very heavy..........................*penicillatus* (p. 45)
Smaller; rostrum slender..........................*angustirostris* (p. 47)
Color blackish brown..*artus* (p. 55)
Hind foot 22 to 24; color broccoli brown.
Dark; rostrum heavy..*pricei* (p. 47)
Paler; rostrum slender......................................*eremicus* (p. 48)

List of species and subspecies, with type localities.

Subgenus Perognathus.

Species and subspecies.	Type locality.
Fasciatus group:	
Perognathus fasciatus Maximilian	Near junction of Missouri and Yellowstone rivers, N. Dak.
fasciatus infraluteus (Thomas)	Loveland, Colo.
flavescens (Merriam)	Kennedy, Nebr.
merriami Allen	Brownsville, Tex.
merriami gilvus nobis	Eddy, N. Mex.
flavus Baird	El Paso, Tex.
flavus bimaculatus (Merriam)	Fort Whipple, Yavapai Co., Ariz.
flavus fuliginosus (Merriam)	San Francisco Mountain, Ariz.
flavus mexicanus Merriam	Tlalpam, Mexico, Mexico.
apache Merriam	Keam Canyon, Apache County, Ariz.
apache melanotis nobis	Casas Grandes, Chihuahua, Mexico.
callistus nobis	Kinney Ranch, Sweetwater Co., Wyo.
Panamintinus group:	
Perognathus panamintinus (Merriam)	Panamint Mountains, Cal.
panamintinus bangsi (Mearns)	Palm Springs, Cal.
panamintinus brevinasus nobis	San Bernardino, Cal.
nevadensis Merriam	Halleck, Nev.
pacificus Mearns	Mexican boundary, shore of Pacific Ocean.
amplus nobis	Fort Verde, Ariz.
longimembris (Coues)	Fort Tejon, Kern Co., Cal.
Parvus group:	
Perognathus parvus (Peale)	Oregon [The Dalles?].
parvus mollipilosus (Coues)	Fort Crook, Shasta Co., Cal.
parvus olivaceus (Merriam)	Kelton, Utah.
parvus magruderensis nobis	Mount Magruder, Nev.
alticola Rhoads	Squirrel Inn, San Bernardino Mountains, Cal.
lordi (Gray)	British Columbia.
lordi columbianus (Merriam)	Pasco, Wash.
Formosus group:	
Perognathus formosus Merriam	St. George, Utah.

Subgenus Chætodipus.

Species and subspecies.	Type locality.
Baileyi group:	
Perognathus baileyi Merriam	Magdalena, Sonora, Mexico.
Hispidus group:	
Perognathus hispidus Baird	Charco Escondido, Tamaulipas, Mexico.
hispidus paradoxus (Merriam)	Trego County, Kans.
hispidus zacatecæ nobis	Valparaiso, Zacatecas, Mexico.
Penicillatus group:	
Perognathus penicillatus Woodhouse	San Francisco Mountain, Ariz.
penicillatus angustirostris nobis	Carriso Creek, Colorado Desert, Cal.
penicillatus pricei (Allen)	Oposura, Sonora, Mexico.
penicillatus eremicus (Mearns)	Fort Hancock, El Paso Co., Tex.
stephensi Merriam	Mesquite Valley, Cal.
arenarius Merriam	San Jorge, Lower California.

Species and subspecies.	Type locality.
Pernix group:	
Perognathus pernix Allen	Rosario, Sinaloa, Mexico.
pernix rostratus nobis	Camoa, Sonora, Mexico.
Intermedius group:	
Perognathus intermedius Merriam	Mud Spring, Mohave Co., Ariz.
nelsoni Merriam	Hacienda La Parada, San Luis Potosi, Mexico.
nelsoni canescens (Merriam)	Jaral, Coahuila, Mexico.
goldmani nobis	Sinaloa, Sinaloa, Mexico.
artus nobis	Batopilas, Chihuahua, Mexico.
fallax Merriam	Reche Canyon, San Bernardino Co., Cal.
anthonyi nobis	Cerros Island, Lower California.
Californicus group:	
Perognathus femoralis Allen	Dulzura, San Diego Co., Cal.
californicus Merriam	Berkeley, Cal.
californicus dispar nobis	Carpenteria, Santa Barbara Co., Cal.
Spinatus group:	
Perognathus spinatus Merriam	Colorado River, Cal.
spinatus peninsulæ Merriam	San Jose del Cabo, Lower California.
bryanti Merriam	San Jose Island, Lower California.
margaritæ Merriam	Santa Margarita Island, Lower California.

Subgenus PEROGNATHUS Maximilian, 1839.

PEROGNATHUS FASCIATUS Maximilian. MAXIMILIAN POCKET MOUSE.

Perognathus fasciatus Maximilian, Nova Acta Acad. Caes. Leop.-Carol., XIX, I, 369–373, Pl. XXXIV, 1839; Reise Nord-Am., I, 449, 1839; Merriam, N. Am. Fauna No. 1, 10, 1889.

Type locality.—Upper Missouri River near its junction with the Yellowstone.

Distribution.—Upper Sonoran and Transition zones of eastern Montana and Wyoming, east into the adjoining parts of North and South Dakota.

General characters.—Size rather small, exceeding *P. flavus*, but not equaling *P. apache;* ears medium, antitragus not lobed; tail subterete, evenly haired, slightly shorter than head and body; proximal half of sole of hind foot hairy.

Color.—Upperparts grayish olivaceous, finely lined with black; hairs clear plumbeous basally, followed by a zone of black-tipped grayish buff; sides not noticeably paler than back; underparts pure white; lateral line bright buff (due to absence of black-tipped hairs), extending from nose to end of tail; tail indistinctly tricolor, dusky above, buffy on sides, and white below; orbital region and ill-defined ring around ears buff; subauricular spot present. *Spring pelage:* General color paler, more buffy and often lacking the olive tinge; contrast with lateral line not marked. *Young:* Dull plumbeous above with slight admixture of buffy.

Skull.—Size small; cranium somewhat arched; interparietal pentagonal, moderately wide; mastoids well developed, slightly projecting; auditul bullæ scarcely meeting anteriorly; coronoid and angular processes of mandible long and slender; lower premolar about equal to or slightly smaller than last molar.

Measurements.—Average of four adults from Tilyou ranch, Montana (near junction of Missouri and Yellowstone): Total length, 134.7; tail vertebræ, 64.5; hind foot, 17. *Skull:* (See table, p. 62.)

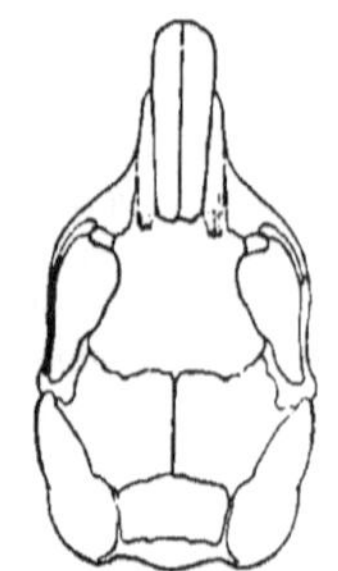

FIG. 3.—Skull of *Perognathus fasciatus*.

Remarks.—In bright new pelage *P. fasciatus* presents a very attractive appearance. Its diminutive size, peculiar greenish back, pure white underparts, and bright buff lateral line make a combination quite unusual among our small mammals. One specimen (No. 65017) from Rosebud Agency, S. Dak., collected May 13, 1894, is at the molting stage, and the incoming hairs show the extreme of this peculiar coloration. The head, back, and rump show small patches of bright, even iridescent, greenish, about which is the duller grayish of the old pelage. Another specimen (No. 65664) is further advanced, the only remains of the old pelage being slight traces on head and back and a large dark-buff rump patch. The species presents little geographic variation. Specimens from Bighorn Basin have somewhat peculiar skulls, but the aberrance is very slight.

Specimens examined.—Total number, 39, from localities as follows:

Montana: Calf Creek, 2; Clark Fork, 2; Big Porcupine Creek, 1; Frenchman River, 3; Lake Basin, Yellowstone County, 2; Powderville, 1; Sage Creek, Bighorn Basin, 1; Tilyou ranch, 27 miles above mouth of Yellowstone River, 6; Wolf Creek, 1; Mouth of Yellowstone, 1.

North Dakota: Forty miles north of Medora, 1.

South Dakota: Cheyenne River, Custer County, 1; Corral Draw, 4; Lugenbeel County, 2; Pine Ridge, 3; Rosebud Agency, 1; Quinn Draw, 3; Smithville, 1.

Wyoming: Kirby Creek, Bighorn Basin 2; Newcastle, 1.

PEROGNATHUS FASCIATUS INFRALUTEUS (Thomas). BUFF-BELLIED POCKET MOUSE.

Perognathus infraluteus Thomas, Ann. and Mag. Nat. Hist., 6th ser., XI, 406, May, 1893.

Type locality.—Loveland, Larimer County, Colo.

Distribution.—Known only from the type locality.

General characters.—Similar to *P. fasciatus*, but smaller and different in color of underparts, which are yellowish buff instead of white; pelage harsher.

Color.—Upperparts as in *P. fasciatus*, but more buffy; underparts buff with traces of white on inguinal and pectoral regions; eye-ring buff, more prominent than in *fasciatus*.

Skull.—Essentially as in *fasciatus*, but smaller and with slightly wider interparietal, as in *flavescens*.

Measurements.—Average of eight young adults from the type locality: Total length, 128; tail vertebræ, 59; hind foot, 17. *Skull:* (See table p. 62.)

Remarks.—The distinguishing character of this form is the buff color of its underparts. This, however, is not invariable, as one specimen in the series from the type locality is pure white below, and thus, but for minor characters, indistinguishable from *fasciatus.* The specimens examined are all young adults, which may partially account for their pecularities. In typical *fasciatus* the underparts are pure white in both young and old.

Specimens examined.—Total number, 10, all from the type locality, Loveland, Colo.

PEROGNATHUS FLAVESCENS (Merriam). PLAINS POCKET MOUSE.

Perognathus fasciatus flavescens Merriam, N. Am. Fauna No. 1, 11, 1889; Allen, Bull. Am. Mus. Nat. Hist., N. Y., VIII, 247, 1896.

Perognathus copei Rhoads, Proc. Acad. Nat. Sci. Phila., 1893, 404.

Type locality.—Kennedy, Nebr.

Distribution.—Upper Austral plains of South Dakota, Nebraska, and Kansas; south possibly to northern Texas, and west to base of Rocky Mountains.

General characters.—Proportions much as in *P. fasciatus;* size slightly smaller; pelage harsher; color buff, lined with black, never showing the strong olivaceous appearance of *fasciatus.*

Color.—*April specimens:* Above, light-grayish buff mixed with dusky; below, white; lateral line, eye-ring, and postauricular spot, clear buff; subauricular spot prominent; large spot on inflexed part of ear white; tail indistinctly bicolor; feet and legs white.

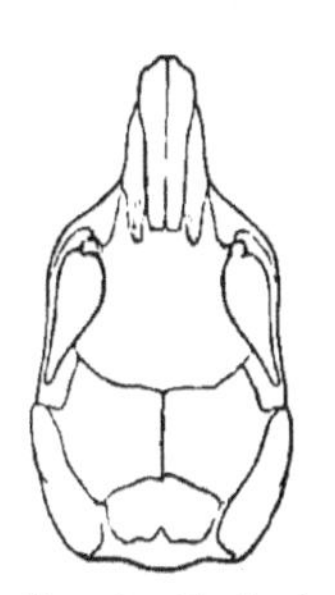

FIG. 4.—Skull of *Perognathus flavescens.*

Skull.—Similar to that of *fasciatus*, but a trifle smaller; interparietal wider; angular process of mandible shorter and broader; lower premolar smaller than last molar.

Measurements.—Average of six adults from Kennedy, Nebr.: Total length, 129.5; tail, 61.5; hind foot, 17.3. *Skull:* (See table, p. 62.)

Remarks.—This species is closely related to *P. fasciatus*, but is entirely distinct. Intergradation between the two is not probable, since typical examples of both have been taken at the same place, Rosebud Indian Agency, S. Dak. They doubtless occur together at other points, but in all cases color alone will be found sufficient to distinguish them.

P. copei from Mobeetie, Tex., was based on a single very imperfect specimen, and its status is accordingly doubtful. Its skull shows no tangible departure from that of true *flavescens.* Possibly it represents a slight southern race of *flavescens*, or it may prove to be an intergrade between that species and *merriami*. Four distorted and desiccated individuals from Santa Fe, N. Mex., have also been doubtfully referred to *flavescens*.

Specimens examined.—Total number, 68, from localities as follows:

Colorado: Boulder County, 1; Greeley, 3; Pueblo, 4; Sterling, 5.

Kansas: Cairo, 4.

Nebraska: Cherry County, 10; Ewing, 1; Kennedy, 6; Lakeside, 1; Lincoln County, 1; Loup Fork, 1; Myrtle, 2; Perch, Rock County, 9; Pole Creek, 40 miles from Fort Riley, 1; Thomas County, 5; Verdigris, 1.

South Dakota: Rosebud Agency, 2; Vermilion, 2.

New Mexico: Santa Fe, 4.

Texas: Mobeetie, 1.

PEROGNATHUS MERRIAMI Allen. MERRIAM POCKET MOUSE.

Perognathus flavus Baird, Mamm. N. Am., 423, 1857 (part); Merriam, N. Am. Fauna No. 1, 12, 1889; Allen, Bull. Am. Mus. Nat. Hist., N. Y., VIII, 58, 1896.

Cricetodipus flavus Thomas, Proc. Zool. Soc. London, 1888, 449.

Perognathus merriami Allen, Bull. Am. Mus. Nat. Hist., N. Y., IV, 45, Mar., 1892.

Perognathus mearnsi Allen, ibid., VIII, 237, Nov., 1896.

Type locality.—Brownsville, Tex.

Distribution.—Subtropical region of southern Texas and northeastern Mexico, and Lower Sonoran of central Texas. The known range extends from Alta Mira, Tamaulipas, northward to Washburn, Tex., and from this point southwestward to the vicinity of Roswell, N. Mex.; on the east it reaches San Antonio, and on the west follows up the Rio Grande as far as Comstock.

General characters.—Size smaller than *flavescens;* tail about equal to or slightly shorter than head and body, very scantily haired; pelage somewhat softer than in *flavescens* but not as in *flavus;* ears small and orbicular; colors bright; proximal half of sole of hind foot hairy.

Color.—Above, ochraceous buff densely mixed with black, forming an imperfectly defined dorsal stripe from the nose to the tail; below, pure white; sides bright buffy ochraceous, lateral line scarcely distinct; ears buff without, dusky within; spot behind ears clear buff; subauricular spot pure white, sharply contrasted with the surrounding black and ochraceous; light orbital area comparatively extensive; transverse nose stripes prominent, intensely black; tail slightly darker above than below; feet and forelegs white. *Late fall* and *winter pelage:* Heavier, softer, and lighter colored.

Skull.—General shape much as in *P. flavescens*, but smaller and slightly more angular; rostrum much heavier; maxillary branches of zygomata often squarely 'elbowed;' zygomata nearly parallel; interparietal more nearly quadrate than in *flavescens*, much wider than in *flavus;* lower premolar about equal to last molar.

Measurements.—Average of twenty adults from Brownsville, Tex.: Total length, 116.3; tail vertebræ, 57; hind foot, 16. *Skull:* (See table, p. 62.)

Remarks.– *P. merriami* is a very distinct species, more closely related to *P. flavescens* and *P. flavus* than to any other form. From *flavescens* it differs in size, color, hairiness of tail, and cranial characters. From *flavus*, to which it has some superficial resemblance, it is distinguishable by its slightly larger size, less hairy tail, smaller mastoids, heavier rostrum, wider interparietal, relatively larger lower premolar, and by other characters. In 1889 Dr. Merriam used a specimen of this species from Mason, Tex., as the basis of his description of *P. flavus*, of which no typical specimens were then extant. His prediction that this specimen would prove different from the El Paso animal was verified when actual topotypes of the latter were obtained. Subsequent authors, however, have continued to use the characters pointed out by Merriam on the basis of this specimen, and slight confusion has occasionally resulted.

The differences due to season are well shown by the large series examined. Early spring specimens (April) still wear the winter coat, which in June and early July often becomes so much worn that the plumbeous bases of the hairs are exposed. In late July and August the summer molt, the only one known, takes place. The new hair comes in rapidly and evenly, progressing from the head backward until the animals are in the bright post-breeding pelage, which is at its height in September and October. In winter there is but a slight change—a greater or less elimination of black and a general thickening of the pelage. The changes after this are evidently only those which result from wear.

Variations in this species are chiefly in size. Specimens from Padre Island, Texas, are smaller than typical ones, and those from some localities in Tamaulipas are abnormally large. *P. mearnsi* is not distinguishable, having been based on *merriami* in winter pelage.

Specimens examined.—Total number, 153, from localities as follows:

New Mexico: Forty miles west of Roswell, 1 (intermediate).

Texas: Austin, 1; Blocker Ranch, 1; Brownsville, 73; Comstock, 2; Kerrville, 6; Mason, 1; Padre Island, 3; Painted Cave, 1; Watson Ranch, San Antonio, 22; Santa Rosa, 10; San Diego, 3; Turtle Creek, Kerr County, 1; Washburn, 2 (intermediate).

Nuevo Leon: Aldama, 1; Doctor Cos, 1; Linares, 1.

Tamaulipas: Alta Mira, 1; Hidalgo, 7; Matamoras, 2; Mier, 5; Reynosa, 2; Victoria, 6.

PEROGNATHUS MERRIAMI GILVUS subsp. nov. DUTCHER POCKET MOUSE.

Type from Eddy, N. Mex. ♂ ad., No. $\frac{35839}{18273}$, U. S. Nat. Mus., Biological Survey Coll. Collected September 18, 1892, by Dr. B. H. Dutcher. Orig. No., 329.

Distribution.—Western Texas and southeastern New Mexico. Lower Sonoran zone.

General characters.—Size and proportions about the same as those of *merriami*, slightly larger than *flavus*; color as in *merriami*, but slightly paler; pelage softer. Skull superficially resembling that of *flavus*, but in detailed characters agreeing more closely with that of *merriami*.

Color.—Paler and more yellowish than *merriami;* back and sides well mixed with black; lateral line wide; postauricular spots rather prominent; tail whitish below, slightly dusky above.

Skull.—Like that of *merriami;* rostrum more slender; maxillary branches of zygomata lighter; mastoids larger. Contrasted with that of *flavus*, it has smaller mastoids, wider interparietal, larger lower premolar, and slightly wider interorbital space.

Measurements.—*Type:* Total length, 118; tail vertebræ, 58; hind foot, 16.5. One topotype: Total length, 122; tail vertebræ, 60; hind foot, 16.5. *Skull:* (See table, p. 62.)

Remarks.—This subspecies combines to some extent the characters of *flavus* and *merriami*. Nevertheless, careful study makes it perfectly evident that these are two distinct species, between which no real connection exists. All the evidence tends to show that *gilvus* has been derived from *merriami*. That its differentiation has been in the direction of *flavus* is probably an accidental circumstance, and does not indicate close relationship. The fact that typical *flavus* occurs with *gilvus* at its type locality (Eddy, N. Mex.), is interesting in this connection. Intergradation of *gilvus* with *merriami* is indicated by specimens from Comstock and Washburn, Tex., and also by a single individual taken 40 miles west of Roswell, N. Mex.

Specimens examined.—Total number, 7, from localities as follows:

New Mexico: Eddy, 4.

Texas: Big Spring, 1; Presidio County, 1; Stanton, 1.

PEROGNATHUS FLAVUS Baird. BAIRD POCKET MOUSE.

Perognathus flavus Baird, Proc. Acad. Nat. Sci. Phila., 1855, 332; Mamm. N. Am., 423, 1857 (part); Allen, Bull. Am. Mus. Nat. Hist., N. Y., VII, 215, 1894.

Type locality.—El Paso, Tex.

Distribution.—Upper and Lower Sonoran zones from northeastern Colorado and western Nebraska to northern Mexico, extending westward into central Arizona and eastward to western Texas. In central Arizona its range meets that of the subspecies *bimaculatus* and in north-central Mexico it merges with that of *mexicanus*.

General characters.—Size very small; ears medium; pelage very soft; tail moderately haired, shorter than head and body; proximal half of hind sole hairy.

Color.—Above, pinkish buff, lightly mixed with black; below, pure white; black-tipped hairs most numerous in median dorsal region, produced anteriorly beneath ears to cheeks; face and orbital region more or less free from dusky; lateral line not sharply contrasted; postauricular spot clear buff, very prominent; subauricular spot present, but inconspicuous; ears light buff outside, blackish inside; tail pale buffy, almost concolor, very faintly dusky above.

Skull.—Mastoid and audital bullæ greatly developed, interparietal very small, pentagonal or subquadrate, nearly as long as wide; rostrum quite slender; maxillary branches of zygomata angular; interorbital space well constricted; lower premolar noticeably smaller than last molar.

Measurements.—Average of ten adults from Fort Huachuca, Ariz.: Total length, 112.5; tail vertebræ, 50; hind foot, 15.8. *Skull:* (See table, p. 62.)

Remarks.—This species exhibits quite a departure from those preceding. Its small size, short tail, and conspicuous postauricular spots serve to mark it externally, while its short, broad skull, with full bulging mastoids and small interparietal distinguish it cranially. In its wide range some local differentiations might well be expected, but none of importance have been found. Its subspecies are not very strongly characterized and perfect intergradation with each is plainly evident.

Specimens examined.—Total number, 131, from localities as follows:

Arizona: Calabasas, 1; Chiricahua Mountains, 2; Dos Cabezos, 4; Fairbank, 2; Fort Grant, 5; Fort Huachuca, 41; Fort Lowell, 2; Mammoth, 1; Tanner Canyon, Huachuca Mountains, 4; Willcox, 15.

Colorado: Burlington, 1; Canyon City, 1; Fort Garland, 2; Greeley, 6; Loveland, 11.

Nebraska: Alliance, 1.

New Mexico: Chico Springs, 1; Deming, 2; Eddy, 1: Dog Spring, Grant County, 1; Taos, 1.

Oklahoma: Beaver River, Beaver County, 1.

Texas: El Paso, 8; Sierra Blanca, 1.

Chihuahua, Mexico: Chihuahua, 10; Escalon, 3; Gallego, 3.

PEROGNATHUS FLAVUS BIMACULATUS (Merriam). YAVAPAI POCKET MOUSE.

Perognathus bimaculatus Merriam, N. Am. Fauna No. 1, 12, 1889; Allen, Bull. Am. Mus. Nat. Hist., N. Y., VII, 216, 1895.

Perognathus apache Allen, ibid, V, 71, 1893 (part).

Type locality.—Fort Whipple, Yavapai County, Ariz.

Distribution.—Central and northeastern Arizona and southeastern Utah.

General characters.—Similar to *P. flavus*, but larger.

Color.—As in *flavus*, but with a greater abundance of black-tipped hairs on dorsum; underparts white with occasional traces of buff; lateral line quite distinct; ears clear buff outside, blackish inside.

Skull.—Much larger than in *flavus;* interparietal relatively smaller; mastoids very large; lower premolar smaller than last molar as in *flavus.*

Measurements.—Average of ten adults from the type locality: Total length, 118; tail vertebræ, 53; hind foot, 17. *Skull:* (See table, p. 62.)

Remarks.—Typical adult specimens of *bimaculatus* are so much larger than *flavus* as to be very easily distinguishable, but immature or undersized examples are apt to give trouble. The average difference in size, however, is considerable and fully warrants recognition. When specimens of equal age are compared, the subspecies may be easily separated from the typical form by its larger ears and feet.

Specimens examined.—Total number, 82, from localities as follows:

Arizona: Fort Whipple, 23; Holbrook, 8; Keam Canyon, 1; Walnut, 1; Winslow, 21.

New Mexico: Fort Wingate, 4.

Utah: Noland Ranch, San Juan River, 9; Riverview, 25.

PEROGNATHUS FLAVUS FULIGINOSUS (Merriam). DUSKY POCKET MOUSE.

Perognathus fuliginosus Merriam, N. Am. Fauna No. 3, 74, 1890.

Type locality.—Cedar belt northeast of San Francisco Mountain, Arizona.

Distribution.—Lava beds in the vicinity of San Francisco Mountain, Arizona.

General characters.—Size and proportions those of *P. bimaculatus;* color very different.

Color.—Upperparts black or nearly black, except buff postauricular spots; lateral line and underparts ochraceous buff, except throat and breast, which are white.

Skull.—As in *bimaculatus.*

Measurements.—*Type:* Total length, 116; tail vertebræ, 58; hind foot, 18.

Remarks.—This form is doubtless a recent offshoot from *flavus* which has acquired dark colors to harmonize with the black lava rock which it inhabits. The fact that the conditions determining its differentiation are so plainly evident should be no reason for not recognizing the subspecies, even though its range be limited.

Specimens examined.—Total number, 3, from localities as follows:

Arizona: Cedar Belt, San Francisco Mountain, 1; Turkey Tanks, 1; Wolf Creek, 1.

PEROGNATHUS FLAVUS MEXICANUS Merriam. MEXICAN POCKET MOUSE.

Perognathus flavus mexicanus Merriam, Proc. Acad. Nat. Sci., Phila., September 27, 1894, 265–266.

Type locality.—Tlalpam, Mexico, Mexico.

Distribution.—Upper and Lower Sonoran zones of the southern half of the table-land of Mexico.

General characters.—Similar to *P. flavus* but larger and darker.

Color.—Similar to that of *flavus*, but averaging much darker, the buff being richer and the fuliginous more extensive; postauricular spots and lateral line ochraceous, well contrasted; underparts white.

Skull.—As in *flavus*, but larger.

Measurements.—Average of 12 young adults from Tlalpam, Mexico: Total length, 115.7; tail, 53.7; hind foot, 17.4.

Remarks.—Some specimens of *mexicanus* are much like true *flavus*, but many are almost as dark as *fuliginosus*. In these the contrast of dusky back and sides with ochraceous lateral line and pure white underparts is very striking. In size *mexicanus* almost equals *bimaculatus*.

Specimens examined.—Total number, 29, from localities in Mexico, as follows:

Guanajuato: Celaya, 2; Guanajuato City, 2.
Hidalgo: Ixmiquilpan, 2.
Jalisco: Huejuquilla, 1.
Mexico: Tlalpam, 13.
San Luis Potosi: Ahualulco, 1; Hacienda La Parada, 3; Jesus Maria, 3.
Zacatecas: Berriozabal, 1; Valparaiso Mountains, 1.

PEROGNATHUS APACHE Merriam. APACHE POCKET MOUSE.

Perognathus apache Merriam, N. Am. Fauna No. 1, 14, 1889; ibid., No. 3, 73, 1890; Allen, Bull. Am. Mus. Nat. Hist., N. Y., V, 71, 1893 (part); ibid., VII, 216, 1895.
Perognathus flavus subsp. Merriam, N. Am. Fauna No. 3, 73, 1890.

Type locality.—Keam Canyon, Apache County, Ariz.

Distribution.—Eastern Arizona, western New Mexico, and southern Utah.

General characters.—Size large, about equaling *longimembris;* pelage rather soft; tail scantily haired; antitragus not lobed; posterior three-fifths of hind sole hairy.

Color.—Above, rich buff, with light admixture of black, effecting a suspicion of olivaceous; lateral line moderately well defined; below, pure white; ears buff, very faintly dusky within, a white spot on inflexed part and on inferior margin; tail white below, buff above with traces of dusky toward tip. In the early spring 'left-over' pelage the color is a beautiful clear buff with very few dusky-tipped hairs.

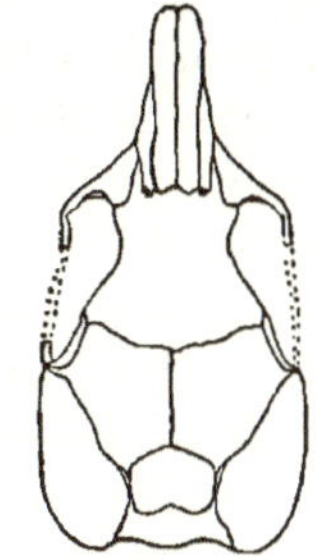

FIG. 5.—Skull of *Perognathus apache.*

Skull.—Size large, equaling *longimembris* and *parvus;* mastoids well developed; audital bullæ apposed anteriorly; interparietal pentagonal, of moderate size; angular process of mandible short and upturned, not long and widespread as in *longimembris;* lower premolar smaller than last molar. Compared with *longimembris* it has larger more bulging mastoids, heavier rostrum, wider interorbital space, shorter nasals, and smaller lower premolar.

Measurements.—Average of four adults from the type locality: Total length, 139.5; tail vertebræ, 67.5; hind foot, 18.5. *Skull:* (See table, p. 62.)

Remarks.—Apart from its subspecies, *P. a. melanotis*, *P. apache* is closely related to no other form except *P. callistus*. In color and size it bears some resemblance to *P. longimembris*, which is quite distant from it geographically. The only other similar form found within its range is *P. flavus bimaculatus*. From this it is distinguished by its larger size, heavy rostrum, and large wide interparietal. Specimens from Walnut, Ariz., are much deeper in color than usual.

Specimens examined.—Total number, 28, from localities as follows:

Arizona: Holbrook, 5; Keam Canyon, Navajo County, 8; Painted Desert, 2; Walnut, Coconino County, 4; Winslow, 1.

New Mexico: Deming, 3; Espanola, 1; Fort Wingate, 1; San Pedro, 1; Santa Fe, 3.

Utah: Noland Ranch, San Juan River, 1; Riverview, 1.

PEROGNATHUS APACHE MELANOTIS subsp. nov. BLACK-EARED POCKET MOUSE.

Type from Casas Grandes, Chihuahua, Mexico. ♀ ad., No. 97416, U. S. Nat. Mus., Biological Survey Coll. Collected May 21, 1899, by E. A. Goldman. Orig. No., 13750.

Distribution.—Known only from the type locality.

Characters.—Similar to *P. apache*, but darker; inside of ears black instead of buff; skull small and otherwise peculiar.

Color.—General color richer buff than that of *P. apache;* upperparts strongly mixed with black, particularly in median dorsal region; inside and inflexed parts of ears black, edges of ears and subauricular spot white; tail bicolor, dusky above, buffy white below; orbital region clear buff; underparts pure white.

Skull.—Similar to that of *P. apache*, but smaller; mastoids and audital bullæ much smaller; interparietal and interorbital space relatively wider.

Measurements.—*Type:* Total length, 133; tail vertebræ, 65; hind foot, 19.5. *Skull:* (See table, p. 62.)

Remarks.—The single specimen upon which this form is based is characterized by both external and cranial peculiarities which are much more than ordinary individual variation. A series of specimens from the type locality would doubtless show the majority of the peculiarities of the type to be constant.

PEROGNATHUS CALLISTUS sp. nov. BEAUTIFUL POCKET MOUSE.

Type from Kinney Ranch, Green River basin, near Bitter Creek, Sweetwater County, Wyo. ♂ yg. ad., No. 88245, U. S. Nat. Mus., Biological Survey Coll. Collected May 14, 1897, by J. Alden Loring. Orig. No., 4122.

Distribution.—Known only from the type locality and vicinity.

General characters.—Size medium, smaller than *apache;* skull quite similar; color very different.

Color.—Above, grayish olive buff uniformly mixed with black; below, pure white; lateral line cream buff, well defined; ears whitish outside, dusky within; postauricular spot creamy buff, quite prominent; tail white below, dusky above.

Skull.—Similar to *apache*, but somewhat heavier and more arched; interparietal slightly wider (though mastoids are larger); audital bullæ scarcely meeting anteriorly.

Measurements.—*Type:* Total length, 135; tail vertebræ, 63; hind foot, 18. *Skull:* (See table, p. 62.)

Remarks.—This species is the most delicately colored of the genus. It has the attractive coloration of *fasciatus*, but softer and more delicate. Its position is evidently between *fasciatus* and *apache*, and its nearest relations are clearly with the latter. Its large size immediately separates it from *fasciatus*, which it resembles externally, especially before maturity.

Specimens examined.—Total number, 7; 6 from Kinney Ranch, Bitter Creek, and 1 from Green River, Wyoming.

PEROGNATHUS PANAMINTINUS (Merriam). PANAMINT POCKET MOUSE.

Perognathus longimembris panamintinus Merriam, Proc. Acad. Nat. Sci. Phila., September 27, 1894, 265.

Type locality.—Perognathus Flat (altitude, 5,200 feet), Panamint Mountains, California.

Distribution.—Panamint Mountains, California, and eastward through southern Nevada to St. George, Utah.

General characters.—Size medium; tail long and moderately hairy; proximal third of hind sole hairy; pelage full, long, and silky; ears moderate.

Color.—Above, grayish buff, often with a pearly appearance caused by a pale buff ground color overlaid by dark-tipped hairs; lateral line pale buff, not sharply defined; subauricular spot small and inconspicuous; forelegs buffy or white; underparts white; tail, above dusky, strongly so distally, below buff or whitish.

Skull.—Size medium; nasals long and narrow; maxillary branches of zygomata gradually narrowing anteriorly; interorbital space wide; lower premolar larger than last molar. Compared with that of *flavus* the skull of *panamintinus* is more elongate, with smaller mastoids, and wider interparietal.

Measurements.—Average of 30 specimens from the type locality: Total length, 143; tail vertebræ, 78; hind foot, 19.7. *Skull:* (See p. 62.)

Remarks.—All the pocket mice without lobed antitragus found in California belong to the *panamintinus* group. *P. panamintinus* itself is easily recognizable by its proportions and dental peculiarities, as well as by its pearly gray color and long soft pelage. Its subspecies are closely related to it; *bangsi* inhabits the arid saline valleys southwest of the Panamint Mountains; *brevinasus* is also found to the southwest; and an incipient form not recognized by name is found in eastern Nevada. From this it appears that strictly typical *panamintinus* is confined to the Panamint Mountains.

Specimens examined.—Total number, 46, from localities as follows:

California: Panamint Mountains, 27.
Nevada: Ash Meadows, 1; Oasis Valley, 1; Oasis Valley (ten miles west), 1; Pahranagat Valley, 3; Pahroc Spring, 6; Panaca, 5; Vegas Valley, 1.
Utah: St. George, 1.

PEROGNATHUS PANAMINTINUS BANGSI[1] (Mearns). BANGS POCKET MOUSE.

Perognathus longimembris bangsi Mearns, Bull. Am. Mus. Nat. Hist., N. Y., X, 300, August 31, 1898.

Type locality.—Palm Springs, Colorado Desert, California.

Distribution.—Desert valleys of southern and southeastern California. Lower Sonoran zone.

General characters.—Similar to *panamintinus*, but smaller and paler.

Color.—Above, pale vinaceous buff, very lightly mixed with black, seldom showing the pearly effect of *panamintinus;* lateral line perfectly blended with sides; lower parts, including feet and fore legs, pure white; ears buffy white, thinly haired, a prominent white spot at the base of each and another on the inflexed portion: tail buff on upper side, rarely showing traces of dusky except at extreme tip, whitish on lower side; transverse nose spots nearly obsolete.

Skull.—Smaller than that of *panamintinus* with relatively smaller mastoids and wider interparietal; otherwise very similar.

Measurements.—*Type:* Length, 138; tail vertebræ, 80; hind foot, 19.

Remarks.—This pallid variety differs from *panamintinus* in color and size only. A convenient character for distinguishing it is the

---

[1] The following subspecies related to *P. panamintinus bangsi* has recently been described in the Proc. Biol. Soc. Wash., XIII, 153, June 13, 1900. Owing to absence in the field, the author has been unable to examine the type.—ED.

PEROGNATHUS PANAMINTINUS ARENICOLA Stephens.

"*Type* from San Felipe Narrows, San Diego County, California. No. 99828, ♂, U. S. Nat. Mus., Biological Survey Coll. Collected April 11, 1892.

"*Characters.*—Similar to *P. panamintinus bangsi* but paler and whiter; mastoids greatly swollen and projecting much further back than the occiput; interparietal very small. Total length, 141; tail vertebræ, 82; hind foot, 19."

color of the upper side of the tail, which is normally dusky in *panamintinus* and buffy in *bangsi*. The specimens from the more eastern localities are larger than those of the Colorado Desert and possibly should be considered intermediate between the latter and true *panamintinus*.

Specimens examined.—Total number, 56, from localities as follows:

California:[1] Argus Mountains (east base), 3; Ash Creek, Owens Lake, 5; Banning, 1; Bishop, Owens Valley, 1; Borax Flat, 4; Cabazon, 10; Hot Springs Valley, 7; Haway Meadows, 1; Little Owens Lake, 5; Moran, 2; Olancha, 1; Palm Springs, 2; Salt Wells Valley, 12; Whitewater, 2.

PEROGNATHUS PANAMINTINUS BREVINASUS subsp. nov. SHORT-NOSED POCKET MOUSE.

Type from San Bernardino, Cal. ♀ ad., No. $\frac{1109}{1581}$, Coll. of C. Hart Merriam. Collected May 2, 1885, by F. Stephens.

Distribution.—Known from a few scattered localities in extreme southwestern California. Upper Sonoran zone.

General characters.—Similar in general to *panamintinus* and *bangsi*; color darker; tail shorter; skull peculiar.

Color.—Above, pinkish buff, much varied with black; below, pure white; lateral line pinkish buff, not very sharply defined; postauricular spot buff, more prominent than in *bangsi*; hairs of back and especially of rump, clear buff nearly to roots, often showing no plumbeous whatever; ears dusky; subauricular spot small; orbital ring buffy; tail buff or buffy white, faintly dusky above; transverse nose stripes blackish, well defined. *Young:* Dull slaty; hairs of back dirty whitish, with plumbeous tips.

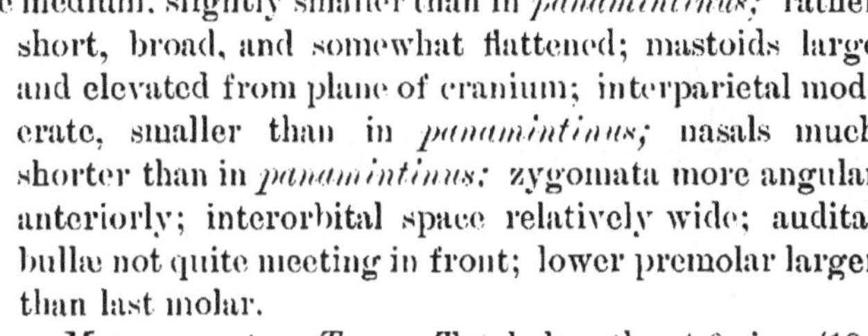

Skull.—Size medium, slightly smaller than in *panamintinus*; rather short, broad, and somewhat flattened; mastoids large and elevated from plane of cranium; interparietal moderate, smaller than in *panamintinus*; nasals much shorter than in *panamintinus*; zygomata more angular anteriorly; interorbital space relatively wide; audital bullæ not quite meeting in front; lower premolar larger than last molar.

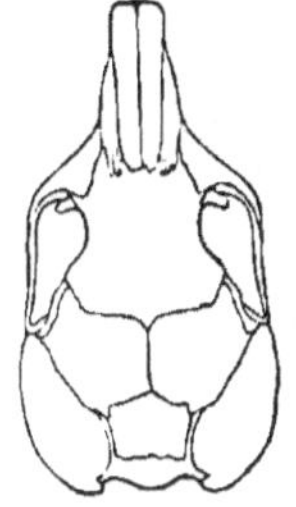

FIG. 6.—Skull of *Perognathus brevinasus*.

Measurements.—*Type:* Total length, 4.9 in. (124 mm.); tail vertebræ, 2.6 in. (66 mm.); hind foot (measured dry), 17.4 mm. Average of three adult males from Ferndale, San Bernardino County, Cal.: Total length, 130; tail, 68; hind foot (measured dry), 18.2. *Skull:* (See table, p. 62.)

Remarks.—This is the *P. longimembris* of recent authors which requires a name, since *longimembris* applies only to the San Joaquin

[1] For details in regard to these localities, and others of the same general region mentioned in this paper, see N. Am. Fauna, No. 7, 361–384.

Valley animal. It ranges near *P. p. bangsi*, but is evidently confined to a higher zone. Whether it intergrades with *panamintinus* or *bangsi* is not satisfactorily shown by the present material. Possibly it should be considered a distinct species.

Specimens examined.—Total number, 61, from localities as follows:

California: Burbank, 1; Ferndale, San Bernardino County, 7; Jacumba, 7; San Bernardino, 44; Summit, Coast Range, San Diego County, 2.

PEROGNATHUS NEVADENSIS Merriam. NEVADA POCKET MOUSE.

Perognathus nevadensis Merriam, Proc. Acad. Nat. Sci. Phila., September 27, 1894, 264.

Type locality.—Halleck, Nev.

Distribution.—Upper Sonoran zone of central Nevada; northward to southern Oregon and northern Utah.

General characters.—Similar in general to *P. panamintinus;* differing in somewhat smaller size, color of underparts, and slight cranial characters.

Color.—Much as in *panamintinus* but darker, and with belly colored like sides.

Skull.—Very similar to that of *panamintinus;* nasals a trifle shorter; zygomata more angular anteriorly; interparietal shorter and broader, occipital side strongly concave; lower premolar larger than last molar.

Measurements.—Average of twenty-four adults from the type locality: Total length, 133; tail vertebræ, 72.4; hind foot, 18.7. *Skull:* (See table, p. 62.)

Remarks.—*P. nevadensis* and *P. panamintinus* are closely related. Whether they are directly connected at the present time remains to be seen. Specimens from Flowing Springs, Nev., are considerably larger than typical, and also interesting as showing a very worn pelage, which is pale grizzled cinnamon with all markings more or less obsolete.

Specimens examined.—Total number, 55, from localities as follows:

Nevada: Austin, 1; Battle Mountain, 5; Devil Gate (twelve miles west of Eureka), 1; Flowing Springs, 10; Golconda, 2; Halleck, 23; Monitor Valley, 2; Osobb Valley, 1; Pyramid Lake, 1; Reese River, 5; Stillwater, 2; Wadsworth, 2.

Oregon: Tumtum Lake, 3.

Utah: Kelton, 1.

PEROGNATHUS PACIFICUS Mearns. PACIFIC POCKET MOUSE.

Perognathus pacificus Mearns, Bull. Am. Mus. Nat. Hist., N. Y., X, 299, August 31, 1898.

Type locality.—Mexican boundary monument No. 258, shore of Pacific Ocean.

Distribution.—Known only from the type locality.

General characters.—Size exceedingly small; similar in color and general characters to *P. p. brevinasus;* tail about equal to or slightly shorter than head and body; proximal third of hind sole hairy; pelage very soft but not long and full as in *panamintinus;* skull much as in the other members of the *panamintinus* group.

Color.—Similar to *P. p. brevinasus* but somewhat darker; sides about like back, between pinkish and salmon buff, very finely and thickly mixed with black; lateral line and slight postauricular spot pinkish buff; ears dusky; subauricular spot present; lower parts white; tail nearly concolor, faintly darker above than below.

Skull.—Size very small; cranium strongly arched; mastoids moderate, not bulging as in *brevinasus;* interparietal much wider than long; zygomata very slender and threadlike; nasals rather short; interorbital space moderately wide; lower premolar plainly larger than last molar.

Measurements.—*Type:* Total length, 113; tail vertebræ, 53; hind foot, 15.5. One adult topotype: Total length, 110; tail vertebræ, 54; hind foot, 15.3. *Skull:* (See table, p. 62.)

Remarks.—This species is by far the most diminutive member of the *panamintinus* group and of the genus. *P. flavus*, which has long been distinguished as the smallest pocket mouse, must now allow its title to pass to this tiny species. There is some superficial resemblance to *flavus*, but the skull is entirely in accord with the characters of the *panamintinus* group. Details which *pacificus* shares with the other members of the group, and which distinguish it from *flavus* and its forms, are small mastoids, wide interparietal, wide interorbital space, and large lower premolar.

Specimens examined.—Total number, 3, all from the type locality.

PEROGNATHUS AMPLUS sp. nov. LORING POCKET MOUSE.

Type from Fort Verde, Ariz. ♂ ad., No. $\frac{34626}{46117}$, U. S. Nat. Mus., Biological Survey Coll. Collected June 26, 1892, by J. Alden Loring. Orig. No., 272.

Distribution.—Known only from the type locality.

General characters.—Size large; tail long, well haired, slightly penicillate; hind sole naked medially to posterior fifth, which is hairy; pelage soft, full, and long; antitragus not lobed; mastoids greatly developed.

Color.—Above, pinkish buff delicately lined with black; basal fifth of hairs plumbeous; underparts white; lateral line buff, rather wide, extending on forelegs nearly to wrist; orbital area pale; white spot present at base of ear above and below; tail buff, mixed with black above.

Skull.—Size large; mastoids excessively developed, bulging in all directions and reaching the maximum shown in the genus; audital bullæ relatively small, about as large as in *P. apache*, weakly apposed anteriorly; interparietal relatively very small, pentagonal, about as long as broad; rostrum long and slender, nasals more slender than in *apache*, nasal branches of premaxillæ wider; zygomata narrowing anteriorly; interorbital width moderate; lower premolar about equal to or very slightly larger than last molar.

Measurements.—Type: Total length, 155; tail vertebræ, 80; hind foot, 20. *Skull:* (See table, p. 62.)

Remarks.—Both externally and cranially *P. amplus* is very peculiar and evidently has no close relation with any previously known species. In proportions (not in size) and character of pelage it is not very unlike *P. panamintinus*, and from some of the forms of this species it is but slightly dissimilar in color, but its remarkable skull and slightly haired hind foot are unique. The great development of mastoids which it shows is not at all correlated with an equal enlargement of the audital bullæ, as these are no larger than in *P. apache*. It has no important characters in common with *apache* and can not be closely related to it.

Specimens examined.—One, the type.

PEROGNATHUS LONGIMEMBRIS (Coues). SAN JOAQUIN POCKET MOUSE.

Otognosis longimembris Coues, Proc. Acad. Nat. Sci. Phila., 1875, 305, under *Cricetodipus parvus*. (Type from Fort Tejon.)

Cricetodipus parvus True, Proc. U. S. Nat. Mus., IV, 474, 1882.

Perognathus inornatus Merriam, N. Am. Fauna No. 1, 15, 1889. (Type from Fresno.)

Type locality.—Fort Tejon, Cañada de las Uvas, Kern County, Cal.

Distribution.—Sonoran zone of the San Joaquin Valley, California, and its immediate extensions.

General characters.—Size large, equaling *P. apache;* color uniform, all markings reduced; antitragus not lobed; pelage rather harsh; proximal third of hind sole hairy.

Color.—Above, buff mixed with more or less black; below, white; bases of hairs on rump slightly or not plumbeous; lateral line poorly defined, concolor with upper sides; tail buff, paler on lower surface, faintly dusky above; upper side of forelegs generally buff to wrist; ears buffy outside, dusky within, a slight stripe of white on inflexed portion and the usual white spot at base. Young adults darker than adults, and showing a slight tinge of olivaceous.

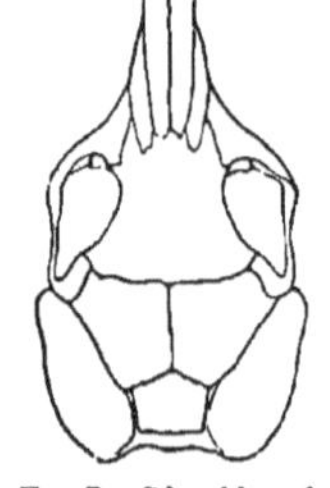
FIG. 7.—Skull of *Perognathus longimembris*.

Skull.—Size large, mastoids and audital bullæ moderate, not bulging as in *brevinasus;* interparietal subquadrate, relatively smaller than in *brevinasus;* interorbital space very narrow, often dorsally concave in old individuals; nasals long; lower premolar larger than last molar.

Measurements.—Average of 4 adult males from Fresno, Cal.: Total length, 145.2; tail vertebræ, 74.5; hind foot, 18.7. Of 4 adult females: Total length, 136; tail vertebræ, 71.5; hind foot, 18.3. *Skull:* (See table, p. 62.)

Remarks.—The above description is based mainly on specimens from Fresno, the type locality of '*inornatus.*' The type of *longimembris* is immature, but its skull shows the narrow interorbital space peculiar to the San Joaquin Valley form. The only available topotype is fortunately a young adult which agrees perfectly with specimens from Fresno and other points in the San Joaquin Valley. Two young specimens from San Emigdio and Rose Station, both very near Fort Tejon, are also clearly the same as those from Fresno, having the harsher pelage and slight olivaceous effect so different from the soft hairs and delicate pearly color of the young of *panamintinus* and subspecies. Thus it seems that the name *longimembris* should be applied to the animal recently called *inornatus* rather than to the San Bernardino form.

The species is very distinct, though its range is limited. It seems to be exclusively confined to the San Joaquin Valley, where it is the only representative of the genus. Young adults may be distinguished from old by their smaller size and darker color. Females are constantly smaller than males. Among adults two phases of color are apparent, one in which the hairs are grayish from the roots and another in which they are buffy.

Specimens examined.—Total number, 111, from localities as follows:

California: Alila, 2; Bakersfield, 5; Delano, 2; Fort Tejon, 2; Fresno, 54; Huron, 3; Livingston, 11; Lodi, 3; Oakdale, 2; Ripon, 2; Rose Station, Kern County, 1; San Emigdio, Kern County, 1; Three Rivers, 2; Tipton, 7; Walker Basin, Kern County, 14.

PEROGNATHUS PARVUS (Peale). OREGON POCKET MOUSE.

Cricetodipus parvus Peale, U. S. Expl. Exp'd., VIII, Mamm. and Ornith., 52–54, 1848.

Perognathus parvus Cassin, U. S. Expl. Exp'd., Mamm. and Ornith., 48–49, 1858; Merriam, N. Am. Fauna No. 1, 28, 1889—Peale's description copied.

Perognathus monticola Baird, Mamm. N. Am., 422, 1857; Merriam, N. Am. Fauna No. 1, 17, 1889.

Type locality.—Oregon. Assumed to be The Dalles, Oreg.

Distribution.—Valley of the Yakima River, Washington, and thence southward to central and southeastern Oregon. Upper Sonoran zone.

General characters.—Size large; tail slightly penicillate, its vertebræ longer than head and body; ears moderate, well haired, antitragus prominently lobed; proximal fourth of hind sole hairy; color variable, presenting two extremes, a gray and a buff.[1]

---

[1] This species is certainly to some degree dichromatic, for the color variation is evidently not due to age, sex, or season. In one phase the buff is reduced to grayish

Color.—Gray phase: Above, pale slaty buff mixed with black, darkest in center of back; below, white, except belly, the hairs of which are normally plumbeous, with pale tips; sides like back, but paler: black-tipped hairs of back running forward across sides and reaching or nearly reaching forearm: lateral line buff; tail tricolor, dusky above, becoming black terminally, buff on sides, generally white below, but sometimes suffused with buffy; ears dusky, lighter on margins; subauricular spot moderate; feet white; inner side of hind legs dusky to heel. *Buff phase:* Everywhere as in gray phase, but general color buff or ochraceous buff instead of slaty. *Young:* Above, clear, light plumbeous, tips of hairs very pale buff, gradually intensifying with increasing age; below, as in adult. In late fall the high pelage which succeeds the breeding pelage becomes much paler as the black tips of the hairs wear off and expose the undercolor.

Skull.—Size large: cranium slightly arched; rostrum somewhat attenuate: audital bullæ and mastoids moderately developed; audital bullæ meeting anteriorly in a well-defined symphysis; interparietal wide, pentagonal, anterior angle strong; lower premolar smaller than last molar.

Measurements.—Average of five adults from Mabton, Wash.:[1] Total length, 171.8; tail vertebræ, 91.8; hind foot, 22.4. *Skull:* (See table, p. 62.)

Remarks.—The group for which *parvus* stands contains seven closely related forms. All are of relatively large size and have the antitragus distinctly lobed, thus requiring but slight comparison with the other members of the subgenus. *P. p. olivaceus* is the most centralized form. It occupies the main part of the Great Basin proper and the others, which are found in the various Great Basin extensions, have evidently been derived from it.

The name *parvus*, though one of the earliest proposed for a pocket mouse, has been usually incorrectly applied. Peale assigns the species to Oregon, and his original description and measurements indicate one of the larger members of the genus.[2] Since but one species is found in the part of Oregon traversed by the Wilkes expedition, and since this agrees in general with Peale's description, there seems to be no reason why the name *parvus* should not now be applied to it. The

---

drab, and in another it is developed into cinnamon, or even bright ochraceous. Between these extremes occur various intermediate stages. As might be expected, one phase is often much more numerous at a given locality than the other, though both are found together. The two are perfectly distinct in both adults and young.

[1] Although numerous specimens from The Dalles have been examined none are sufficiently adult to afford satisfactory measurements, so that it has been necessary to use the Mabton series for this purpose.

[2] The measurements alone are sufficient to prove that the name should never be used for a five-toed kangaroo rat. Cf. Rhoads, Proc. Acad. Nat. Sci. Phila., 1893, 407–410.

form found at The Dalles is here considered typical. The chances that the type was taken there are considerable since the species is very abundant there and members of the Wilkes expedition camped at or near that place on several different occasions.[1]

It is also not improbable that the type of Baird's '*monticola*' was also taken at The Dalles. Baird's queried statement that it came from St. Mary's Mission, Mont., is rendered much more doubtful by the unsuccessful efforts of recent collectors to obtain additional specimens from that locality. Dr. Suckley, who collected this type, stopped for some time at The Dalles and may have obtained it there, as pocket mice are probably more abundant there than at any other point at which he stopped. Its skull agrees more nearly with that of *parvus* than with that of any other form.

Specimens examined.—Total number, 103, from localities as follows:

Oregon: Antelope, 1; Burns, 5; Crown Rock, John Day River, 3; Harney, 1; Heppner, 2; Lost River, Klamath Basin, 5; Narrows, Malheur Lake, 6; North Dalles, 11; Prineville, 1; Rock Creek Sink, 2; Shirk, 5; The Dalles, 13; Tule Lake, 5; Tumtum Lake, 7; Twelve-mile Creek, 1; Umatilla, 2; Willows Junction, 2.

Washington: Mabton, 25; North Yakima, 6.

PEROGNATHUS PARVUS MOLLIPILOSUS (Coues). Coues Pocket Mouse.

Perognathus mollipilosus Coues, Proc. Acad. Nat. Sci. Phila., 1875, 296 (under *P. monticola*).

Perognathus monticola Townsend, Proc. U. S. Nat. Mus., X, 177, 1888.

Type locality.—Fort Crook, Shasta County, Cal.[2]

Distribution.—Great Basin extension of northeastern California, north to Klamath Basin, Oregon. Upper Sonoran zone, except on Mount Shasta, where it ascends to the Boreal.

General characters.—Size somewhat smaller than *parvus;* ears much larger, antitragal lobe prominent; coloration dark; markings intense.

Color.—Above, rich ochraceous buff, black-tipped hairs very abundant; lateral line prominent; white subauricular spot very faint or not evident; below, white, varying to tawny ochraceous on belly.

Skull.—Size relatively rather small; very similar to *P. olivaceus*, but with the ascending branches of the premaxillæ abruptly truncated, not exceeding the nasals.

Measurements.—Average of three adults from the type locality: Total length, 168.3; tail vertebræ, 88; hind foot, 22.3; ear from meatus (dry), 8.2. *Skull:* (See table, p. 62.)

Remarks.—The specimens from Fort Crook and Fall River Valley are the only ones that may be considered strictly typical. They are

[1] Wilkes, Narrative U. S. Expl. Exp'd, IV, 403–432, 1845.

[2] Fort Crook, now abandoned, was located about 2 miles northeast of the present site of Burgettville, or Swasey.

well characterized by large ears, rich color, obsolescent subauricular spots, and truncated premaxillæ. Nearly all the others here referred to *mollipilosus* show greater or less tendency toward *olivaceus*. The form seems to be one like *magruderensis*, which is rather ill defined, but of a type too strongly characterized to be left unrecognized. Specimens from the Boreal zone on Mount Shasta do not seem to be separable, notwithstanding their very anomalous distribution.[1]

Specimens examined.—Total number, 44, from localities as follows:

> **California:** Alturas, 1; Cassel, 6; Edgewood, 3; Fall Lake, Fall River Valley, 1; Fort Crook, 5; Likely, 1; Madeline Plains, 2; Mount Shasta (head of Panther Creek, altitude 7,800 feet, 8; pine belt, south base 4), 12; Sisson, 2; Susanville, 2.
>
> **Oregon:** Summer Lake, 2; Swan Lake Valley, 4; Williamson River, 3.

PEROGNATHUS PARVUS OLIVACEUS (Merriam). GREAT BASIN POCKET MOUSE.

Perognathus olivaceus Merriam, N. Am. Fauna No. 1, 15, 1889; ibid., No. 5, 71, 1891; Elliott, Field Columbian Mus., Zool. Ser., I, No. 10, 211, 1898.
Perognathus olivaceus amœnus Merriam, N. Am. Fauna No. 1, 16, 1889.

Type locality.—Kelton, Utah.

Distribution.—Upper Sonoran zone throughout the Great Basin, from northern Utah and southern Idaho southwest to Owens Valley, California, and west to southern Oregon and northeastern California.

General characters.—Similar to *P. parvus*; differing in softer pelage, lighter color, and slight cranial characters.

Color.—Similar to the buff phase of *P. parvus*, but with clearer, softer colors; above, bright cinnamon buff finely mixed with black; lateral line distinct; subauricular spot conspicuous; hairs of belly pure white or with plumbeous bases and buff tips; inner side of foreleg white or buff. Late fall pelage paler.

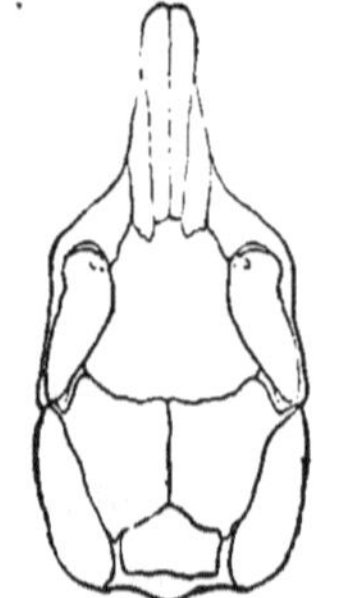

FIG. 8.—Skull of *Perognathus olivaceus*.

Skull.—Similar to that of *parvus* but slightly larger; mastoids more inflated; interparietal slightly smaller (ratio of interparietal width to basilar length of Hensel, 27.8); audital bullæ meeting anteriorly in a very weak symphysis or not meeting; ascending branches of premaxillæ generally exceeding nasals.

Measurements.—*Type:* Total length, 184; tail vertebræ, 101; hind foot, 23. Average of three males from Salt Lake City, Utah: Total length, 175.6; tail vertebræ, 95.6; hind foot, 22. Average of three females from Ogden, Utah: Total length, 167.7; tail vertebræ, 88; hind foot, 21.7. *Skull:* (See table, p. 62.)

Remarks.—In the wide range of this form are found numerous

[1] See N. Am. Fauna No. 16, 98, 1899.

more or less trivial deviations from the type. Most of these are of size only and probably represent nothing more than individual variation, which in this respect is often considerable. A difference in size between the sexes is also quite noticeable. The dark undercolor shown by the type of '*amœnus*' has been observed in many specimens from various localities, and in the series now available from Nephi are individuals with pure white belly hairs, as in the type of *olivaceus*.

Specimens examined.—Total number, 126, from localities as follows:

California: Benton, 1; Bishop Creek, 1; Long Valley, 4; Lower Alkali Lake, 1; Moran, 4.

Idaho: Bear Lake (east side), 10; Big Butte, 1; Birch Creek, 3; Blackfoot, 2; Lemhi, 1; Pahsimeroi Valley, 3.

Nevada: Anderson, 1; Bull Run Mountains, 3; Carson Valley, 1; Cottonwood Range, 5; Elko, 6; Golconda, 1; Granite Creek, 5; Halleck, 5; Monitor Valley, 5; Mountain City, 3; Pyramid Lake, 3; Reese River, 6; Ruby Valley, 9; Winnemucca, 1.

Utah: Blacksmith Fork, Cache County, 2; Kelton, 2; Laketown, 2; Nephi, 9; Ogden, 17; Otter Creek, 2; Salt Lake City, 4.

Wyoming: Fort Bridger, 1.

PEROGNATHUS PARVUS MAGRUDERENSIS subsp. nov. MOUNT MAGRUDER POCKET MOUSE.

Type from Mount Magruder, Nev. (altitude 8,000 feet). ♂ ad., No. $\frac{28427}{10531}$, U. S. Nat. Mus., Biological Survey Coll. Collected June 6, 1891, by Vernon Bailey. Orig. No., 2899.

Distribution.—Upper Sonoran and Transition zones of the desert ranges of southern Nevada and adjoining portion of California.

General characters.—Similar to *P. p. olivaceus*, but very much larger, being the largest member of the *parvus* group.

Color.—As in *P. p. olivaceus*.

Skull.—Very much as in *olivaceus*, but considerably larger and heavier; interparietal relatively narrower (ratio of interparietal width to basilar length of Hensel, 25.1).

Measurements.—*Type:* Total length, 198; tail vertebræ, 107; hind foot, 26. Average of five adult topotypes: Total length, 191; tail vertebræ, 102.2; hind foot, 24.2. *Skull:* (See table, p. 62.)

Remarks.—*P. p. magruderensis* is a large incompletely differentiated mountain form closely related to *olivaceus* which is found near it at a lower altitude. The form found on the Panamint Mountains shows trifling differences from typical *magruderensis*, but is here considered the same.

Specimens examined.—Total number, 27, from localities[1] as follows:

California: Coso, 8; Inyo Mountains, 2; Panamint Mountains, 7; White Mountains, 2.

Nevada: Mount Magruder, 7; Grapevine Mountains, 1.

---

[1]See N. Am. Fauna, No. 7, 361–384, 1893.

PEROGNATHUS ALTICOLA Rhoads. WHITE-EARED POCKET MOUSE.

Perognathus alticolus Rhoads, Proc. Acad. Nat. Sci., Phila., December, 1893, 412.

Type locality.—Squirrel Inn, San Bernardino Mountains, California.

Distribution.—Known only from the type locality.

General characters.—Similar to *P. p. olivaceus*, from which it differs in somewhat smaller size, in color of ears and tail, and in slight cranial characters.

Color.—Above, as in *P. p. olivaceus;* sides like back, lateral line not prominent; below, white; ears clothed within and without with clear white hairs; tail faint buff above, terminal fourth slightly dusky, white below.

Skull.—Essentially as in *P. p. olivaceus;* ascending branches of supraoccipital very broad and heavy; interparietal rather narrow.

Measurements.—Average of two adult topotypes: Total length, 165; tail vertebræ, 83.5; hind foot, 22.2. *Skull:* (See table, p. 62.)

Remarks.—This isolated species may be immediately distinguished from the other members of the *parvus* group by its light ears and tail. The type agrees perfectly with the topotypes upon which the description is based.

Specimens examined.—Total number, 4, all from the type locality.

PEROGNATHUS LORDI (Gray). NORTHWEST POCKET MOUSE.

Abromys lordi Gray, Proc. Zool. Soc. London, 1868, 202.
Perognathus lordi Rhoads, Proc. Acad. Nat. Sci. Phila., 1893, 405.

Type locality.—British Columbia.

Distribution.—Upper Sonoran and Transition zones of the plains of the Columbia River, Washington, and suitable adjacent territory in southern British Columbia.

General characters.—Similar to *P. parvus;* size large (nearly equaling *magruderensis*); tail long; feet and ears moderate; antitragus lobed; color dark; interparietal narrow.

Color.—Above, pale slaty buff, strongly mixed with black; general color as in the gray phase of *P. parvus;* hairs of belly generally with plumbeous bases and buffy tips, leaving a small inguinal and a large pectoral patch pure white; subauricular spot small but distinct; tail tricolor, as in *parvus*.

Skull.—Size large; audital bullæ and mastoids inflated; audital bullæ always connected anteriorly; interparietal squarish pentagonal, deeply notched by occipital.

Measurements.—Average of seven adults from Oroville, Wash.: Total length, 183, tail vertebræ, 97.7; hind foot, 23.2. *Skull:* (See table, p. 62.)

Remarks.—The numerous specimens examined from various parts of the country in which John Keast Lord collected leave little doubt that this was the pocket mouse to which his name was given by Gray in 1868; but in order to remove all uncertainty, specimens were sent to Mr. Oldfield Thomas, curator of mammals in the British Museum, who kindly compared them with the type and found that they agreed in every essential particular. In color *lordi* is almost identical with the gray phase of *P. monticola*, but its large size and small interparietal show it to be a very different species. Apparently it does not occur on the west side of the Columbia at Wenatchee or south of that point. Specimens from Coulee City, Douglas, and vicinity are grading toward *columbianus*.

Specimens examined.—Total number, 131, from localities as follows:

> **British Columbia:** Ashcroft, 14; Kamloops, 6; Okanagan, 12; Vernon, 2.
> **Idaho:** Lewiston, 1.
> **Washington:** Almota, 16; Asotin, 11; Chelan, 2; Cheney, 3; Conconully, 3; Coulee City, 6; Douglas, 11; Fort Spokane, 7; Marcus, 1; Orondo, 7; Oroville, 9; Spokane Bridge, 11; Wenatchee (east bank of Columbia), 9.

PEROGNATHUS LORDI COLUMBIANUS (Merriam). COLUMBIAN POCKET MOUSE.

Perognathus columbianus Merriam, Proc. Acad. Nat. Sci. Phila., September 27, 1894, 236.

Type locality.—Pasco, Wash.

Distribution.—Vicinity of type locality.

General characters.—Similar to *P. lordi*, from which it differs in slight cranial characters.

Color.—As in *P. lordi*.

Skull.—Audital bullæ and mastoids highly developed: interparietal width much reduced; otherwise as in *P. lordi*.

Measurements.—Average of five adults from Pasco, Wash.: Total length, 179.8; tail vertebræ, 92; hind foot, 22.8. *Skull:* (See table, p. 62.)

Remarks.—This form is found only on the hot plains about the Great Bend of the Columbia. The great development of audital bullæ and mastoids and consequent reduction of interparietal width exhibited by it is the extreme shown in the *parvus* group.

Specimens examined.—Total number, 26, from localities as follows:

> **Washington:** Pasco, 12; Touchet, 14.

PEROGNATHUS FORMOSUS Merriam. LONG-TAILED POCKET MOUSE.

Perognathus formosus Merriam, N. Am. Fauna No. 1, 17, October 25, 1889.

Type locality.—St. George, Utah.

Distribution.—Southwestern Utah, southern Nevada, and the adjoining portion of California. Lower Sonoran zone.

General characters.—Size large (about equal to *P. p. magruderensis*); tail much longer than head and body, heavily crested penicillate; ears large, somewhat attenuate, scantily haired; antitragus prominently lobed; soles naked.

Color.—Above, grizzled sepia; below, white; sides not noticeably lighter than back; dark hairs generally extending down front leg to forearm; ears dusky black, tuft of bristly hairs at base mixed black and whitish; subauricular spot small, noticeable only in very high pelage; feet white; tail buff to pencil below, buff mixed with dusky above, intensifying toward pencil, which is brownish black. Worn pelage, drab instead of sepia. *Young:* Smoky gray above, white below.

Skull.—Size medium; cranium slightly arched; mastoids well developed, bulging very slightly behind, rather smaller than in the *parvus* group; interorbital space wide; interparietal large and wide, pentagonal; nasals shorter than in *magruderensis;* audital bullæ slightly touching anteriorly; lower premolar larger than last molar.

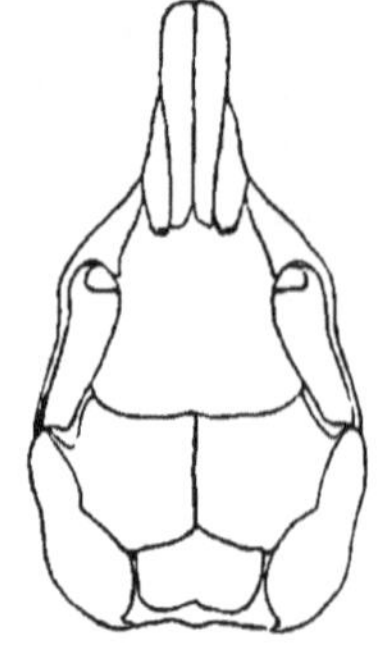

Fig. 9.—Skull of *Perognathus formosus.*

Measurements.—Average of five adults from St. George, Utah: Total length, 189.6; tail vertebræ, 106.4; hind foot, 24. *Skull:* (See table, p. 62.)

Remarks.—This peculiar species is the only member of the subgenus *Perognathus* which has a heavily crested tail. In this respect it is like *Chætodipus*, but its skull shows the characters of true *Perognathus*. It inhabits remote western deserts little frequented by collectors. With the exception of the type, all the specimens known were taken by the Death Valley Expedition in 1891.

Specimens examined.—Total number, 136, from localities[1] as follows:

California: Argus Mountains, 6; Bennett Wells, 2; Emigrant Spring, 12; Funeral Mountains, 7; Furnace Creek, 4; Grapevine Springs, 11; Little Owens Lake, 3; Lone Pine, 2; Lone Willow Spring, 2; Panamint Mountains, 15; Resting Springs, 1; Saline Valley, 6; Saratoga Springs, 6.

Nevada: Ash Meadows, 4; bend of Colorado River near Callville, 12; Bunkerville, 2; Charleston Mountains, 1; Grapevine Mountains, 6; Oasis Valley, 2; Pahranagat Valley, 2; Pahroc Spring, 2; Pahrump Valley, 17; Thorp Mill, 2.

Utah: St. George, 9.

Subgenus CHÆTODIPUS Merriam, 1889 (see p. 14).

Chætodipus Merriam, N. Am. Fauna No. 1, 5, 1889. Type, *Perognathus spinatus* Merriam, 1889, from Colorado River, California.

PEROGNATHUS BAILEYI Merriam. Bailey Pocket Mouse.

Perognathus baileyi Merriam, Proc. Acad. Nat. Sci. Phila., September 27, 1894, 262.

Type locality.—Magdalena, Sonora, Mexico.

---

[1] See N. Am. Fauna No. 7, 361–384, 1893.

Distribution.—South central Arizona and thence south into Sonora and northern Lower California, Mexico.

General characters.—Size, very large; tail very long and penicillate: color similar to that of *P. formosus;* skull large and heavy.

Color.—As in *formosus*, but paler, being grayish rather than buffy; under side of tail whitish instead of buffy.

Skull.—Large and massive; mastoids relatively smaller than in *formosus;* mastoid side of parietal scarcely longest, about equaling other long sides; audital bullæ very weakly apposed in front; interparietal large, pentagonal, relatively wider than in *formosus*, interparietal width about equal to interorbital width; lower premolar smaller than or about equal to last molar.

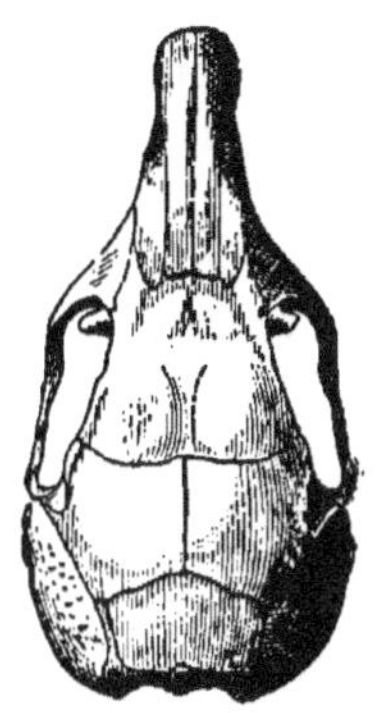

FIG. 10.—Skull of *Perognathus baileyi.*

Measurements.—Average of five adults from the type locality: Total length, 214.6; tail vertebræ, 120.6; hind foot, 27. *Skull:* (See table, p. 62.)

Remarks.—*P. baileyi* stands somewhat alone. It seems most nearly related to *formosus*, although the sum of its characters places it in a different subgenus. The size and massiveness of its skull suggest relationship to some of the larger species of *Chætodipus*, like *paradoxus* or *femoralis*, but detailed characters indicate little affinity in this direction.

Specimens examined.—Total number, 17, from localities as follows:

Arizona: Mammoth, 1; New River, 5; Tucson (75 miles southwest), 1; Santa Catalina Mountains, near Tucson, 1.
Sonora: Magdalena, 8.
Lower California: Comondu, 1.

PEROGNATHUS HISPIDUS Baird. HISPID POCKET MOUSE.

Perognathus fasciatus Baird, Mamm. N. Am., 420, 1857; Thomas, Proc. Zool. Soc. Lond., 1888, 449.

Perognathus hispidus Baird, Mamm. N. Am., 421, 1857; Merriam, N. Am. Fauna No. 1, 23, 1889.

Perognathus paradoxus spilotus Merriam, N. Am. Fauna No. 1, 25, 1889; Allen, Bull. Am. Mus. Nat. Hist., N. Y., VI, 172, 1894; ibid., VIII, 58, 1896.

Perognathus paradoxus Allen, Bull. Am. Mus. Nat. Hist., N. Y., VI, 172, 1894.

Type locality.—Charco Escondido, Tamaulipas, Mexico.

Distribution.—Southern and western Texas, north to Oklahoma and south into border States of Mexico. Lower Sonoran zone.

General characters.—Size large; tail equal to or slightly shorter than head and body, not crested or penicillate; pelage harsh, no spines or bristles anywhere; ear small, antitragus lobed, tragus quite evident; soles of hind feet naked in median line; skull heavy and somewhat ridged.

Color. Above, ochraceous much mixed with black; sides scarcely paler than back; lateral line clear ochraceous, extending on fore and hind legs for half their length; face and orbital region light, lower cheeks continuous with lateral line; underparts white; ears dusky inside, buffy white on margins and on outer side, except an elliptical black spot on inflexed portion; feet white; tail whitish below, buffy on sides, sharply black above. *Spring pelage:* Much paler.

Skull.—Size large; rostrum heavy, somewhat arched; interorbital space wide; supraorbital bead very evident; mastoids relatively small, not bulging behind; mastoid side of parietal short; interparietal large, imperfectly pentagonal, all angles much rounded, anterior one sometimes entirely annihilated; ascending branches of supraoccipital short and heavy; audital bullæ normally separated anteriorly by breadth of basisphenoid, occasionally approaching each other; lower premolar about equal to last molar.

Measurements.—Average of six adults from Brownsville, Tex.: Total length, 204.5; tail vertebræ, 100.5; hind foot, 25. *Skull:* (See table, p. 62.)

Remarks.—This species typifies one of the most peculiar groups of the genus. It is characterized by its large size, short uncrested tail, and heavy ridged skull. Its skull, though peculiar, is plainly that of a *Chætodipus*, but external characters, excepting size, do not prohibit its being classed with restricted *Perognathus*, thus reversing the conditions presented by *formosus*. Baird's type agrees in essential characters with specimens from Brownsville, Tex., and other points near the type locality. In examining this type it was discovered that the broken skull supposed to belong to it is composite. The posterior section is the only part which may be safely assumed to have been originally within the skin. The anterior part and the mandible seem to have belonged with some other skin. Besides many differences of proportion which show this to be the case, there is a distinct difference in the texture and surface appearance of the bone in the two parts, indicating that they were cleaned and used differently. The skull of Baird's second specimen (No. 1695), which he figured, is nearly perfect and agrees in detail with many recently collected ones. The posterior section of the skull of the type agrees with this one, and also with the same parts of numerous others from the same vicinity. The skin of the type is also easily recognizable, so that when everything is considered there is no good excuse for allowing the name *Perognathus hispidus* to remain doubtful.

The form described as *P. p. spilotus* is here considered synonymous with *hispidus*, though there is some difference between the two. In a general way the southern animals are smaller, and with harsher pelage and higher color than the northern. The difference, which is chiefly of size, is fairly marked, and the increase quite gradual from typical *hispidus* to typical *paradoxus*, leaving '*conditi*' and '*spilotus*' exactly intermediate, in character as well as geographic situation. Individual variation in size is often considerable, as is well shown by the Brownsville series, in which the length of the hind foot varies from 22 mm to 25 mm.

Specimens examined.—Total number, 175, from localities as follows:

Texas: Bee County, 5; Beeville, 2; Blocker Ranch, 1; Brazos, 1; Brownsville, 40; Chileipin Creek, San Patricio County, 1; Colorado, 1; Corpus Christi, 2; Cuero, 1; Gainesville, 6; Llano, 2; Lomita Ranch (near Rio Grande City), 3; Long Point, 1; Los Indios Ranch, Nueces County, 1; Nueces Bay, 5; Oconnorport, 1; Padre Island, 1; Rio Grande City, 3; Rockport, 30; Roma, 1; Saginaw, 1; San Antonio, 46; Santa Rosa, 1; San Thomas, 2; Sauz Ranch, Cameron County, 1; Sycamore Creek (mouth), 2.
Nuevo Leon: Linares, 2.
Tamaulipas: Matamoras, 1; Mier, 3; Victoria, 1.

PEROGNATHUS HISPIDUS PARADOXUS (Merriam). KANSAS POCKET MOUSE.

Perognathus paradoxus Merriam, N. Am. Fauna No. 1, 24, October 25, 1889.
Perognathus latirostris Rhoads, Am. Nat., XXVIII, 185, February, 1894.
Perognathus conditi Allen, Bull. Am. Mus. Nat. Hist., N. Y., VI, 318, November, 1894.

Type locality.—Trego County, Kans.

Distribution.—Upper Sonoran zone of the Great Plains from the Dakotas to Texas, westward to base of Rocky Mountains.

General characters.—Very similar to *P. hispidus*, but larger and with softer pelage; skull much heavier and more ridged.

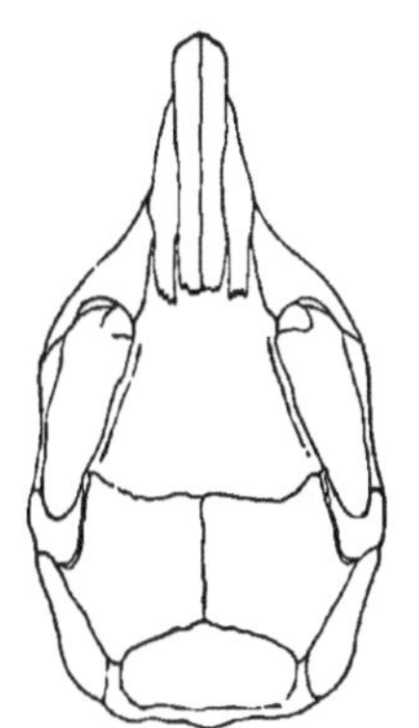

FIG. 11.—Skull of *Perognathus paradoxus*.

Color.—Much as in *hispidus*, but duller and paler.

Skull.—As in *hispidus*, but much larger, heavier, more angular and more ridged; otherwise not tangibly different.

Measurements.—Average of six adults from Kansas and Nebraska: Total length, 222.3; tail vertebræ, 108; hind foot, 26.5. *Skull:* (See table, p. 62.)

Remarks.—The average difference in size between *paradoxus* and typical *hispidus* is considerable, but apart from this there are no very important distinctive characters. The skull varies individually more than is usual in the genus and affords scarcely any reliable differences. *P. paradoxus* has few characters in common with *P. femoralis*, which it rivals in size. It is heavier and more robust than *femoralis* and different in many other ways. The type of '*P. latirostris*' Rhoads is slightly larger than any other specimen examined, but, in view of the variation shown in the group, the chances of its being even subspecifically distinct seem very slight. Specimens from Arizona and western Texas are here referred to *paradoxus*, as they seem slightly nearer to that form than to *hispidus*.

Specimens examined.—Total number, 61, from localities as follows:

Arizona: Fort Huachuca, 1; San Bernardino Ranch, 2.
Colorado: Boulder County, 2; Sterling, 2.
Kansas: Colby, 1; Ellis, 2; Garden Plain, 1; Pendennis, 1; Trego County, 3.
Nebraska: Callaway, 1; Cherry County, 1; Myrtle, 2; Red Cloud, 1.
New Mexico: Las Vegas, 1; Roswell, 1.
Oklahoma: Alva, 11; Orlando, 3; Ponca, 1.
South Dakota: Corral Draw, Pine Ridge Indian Reservation, 8; Quinn Draw, Cheyenne River, 3; Smithville, 1.
Rocky Mountains: 1 (type of '*latirostris*').
Texas: Amarillo, 1; Marfa, 3; Presidio County, 1.
Chihuahua: Chihuahua, 1; Santa Rosalia, 2; Casas Grandes, 12.

PEROGNATHUS HISPIDUS ZACATECÆ subsp. nov. ZACATECAS POCKET MOUSE.

Type from Valparaiso, Zacatecas, Mexico. ♀ yg. ad., No. 91877, U. S. Nat. Mus., Biological Survey Coll. Collected December 16, 1897, by E. A. Goldman. Orig. No., 11968.

Distribution.—Upper Sonoran zone from Valparaiso, Zacatecas, to Celaya, Guanajuato, Mexico.

General characters.—Somewhat larger and darker-colored than *hispidus;* otherwise similar.

Color.—Much darker and more olivaceous than in *hispidus;* general color of upperparts between the hair-brown and olive of Ridgway; bases of hairs very dark plumbeous; lateral line pure ochraceous, well defined, slightly paler than in *hispidus;* spots at base of whiskers intensely black and very conspicuous; tail sharply black above; underparts white.

Skull.—As in *hispidus*, but somewhat larger.

Measurements.—*Type:* Total length, 211; tail vertebræ, 105; hind foot, 27.5. *Skull:* (See table, p. 62.)

Remarks.—This form seems to be related most nearly to *paradoxus* and, like it, inhabits the Upper Sonoran zone. Its dark olivaceous color makes it easily recognizable.

Specimens examined.—Total number, 10, from localities in Mexico, as follows:

Guanajuato: Celaya, 1.
Zacatecas: Valparaiso, 9.

PEROGNATHUS PENICILLATUS Woodhouse. DESERT POCKET MOUSE.

Perognathus penecillatus Woodhouse, Proc. Acad. Nat. Sci. Phila., 1852, 200.
Perognathus penicillatus Woodhouse, Sitgreaves Exp'd. Zuñi and Colorado River, 49, pl. 3, 1854; Merriam, N. Am. Fauna No. 1, 22, 1889.

Type locality.—San Francisco Mountain, Arizona.[1]

---

[1] Woodhouse does not specify exactly where the type was taken. It seems to have been between his camps 15 and 18, which were on the northeast side of the mountain. It is not unlikely that the type came from the Little Colorado Desert, a few miles farther to the northeast.

Distribution.—Vicinity of Colorado River, from Bunkerville, Nev., to Yuma, Ariz., where it meets the range of its subspecies *angustirostris.* The type is the only specimen known from the type locality. Lower Sonoran zone.

General characters.—Size rather large, about equal to *formosus;* tail long, heavily crested, penicillate; sole of hind foot naked to heel; ears scantily haired, shorter and rounder than in *formosus*, antitragus lobed; pelage rather soft; no spines on rump; color very uniform, markings almost obsolete.

Color.—Above, vinaceous buff very finely sprinkled with black; sides exactly like back; lateral line obsolete; subauricular spot present; face and cheeks like back except for a slight darkening under ears; no black spots at base of whiskers; ears outside like back, inside slightly dusky; tail white below to pencil, upper surface and pencil dusky brownish. In the 'left-over' winter pelage the general color is écru drab instead of vinaceous buff.

Skull.—Size medium or rather large; rostrum heavy and high; parietals somewhat flattened; mastoid side of parietal about equaling squamosal side, much exceeded by others; interparietal moderate, all angles rounded, especially posterior ones, anterior angle rounded but distinctly evident; ascending branches of supraoccipital quite heavy; audital bullæ widely separated anteriorly; lower premolar larger than last molar.

Measurements.—Average of four adults from bend of Colorado River, Nevada: Total length, 205; tail vertebræ, 109; hind foot, 25.5. *Skull:* (See table, p. 62.)

Remarks.—The members of the *penicillatus* group are true *Chætodipus*, but none of them have rump spines. Characters marking the typical form are large size, uniform color, subdued markings, and heavy skull. The skull of the type which is now available for examination does not agree perfectly with any of the series from the bend of the Colorado River. It is larger and heavier than these, the anterior part is much elevated, and the rostrum broad. These characters, however, are quite pronounced in the Colorado River specimens, and it seems safe to consider them *penicillatus*, even though no exact duplicates of the type are among them. Even the most northern of the Colorado River specimens is somewhat intermediate between true *penicillatus* and *angustirostris.*

Specimens examined.—Total number, 55, from localities as follows:

Arizona: Ehrenberg, 5; Harper Ferry, 3; Fort Mohave, 9; Norton, 4; San Francisco Mountain, 1 (type).

California: Mohave Mountains, 1.

Nevada: Bunkerville, 3; Colorado River, Lincoln County, 8; Colorado River, near Callville, 8; Vegas Valley, 13.

PEROGNATHUS PENICILLATUS ANGUSTIROSTRIS subsp. nov. CALIFORNIA DESERT POCKET MOUSE.

Type from Carriso Creek, Colorado Desert, Cal. ♂ ad., No. 73881, U. S. Nat. Mus., Biological Survey Coll. Collected March 31, 1895, by A. W. Anthony. Orig. No., 22.

Distribution.—Colorado Desert; south to northern Lower California and east to the Colorado River and southwestern Arizona, where it meets the range of *penicillatus* and *pricei.* Lower Sonoran zone.

General characters.—Similar to *P. penicillatus*, but smaller; color about the same; skull lighter and with longer and more slender rostrum.

Color.—As in *P. penicillatus.*

Skull.—Similar in general to *P. penicillatus;* nasals and ascending premaxillæ long and narrow, much more slender than in *penicillatus;* interparietal averaging larger and more angular.

Measurements.—*Type:* Total length, 191; tail vertebræ, 105; hind foot (measured dry), 24.4. Average of five topotypes: Total length, 181; tail vertebræ, 103; hind foot, 24. *Skull:* (See table, p. 62.)

Remarks.—The numerous specimens of this subspecies which have been examined include many which are not strictly typical. This is true of the large series from the Colorado River at monument No. 204 and the several localities in the vicinity of Yuma, all of which tend in differing degrees toward true *penicillatus.* From Yuma eastward the tendency is toward *pricei.* The characters of small size and slender rostrum are very constant in the many specimens from the Colorado Desert, California.

Specimens examined.—Total number, 253, from localities as follows:

Arizona: Bradshaw City, 1; Gila City, 3; Yuma, 9.

California:[1] Agua Caliente, 3; Baregas Springs, 4; Carriso Creek, 15; Colorado Desert, 7; Coyote Wells, 3; Indian Wells, 1; Laguna, 5; Mexican Boundary monument No. 204, near Colorado River, 78; Palm Springs, 55; Salt Creek, 1; San Felipe Canyon, 6; Unlucky Lagoon, 9; Vallecitas, 10; Walters, 7; Whitewater, 2; Fort Yuma, 15.

Lower California: Gardner Lagoon, 5; Hardee River (head, near mouth of Colorado River), 2; Poso Vicente, 2; Seven Wells, 10.

PEROGNATHUS PENICILLATUS PRICEI (Allen). PRICE POCKET MOUSE.

Perognathus pricei Allen, Bull. Am. Mus. Nat. Hist., N. Y., VI, 318, November, 1894.
Perognathus obscurus Allen, Bull. Am. Mus. Nat. Hist., N. Y., VII, 216, June, 1895.

Type locality.—Oposura, Sonora, Mexico.

Distribution.—South central Arizona and Northwestern Mexico, west of the Sierra Madre.

General characters.—Similar to *penicillatus*, but smaller; pelage

[1] Nearly all these localities are in the Colorado Desert.

harsher, no spines on rump; upperparts more strongly mixed with black; skull short and heavy.

Color.—Above, general effect drab or broccoli brown, produced by vinaceous buff strongly lined with black; sides like back, lateral line faintly evident; ears very scantily haired, same color as back; underparts white; tail bicolor, white below, dusky above.

Skull.—Size medium, much smaller than in *penicillatus;* rostrum short and heavy; nasal branches of premaxillæ barely exceeding nasals; interparietal moderately wide, anterior angle often obliterated; lower premolar larger than last molar. Contrasted with *penicillatus* the skull of *pricei* is much smaller, smoother, or less angular, and has very much shorter nasals. In comparison with *intermedius* it is heavier and less arched, the rostrum is broader, and the nasals are shorter, the mastoids are smaller, and the interparietal is narrower.

Measurements.—Average of seventeen adults from Hermosillo, Sonora, Mexico: Total length, 172.5; tail vertebræ, 92.8; hind foot, 22.3. *Skull:* (See table, p. 62.)

Remarks.—The type of *P. pricei* is very immature, but its skull shows characters amply sufficient to prove that it belongs to the *penicillatus* rather than the *intermedius* group. Although these groups inhabit the same general region and resemble each other so closely in superficial characters, the skulls are so markedly different as to indicate that they bear no close relation to one another. The only external difference is found in the rump spines. This is not to be relied upon absolutely, however, for though never present in *penicillatus* and its forms, they are sometimes, though very rarely, absent or undeveloped in *intermedius*. In local habitat the two also differ in an interesting way, *pricei* being found in sandy places, while *intermedius* prefers the rocks.

The extreme form of *pricei* is found in southern Sonora, where it is so different from typical *penicillatus* as to suggest full specific rank.

Specimens examined.—Total number, 187, from localities as follows:

Arizona: Calabasas, 6; Dos Cabezos, 1; Fairbank, 28; Fort Bowie, 2; Fort Huachuca, 1; Fort Lowell, 39; La Osa, 2; Mammoth, 12; New River, 5; Phœnix, 5; Santa Cruz River (west of Patayone Mountain), 3; Sentinel, 2; Tubac, 3; Tucson (twenty miles south), 3; Willcox, 6.

Sonora: Batomotal, 13; Hermosillo, 17; Magdalena, 6; Oposura, 8; Ortiz, 10; Quitobaquita, 10; Sonora, 1; Sonoyta, 4.

PEROGNATHUS PENICILLATUS EREMICUS (Mearns). EASTERN DESERT POCKET MOUSE.

Perognathus (*Chætodipus*) *eremicus* Mearns, Bull. Am. Mus. Nat. Hist., N. Y., X, 300, August 31, 1898.

Type locality.—Fort Hancock, El Paso County, Tex.

Distribution.—Extreme western Texas, thence south into north central Mexico east of the Sierra Madre at least to La Ventura, Coahuila.

General characters.—Size about equal to *pricei;* color slightly paler; pelage softer; nasals longer and more slender; skull otherwise peculiar.

Color.—Essentially as in *pricei*, but paler; general effect fawn lightly mixed with black; dark area below ears quite prominent; spot at base of whiskers faint.

Skull.—Similar to *pricei;* cranium somewhat arched; nasals long and slender; nasal branches of premaxillæ widened at extremities, extending much beyond nasals; supraoccipital slightly bulging behind.

Measurements.—*Type:* Total length, 163; tail vertebræ, 83; hind foot, 22.1. *Skull:* (See table, p. 62.)

Remarks.—The average difference between this eastern form of the *penicillatus* group and its western relative *pricei* is considerable. The long slender nasals and high arched skull of this form are never found in specimens from west of the Sierra Madre. Specimens from Chihuahua and Coahuila appear to be quite typical. *P. eremicus* differs from *pricei* much as *angustirostris* does from true *penicillatus*. In fact, its skull is not very unlike that of *angustirostris*, but the two are not likely to be confused, on account of the difference in size and color. Specimens from San Bernardino ranch, Arizona, are not typical, being dark-colored and otherwise intermediate.

Specimens examined.—Total number, 93, from localities as follows:

Arizona: San Bernardino Ranch, Cochise County, on Mexican boundary, 27.
Texas: El Paso, 5; Fort Hancock, 3.
Chihuahua: Ciudad Juarez, 2; Escalon, 1; Samalayuca, 3; Santa Rosalia, 24.
Coahuila: Jimulco, 1; La Ventura, 12; Torreon, 14.
Durango: Mapimi, 1.

PEROGNATHUS STEPHENSI Merriam. STEPHENS POCKET MOUSE.

Perognathus stephensi Merriam, Proc. Acad. Nat. Sci. Phila., September 27, 1894, 267.

Type locality.—Mesquite Valley, northwest arm of Death Valley, Inyo County, Cal.

Distribution.—Known only from the type locality.

General characters.—Similar to *penicillatus;* size very much smaller; tail long, well crested; hind feet naked below; very little or no black in color.

Color.—'*Left-over' winter pelage:* Above, between pinkish buff and vinaceous buff; effect perfectly uniform, no traces of black anywhere; ears sparsely haired, same color as back; lateral line entirely obliterated; face slightly lighter than back and sides; below, white; tail below white, above like back. The post-breeding pelage is doubtless darker and may have more or less black in it.

Skull.—Size small; general form much like that of *penicillatus;*

cranium slightly arched; mastoids rather small; interparietal correspondingly large; ascending branches of supraoccipital relatively heavy; lower premolar very large, nearly twice as large as last molar.

Measurements.—Type: Total length, 177; tail vertebræ, 96; hind foot, 21. *Skull:* (See table, p. 62.)

Remarks.—P. stephensi is a miniature of *penicillatus* and but slightly larger than *arenarius*. It is at once separated from the former by its small size and from the latter by its cranial characters. Further collections from the desert region of California will doubtless yield more of this interesting species, but at present it is known only from the two specimens which Mr. Stephens caught in the extension of Death Valley known as Mesquite Valley.

Specimens examined.—Total number, 2, the type and one topotype.

PEROGNATHUS ARENARIUS Merriam. LITTLE DESERT POCKET MOUSE.

Perognathus arenarius Merriam, Proc. Cal. Acad. Sci., 2d ser., IV, 461, September 25, 1894.

Type locality.—San Jorge, near Comondu, Lower California.

Distribution.—Known only from the type locality.

General characters.—Size very small; tail short, slightly exceeding head and body; pelage rather soft, no bristles anywhere; color plain and uniform, lateral line obsolete; skull short and broad.

Color.—Very similar to *penicillatus;* dorsum buffy drab, finely mixed with black; sides somewhat paler, lateral line not evident; ears dusky, a minute white spot on lower margins; underparts white; tail bicolor.

Skull.—Size very small; cranium slightly arched; interorbital and mastoid width relatively great; mastoids moderate, relatively larger than in *penicillatus;* interparietal broadly pentagonal; nasals rather slender, slightly emarginate at frontal endings; zygomata extremely frail and light; lower premolar larger than last molar.

Measurements.—Type (from dry skin): Total length, 136; tail vertebræ, 70; hind foot, 20. *Skull:* (See table, p. 62.)

Remarks.—P. arenarius is a very aberrant member of the *penicillatus* series. It is about the same color as *stephensi*, but differs from it in size and cranial details, such as more slender nasals, wider interorbital space, larger mastoids, and shorter premaxillæ. As far as known it is the smallest member of the subgenus *Chætodipus*.

Specimen examined.—The type.

PEROGNATHUS PERNIX Allen. SINALOA POCKET MOUSE.

Perognathus pernix Allen, Bull. Am. Mus. Nat. Hist., N. Y., X, 149, April, 1898.

Type locality.—Rosario, Sinaloa, Mexico.

Distribution.—Coast of western Mexico in the States of Sinaloa and Jalisco.

General characters.—Size small; tail rather long, thinly haired, slightly crested; colors dark; pelage slightly hispid, no spines or bristles anywhere; ears medium; feet naked below.

Color.—General color above, hair-brown, uniform over all parts above the lateral line; lateral line distinct, between pinkish buff and ochraceous buff; underparts soiled white; ears dusky, a minute white spot on inferior margins; tail brownish black above, whitish below.

Skull.—Size rather small; form narrow and elongate; mastoids quite small; interorbital space much constricted; nasals rather broad and flattened, of medium length; naso-frontal suture not emarginate; interparietal wide, somewhat produced anteriorly; posterior angles much rounded; molar teeth small and weak; lower premolar larger than last molar.

Measurements.—Average of four adult topotypes: Total length, 175; tail vertebræ, 97; hind foot, 22.3. *Skull:* (See table, p. 63.)

Remarks.—*Perognathus pernix* differs from other Mexican species in much smaller size. Its dark color, narrow interorbital space and long nasals distinguish it from all other *Chætodipus* not having rump spines.

Specimens examined.—Total number, 48, from localities in Mexico, as follows:

Sinaloa: Altata, 2; Culiacan, 17; Mazatlan, 11 (not typical); Rosario, 10.
Tepic: Acaponeta, 8.

PEROGNATHUS PERNIX ROSTRATUS subsp. nov. BROAD-NOSED POCKET MOUSE.

Type from Camoa, Rio Mayo, Sonora, Mexico. ♂ yg. ad., No. 95818, U. S. Nat. Mus., Biological Survey Coll. Collected October 28, 1898, by E. A. Goldman. Orig. No., 13167.

Distribution.—Coast plains of southern Sonora and northern Sinaloa, Mexico.

General characters.—Size, proportions, and general color about as in *P. pernix;* skull quite different.

Color.—Above, slightly lighter and grayer than *pernix;* general color oftener broccoli brown than hair-brown; facial area distinctly paler than back and sides; lateral line pinkish buff; lower parts soiled white.

Skull.—Similar to *pernix*, but shorter and broader; rostrum very heavy; nasals, premaxillæ, and premaxillary branches of zygomata all heavier than in *pernix;* nasals shorter; interorbital space wider; interparietal, mastoids, and audital bullæ not tangibly different.

Measurements.—*Type:* Total length, 162; tail vertebræ, 94; hind foot, 23.5. Average of four topotypes: Total length, 161; tail vertebræ, 88; hind foot, 22.5. *Skull:* (See table, p. 63.)

Remarks.—This form is quite a departure from *pernix*, but intergradation with that species is evidenced by a single specimen from

Sinaloa. The series of topotypes from Camoa are constant in their cranial differences from *pernix*, and though no external characters are evident the form seems well worth recognition.

Specimens examined.—Total number, 10, from localities in Mexico, as follows:

Sinaloa: Sinaloa, 1.
Sonora: Camoa, 9.

PEROGNATHUS INTERMEDIUS Merriam. INTERMEDIATE POCKET MOUSE.

Perognathus intermedius Merriam, N. Am. Fauna No. 1, 18–19, 1889; ibid., No. 3, 74, 1890.
Perognathus obscurus Merriam, ibid., No. 1, 20–21, 1889.

Type locality.—Mud Spring, Mohave County, Ariz.

Distribution.—Known from several scattered localities in the Sonoran zone of Arizona, New Mexico, and northern Mexico.

General characters.—Size medium, smaller than *penicillatus;* color much darker, with well-defined markings; rump spines rather weak; skull rather small and light.

Color.—*Winter pelage:* Above, general effect drab, with a strong mixture of black on back and rump; sides paler than back; lateral line pale fawn, quite narrow; ears dusky; tail dusky above, becoming black toward pencil, whitish below, faintly buffy on sides; underparts white, with suggestions of buff.

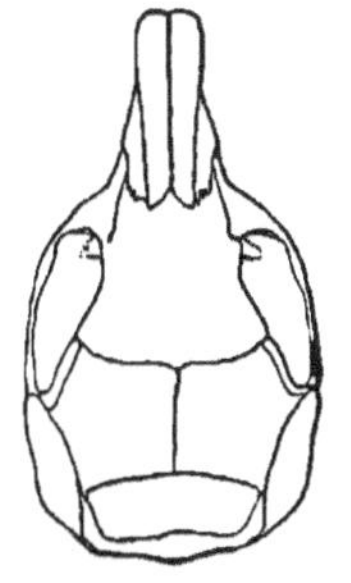

FIG. 12.—Skull of *Perognathus intermedius.*

Skull.—Size medium; cranium well arched; rostrum slender, somewhat depressed; interparietal very wide and strap-shaped, anterior angle normally obliterated, others but slightly rounded; lower premolar larger than last molar. Compared with *penicillatus* it is smaller and less angular; rostrum and nasals much more slender; zygomata more sloping; mastoids relatively larger and fuller; ascending branches of supraoccipital much lighter; interorbital space wider.

Measurements.—Average of four adults from the Grand Canyon of the Colorado, Arizona: Total length, 179.5; tail vertebræ, 102.7; hind foot, 22.7; ear from anterior base, 7. *Skull:* (See table, p. 62.)

Remarks.—Specimens of typical *intermedius* are not numerous at present, and the few that are available are in the winter pelage. This makes the determination of '*P. obscurus*' a little difficult. The latter is identical with *intermedius* in cranial characters, but slightly more ruddy in color.

P. intermedius is much rarer than *penicillatus*, some form of which is often found near it. In the vicinity of El Paso, Tex., Mr. Vernon Bailey collected both *intermedius* and *eremicus*, the one being found in the rocks and the other in the sandy places. At other localities where both occur the same conditions seem to obtain.

Specimens examined.—Total number, 46, from localities as follows:

Arizona: Grand Canyon, 4; Harper Ferry, 1; Fort Bowie, 1; Fort Huachuca, 1; Little Colorado River, Painted Desert, 2; Mud Spring, 2; Willow Spring, 1.
New Mexico: Camp Apache, Grant County, 14.
Texas: Alpine, 1; El Paso, 2.
Chihuahua: Casas Grandes, 4; Chihuahua, 13.

PEROGNATHUS NELSONI Merriam. NELSON POCKET MOUSE.

Perognathus nelsoni Merriam, Proc. Acad. Nat. Sci. Phila., September 27, 1894, 266.

Type locality.—Hacienda La Parada, San Luis Potosi, Mexico.

Distribution.—Upper and Lower Sonoran zone of central Mexico, covering the table-land from Inde, Durango, south to Lagos, Jalisco, and east to Jaumave, Tamaulipas.

General characters.—Similar to *intermedius*, but larger, darker, and harsher pelaged: tail heavily crested: rump bristled.

Color.—Above, general effect hair-brown; hairs dark plumbeous, basally followed by a narrow grayish fawn zone and a heavy black tip; sides like back, orbital region scarcely lighter; lateral line fawn, well defined: underparts dirty whitish; ears dusky, slightly hoary on margins; tail bicolor, black above, whitish below. Worn pelage much paler, becoming drab or ecru drab.

Skull.—Similar to *intermedius*, but larger and heavier, rostrum and nasals particularly so; interparietal smaller; nasal branches of premaxillæ exceeding nasals; ascending branches of supraoccipital heavy.

Measurements.—Average of ten adults from the type locality: Total length, 182; tail vertebræ, 104; hind foot, 23; ear from anterior base, 8. *Skull:* (See table, p. 62.)

Remarks.—This is the commonest pocket mouse of Mexico. It is found in suitable localities over the entire table-land. It is closely related to *intermedius* and possibly intergrades with it. There are some slight variations in the species, but none are marked enough to warrant separation.

Specimens examined.—Total number, 65, from localities in Mexico, as follows:

Aguas Calientes: Chicalote, 5.
Coahuila: Jimulco, 1; La Ventura, 1; Sierra Encarnacion, 1.
Durango: Durango City, 10; Inde, 3; Mapimi, 1.
Jalisco: Lagos, 9.
San Luis Potosi: Hacienda La Parada, 19; Jesus Maria, 3.
Zacatecas: Berriozabal, 9; Cañitas, 1; Hacienda San Juan Capistrano, 1; Valparaiso Mountains, 1.

PEROGNATHUS NELSONI CANESCENS (Merriam). GRAY POCKET MOUSE.

Perognathus intermedius canescens Merriam, Proc. Acad. Nat. Sci. Phila., September 27, 1894, 267.

Type locality.—Jaral, Coahuila, Mexico.

Distribution.—Known only from the type locality.

General characters.—Size larger than *intermedius;* color much paler and more grayish; skull similar to that of *P. nelsoni.*

Color.—General color of upperparts drab gray; lateral line pinkish buff, rather narrow; underparts pure white; tail bicolor, mouse gray above, white below.

Skull.—Similar to that of *nelsoni;* differs in more slender nasals, constricted interorbital space, and slightly smaller mastoids.

Measurements.—*Type:* Total length, 193; tail vertebræ, 117; hind foot, 22. One topotype: Total length, 184; tail vertebræ, 105; hind foot, 22. *Skull:* (See table, p. 63.)

Remarks.—This form seems to be quite localized. Its habitat is similar to that of the other members of the group. The type and cotypes were caught in the cliffs of a rocky canyon.

Specimens examined.—Total number, 3, from the type locality.

PEROGNATHUS GOLDMANI sp. nov. GOLDMAN POCKET MOUSE.

Type from Sinaloa, Sinaloa, Mexico. ♀ ad., No. 96673, U. S. Nat. Mus., Biological Survey Coll. Collected February 15, 1899, by E. A. Goldman. Orig. No., 13428.

Distribution.—Coast plains of northern Sinaloa and southern Sonora, Mexico.

General characters.—Size large; tail moderately long and heavily crested; pelage somewhat hispid, rump with a few short bristles; ears relatively large, much larger than those of *nelsoni;* antitragal lobe prominent, wider at base than at apex; in color and markings similar to *nelsoni;* skull relatively large and heavy.

Color.—Similar in general to *nelsoni;* general color across shoulders and anterior portion of upperparts, broccoli brown; posterior half of dorsum much darkened by admixture of black; lateral line pinkish buff; ears blackish with hoary margins, externally whitish for distal half; subauricular spot present; tail sharply bicolor, blackish above, white below.

Skull.—Size large, much heavier than in *nelsoni;* mastoids somewhat smaller and more ridged; nasals much larger and heavier; skull noticeably higher and not so wide posteriorly, thus making the zygomata more nearly parallel.

Measurements.—*Type:* Total length, 202; tail vertebræ, 108; hind foot, 28. Average of five topotypes: Total length, 202; tail vertebræ, 112; hind foot, 28; ear from anterior base, 11. *Skull:* (See table, p. 63.)

Remarks.—The large orbicular ears of this species easily distinguish it from *nelsoni*, its nearest relative. It is one of the several forms peculiar to western Mexico, and, like the others, its known range is quite limited. Specimens from Camoa and Alamos are slightly smaller than those from Sinaloa.

Specimens examined.—Total number, 36, from localities in Mexico, as follows:

> **Sinaloa:** Sinaloa, 7.
> **Sonora:** Alamos, 18; Camoa, 11.

PEROGNATHUS ARTUS sp. nov. BATOPILAS POCKET MOUSE.

Type from Batopilas, Chihuahua, Mexico. ♀ ad., No. 96298, U. S. Nat. Mus., Biological Survey Coll. Collected October 6, 1898, by E. A. Goldman. Orig. No., 13090.

Distribution.—Known only from a few scattered localities in western Mexico.

General characters.—Externally similar to *goldmani;* rump bristles weak or undeveloped; skull distinctive.

Color. As in *goldmani.*

Skull.—Similar to that of *goldmani*, but smaller and narrower; mastoids much smaller with more strongly marked transverse ridges; audital bullæ smaller; nasals moderate, exceeded by ascending premaxillæ; interparietal nearly elliptical, slightly produced anteriorly; zygomata nearly parallel.

Measurements.—Average of five adult topotypes: Total length, 191; tail vertebræ, 106; hind foot, 24.6. *Skull:* (See table, p. 63.)

Remarks.—The large size of this species at once distinguishes it from *perniæ* and *rostratus*, and its very small mastoids separate it from other Mexican species. Externally it is very similar to *goldmani*, but it has less prominent rump bristles; in fact, they are not at all evident in the majority of specimens. *P. perniæ* was generally found by Mr. Goldman at the same localities as *P. goldmani*, but at Culiacan he found it in company with *P. artus*.

Specimens examined.—Total number, 15, from localities in Mexico, as follows:

> **Chihuahua:** Batopilas, 8.
> **Durango:** Chacala, 3.
> **Sinaloa:** Culiacan, 4.

PEROGNATHUS FALLAX Merriam. SHORT-EARED CALIFORNIA POCKET MOUSE.

Perognathus fallax Merriam, N. Am. Fauna No. 1, 19, 1889; Allen, Bull. Am. Mus. Nat. Hist., N. Y., V, 184, 1893.

Type locality.—Reche Canyon, 3 miles southeast of Colton, San Bernardino County, Cal.

Distribution.—Extreme southwestern California, occupying the region west of the San Bernardino and San Jacinto ranges and extending south into northern Lower California.

General characters.—Size medium, somewhat larger than *intermedius;* general color similar but darker; wider and brighter lateral line; rump bristles heavier; tail long and crested; ears moderate.

Color.—Above, general effect bister, middle of back and rump with a strong element of black; lateral line and subterminal zone of hairs of upperparts pinkish buff; underparts creamy white; ears dusky on inflexed portions, hoary on inner sides; tail bicolor.

Skull.—Similar to *intermedius;* cranium arched; nasals slender; mastoids rather large and full; interparietal wide, anterior angle slightly developed; naso-frontal suture slightly or not emarginate.

Measurements.—Average of six adults from the type locality: Total length, 192; tail vertebræ, 11: hind foot, 23; ear from anterior base, 9. *Skull:* (See table, p. 63.)

Remarks.—This species falls readily into the small group typified by *intermedius.* It differs from the other members in size, color, and shape of interparietal. It has been much confused with *femoralis* on account of its similar color, but its much smaller ear is a convenient external character for distinguishing it. Two specimens from Turtle Bay, Lower California, are similar in color to *anthonyi*, but cranially the same as *fallax*, to which they are here referred.

Specimens examined.—Total number, 120, from localities as follows:

California:[1] Ballenas, 1; Bergmann, Riverside County, 1; Carlsbad, 1; Chihuahua Mountains, 1; Dulzura, 24; El Nido, 3; Encinitas, 1; Herron, San Bernardino County, 5; Jacumba, 8; Lajolla, 1; Mountain Spring, 11; Radec, 5; Reche Canyon, Riverside County, 10; Riverside, Riverside County, 1; Rose Canyon, 10; San Felipe Valley, 4; San Pasqual Valley, 4; Santa Ysabel, 10; San Ygnacio Valley, 1; Summit (Coast Range), San Bernardino County, 4; Temescal, Riverside County, 1.

Lower California: Cape Colnett, 2; Ensenada, 1; Gato Creek, 1; Jamul Creek, 1; San Isidro Ranch, 2; Sanos Cedros, 1; San Quintin Bay, 1; Tia Juana, 2; Turtle or San Bartolome Bay, 2.

PEROGNATHUS ANTHONYI sp. nov. CERROS ISLAND POCKET MOUSE.

Type from South Bay, Cerros Island, Lower California. ♀ ad., No. 81058, U. S. Nat. Mus., Biological Survey Coll. Collected July 29, 1896, by A. W. Anthony. Orig. No., 71.

Distribution.—Known only from the type locality.

General characters.—Similar in general to *P. fallax;* differing in slightly smaller size, more ruddy color, and cranial characters.

Color.—Above, grayish fawn mixed with black; lateral line brownish fawn, poorly defined; ears dusky; white subauricular spot present; tail dusky above, whitish below.

Skull.—Similar to *P. fallax;* cranium less arched; rostrum heavier; mastoids smaller; interparietal smaller and shorter; zygomatic breadth greater anteriorly.

[1] Most of these localities, unless otherwise stated, are in San Diego County.

Measurements.—Type: Total length, 168; tail vertebræ, 92; hind foot, 23.5. *Skull:* (See table, p. 63.)

Specimens examined.—One, the type.

PEROGNATHUS FEMORALIS Allen. GREAT CALIFORNIA POCKET MOUSE.

Perognathus femoralis Allen, Bull. Am. Mus. Nat. Hist., N. Y., III, 281, June 30, 1891; Rhoads, Proc. Acad. Nat. Sci. Phila., 1893, 407.

Type locality.—Dulzura, San Diego County, Cal.

Distribution.—Known from a few localities in San Diego County, in extreme southern California, and the adjoining part of Lower California.

General characters.—Size very large; tail long, heavily crested penicillate; color dark; ears large and elongate; pelage harsh; rump and flanks furnished with strong bristles or spines; skull large and heavy.

Color.—Similar to *fallax*, but quite intensified; above, general color bister, hairs heavily tipped with intense black; lateral line rich pinkish buff; underparts dirty whitish, sometimes washed or flecked with buffy; tail bicolor.

Skull.—Large and heavy; less arched than in *fallax;* rostrum and nasals much heavier; mastoids relatively smaller; molar teeth relatively weaker; interparietal subquadrate, rarely developing a fifth angle; naso-frontal suture slightly emarginate.

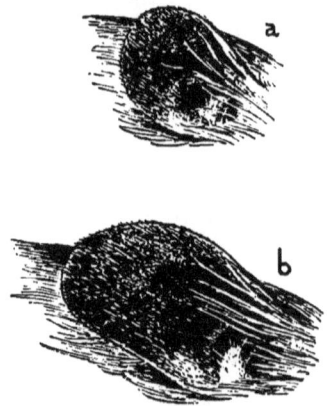

FIG. 13.—Ear of (a) *Perognathus fallax;* (b) *Perognathus femoralis.*

Measurements.—Average of six adults from the type locality: Total length, 223; tail vertebræ, 126; hind foot, 27.5; ear from anterior base, 12. *Skull:* (See table, p. 63.)

Remarks.—This species has the longest tail and largest hind foot found in the genus, but its body is light in comparison with that of *paradoxus.* In color it has a remarkable resemblance to *fallax*, which is found within its range, but its large size, long ears, and heavy skull are amply sufficient to distinguish it.

Specimens examined.—Total number, 60, from localities as follows:

California (San Diego County): Dulzura, 32; Santa Ysabel, 9; Twin Oaks, 16.

Lower California: Nachoguero Valley, 3.

PEROGNATHUS CALIFORNICUS Merriam. CALIFORNIA POCKET MOUSE.

Perognathus californicus Merriam, N. Am. Fauna No. 1, 26, 1889; Allen, Bull. Am. Mus. Nat. Hist., N. Y., 263, 1896; Elliott, Field Columbian Mus., Zool. Ser., I, No. 10, 211, 1898.
Perognathus armatus Merriam, l. c., 27.

Type locality.—Berkeley, Cal.

Distribution.—Vicinity of San Francisco Bay and south to Bear Valley, San Benito County, where it meets the range of its subspecies *dispar.*

General characters.—Similar to *P. femoralis*, but smaller; about equal in size to *fallax;* ears quite elongate: rump and flanks well supplied with bristles; skull very peculiar.

Color.—Nearly the same as *femoralis*, much darker than *fallax;* general effect of upperparts bister: hairs pale plumbeous basally, darkening distally; subterminal zone pinkish buff followed by heavy black tips; tail bicolor; underparts and feet yellowish white.

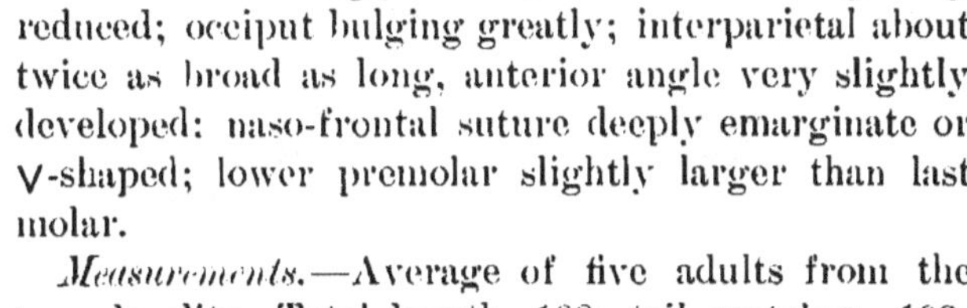

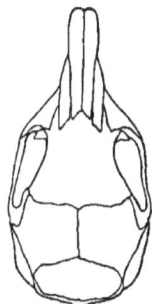

Skull.—Size medium; cranium considerably arched; mastoids exceedingly small; mastoid width greatly reduced; occiput bulging greatly; interparietal about twice as broad as long, anterior angle very slightly developed: naso-frontal suture deeply emarginate or V-shaped; lower premolar slightly larger than last molar.

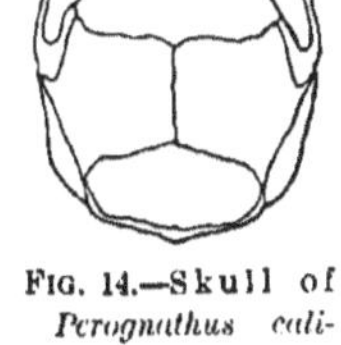

FIG. 14.—Skull of *Perognathus californicus.*

Measurements.—Average of five adults from the type locality: Total length, 192; tail vertebræ, 103; hind foot, 24; ear from anterior base, 10.5. *Skull:* (See table, p. 63.)

Remarks.—*P. californicus* is remarkable for its very small mastoids. It has no close relation to *fallax*, with which it has sometimes been confused. Its long ears and its cranial characters indicate that its closest affinities are with *femoralis.* Even within its very limited range it is quite a rare animal, and but few specimens are in collections.

Specimens examined.—Total number, 18, from localities as follows:

California: Berkeley, 7; Bear Valley, San Benito County, 2; Gilroy, 3; Portola, San Mateo County, 2; Stanford University, 2.

PEROGNATHUS CALIFORNICUS DISPAR subsp. nov. ALLEN POCKET MOUSE.

Type from Carpenteria, Santa Barbara County, Cal. ♂ ad., No. $\frac{32116}{43928}$, U. S. Nat. Mus., Biological Survey Coll. Collected December 19, 1891, by E. W. Nelson. Orig. No., 1655.

Distribution.—Coast valleys of California from San Bernardino to San Benito County and north along the foothills of the west slope of the Sierras to Placer County.

General characters.—Larger and paler colored than *californicus;* pelage somewhat softer; skull quite different.

Color.—Similar to *fallax*, paler than *californicus* or *femoralis;* above, general color bister; facial area slightly lightened; lateral line pinkish buff, sometimes approaching ochraceous buff; underparts buffy white; tail bicolor.

Skull.—Similar to that of *californicus*, but larger and heavier; in general form resembling that of *femoralis;* mastoids quite small; nasals heavy, somewhat elongate; interorbital space narrow.

Measurements.—*Type:* Total length, 218; tail vertebræ, 120; hind foot, 27. Average of six typical adults: Total length, 210; tail vertebræ, 117; hind foot, 26; ear from anterior base, 12. *Skull:* (See table, p. 63.)

Remarks.—Although this subspecies is somewhat intermediate in character between *californicus* and *femoralis* there seems to be no good evidence of any connection with the latter. It intergrades with *californicus* in the vicinity of Bear Valley, San Benito County. In typical form, its skull presents the characters of small mastoids and narrow interorbital space found in *californicus* at the same time almost attaining the large size of the skull of *femoralis*.

Specimens examined.—Total number, 56, from localities as follows:

California: Auburn, 1; Bitter Water, 3; Carpenteria, 4; Fort Tejon, 2; Hueneme (10 miles west), 1; Kern River (25 miles above Kernville), 1; Las Virgines Creek, Los Angeles County, 1; Milo, 1; Nordhoff, 4; Raymond, 1; San Bernardino Peak, 3; San Emigdio, 4; San Fernando, 3; San Luis Obispo, 8; San Simeon, 1; Santa Monica, 1; Santa Paula, 1; Three Rivers, 9; Ventura River, 7.

PEROGNATHUS SPINATUS Merriam. SPINY POCKET MOUSE.

Perognathus spinatus Merriam, N. Am. Fauna No. 1, 21, October 25, 1889.

Type locality.—Twenty-five miles below the Needles, Colorado River, California.

Distribution.—Desert region of southern California and northern Lower California.

General characters.—Size medium, tail moderately long and crested; ears small and orbicular; pelage hispid, spines large and prominent on rump, scattered on flanks and sides and often extending to shoulders; lateral line very faint or wanting.

Color.—Above, general effect drab brown; hairs plumbeous basally, ecru drab subterminally and black-tipped; sides and orbital region slightly paler than back; underparts buffy white; lateral line generally faint, in very bright pelage showing as a slender line of ecru drab; ears dusky, subauricular spot small; tail hair-brown above, whitish below; spines white with dusky tips, except on sides where the tips are also white. Many of the hairs of the back often end with a broad zone of ecru drab without the usual black tip. These, when combined with those having black tips, cause a peculiar mottled appearance. The 'left-over' winter pelage is much paler and grayer, the general effect being pale drab.

Skull.—Size medium; cranium rather slender and much flattened; parietals on nearly same plane as interparietal; mastoids small, not so full as in *fallax* and *intermedius;* interparietal broad but normally with slight evidence of an anterior or fifth angle; supraorbital ridge slightly trenchant; lower premolar about equal to last molar.

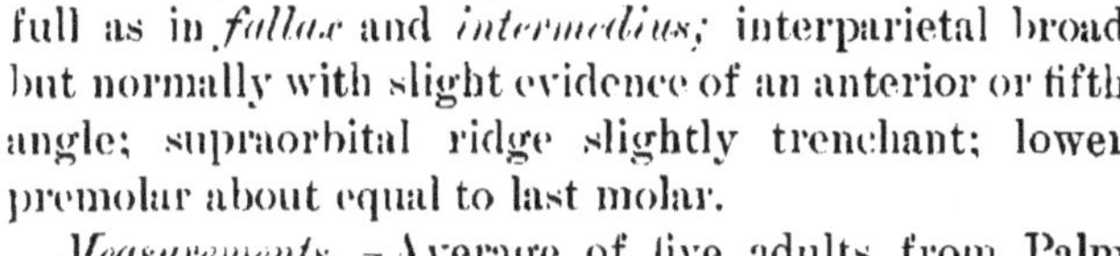

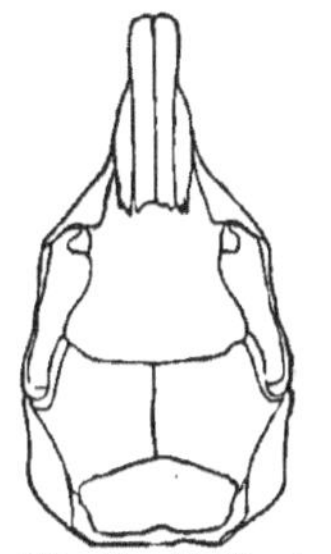

Fig. 15.—Skull of *Perognathus spinatus.*

Measurements.—Average of five adults from Palm Springs, Cal.: Total length, 181; tail vertebræ, 101; hind foot (measured dry), 21.5. *Skull:* (See table, p. 63.)

Remarks.—*Perognathus spinatus* has a limited range, and aside from the excellent series from Palm Springs in the Bangs collection but few specimens have found their way into collections. It is the type of the subgenus *Chætodipus* and the representative of a small group quite distinct from all others. Young adults of this group differ from old in being of a grayish color and in having weaker and less numerous spines.

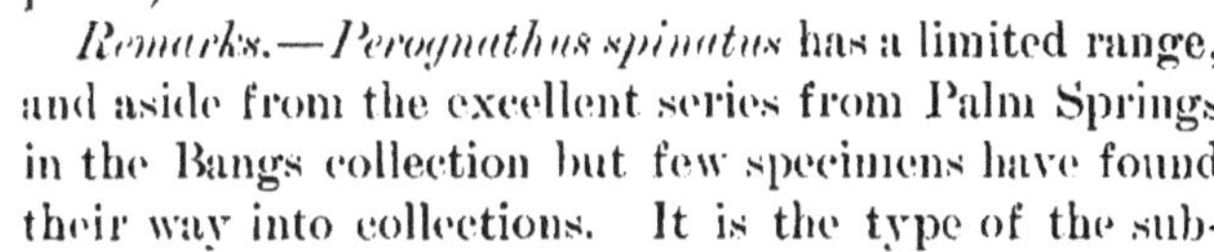

In juveniles the first pelage is soft and without spines, which do not appear until an entire new pelage is acquired. In this species, as in most others, the males average slightly larger than the females.

Specimens examined.—Total number, 46, from localities as follows:

California: Colorado River (twenty-five miles below Needles), 1 (type); Coast Range, San Diego County, 3; La Puerta, San Diego County, 6; Palm Springs, 21; San Felipe Cañyon, 12; Vallecitas, San Diego County, 2.

Lower California: Cocopah Mountains, 1.

PEROGNATHUS SPINATUS PENINSULÆ Merriam. CAPE ST. LUCAS POCKET MOUSE.

Perognathus spinatus peninsulæ Merriam, Proc. Cal. Acad. Sci., 2d ser., IV, 460, September 25, 1894.

Type locality. San Jose del Cabo, Lower California.

Distribution.—Cape region of Lower California.

General characters.—Similar to *P. spinatus*, but much larger; pelage a trifle more hispid; tail more scantily haired and relatively shorter; ears large and rounded.

Color.—As in *P. spinatus.*

Skull.—Similar to that of *P. spinatus*, but averaging much larger; supraorbital ridges flattened and shelf-like, with very trenchant edges, forming a nearly straight line from mastoids to lachrymals.

Measurements.—Average of five adults from San Jose del Cabo and Cape St. Lucas: Total length, 188; tail vertebræ, 101; hind foot, 24. *Skull:* (See table, p. 63.)

Specimens examined. Total number, 23, from localities as follows:

Lower California: Cape St. Lucas, 7; Comondu, 2 (intermediate); San Jose del Cabo, 5; Santa Anita, 9.

PEROGNATHUS BRYANTI Merriam. BRYANT POCKET MOUSE.

Perognathus bryanti Merriam, Proc. Cal. Acad. Sci., 2d ser., IV, 458, September 25, 1894.

Type locality.—San Jose Island, Lower California.

Distribution.—Known only from the type locality.

General characters.—Larger and longer-tailed than *peninsulæ;* skull slightly characterized; otherwise similar.

Color.—Apparently as in *peninsulæ.*

Skull.—Slightly larger and heavier than in *peninsulæ;* somewhat more elongate; nasals longer and more slender; interparietal wide and subquadrate; lower premolar equal to or slightly larger than last molar.

Measurements.—*Type:* Total length, 216; tail vertebræ, 127; hind foot, 25. One adult topotype: Total length, 225; tail vertebræ, 128; hind foot, 25. *Skull:* (See table, p. 63.)

Remarks.—This insular form is well characterized by its large size and long tail. In color it is probably the same as *peninsulæ*, though the material examined is not sufficient to determine with certainty.

Specimens examined.—Total number, 8, all from the type locality.

PEROGNATHUS MARGARITÆ Merriam. MARGARITA POCKET MOUSE.

Perognathus margaritæ Merriam, Proc. Cal. Acad. Sci., 2d ser., IV, 459, September 25, 1894.

Type locality.—Santa Margarita Island, Lower California.

Distribution.—Known only from the type locality.

General characters.—Size medium; tail longer than head and body; ears moderate; pelage rather harsh, rump and flanks with a few bristles; skull peculiar.

Color.—Above, much as in *spinatus*, pale fawn mottled and lined with hair-brown and black; lateral line scarcely evident; subauricular spot present, but very small; underparts and feet dirty white.

Skull.—Size rather small; cranium somewhat arched; mastoids exceedingly small, fully as small as in *californicus;* nasals moderate, naso-frontal suture emarginate; occiput not projecting posteriorly; interparietal wide, anterior angle evident, others very much rounded; interorbital space moderate, about as in *californicus;* lower premolar larger than last molar.

Measurements.—*Type* (from dry skin): Total length, 180; tail vertebræ, 102; hind foot, 22.5. *Skull:* (See table, p. 63.)

Remarks.—So far as known, this species has no near relative on the mainland adjacent to its habitat. In cranial characters it seems to be somewhat similar to *californicus*, while externally it is a combination of *fallax* and *spinatus*.

Specimen examined.—The type.

Cranial measurements of Perognathus.

[All measurements are in millimeters.]

Species.	Locality.	Basilar length of Hensel.[1]	Occipito-nasal length.	Greatest mastoid breadth.	Length of interparietal.	Greatest width of interparietal.	Interorbital width at narrowest point.	Length of nasals.	Number of specimens averaged.
Perognathus fasciatus	Tilyou ranch, Mont	16.5	23	11.8	......	4.3	......	......	3
P. f. infraluteus	Loveland, Colo	15.4	21.8	11.7	......	4.8	......	......	3
P. flavescens	Kennedy, Nebr	15.1	22.1	11.6	......	4.8	......	......	3
P. merriami	Brownsville, Tex	14.8	20.4	11.2	2.3	3.6	......	......	4
P. m. gilvus	Eddy, N. Mex	15	20.7	11.7	2.2	3.3	4.7	7.5	2
P. flavus	El Paso, Tex	14.6	21	12	2.6	2.9	......	......	1
P. flavus	Fort Huachuca, Ariz	14.5	20.3	11.7	2.4	2.8	......	......	4
P. f. bimaculatus	Fort Whipple, Ariz	15.9	21.6	13	2.5	3.1	4.5	7.6	3
P. apache	Keams Canyon, Ariz	16.3	23	13	2.9	4	5	8.6	4
P. apache melanotis[2]	Casas Grandes, Chihuahua, Mexico.	15.3	21.5	11.8	2.8	4.3	5	8	1
P. callistus	Kinney ranch, Wyo	16	22.9	13	2.8	4.6	4.8	8.2	1
P. panamintinus	Panamint Mountains, California.	14.9	21.4	11.9	2.5	3.5	5.2	8.3	3
P. p. brevinasus	San Bernardino, Cal	14.2	20.1	11.8	3	3.8	5	7.1	3
P. nevadensis	Halleck, Nev	14.8	20.6	12	2.2	3.7	5.2	8.2	2
P. pacificus[2]	Edge of Pacific Ocean, Mexican boundary.	13	19	11	2.2	3.4	4.9	7	1
P. longimembris	Fresno, Cal	17.5	23.6	13	3	3.8	4.8	9.2	3
P. amplus[2]	Fort Verde, Ariz	16.8	23.6	13.9	3	3.3	5	9.2	1
P. parvus	Mabton, Wash	17.6	25.1	13	3.1	5.4	......	......	5
P. p. olivaceus	Salt Lake City, Utah	17.6	......	12.6	3.7	4.9	......	......	4
P. p. magruderensis	Mount Magruder, Nev	19.1	27	13.6	3.1	4.8	......	......	4
P. p. mollipilosus	Fort Crook, Cal	17.8	25.3	12.8	3.1	5.1	......	......	3
P. alticola	San Bernardino Mts., Cal	18	25.4	12.6	3.2	4.6	......	......	2
P. lordi	Oroville, Wash	18.7	26.7	13.6	3.3	4.7	......	......	5
P. l. columbianus	Pasco, Wash	18.6	25.9	14.1	3.1	4.1	......	......	4
P. formosus	St. George, Utah	18.6	26.7	14	3.7	5.8	6.6	10.4	3
P. baileyi	Magdalena, Mexico	21.5	30	15.6	4.2	6.8	6.8	12.2	3
P. hispidus	Mier, Mexico	20.2	28	13.8	4.1	7.2	7	10.6	3
P. h. paradoxus	Kansas and Nebraska	24	32	15	4.7	8	7.5	13.2	3
P. h. zacatecæ[2]	Valparaiso Mountains, Mexico.	22.5	30.2	15	4	8	7	12	1
P. penicillatus[2]	Little Colorado Desert, Arizona.	20	28.3	14.5	3.5	6.9	6.8	10	1
P. penicillatus	Colorado River, near Callville, Nev.	19.5	27.8	13.4	3.2	7.6	6.6	10.8	3
P. p. angustirostris	Carriso Creek, California	18	26	13	3.1	7	6.4	10.3	3
P. p. pricei	Oposura, Sonora, Mexico	18.4	26	13	3.4	6.7	6.2	9.4	3
P. p. eremicus[2]	Fort Hancock, Tex	17.5	......	12.6	3	7	6.4	9.8	1
P. stephensi[2]	Mesquite Valley, California	16	22.7	12	3	6.7	6	9	1
P. arenarius[2]	San Jorge, Lower California	15.3	23	12	3.5	6.4	6.2	8.8	1
P. intermedius[2]	Mud Spring, Ariz	17	24.5	13.5	3	8	6.3	9.4	1
P. nelsoni	La Parada, Mexico	18	26	13.8	3.5	7.6	6.7	10	3

[1] See note, p. 14.

[2] Type.

Cranial measurements of Perognathus—Continued.

[All measurements are in millimeters.]

Species.	Locality.	Basilar length of Hensel.	Occipito-nasal length.	Greatest mastoid breadth.	Length of interparietal.	Greatest width of interparietal.	Interobital width at narrowest point.	Length of nasals.	Number of specimens averaged.
P. n. canescens[1]	Jaral, Coahuila, Mexico	17.5	25	13.5	3.7	7.2	6.1	9.3	1
P. goldmani	Sinaloa, Sinaloa, Mexico	20.6	27.7	14.3	3.8	7.4	6.5	11.1	3
P. artus	Batopilas, Chihuahua, Mexico.	18.8	25.4	12.4	3.3	7.1	6.1	9.7	3
P. fallax	Reche Canyon, California	18	26	11	3.8	7.8	6.6	10	3
P. anthonyi[1]	Cerros Island, Lower California.	17.4	25.4	12.9	2.6	5.8	6	10.2	1
P. femoralis	Dulzura, Cal	20.3	29.6	14.3	4	8.1	7.1	11.4	3
P. californicus	Berkeley, Cal	18.9	27.4	13	4.4	8.1	6.4	10.2	3
P. c. dispar	Carpenteria, Cal	19.8	28	13.1	4	8.3	6.5	11.2	3
P. spinatus	Palm Springs, Cal	17	24.3	12.3	3.4	7.7	6.2	9.4	4
P. s. peninsulæ	San Jose del Cabo, Lower California.	18	26.5	13	3.7	7.6	6.6	9.8	4
P. bryanti	San Jose Island, Lower California.	18.9	27.3	13.1	3.5	8.1	6.8	10.3	3
P. margaritæ[1]	Margarita Island, Lower California.	18	25.9	12	3.7	8	6.5	10.3	1
P. pernix	Rosario, Sinaloa, Mexico	17.4	24.4	12.2	3.3	7.2	5.4	9.2	3
P. p. rostratus	Camoa, Sonora, Mexico	16.5	22.7	11.7	3.4	7	5.5	8.6	3

[1] Type.

INDEX.

[Names of new species in **black-face** type, synonyms in *italics*.]

PLATE I.

[One and one-half times natural size.]

FIG. 1. *Perognathus flavus* Baird. Topotype. El Paso, Tex. (No. 25029, U. S. Nat. Mus.)

2. *Perognathus amplus* Osgood. Type. Fort Verde, Ariz. (Type No. 46711, U. S. Nat. Mus.)

3. *Perognathus merriami* Allen. Topotype. Brownsville, Tex. (No. 41764, U. S. Nat. Mus.)

4. *Perognathus flavus bimaculatus* (Merriam). Topotype. Fort Whipple, Ariz. (No. 46478, U. S. Nat. Mus.)

5. *Perognathus* (*Chætodipus*) *californicus* Merriam. Topotype. Berkeley, Cal. (No. 55560, U. S. Nat. Mus.)

6. *Perognathus* (*Chætodipus*) *pernix* Allen. Topotype. Rosario, Sinaloa, Mexico. (No. 91324, U. S. Nat. Mus.)

7. *Perognathus* (*Chætodipus*) *penicillatus* Woodhouse. Type. Near San Francisco Mountain, Ariz. (Type No. 2676, U. S. Nat. Mus.)

8. *Perognathus* (*Chætodipus*) *pernix rostratus* Osgood. Type. Camoa, Sonora, Mexico. (Type No. 95818, U. S. Nat. Mus.)

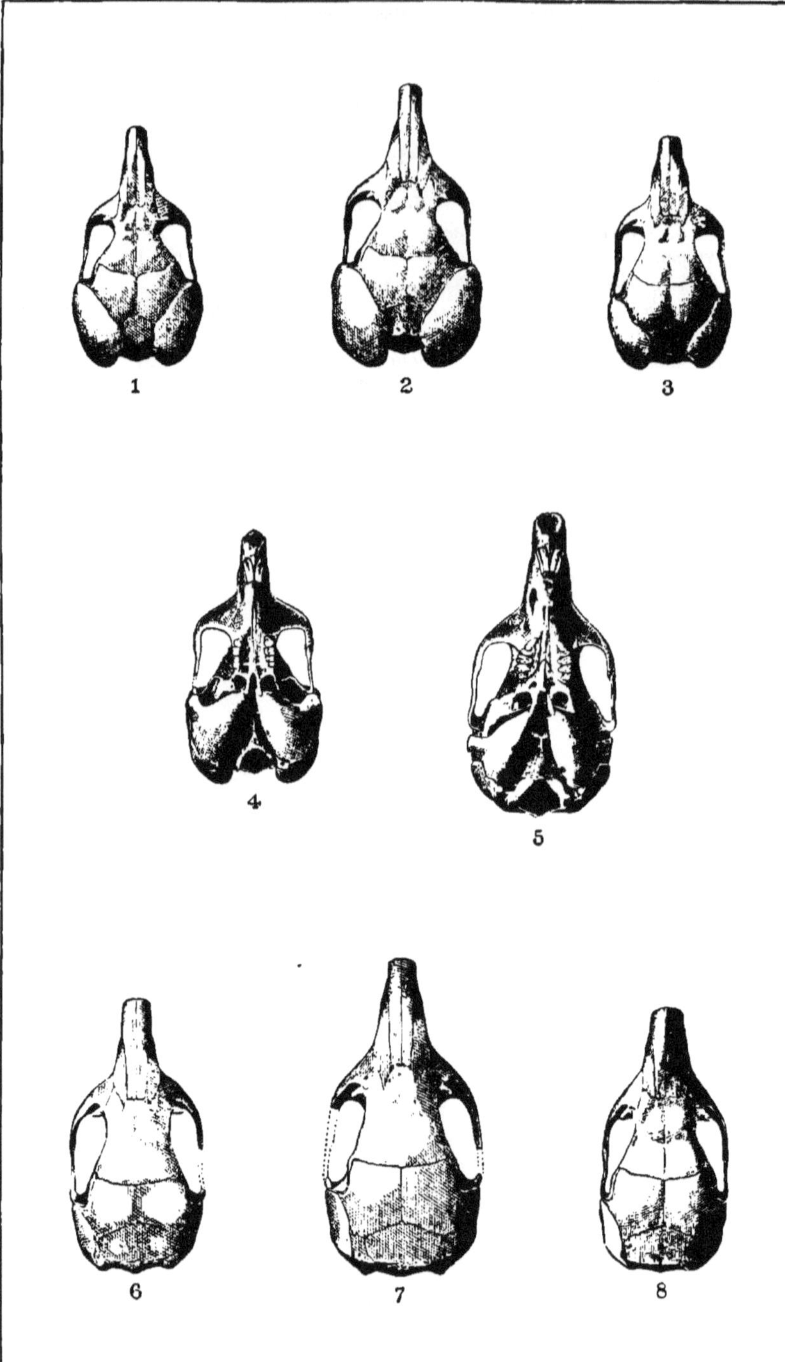

SKULLS OF PEROGNATHUS.

1. *Perognathus flavus.*
2. *P. amplus.*
3. *P. merriami.*
4. *P. flavus bimaculatus.*
5. *P. (Chætodipus) californicus.*
6. *P. (Chætodipus) pernix.*
7. *P. (Chætodipus) penicillatus.*
8. *P. (Chætodipus) pernix rostratus.*

PLATE II.

[One and one-half times natural size.]

FIG. 1. *Perognathus panamintinus* (Merriam). Type. Panamint Mountains, Cal. (Type No. 39866, U. S. Nat. Mus.)

2. *Perognathus lordi columbianus* (Merriam). Type. Pasco, Wash. (Type No. 39450, U. S. Nat. Mus.)

3. *Perognathus nevadensis* Merriam. Topotype. Halleck, Nev. (No. 54565, U. S. Nat. Mus.)

4. *Perognathus* (*Chætodipus*) *bryanti* Merriam. Type. San Jose Island, Lower California, Mexico. (No. 550, Coll. Calif. Acad. Sci.)

5. *Perognathus* (*Chætodipus*) *margaritæ* Merriam. Type. Santa Margarita Island, Lower California, Mexico. (No. 90, Coll. Calif. Acad. Sci.)

6. *Perognathus* (*Chætodipus*) *spinatus peninsulæ* Merriam. Type. San Jose del Cabo, Lower California, Mexico. (No. 274, Coll. Calif. Acad. Sci.)

7. *Perognathus* (*Chætodipus*) *arenarius* Merriam. Type. San Jorge, near Comondu, Lower California, Mexico. (No. 99, Coll. Calif. Acad. Sci.)

8. *Perognathus* (*Chætodipus*) *stephensi* Merriam. Topotype. Mesquite Valley, Cal. (No. 39874, U. S. Nat. Mus.)

9. *Perognathus* (*Chætodipus*) *nelsoni* Merriam. Type. Hacienda La Parada, San Luis Potosi, Mexico. (Type No. 50214, U. S. Nat. Mus.)

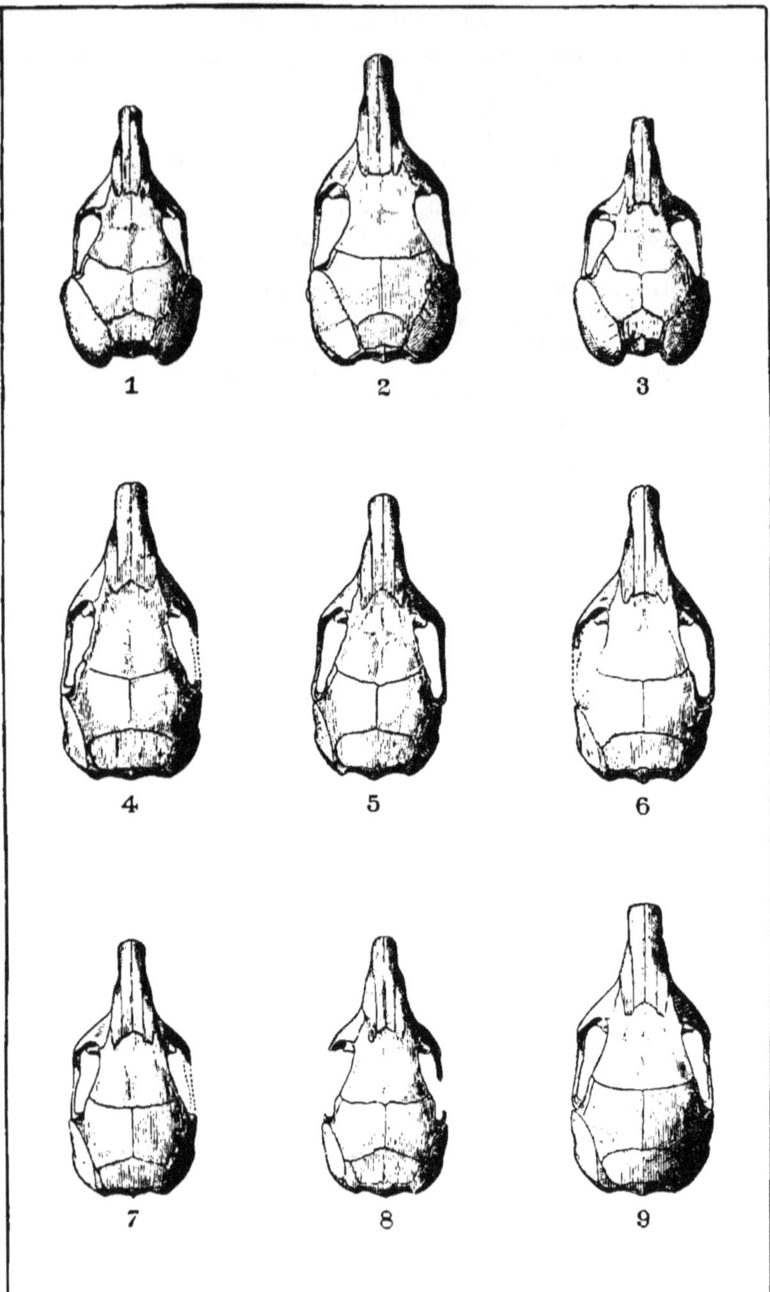

SKULLS OF PEROGNATHUS.

1. *Perognathus panamintinus.*
2. *P. lordi columbianus.*
3. *P. nevadensis.*
4. *P. (Chætodipus) bryanti.*
5. *P. (Chætodipus) margaritæ.*
6. *P. (Chætodipus) spinatus peninsulæ.*
7. *P. (Chætodipus) arenarius.*
8. *P. (Chætodipus) stephensi.*
9. *P. (Chætodipus) nelsoni.*

PLATE III.

Map showing distribution of the subgenus *Perognathus*.

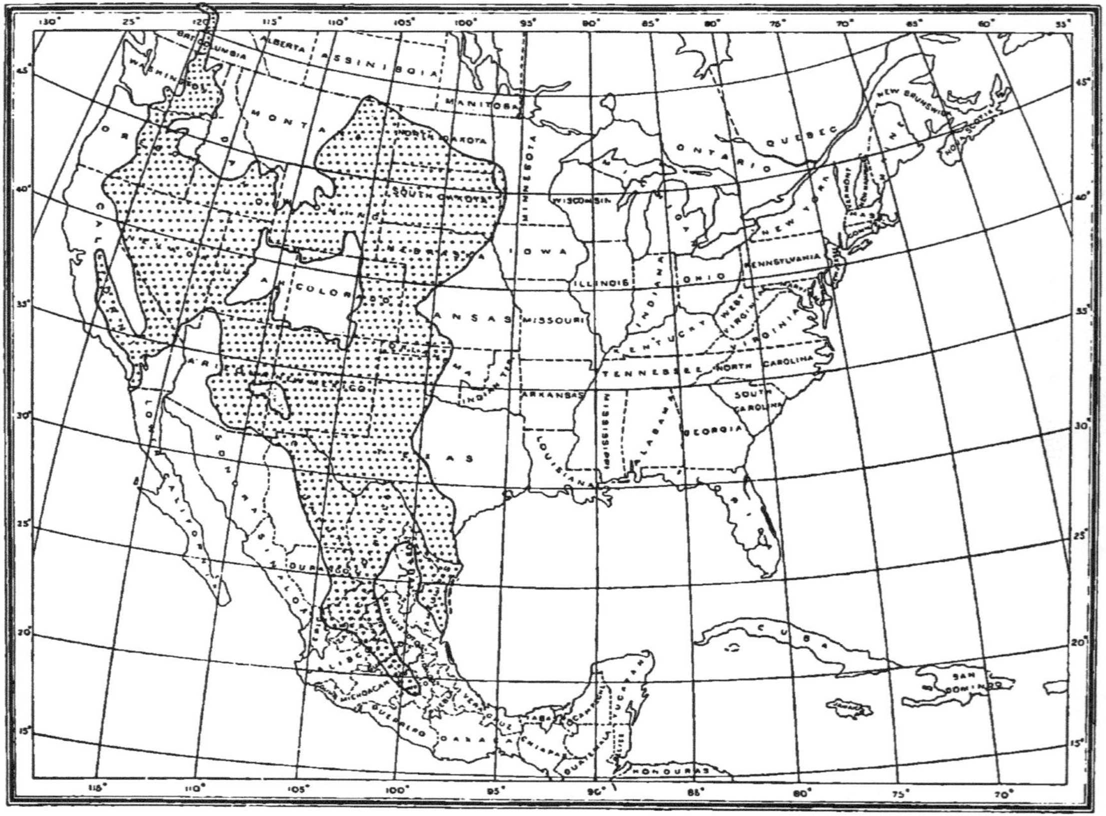

MAP SHOWING DISTRIBUTION OF THE SUBGENUS PEROGNATHUS.

PLATE IV.

Map showing distribution of the subgenus *Chætodipus*.

O

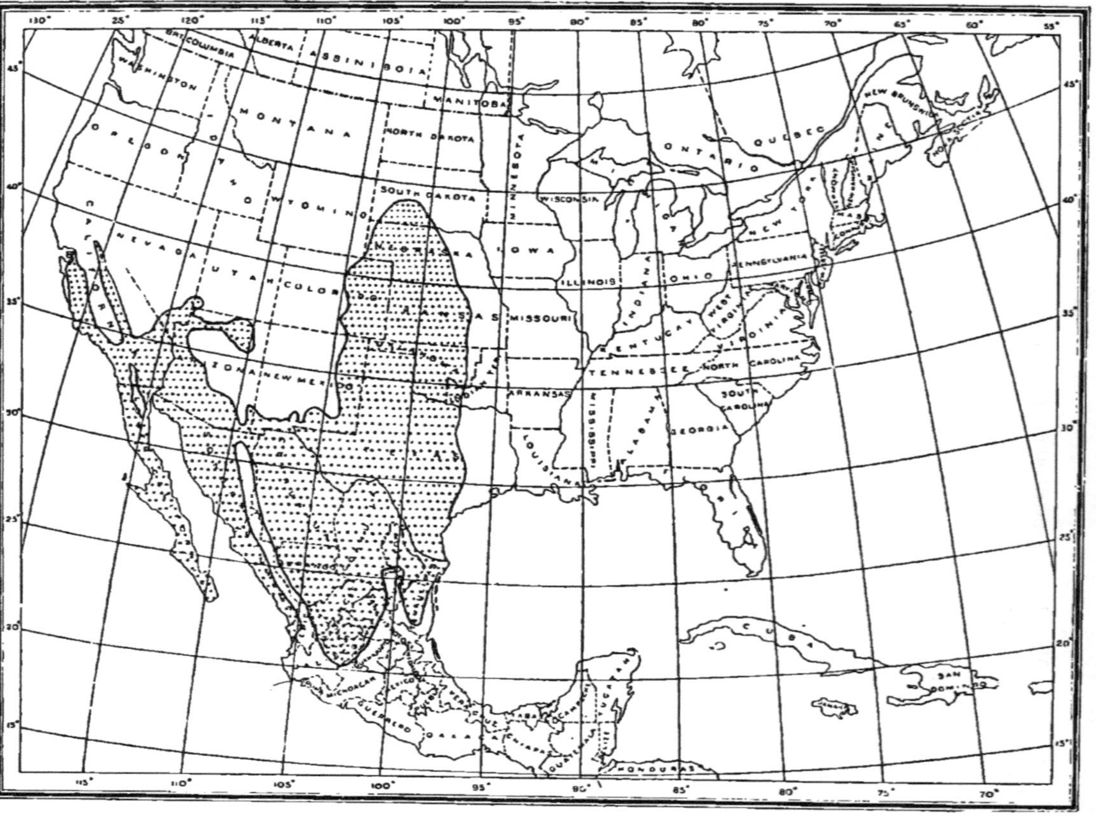

MAP SHOWING DISTRIBUTION OF THE SUBGENUS CHÆTODIPUS.

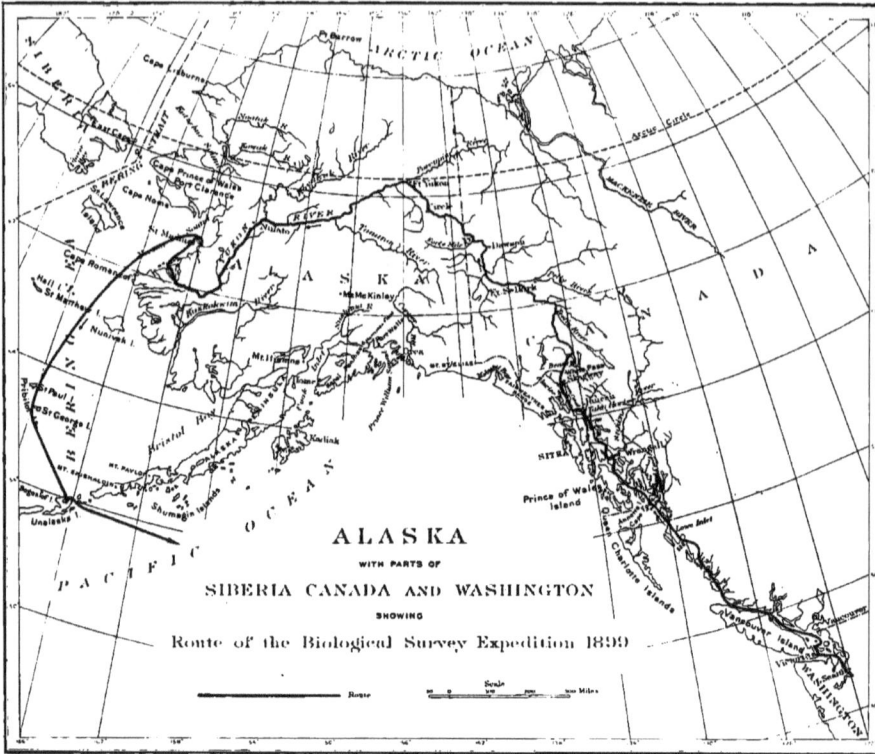

ALASKA

WITH PARTS OF

SIBERIA CANADA AND WASHINGTON

SHOWING

Route of the Biological Survey Expedition 1899

Route

Scale 50 0 100 200 300 Miles

U. S. DEPARTMENT OF AGRICULTURE

DIVISION OF BIOLOGICAL SURVEY

NORTH AMERICAN FAUNA

No. 19

[Actual date of publication, October 6, 1900]

RESULTS OF A BIOLOGICAL RECONNOISSANCE OF THE YUKON RIVER REGION

General Account of the Region
Annotated List of Mammals

By Wilfred H. Osgood

Annotated List of Birds

By Louis B. Bishop, M. D.

Prepared under the direction of

Dr. C. HART MERRIAM

CHIEF OF DIVISION OF BIOLOGICAL SURVEY

WASHINGTON
GOVERNMENT PRINTING OFFICE
1900

STATISTICAL DEPARTMENT
OCT 16 1900
BOSTON PUBLIC LIBRARY

LETTER OF TRANSMITTAL.

U. S. DEPARTMENT OF AGRICULTURE,
WASHINGTON, D. C., *July 28, 1900.*

SIR: I have the honor to transmit for publication, as No. 19 of North American Fauna, a report entitled 'Results of a Biological Reconnoissance of the Yukon River Region,' by Wilfred H. Osgood and Louis B. Bishop.

Under instructions dated May 11, 1899, Wilfred H. Osgood, an assistant in the Biological Survey, proceeded to Skagway, Alaska, and thence over White Pass to the headwaters of the Yukon and down the entire length of the Yukon River to St. Michael. He was accompanied by Dr. Louis B. Bishop, of New Haven, as volunteer assistant; Dr. Bishop has prepared the report on the birds observed during the trip. These are the first investigations of the kind undertaken on the Upper Yukon, and the results herewith presented will be found to contain many important facts concerning the distribution of mammals, birds, and trees in this region.

Respectfully,

C. HART MERRIAM,
Chief, Biological Survey.

HON. JAMES WILSON,
Secretary of Agriculture.

CONTENTS.

ILLUSTRATIONS.

RESULTS OF A BIOLOGICAL RECONNOISSANCE OF THE YUKON RIVER REGION.

GENERAL ACCOUNT OF THE REGION.

By Wilfred H. Osgood.

Nowhere else in North America is such a vast extent of boreal country so easily accessible as along the Yukon. The navigable waters of the river begin at Lake Bennett, only 35 miles from the port of Skagway, on the coast of southeast Alaska, and with but one short interruption, extend northward as far as the Arctic Circle and then westward to Bering Sea; in all, a distance of more than 1,800 miles. The recent developments resulting from the discovery of gold in this region include a modern railroad from Skagway to Bennett and a tram-car service around the dangerous White Horse Rapids. The chief obstacles to ready access to the territory have thus been removed, and an opportunity is afforded for obtaining specimens and information from a region much of which was previously unknown to naturalists. Accordingly, with Dr. Louis B. Bishop as voluntary companion and A. G. Maddren as assistant, I was detailed to make a hasty biological reconnoissance of this region during the summer of 1899.

ITINERARY.

After outfitting at Seattle, Wash., we sailed via the Inside Passage direct to Skagway, Alaska, where we arrived on May 30. From Skagway we worked slowly over White Pass and down to Lake Bennett, at the head of navigation on the Yukon. Here we embarked in a small flat-bottomed boat suited to our needs and sailed down the series of lakes that follow one another for nearly 200 miles. From the lakes we passed into Thirty-Mile River, thence into Lewes River, and finally into the Yukon proper, stopping frequently and making collections at favorable points. With the aid of the swift, even current we were able to make easy and rapid progress. Thus we continued until an unfortunate capsize between Fort Yukon and Fort Hamlin prevented further detailed work on the river, and we were obliged to proceed direct to St. Michael, where a month was spent in collecting on the coast and tundra. Finally, late in September, our work was brought to a close by the approach of the long arctic winter. We returned to Seattle on the U. S. revenue cutter *Corwin*, which stopped on her way for a few hours at St. George Island and at Unalaska, at each of which places we collected a few birds.

A relatively large part of our time was spent in the White Pass region and about the headwaters of the Yukon, as this was an absolutely virgin field, whereas part of the lower river had been previously visited by naturalists. We were unable to do any collecting in the mountains which lie back from the river, owing to the great distance to be covered and the shortness of the season. Legions of mosquitoes were attendant upon us almost constantly. At first they seemed positively unbearable and were a real hindrance to the work, but we gradually became accustomed to them, and by the use of gloves, head nets, and canopies to sleep under, managed to exist in comparative comfort. Aside from insect pests, however, outdoor life on the Yukon in June and July is very enjoyable; good camping places are abundant, and the weather is mild and beautiful. During the latter part of August and in September strong winds sweep up the river and frequent rains occur.

FAUNAL DISTRICTS.

The country traversed may be divided for convenience into five districts: (1) The Lynn Canal district, (2) the White Pass district, (3) the Canadian Yukon district, (4) the Hudsonian Yukon district, and (5) the Alaska Tundra district. These districts are limited in a general way by their respective life zones, but they are not of equal extent or importance, and the names applied to them are used not to specially designate restricted parts of zones already recognized, but purely as a matter of convenience. They are longitudinal districts—that is, they are very much longer than wide, and each is merely a narrow tract covered by our route through some larger faunal region.

Lynn Canal district.—Skagway and the country bordering Lynn Canal are in the northern part of that faunal area which Nelson has called the 'Sitkan district' and which has often been included in the Northwest Coast district. The trees and shrubs are much the same as those at Juneau, Wrangell, and other points farther south, but the vegetation is not quite so dense and luxuriant. The shores of Lynn Canal are steep, rocky, and comparatively sparsely timbered, but in some places, as at Haines, low, swampy ground and heavy saturated forests are found. At Skagway, poplars (*Populus tremuloides* and *Populus balsamifera*) are very common; they share the river bottom with willows and extend well up the steep canyon sides, where they occupy large areas adjacent to the pines, firs, and spruces. Skagway is surrounded by high mountains, and its fauna is limited chiefly by altitude. Glacier Station, 14 miles distant, and about 1,900 feet higher, is near the boundary between the Lynn Canal and White Pass districts. The station is situated on the side of a wooded gulch through which a fork of Skagway River flows. The immediate vicinity is similar to the country about Skagway, but shows the influence of the Hudsonian zone of the White Pass district, which begins only a short distance beyond. On either side of the gulch are glaciated granite cliffs supporting an irregularly distributed vegetation, chiefly groves of poplars and dense thickets of alders, while in the bottom of the gulch conifers are the prevailing trees. The most common trees and shrubs are lodgepole pines (*Pinus murrayana*), alpine firs (*Abies lasiocarpa*), tidewater spruces (*Picea sitchensis*), poplars or aspens (*Populus tremuloides* and *Populus balsamifera*), alders (*Alnus sinuata*), dwarf birches (*Betula glandulosa*), currants (*Ribes laxiflorum*), and huckleberries (*Vaccinium ovalifolium*). The black crowberry (*Empetrum nigrum*) and several other heather-like plants occur in the gulch but are more common higher up. Along the trickling streams are many ferns and mosses, as well as occasional patches of the lichen known as 'reindeer moss.' Among the mammals of this region are the Streator shrew (*Sorex p. streatori*), the Bangs white-footed mouse (*Peromyscus oreas*), the Dawson red-backed mouse (*Evotomys dawsoni*), the long-tailed vole (*Microtus mordax*), and the red squirrel (*Sciurus h. petulans*). Characteristic birds are the sooty song sparrow (*Melospiza m. rufina*), the Townsend fox sparrow (*Passerella i. townsendi*), the Oregon snowbird (*Junco h. oregonus*), and the varied thrush (*Hesperocichla nævia*).

White Pass district.—The summits of the mountains that rise directly east of Skagway are covered with glaciers and perpetual snow, which feed numerous streams that flow down between massive walls of granite. The sides of the wider canyons have been smoothed and scored by glaciation, and the smaller and more recent ones are but jagged rock-bound chasms. These unfavorable conditions cause a rapid change in the character of the plant and animal life, and from Glacier to the summit of White Pass the zones are Hudsonian and Arctic-alpine. A few hundred feet above Glacier the trees become smaller and more scattered, and at Summit only the alpine juniper (*Juniperus nana*), the bearberry (*Arctostaphylos uvaursi*), and depauperate alpine hemlocks (*Tsuga mertensiana*) occur. Heathers and mosses prevail and large areas of reindeer moss are conspicuous. For some distance on the summit of White Pass (Plate II, fig. 1) the elevation and physiography are much the same; the country is slightly rolling and consists entirely of granite rock, about which cling many mosses and heathers, while small alpine junipers and hemlocks struggle for existence in favorable places. The breeding birds found with these Hudsonian plants were ptarmigan (*Lagopus rupestris* and *L. leucurus*), pipits (*Anthus pensilvanicus*), rosy finches (*Leucosticte t. littoralis*), and golden-crowned sparrows (*Zonotrichia coronata*). Characteristic mammals noted were pikas (*Ochotona collaris*), hoary marmots (*Arctomys caligatus*), and mountain goats (*Oreamnos montanus*).

Canadian Yukon district.—Lake subdivision: On the north side of the divide the hemlocks are soon replaced by pines and spruces, and in the vicinity of Shallow Lake the boundary of the interior fauna and flora is reached. The change is complete at Log Cabin, British Columbia, nine miles from the head of Lake Bennett, where the characteristic features of the Canadian zone are again established and the general aspect of the country is very different. The most abundant tree is the white spruce (*Picea canadensis*), and among shrubs seen for the first time the buffalo berry (*Lepargyræa canadensis*) is very common. Birds marking a change of fauna are the slate-colored junco (*Junco hyemalis*), the Alaska jay (*Perisoreus canadensis fumifrons*), the intermediate sparrow (*Zonotrichia l. gambeli*), and the black-poll warbler (*Dendroica striata*). A new chipmunk (*Eutamias caniceps*) is very conspicuous. At the head of Lake Bennett another change occurs; the country becomes more arid and rocky and there is a tinge of Hudsonian.

Lake Bennett is a long, narrow sheet of water inclosed by high granite cliffs, the sides of which are often so steep as to be unfavorable for plant and animal life, and whose summits are doubtless similar to White Pass in fauna and flora. Cold winds sweep down the lake much of the time, and cool shadows envelop the east side most of the forenoon and the west side most of the afternoon, so that opportunity for warmth by direct sunlight is limited. Hence there is quite a strong Hudsonian element about the lake. Among the plants[1] collected here are the pale dwarf laurel (*Kalmia glauca*), the Greek valerian (*Polemonium humile*), the forget-me-not (*Myosotis sylvatica alpestris*), the alpine juniper (*Juniperus nana*), the bush cranberry (*Viburnum pauciflorum*), the dwarf birch (*Betula glandulosa*), the bearberry (*Arctostaphylos uvaursi*), the buffalo berry (*Lepargyræa canadensis*), the shadbush (*Amelanchier alnifolia*), the Labrador tea (*Ledum grœnlandicum*), and the black crowberry (*Empetrum nigrum*). Where trickling streams come down to the lake alder thickets abound, and along terraces of rock clumps of pines and spruces as well as poplars find support. Among Hudsonian mammals were found pikas (*Ochotona collaris*), hoary marmots (*Arctomys caligatus*), and Dall sheep

[1] Identified by F. V. Coville, chief botanist, U. S. Department of Agriculture.

FIG. 1.—SUMMIT OF WHITE PASS.

FIG. 2.—CANADIAN POLICE STATION AT CARIBOU CROSSING.

(*Ovis dalli*). Although the lake widens slightly at its lower end, its outlet is a narrow stream about 2 miles long, called Caribou Crossing (Plate II, fig. 2), on the north side of which is an open, grassy swamp bordered by willow thickets. This low country, though very limited and not extending to the next lake, affords a breeding place for a few mammals and birds not found about Bennett.

Lake Tagish, which receives the waters of Bennett through Caribou Crossing, is like Bennett in character, though not so closely walled, and is characterized by practically the same plants and animals. The surrounding mountains are covered with dense forests, which in many places are almost impenetrable.

Connected with Lake Tagish by a short, narrow stream, known as Six-Mile River, is Lake Marsh, a long shallow lake on each side of which extends low country, with rolling hills farther back. The valley widens here quite appreciably, and the open country is like that at Caribou Crossing. On the east side are sedgy bogs surrounded by willow thickets, and in many places a wide margin of beautiful green sedge meets the edge of the water. Rocky shores are found at some points on the northwest side, but in general the country is low and moist, in marked contrast to that about Bennett and Tagish. The mountain animals of those lakes are of course absent, and the bird life is also somewhat different.

Fifty-Mile River, into which the Yukon waters proceed from Lake Marsh, is rather narrow, and for a short distance at White Horse Rapids very swift. Its banks are chiefly abrupt bluffs of sandy clay (from 50 to 100 feet high) but at Miles Canyon it is confined between walls of basalt. Below the rapids the stream widens somewhat and the high banks become less frequent, often being replaced by low ones thickly grown with willows. The timber is somewhat scattered, and on the rolling hills back from the river bare granite spaces may be frequently seen. At the head of Fifty-Mile River, we first met with birch trees (*Betula papyrifera ?*), and from that time on they were seen daily. They do not grow to large size—trees more than 8 inches in diameter were seldom seen. Several small streams flow into Fifty-Mile River, which favor the growth of thickets of alders along their banks and large clumps of willows about their mouths. The little boreal sagebrush (*Artemisia frigida*) grows abundantly on the warm exposed slopes that occasionally alternate with the sandy bluffs. Lodgepole pines are also abundant and frequently occupy large areas to the exclusion of all other trees. Spruce and poplar, however, are still the strongest elements in the forest.

From Fifty-Mile River we enter Lake Lebarge, the last and largest of the lakes. All about its clear, cold waters are low granite mountains (Plate III, fig. 1). Occasionally patches of heavy spruce forest are found near the water, but in many places cliffs rise abruptly from the water's edge, and the timber is very sparsely sprinkled over them. The rocks found here and a few in Thirty-Mile River are the last we saw showing signs of glaciation. Lake Lebarge is quite different from Lake Marsh, and is more similar to Lakes Tagish and Bennett, though all the Hudsonian elements of these are not present.

River subdivision: This area includes the section from the foot of Lake Lebarge to the mouth of the Pelly River at Fort Selkirk. There is very little variety in the character of the country between these points. Thirty-Mile River, which proceeds from Lake Lebarge, is a swift, narrow stream, and at low water is barely navigable for small steamers. A conspicuous feature of its banks, which are cut abruptly like those of Fifty-Mile River, is a narrow ribbon-like stratum of volcanic ash about 6 inches below the surface that may be seen wherever the bank is exposed. On the mountains a short distance from the river the forest of spruce is heavier and purer than any previously noted. The poplars and willows are more confined to the brink of the river, and the birches are scattered. Thirty-Mile River is simply that portion of the Yukon between Lake Lebarge and the mouth of the Hootalinqua or Teslin River. The stream is greatly augmented by the waters of the Hootalinqua, and from this point on to Fort Selkirk is known as the Lewes River. Below the Hootalinqua it cuts through the Semenow Hills, for the most part abrupt, rocky, and rather barren mountains from 2,000 to 3,000 feet high. Near their bases and at the water's edge are forested areas, but the exposed hillsides are covered with boreal sagebrush (*Artemisia frigida*), with here and there a prostrate juniper or a small clump of spruces. The river now widens rapidly, receiving in succession the waters of the Big Salmon, the Little Salmon, and the Nordenskiold. The rolling hills are sometimes a mile or several miles from the river bank, with low willow swamps intervening. Islands varying from 1 to 100 acres in extent and covered with luxuriant vegetation are abundant. The distribution of trees on the small, regular-shaped islands is very uniform, the different kinds being grouped in concentric belts. Alders generally form the outer margin; next come the willows; next the poplars, rising somewhat higher; and finally the dark-green spruces, which occupy the central area. The whole effect is quite picturesque. On the larger islands the spruces are larger, and usually predominate to such an extent that almost everything else should be classed as undergrowth (including trees and shrubs belonging to the genera *Alnus*, *Salix*, *Populus*, *Lepargyraea*, *Cornus*, *Viburnum*, *Rosa*, *Ledum*, *Vaccinium*, *Ribes*, and others). Lodgepole pines still occur, though unlike the spruces they nowhere form continuous forest and disappear entirely a short distance beyond Fort Selkirk.

The Canadian Yukon district as a whole is very well marked. Characteristic mammals are the gray-headed chipmunk (*Eutamias caniceps*),

Fig. 1.—Cliffs on East Side of Lake Lebarge.

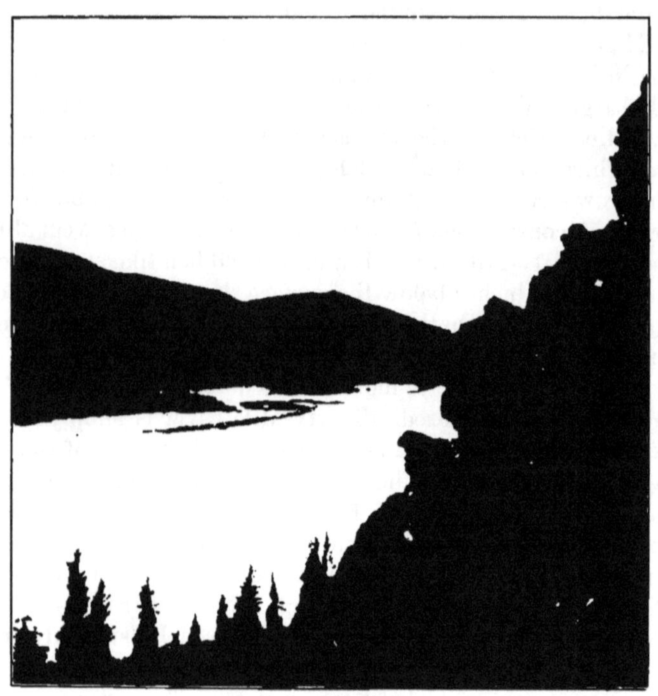

Fig. 2.—Yukon River, 50 Miles below Fort Selkirk.

the Bennett ground squirrel (*Spermophilus empetra plesius*), the Northern bushy-tailed rat (*Neotoma saxamans*), white-footed mice (*Peromyscus oreas* and *Peromyscus maniculatus arcticus*), and the varying hare (*Lepus saliens*). All of these species and three of the genera, *Eutamias*, *Neotoma*, and *Peromyscus*, find their northern limits in this district. Among birds that are known from the Yukon only in this district may be mentioned the sparrow hawk (*Falco sparverius*), the screech owl (*Megascops asio kennicotti?*), the night hawk (*Chordeiles virginianus*), the tree swallow (*Tachycineta bicolor*), the Tennessee warbler (*Helminthophila peregrina*), the pileolated warbler (*Wilsonia pusilla pileolata*), and the mountain bluebird (*Sialia arctica*). Of these, *Chordeiles* is perhaps the most noteworthy, as it is decidedly a southern genus. It is very common, and was seen nightly from Caribou Crossing to Rink Rapids, but after we had passed that point it disappeared. Its range in this region as observed by us is probably accurate and corresponds with the limits of the district. Among trees, the lodgepole pine (*Pinus murrayana*) is common throughout the district, but does not extend beyond it.

Hudsonian Yukon district.—This district, as here considered, includes all of the Yukon region from Fort Selkirk to the limit of trees. The Lewes River is joined at Fort Selkirk by the Pelly, after which the increased volume of water flows on between heavily forested slopes and jutting cliffs (Plate III, fig. 2), which replace the sandy banks of the upper river. From the mouth of the Selwyn northward the topography of the river banks is but slightly different. The number of poplars in the forest is much increased; the spruces are correspondingly decreased not only in number but also in size; while the birches about hold their own, and the pines are not present at all, having disappeared between Fort Selkirk and the mouth of the Selwyn River. As we approach Dawson spruces become dwarfed and entirely subordinate to the poplars, which crowd their bushy tops together for miles and miles. The spruces are in the gulches and in small clumps elsewhere, and a few are scattered about, their dark-green spike tops showing off well against the billowy mass of the lighter foliage of poplar and birch. The undergrowth remains much the same, and deep moss covers the ground and rocks. In damp sandy places along shore and on islands occasionally overflowed a bright-green scouring rush (*Equisetum*) grows so abundantly as to be a characteristic plant. The alpine juniper (*Juniperus nana*) is found occasionally on hillsides not too thickly grown with poplars, and on the more open hillsides the landscape is brightened by masses of fireweed (*Chamænerion angustifolium*), for even here forest fires are not a novelty.

Two more large rivers, the White and the Stewart, empty into the Yukon in this vicinity. About the mouths of these and other tributaries is more or less low country covered with willows. Islands become still more numerous and larger, and have a forest growth that is more uniform in character than that of the river banks. High cliffs overhanging the river are of frequent occurrence.

From Dawson to the Alaska boundary and thence to Circle the country is about the same. For a long distance in the vicinity of the boundary a range of high mountains is visible to the northward on the right bank of the river. The low, rolling hills which border the upper river do not quite reach Circle, but are replaced by a broad, flat country known as the 'Yukon Flats,' which extends from near Circle to Fort Hamlin, a distance of about 200 miles. Through the 'Flats' the course of the river breaks up into a great many channels, and the islands still further increase in size and number. These are composed of sand and silt, in which poplars thrive better than spruces, though the latter are by no means eliminated. A wild rose (*Rosa cinnamomea?*) is the most abundant shrubby plant, and on the ground below it the *Equisetum* is rampant. The larger islands are identical in character with the mainland, and on them the spruces form quite a heavy forest, with deep moss beneath. At Fort Hamlin the river narrows again and flows between rolling wooded hills, which are similar to many farther up the river. Small streams enter the main river frequently, and the timber is much the same; poplars, alders, and birches cover the hills in dense thickets, through which spruces are sprinkled. Alders are more numerous than before. The hills vary in height from 500 to 3,000 feet, and the highest have a distinct timberline at about 2,000 feet. At the mouth of the Tanana the hills become smaller and the river very much wider. Here, at Fort Gibbon, Dr. Bishop found the larch (*Larix americana*) quite abundant. This was the only point at which it was seen by any of our party.

The Lower Yukon beyond the Tanana is very uniform in character. The banks are low and rolling and overgrown with willows and alders; farther back are higher hills covered with poplars and birch; occasionally the summits of a few hills higher than the rest are devoid of trees. On the sandy islands the willow thickets are impenetrable, and where a cut bank exposes a section of them their slender perpendicular trunks stand so closely as to present a solid front like a thick hedge or canebrake. Thus it continues until the limit of timber is reached at Andraefski, 90 miles above the mouth of the Yukon.

This district as a whole is characterized by absence of southern plants and animals. Among migratory birds a few have their center of abundance farther south, but all the mammals are northern forms, and nearly all belong to genera of circumpolar distribution.[1] Plant life, though quite luxuriant, is made up of only a small number of hardy species. Characteristic mammals are the Fort Yukon ground

---

[1] The only exceptions are *Synaptomys*, *Fiber*, and *Erethizon*.

squirrel (*Spermophilus osgoodi*), Dawson red-backed mouse (*Evotomys dawsoni*), yellow-cheeked vole (*Microtus xanthognathus*), Yukon lemming (*Lemmus yukonensis*), Dall varying hare (*Lepus americanus dalli*), and tundra weasel (*Putorius arcticus*). Of the birds, the most characteristic are the duck hawk (*Falco peregrinus anatum*), pigeon hawk (*Falco columbarius*), Alaska longspur (*Calcarius lapponicus alascensis*), hoary redpoll (*Acanthis hornemanni exilipes*), fox sparrow (*Passerella iliaca*), Bohemian waxwing (*Ampelis garrulus*), and wheatear (*Saxicola œnanthe*).

Alaska Tundra district.—The Yukon from Andraefski to the coast of Norton Sound is bounded on both sides by typical tundra. The country is low and gently undulating, and its surface a short distance away appears to be thickly carpeted with grass. That this is not the real condition a short walk ashore soon demonstrates; but the delusion is so complete that were it not for the presence of the great river one might fancy himself looking out over the undulating plains at the eastern base of the Rocky Mountains in the western United States.

The flora of the tundra, though devoid of trees deserving of the name, is found on careful examination to be quite varied. Besides the numerous mosses and heathers and many small berry-bearing plants are dwarf willows, birches, and alders. The alders attain the greatest size, but are usually found in isolated clumps in favorable spots, where they often grow from 6 to 8 feet high. The ground is frozen a few inches below the surface, and the heavy, spongelike covering of vegetation is kept constantly saturated. Occasional high bluffs on the coast in exposed situations are bleak and bare, but besides these there is scarcely a spot not covered with low, matted vegetation. Numerous small ponds are irregularly distributed over the tundra, and around them the vegetation is ranker than elsewhere. Broken lava borders the shores of St. Michael Island, and small moss-covered heaps of it, which form practically the only solid footing on the island, are scattered about over the tundra.

Characteristic mammals at St. Michael are the Hall Island fox (*Vulpes hallensis*), Nelson vole (*Microtus operarius*), tundra red-backed mouse (*Evotomys dawsoni alascensis*), Nelson pied lemming (*Dicrostonyx nelsoni*), Alaska lemming (*Lemmus alascensis*), and Alaska Arctic hare (*Lepus othus*). Land birds known to breed are the hoary redpoll (*Acanthis hornemanni exilipes*), common redpoll (*Acanthis linaria*), Alaska longspur (*Calcarius lapponicus alascensis*), western tree sparrow (*Spizella monticola ochracea*), golden-crowned sparrow (*Zonotrichia coronata*), and Siberian yellow wagtail (*Budytes flavus leucostriatus*). Common tundra plants[1] are *Cassiopea tetragona*, *Andromeda polifolia*, *Vaccinium vitisidæa*, *Mairania alpina*, *Ledum palustre*, *Artemisia*

[1] Nelson, Report upon Natural History Collections in Alaska, 30, 1887.

arctica, *Rubus chamæmorus*, *Rubus arcticus*, *Betula nana*, *Alnus sinuata*, *Chamæcistus procumbens*, and *Tussilago frigida*.

SUMMARY OF FAUNAL DISTRICTS.

All the country here considered is in the boreal zones, the Tundra district and a small part of the White Pass district belonging to the Arctic subdivison, and the Yukon Valley principally to the Hudsonian, though it has also a well-marked Canadian section. Birds are comparatively rare in all the interior region, and it is difficult to determine the exact range of many species. Some were seen but once or twice; others appeared sporadically at rather long intervals; while still others that are known from the region were not seen at all; so it is hardly safe, in making generalizations, to rely too much on the ranges observed by us. The distribution of trees and shrubby plants and of many of the mammals, however, could be determined with much greater accuracy and constitute reliable guides in fixing the limits of the districts. These districts are in general the same as those recognized by Nelson, but with more definite and somewhat modified limits. Names slightly different from those he used are adopted here in order to agree with the commonly accepted names of the primary zones of North America. Thus the part of his 'Alaskan-Canadian' district here considered is called the 'Hudsonian Yukon' district, since it lies entirely within the transcontinental Hudsonian zone. Owing to fluviatile conditions, the boundaries of the Yukon districts doubtless do not agree in latitude with those which might be made away from the rivers.

The zones which we successively traversed in going from Skagway to St. Michael via White Pass and the Yukon are: (1) Canadian; (2) Hudsonian; (3) Arctic-Alpine; (4) Canadian; (5) Hudsonian, and (6) Arctic. The Lynn Canal district is in the Canadian zone, but it has some slight peculiarities such as are to be expected in a coast district. Though it does not have the Hudsonian animals of the northern coast, it lacks several of the typical Canadian forms of the coast farther south.[1] It is really near the northern limit of the Canadian zone on

---

[1]The coast of Alaska south of the peninsula, or what has been known as the 'Sitkan district,' may be easily divided into two districts corresponding to the Canadian and Hudsonian zones. Lynn Canal is situated near the boundary between these districts. Among Canadian mammals which have their center of abundance in the restricted Sitkan district on the coast south of Lynn Canal are: *Odocoileus sitkensis*, *Sciurus vancouverensis*, *Peromyscus macrorhinus*, *Evotomys wrangeli*, *Microtus macrurus*, *Synaptomys wrangeli*, *Zapus saltator*, and *Myotis alascensis*. Among the Hudsonian forms found on the coast only north of Lynn Canal are: *Rangifer* sp., *Ovis dalli*, *Sciurus hudsonicus*, *Spermophilus c. plesius*, *Zapus h. alascensis*, *Ochotona collaris*, and *Myotis lucifugus*. Among trees which find their northern limit in the vicinity of Lynn Canal are: *Thuja plicata*, *Abies lasiocarpa*, and *Pinus murrayana*. The northern district from Lynn Canal to Kadiak is so similar to the great interior Hudsonian region that it hardly merits recognition as a distinct district, but it certainly should not be included in the Sitkan district.

the Pacific coast. The occurrence at Skagway of mammals of the interior, such as *Microtus mordax*,[1] *Evotomys dawsoni*, and *Peromyscus oreas*, seems to show an approach to the condition farther north where the coast and interior forms are practically the same. The Canadian zone of the Lynn Canal district gives way to the Hudsonian and Arctic-Alpine in the White Pass district. The character of this district is essentially the same as that of other mountain regions in western North America. This is well indicated by the fact that its mammals include the hoary marmot (*Arctomys caligatus*), the Alaska pika (*Ochotona collaris*), and the mountain goat (*Oreamnos montanus*), and its birds the ptarmigan (*Lagopus leucurus* and *L. rupestris*), the pipit (*Anthus pensilvanicus*), and the rosy finch (*Leucosticte t. littoralis*).

The Canadian Yukon district from Bennett to Fort Selkirk merely represents the extent to which our route entered the extreme northern part of the Canadian zone; that is, its limits are those of the section that our route made across the end of a narrow tongue which extends northward from the great areas occupied by the Canadian zone farther south. Owing to its being so near the border of the Hudsonian zone, its character is not purely Canadian, but the presence of such forms as chipmunks (*Eutamias*) and white-footed mice (*Peromyscus*) among mammals, night hawks (*Chordeiles*) among birds, and lodgepole pines (*Pinus murrayana*) among trees, makes it evident that the Canadian element is very strong. The Hudsonian Yukon district represents the complete section which the Yukon River makes through the great northern forest belt of the Hudsonian zone. This belt corresponds to the Alaskan-Canadian district outlined by Nelson. It is bounded on the south by the Canadian zone and the extreme northern limit of southern forms, and on the north by the treeless tundra. On the west it probably reaches and includes the coast from Kadiak to Lynn Canal; on the east its limits are unknown. The Alaska Tundra district defines itself. Its character is the combined result of latitude and rigorous coast climate. Our experience in this treeless district was limited to St. Michael Island and the ninety miles between Andraefski and the mouth of the Yukon. The animals of this region are not all absolutely confined to it, many of them ranging some distance into the forest belt. Small mammals, such as the Nelson vole (*Microtus operarius*), occupy so-called 'islands'—local spots offering what are practically tundra conditions—as far within the forest belt as Circle.

---

[1]The case of *Microtus mordax* at the head of Lynn Canal is particularly interesting, since a closely related form, *M. macrurus*, has been found at Glacier Bay on the north side of the mouth of the canal and also at Juneau on the south side, and would therefore be expected at Skagway, which is halfway between. Assuming that *macrurus* has been dispersed from the south northward on the coast, it seems that it did not reach Glacier Bay by way of the present mainland, otherwise it would be found at Skagway. *M. mordax* doubtless reached Skagway from the interior.

The fauna of the Yukon basin as a whole is composed of two principal elements, one containing absolutely circumpolar forms, evidently derived from the north, the other containing forms which have their center of abundance farther south. This is particularly true of the mammals. Among the genera belonging to the north may be mentioned *Rangifer*, *Evotomys*, *Lemmus*, and *Dicrostonyx*, all of which are circumpolar in distribution; those from farther south are *Alces*, *Sciuropterus*, *Eutamias*, *Peromyscus*, *Neotoma*, *Fiber*, and *Synaptomys*. With the exception of alpine species and a few wide-ranging forms, chiefly carnivores, the variations of which are not sufficiently known to be of use in defining faunal regions, no species of mammals are common to the Yukon region and the Sitkan coast district. From this it seems that all the southern forms which reach the Yukon region have been derived from the interior rather than from the coast. This is also true of the trees and to a great degree of the birds. But, on the other hand, some species of land birds are common to the lower Yukon and the Sitkan district while a large intervening area in the interior is uninhabited by them.[1] *Selasphorus rufus*, *Dendroica townsendi*, and *Hylocichla aonalaschkæ* were found on both sides of White Pass, but only rarely and for a very short distance on the interior side.

PREVIOUS WORK.

Our knowledge of the natural history of the Yukon region has been derived chiefly from two sources—the members of the Russo-American Telegraph Expedition and the Signal Service officers formerly stationed at St. Michael. The first information was gathered by the scientific corps of the Telegraph Expedition of 1865 to 1868. Prominent among the members of this corps were Robert Kennicott, William H. Dall, and J. T. Rothrock, all of whom secured valuable specimens and information. The notes of Kennicott were not published, owing to his untimely death at Nulato, May 13, 1866, but numerous specimens, particularly from the vicinity of Fort Yukon, are now in the National Museum, a monument to his faithful pioneer work. Among the numerous papers on various subjects relating to Alaska published by Dail are lists of birds and mammals.[2] A list of plants including some records from Fort Yukon was published by Rothrock.[3]

---

[1] One of these species is the varied thrush (*Hesperocichla nævia*) which was found in the Lynn Canal district, but not in any numbers in the Yukon Valley above Dawson. Below Dawson it is quite common along the Yukon and undoubtedly breeds there. It is well known to range along the Pacific coast to Kadiak, and thence to the shores of Kotzebue Sound and up the Kowak River. Its absence on the Upper Yukon and its occurrence all along the coast make it extremely probable that in reaching the Yukon its course of migration is up the river from its mouth.

[2] List of Birds of Alaska, by W. H. Dall and H. M. Bannister < Trans. Chicago Academy of Sciences, I, pt. II, 267–310, 1869; also Alaska and its Resources, by W. H. Dall, Boston, Lee & Shepard, 1870.

[3] Sketch of the Flora of Alaska, by J. T. Rothrock < Annual Report Smithsonian Institution, 1867, 433–463.

In 1874, with the establishment of a meteorological station at St. Michael, work was begun by Lucien M. Turner. He collected about St. Michael and secured specimens from the fur traders and natives on the Yukon as far up as Fort Yukon. The results of his work were published in the Arctic Series of the Signal Service.[1]

Turner was relieved in 1877 by E. W. Nelson, who continued to collect specimens until 1881. His work was more extensive than that of Turner, but was carried out along the same lines. He made several trips up and down the coast from St. Michael, and also worked about the Yukon delta and up the river as far as Anvik. L. N. McQuesten, who conducted a trading post at Fort Reliance, furnished him with numerous specimens and much valuable information. The results of Nelson's work were also published in the Arctic Series of the Signal Service[2] and form by far the most valuable contribution to the natural history of Alaska yet made. In 1889 an important report[3] was published by Dr. George M. Dawson, covering the region of the sources of the Yukon and down as far as Fort Selkirk. This report contains detailed descriptions of the physical features of the upper river, notes on natural history, particularly on the distribution of trees, and a list of plants by John Macoun.

NEW SPECIES.

Nine new species and subspecies of mammals are described in the present report. They are as follows:

Sciuropterus yukonensis.
Sciurus hudsonicus petulans.
Eutamias caniceps.
Spermophilus empetra plesius.
Neotoma saxamans.
Fiber spatulatus.
Lepus saliens.
Lutreola vison ingens.
Mustela americana actuosa.

In the collection of birds, three new forms were found. These have been described by Dr. Bishop[4] as follows:

Canachites canadensis osgoodi.
Sayornis saya yukonensis.
Contopus richardsoni saturatus.

[1] Contributions to the Natural History of Alaska, by L. M. Turner, Arctic Series, Signal Service, No. II, Washington, 1886.

[2] Report upon Natural History Collections made in Alaska, by E. W. Nelson, Arctic Series, Signal Service, No. III, Washington, 1887.

[3] Geological and Natural History Survey of Canada, Annual Report, III (1887-88), Pt. I, 6 B-277 B, 1889.

[4] Auk, XVII, 113-120, April, 1900.

Besides the above, several new mammals which come within the scope of the present report have been recently described by Dr. Merriam.[1] These are as follows:

Spermophilus osgoodi.	*Lepus americanus dalli.*
Lemmus yukonensis.	*Lepus othus.*
Lemmus alascensis.	*Vulpes hallensis.*
Dicrostonyx nelsoni.	*Sorex personatus arcticus.*
Erethizon epixanthus myops.	*Sorex tundrensis.*

In addition to these, three new forms recently described by Witmer Stone[2] should be mentioned:

Dicrostonyx hudsonius alascensis (equals *D. nelsoni* Merriam).
Putorius rixosus eskimo.
Lynx canadensis mollipilosus.

---

[1] Proc. Wash. Acad. Sci., II, 13–30, March 14, 1900.

[2] Proc. Acad. Nat. Sci. Phila., March 24, 1900, 33–49.

MAMMALS OF THE YUKON REGION.

By Wilfred H. Osgood.

INTRODUCTION.

The following list, primarily based on collections made during the past year, includes all the known mammals of the Yukon region. Besides the species which belong strictly to the Yukon, are included those found in the Lynn Canal and White Pass districts and those known to occur at St. Michael. This makes a list containing the majority of the mammals known from Alaska, which is not, however, intended to be comprehensive, but should be considered as supplementary to the list published by Nelson in 1887. As may be seen from the itinerary, our collections were made during a hasty trip from the coast of southeastern Alaska to the head waters of the Yukon and thence down the river to St. Michael. Good series of all the common small mammals were secured, but the larger and rarer species were not often obtained. It was not only difficult to secure specimens of the larger mammals, but it was hard to gain much accurate information in regard to them. Most of the miners we met had been in the country but a short time and their knowledge of animals was limited; natives were seldom met on the upper river and the few that were interviewed seemed disinclined to talk. The fur trade on the Yukon has dwindled to comparatively meager proportions. The Indians still bring a few furs to the traders every year and receive pittances of flour and tea in return; but the trade is apparently very small and were it not for the transportation business which has recently become so important, the large companies would doubtless find it difficult to maintain themselves.

In identifying the recently collected specimens and studying their geographical distribution, it has been necessary to refer constantly to the specimens collected by Kennicott, Dall, Nelson, and Turner. Many of these, which are in the National Museum, were found to be in poor condition and required considerable renovating to make them comparable with modern specimens. For the free use of these specimens I am indebted to Gerrit S. Miller, jr., assistant curator of mammals in the National Museum. I am also indebted to Outram Bangs for the use of specimens, and E. W. Nelson for much valuable information. The identifications of some of the mammals have been verified by specialists as follows: The species of *Sorex* by Dr. C. Hart Merriam; of *Microtus* by Vernon Bailey, and of *Zapus* by Edward A. Preble. All measurements are in millimeters.

LIST OF SPECIES AND SUBSPECIES.

1. *Rangifer montanus* Seton-Thompson.
2. *Rangifer arcticus* (Richardson).
3. *Rangifer tarandus* (Linnæus).
4. *Alces gigas* Miller.
5. *Ovis dalli* Nelson.
6. *Oreamnos montanus* (Ord).
7. *Sciuropterus yukonensis* sp. nov.
8. *Sciurus hudsonicus* Erxleben.
9. *Sciurus hudsonicus petulans* subsp. nov.
10. *Eutamias caniceps* sp. nov.
11. *Spermophilus empetra plesius* subsp. nov.
12. *Spermophilus osgoodi* Merriam.
13. *Arctomys caligatus* Eschscholtz.
14. *Castor canadensis* Kuhl.
15. *Mus decumanus* Pallas.
16. *Peromyscus oreas* Bangs.
17. *Peromyscus maniculatus arcticus* (Mearns).
18. *Neotoma saxamans* sp. nov.
19. *Evotomys dawsoni* Merriam.
20. *Evotomys dawsoni alascensis* (Miller).
21. *Microtus mordax* (Merriam).
22. *Microtus drummondi* (Aud. and Bach.).
23. *Microtus xanthognathus* (Leach).
24. *Microtus operarius* (Nelson).
25. *Fiber spatulatus* sp. nov.
26. *Synaptomys dalli* Merriam.
27. *Lemmus yukonensis* Merriam.
28. *Lemmus alascensis* Merriam.
29. *Dicrostonyx nelsoni* Merriam.
30. *Zapus hudsonius alascensis* Merriam.
31. *Erethizon epixanthus myops* Merriam.
32. *Ochotona collaris* (Nelson)
33. *Lepus saliens* sp. nov:
34. *Lepus americanus dalli* Merriam.
35. *Lepus othus* Merriam.
36. *Lynx canadensis mollipilosus* Stone.
37. *Canis occidentalis* Richardson.
38. ? *Vulpes fulvus* (Desmarest).
39. *Vulpes hallensis* Merriam.
40. *Ursus americanus* Pallas.
41. *Ursus horribilis alascensis* Merriam.
42. *Lutra canadensis* (Schreber).
43. *Lutreola vison ingens* subsp. nov.
44. *Putorius arcticus* Merriam.
45. *Putorius cicognani alascensis* (Merriam.)
46. *Putorius rixosus eskimo* Stone.
47. *Mustela americana actuosa* subsp. nov.
48. *Mustela pennanti* Erxleben.
49. *Gulo luscus* (Linnæus).
50. *Sorex personatus streatori* Merriam.
51. *Sorex personatus arcticus* Merriam.
52. *Sorex obscurus* Merriam.
53. *Sorex tundrensis* Merriam.
54. *Myotis lucifugus* (Le Conte).

ANNOTATED LIST OF SPECIES.

Rangifer montanus Seton-Thompson. Mountain Caribou.

Rangifer montanus Seton-Thompson, Ottawa Naturalist, XIII, No. 5, 6, Aug. 11, 1899; Allen, Bull. Am. Mus. Nat. Hist., N. Y., XIII, 1–18, April 3, 1900.

This large woodland caribou is reported as quite common in northern British Columbia about the head waters of the Yukon and for an indefinite distance northward. It does not occur on the coast south of Cook Inlet, but is reported from many points immediately beyond the summit of the coast mountains. It prefers the higher ground in summer and is not found along river bottoms like the moose, for which reason few are killed by parties descending the river. Its flesh is smoked and dried by the Indians for winter food, and when so cured is preferred to all other meat of the country. The hides, like those of the moose, serve the natives for various articles of clothing and are utilized especially for sleeping robes.

Rangifer arcticus (Richardson). Barren Ground Caribou.

The barren ground caribou ranges over nearly all of extreme northern North America from northwestern Labrador to the Aleutian Islands. It was formerly abundant over this great territory, but is now quite rare. Even at the time of Nelson's work in 1877 it had become comparatively uncommon, though it was once common all about Norton Sound and for some distance up the river. The southern and interior limits of its range are uncertain. During our stay in St. Michael, I saw half a dozen skins which had been secured near Andraefski, 90 miles above the mouth of the Yukon. There are specimens in the National Museum from Nushagak and Unalakleet, Alaska; and from Rampart House and La Pierre House, Northwest Territory.

Rangifer tarandus (Linnaeus). Domesticated Reindeer.

During the past few years, as is well known, an effort has been made to introduce domesticated reindeer from Siberia into Alaska. The animals as a rule have been carefully herded, but in a few cases they have had opportunities to stray away and run wild. The herd that had perhaps the best chance to stray was one which was brought from Lapland to Haines in 1898, and driven inland over the Dalton trail. A short time after it started several of the animals were seen in the forest near Haines, and one of them was killed. This was the only instance of the kind brought to my attention, but I have no doubt that reindeer have occasionally wandered from the care of the herders at other times and in other places.

Alces gigas Miller. Alaska Moose.

Alces gigas Miller, Proc. Biol. Soc. Wash., XIII, 57–59, May 29, 1899.

The Alaska moose, as has frequently been stated, is the largest of the deer family in North America. Its distribution along the Yukon extends from Lakes Atlin and Tagish at least to the mouth of the Tanana and probably somewhat farther. Whymper[1] says that it was 'never known as low as Nulato,' even in the time of its greatest abundance. It is evident, however, from the record of Nelson[2] at the Yukon delta, and that of Richardson[3] at the mouth of the Mackenzie, that it does occasionally leave its favorite woodlands of the interior and wander as far as the Arctic coast. At present it is still quite numerous, but is chiefly confined to the small streams tributary to the Yukon. According to reports which came to me it is abundant in the region about the upper waters of the Stewart, Pelly, and MacMillan rivers. Along the great river itself numbers have been killed during

[1] Travels in Alaska and on the Yukon, 245, 1869.

[2] Report upon Natural History Collections in Alaska, 287, 1887.

[3] Fauna Boreali-Americana, 233, 1829.

the recent influx of prospectors. At the beginning of the Klondike rush, it was not uncommon for a party to secure one or two moose while descending the river, but such is rarely the case at present. Our party failed to see any, though we spent nearly three months in the region; during this time we heard of but two animals being killed, one near the foot of Lake Lebarge and another on upper Charlie Creek, a short distance above Circle; both were secured by Indians. We saw comparatively few fresh tracks.

In winter, moose meat is the staple diet of both Indians and whites and has readily sold in the mining camps at $1 to $2 per pound. Such a price, even in this country of high wages, has been a great incentive to hunting, and many a miner has left his claim to pursue the moose. The hides also are a source of profit, particularly to the Indians, who tan them and make them into mittens and moccasins. What the Indians do not need they sell readily to miners and prospectors. It is difficult to estimate the number of animals that have been killed, but it must be very large, for the demand has been steady and a comparatively large population has been supplied with meat. On one hunt, an account of which has been given by Tappan Adney,[1] 44 moose were killed in about one month, and a single party of Indians was credited with a total of 80 moose and 65 caribou in one winter.

Ovis dalli (Nelson). Dall Mountain Sheep.

Most of the specimens of the Dall sheep which have reached our museums were secured in the vicinity of Cook Inlet, but the animal occurs in nearly all the high mountains of Alaska, and in the north ranges to the Arctic coast. Since we were at a distance from the mountains during the greater part of our trip, I was unable to secure much information in regard to the distribution of the species. Sheep are said to occur about the West Arm of Lake Bennett, and Windy Arm of Lake Tagish. A prospector with whom I talked at Lake Tagish said he had seen and killed them at both of these places. Lake Bennett is not far from the type locality of *O. stonei*, and it is possible that this species occurs there with *O. dalli*. Both white and gray sheep are reported, though all are said to be white in winter. I was told that white sheep were killed some years ago on the cliffs about Lake Lebarge, but I failed to find signs of them there. Prospectors at Fort Selkirk say that sheep are always to be found in the mountains along Pelly River, particularly in the MacMillan Mountains[2] near the mouth

[1] Harper's Magazine, C, 495–507, March, 1900.

[2] The sheep from the MacMillan Mountains are said to be the 'black sheep,' which name could hardly apply to *O. dalli*, but is the name commonly given to *O. stonei*. If *stonei* really does occur in these mountains the record is a very interesting one, and the locality much farther north than any from which the species has been previously recorded.

of MacMillan River, and they were also reported from the head waters of the Stewart and from the Tanana Hills.

Oreamnos montanus (Ord). Mountain Goat.

Goats occur on the high granite cliffs which inclose the upper part of Lynn Canal; they are also common on the mountains near White Pass and about the rocky walls of Lake Bennett. I was told that they had been killed recently at the upper end of Little Windy Arm on Lake Tagish, but I could obtain no reliable report of their occurrence in the interior beyond this point. At Lake Lebarge they were very doubtfully reported. Their range is known to extend north to White Pass in the coast mountains at least to Copper River,[1] but does not reach far into the interior. Hunters from the mountains about the upper waters of the Pelly and Stewart rivers asserted positively that none had been heard of in that region.

The station agent at Glacier, near White Pass summit, told me that goats frequently appear on the cliffs within easy view of his house. He also showed me the hide of one that had been killed near there a short time before our arrival. I made one short trip into these mountains, but failed to see any goats. The character of the cliffs is ideal for them, but they had evidently gone farther back to their summer feeding grounds, as the abundant tracks and dung were a few weeks old.

Sciuropterus yukonensis sp. nov. Yukon Flying Squirrel.

Type from Camp Davidson, Yukon River, near Alaska-Canada boundary. No. $\frac{12008}{35326}$, U. S. Nat. Mus. Collected December 8, 1890, by R. E. Carson.

Characters.—Size largest of North American flying squirrels; tail exceedingly long; color rather dark, underparts suffused with fulvous; skull slightly characterized.

Color.—Top of head, neck, and upperparts to base of tail pale cinnamon or between the wood brown and cinnamon of Ridgway; underfur bluish black, partially exposed on legs and membranes; underparts dull whitish, irregularly suffused with cinnamon fawn; feet dusky above, lightly edged with creamy white, buffy white below; cheeks and sides of head ashy, lightly mixed with cinnamon; end of nose slightly paler than top of head, not light ashy as in *S. sabrinus;* black eye-ring prominent; tail light fawn below, with a light edging of dusky, becoming broader toward tip; tail above fawn heavily mixed with black, which predominates for terminal fifth.

Skull.—Size large, slightly larger than in *S. alpinus;* audital bullæ larger; width at postorbital constriction greater; molars heavier, particularly the mandibular series.

---

[1] H. T. Allen, Science, VII, 57, 1886.

Measurements. Total length 365; tail 180; hind foot (measured dry) 41. Skull: Occipitonasal length 40; zygomatic breadth 25; postorbital constriction 10.

Remarks.—This species is distinguished from both *S. sabrinus* and *S. alpinus* by its large size and very long tail, but it is also very different from either in color. It is evidently a very rare squirrel, as the type and one topotype are the only specimens known. A specimen from Chilkoot Inlet which may possibly be this species has been recorded by Dr. George M. Dawson.[1] Camp Davidson is the northernmost point at which the genus *Sciuropterus* is known to occur. The type and one other specimen were secured by R. E. Carson, who was a member of the boundary survey party of the U. S. Coast and Geodetic Survey under J. E. McGrath, in 1890. Dr. W. W. Kingsbury, also a member of the party, writes me as follows in regard to these specimens:

> I send you the following notes taken from my journal regarding two Flying Squirrels which were captured by a member of our party while in Alaska, in 1890; their skins were sent to the National Museum at Washington.
>
> The female was caught Dec. 8th, 1890, and the male Dec. 9th, 1890. Both squirrels were caught in a trap known as the 'dead fall,' which was set by R. E. Carson for marten. The traps were set in the bed of a frozen stream, where it ran through a clump of spruce trees about one-fourth of a mile back from the Yukon river. This clump of trees is about 2½ miles east of the International boundary line, and on the east bank of the Yukon river.
>
> We showed these skins to both McQuesten and Mayo, two traders who had been in that country over twenty years, and who said that they had seen Flying Squirrels along the Yukon river quite a number of times before, and had also seen them at Ft. Reliance and Ft. Yukon; but had not seen any of them for a number of years before this date. We also showed the skins to an Indian, who said these squirrels would attack a man by flying in his face, and the Indians would not eat them because the squirrels ate dirt.
>
> During the winter and spring of that year, I hunted very carefully in the vicinity where these squirrels were captured, but failed to find further trace of them. The stomachs of both of these squirrels were empty. The traps in which they were caught were set for martens, and two or three had been caught, but none were caught in these traps after the squirrels were captured.

Sciurus hudsonicus Erxleben. Hudson Bay Red Squirrel.

All the red squirrels from the Yukon basin and northern Alaska, as far as can be determined at present, are referable to *Sciurus hudsonicus* 'proper,' although those from the Upper Yukon show considerable tendency toward *S. h. streatori*. Most of the Yukon specimens are in summer pelage, while the few available specimens from eastern Canada and the vicinity of Hudson Bay are in winter pelage, so that close comparison is not possible. Specimens from various points along the Yukon from Bennett to Nulato have been examined. The

[1] Geol. and Nat. Hist. Survey of Canada, Annual Report, III (1887–88), pt. I, 50 A, 1889.

FIG. 1.—NESTS OF RED SQUIRRELS IN SPRUCE THICKET.

FIG. 2.—BURROWS MADE BY RED SQUIRRELS IN LOOSE SCALES STRIPPED FROM SPRUCE CONES.

animal is exceedingly abundant in all the spruce forest, and doubtless ranges northward to the limit of trees.

Evidences of its activity are to be found all through the spruce forest. Its globular nests of grass, moss, bark, and refuse are common (Plate IV, fig. 1), and are usually situated near the trunk of some slender spruce, 10 or 20 feet from the ground. Sometimes several will be found in the same tree, and a half dozen or more are very often to be seen at the same time. Little excavations in the moss show where the chickarees have been digging for roots; and spruce cones tucked away in these and other out-of-the-way places are further evidence of their sagacity. The ground is often strewn for some distance with the scales of spruce cones which they have stripped (Plate IV, fig. 2). Near Lake Marsh I found one such place 20 feet square which was covered 6 inches deep with scales.

Sciurus hudsonicus petulans subsp. nov.

Type from Glacier, White Pass, Alaska (altitude 1,870 feet). No. 97457, U. S. Nat. Mus., Biological Survey Collection, ♀ ad. Collected June 4, 1899, by W. H. Osgood. Original No., 370.

Characters.—Similar to *S. hudsonicus*, but larger and darker; central portion of tail darker and with slight mixture of black; submarginal black in tail wider; edging of tail much darker; underparts not pure white in summer. Similar to *Sciurus h. streatori*, but more reddish; central portion of tail with much less admixture of black; subterminal black in tail much narrower. Somewhat similar to *S. vancouverensis*, but paler and cranially different; lateral stripe much more prominent; submarginal and subterminal black in tail narrower; median dorsal stripe less suffused; median dorsal hairs of tail with much less black.

Color.—*Summer pelage:* Upperparts between the raw umber and Prout's brown of Ridgway; top of head slightly darker than back; lateral line prominent, intense black; forelegs and feet russet; underparts lightly washed with fulvous; median dorsal portion of tail hazel, slightly mixed with black-tipped hairs; submarginal and subterminal black in tail rather limited; edging of tail ochraceous; under surface of tail paler than upper, the grayish roots of the hairs showing through. *Winter pelage:* Similar to the corresponding pelage of *S. hudsonicus*, but considerably darker; median dorsal line more diffuse; tail darker and with greater admixture of black in central portion.

Skull.—Similar to that of *hudsonicus* and its other subspecies; nasals longer and posteriorly more compressed than in *S. vancouverensis;* orbital arch with a sharp indentation between lachrymal and postorbital process. (See Plate V, fig. 2.[1])

Measurements.—Average of two specimens from type locality: Total length 303; tail 120; hind foot 50.

---

[1] Topotype No. 97460, U. S. Nat. Mus. Compare with fig. 1, *S. vancouverensis*, No. 71889, U. S. Nat. Mus., from Goldstream, Vancouver Island.

Remarks.—The closest relationship of this red squirrel is evidently with *hudsonicus* of northern Alaska.[1] A single specimen from Yakutat Bay shows a decided tendency toward the northern form, and those from Cook Inlet are clearly referable to it. A more or less imperfect specimen from Inverness, British Columbia, indicates a possible intergradation with *Sciurus h. streatori.* There is ample material demonstrating by skulls as well as by color that it has no very close relationship to *S. vancouverensis.* My specimens of *petulans* taken early in June are in new summer pelage or in old winter pelage just previous to or in process of change. The latter doubtless does not fairly represent the winter pelage; but in making comparison with eastern specimens, I have chosen those in a similar condition.

About Lynn Canal and on the southwest side of White Pass I found these red squirrels abundant. Several at Glacier had become quite tame, and came every day to the cabin of one of the railroad hands to be fed. They have all the vivacious energy, curiosity, and vocal accomplishments of their Eastern cousins, and fully maintain their reputation for rollicking good nature and fearlessness.

Eutamias caniceps sp. nov. Gray-headed Chipmunk.

Type from Lake Lebarge, Northwest Territory, Canada. No. 99200, U. S. Nat. Mus., Biological Survey Collection, ♀ ad. Collected July 13, 1899, by W. H. Osgood. Original No., 603.

Characters.—Similar to *E. borealis*, but grayer, particularly the head, tail, and feet; postauricular spots more prominent; underparts pure white.

Color.—*Summer or postbreeding pelage:* Sides bright ochraceous, extending from flanks forward and stopping immediately below ears, but interrupted at shoulders by the extension of gray from arm; five black stripes on back very distinct and, except outer ones, entirely unmixed with ochraceous; outer pair of light stripes pure white, prominent, not continuous with postauricular spots; inner light stripes bluish white mixed with ochraceous; top of head brownish gray; postauricular spots bluish white, connected with throat by a continuous light stripe running below ear; light stripes on sides of head prominent, almost pure white; dark stripes rufous mixed with blackish, narrower and darker than in *E. borealis;* underparts pure white; feet yellowish white. *Worn pelage:* General effect of upperparts olive gray relieved by the black and white stripes of the back and faint traces of the fulvous, which has been worn away; feet grayish white; tail above black, grizzled and overlaid with white, below clay color submargined by black and margined by white.

[1] The *hudsonicus* of northern Alaska is here considered the same as that from eastern Canada, but will doubtless prove separable when an abundance of material in all pelages is available.

Skull.—Similar to that of *E. borealis*, but with a slightly fuller brain-case and larger audital bullæ.

Measurements.—Type (from dry skin): Total length 223; tail vertebræ 103; hind foot 32.

Remarks.—The type[1] of *E. borealis* from Fort Liard, British Columbia, is missing, but specimens from Fort Simpson, which is not far from Fort Liard, and other points east of the Rocky Mountains, are available for comparison. These are all much suffused with fulvous, and are very easily distinguished from those of the Upper Yukon. *E. caniceps* is characterized not only by gray head and cheeks, but by gray feet, gray edging to tail, and pure white underparts.

This species is found from the headwaters of the Yukon about Lake Lindeman to the vicinity of Fort Selkirk, where it was last seen by our party. I found it most common in the dry and open rocky country about Lake Bennett and Lake Lebarge, and a few were taken in the thickets of *Lepargyræa* about Lake Marsh and Fifty-Mile River. It is not abundant anywhere in the region, but is remarkably tame and unsuspicious. I seldom saw more than two or three in a half day's tramp, but these would often frisk about within a few feet of me as if entirely oblivious of my presence.

Spermophilus empetra plesius subsp. nov. Bennett Ground Squirrel.

Type from Bennett City, head of Lake Bennett, British Columbia. No. 98931, U. S. Nat. Mus., Biological Survey Collection, ♀ ad. Collected June 19, 1899, by W. H. Osgood. Original No., 465.

Characters.—Similar to *S. empetra* and *S. kadiacensis*, but smaller; general color less fulvous; under side of tail always clear bright cinnamon rufous; molar teeth relatively much larger than in *kadiacensis;* skull small and light and otherwise slightly peculiar.

Color.—*Postbreeding pelage:* Above, mottled as in *S. empetra*, but general color less fulvous; upperparts, mixed black, white and yellowish gray extending forward to top of head, becoming narrower and slightly grayer between shoulders; top of head chestnut mixed with black; nose and forehead clear hazel; under side of body cinnamon rufous paling to nearly white around chin and extending to sides of body, neck and cheeks, and both sides of legs; under side of tail somewhat deeper cinnamon rufous margined by yellowish white; subterminal black in tail less extensive than in *empetra* and *kadiacensis;* median part of upper side of tail grizzled black and yellowish, narrow submargin and subterminal zone black, the whole edged and overlaid with yellowish white. The hairs of the back in *S. plesius* are of two kinds, some being of several colors arranged in zones and some pure black for their entire length. The former, which are most abundant, are dark sooty plumbeous at the base followed by a zone of light gray,

[1] See Allen, Bull. Am. Mus. Nat. Hist., N. Y., III, 109, 1890.

then one of black, then yellowish white, and finally a black tip. In *S. empetra*, the arrangement is practically the same, but the upper part of the light gray zone blends into fulvous. As this is the widest zone, it gives a fulvous suffusion to the entire upperparts of the animal. In *plesius* the black submargin of the tail never shows through on the under side. *Worn spring pelage:* Upperparts yellowish gray; top of head, forehead, and nose cinnamon rufous; thighs with faint suggestions of rufous; shoulders and neck hoary; sides and underparts grayish white washed with yellowish and flecked with ochraceous; feet pale buffy ochraceous; tail paler than in postbreeding pelage.

Skull.—Similar to that of *S. empetra* from Unalaska, but smaller and lighter; nasals shorter and wider in proportion to their length; postpalatal notch extending farther forward, being almost on a plane with the last molar; molar teeth decidedly larger than those of *kadiacensis*.

Measurements.—Type (from dry skin): Total length 345; tail vertebræ 93; hind foot 50. Skull of type: Basal length 45; zygomatic breadth 35; postorbital constriction 13; length of nasals 18; least width of nasals 6; alveolar length of molar series 13.

Remarks.—The material representing *Spermophilus empetra* is still very scanty and imperfect. Specimens from the Arctic coast are few in number and poor in quality, while from Hudson Bay one flat skin, unaccompanied by a skull, is all I am able to find. I have considered this (No. 13932, U.S.N.M.) to be typical of *empetra* and have used it in making skin comparisons. Since it agrees fairly well with specimens of the ground squirrel which has been introduced on Unalaska, I have used the skulls of these for skull comparisons. Specimens from Bristol Bay and the Alaska peninsula are apparently intermediate between *empetra* and *plesius*. *S. kadiacensis* is apparently confined to Kadiak Island, as specimens from the mainland immediately opposite the island are cranially and dentally distinct. The southern members of the group, *columbianus* and *erythroglutæus*, also need not be considered, as they are very different from *empetra* and *plesius*.

S. plesius was first met with on the south side of White Pass near Glacier, where a small colony was found on a steep rocky slope above the canyon. They were active here in early June while patches of snow still lay on the ground. On the summit of White Pass another small colony was found, and at Lake Bennett they were very abundant. Here their burrows are to be found wherever the conformation of the rocks affords lodgment of sufficient soil. From Bennett on to Fort Selkirk they are exceedingly abundant. We saw them daily about all the lakes, and as we floated down Fifty-Mile and Thirty-Mile rivers, we often saw them bobbing in and out of their burrows or scurrying along their little trails which score the banks.

From sunrise till late in the afternoon, their sharp clicking cries rang out across the water, so that, if not to be seen, they at least reminded us of their presence nearly all the time. When alarmed, they stand erect on their haunches near their burrows and violently utter their sharp, high-pitched *clickety click* as long as the exciting cause is in sight, always emphasizing each cry by vigorously slapping their short tails against the ground behind them. As a rule they were quite wary, and it was not possible to get within gunshot without some concealment and careful stalking. The limit of the range of the species along the river is near Fort Selkirk. The last specimen secured was caught near Rink Rapids, but I learned that quite a colony of ground squirrels exists on the west bank of the river just below Fort Selkirk.

Spermophilus osgoodi Merriam. Fort Yukon Ground Squirrel.

Spermophilus osgoodi Merriam, Proc. Wash. Acad. Sci., II, 18, March 14, 1900.

From Fort Selkirk, near the limit of *Spermophilus plesius* in the interior, nearly to Circle, we saw no signs of ground squirrels of any kind. Just before reaching Circle, however, we began to see unmistakable signs of them and were soon attracted to a small colony by their clicking calls which reached our ears as we floated down in midstream. The call is executed in about the same time as that of *S. plesius*, but its pitch is much lower and its effect on the ear is utterly different. It suggests the click of castanets. On going ashore we found their burrows and connecting paths scattered over quite an area on the hillside. The colony occupied the open hillside and a few ledges of loose rock, and even extended down into a thicket of alder and willow at the foot of the hill. The animals were very shy and became much excited at our approach. Their long tails were very noticeable in marked contrast to the short ones of *S. plesius*, which we had been accustomed to seeing. Fifteen specimens were secured.[1] At this time (Aug. 14) they were all very fat and in splendid postbreeding pelage; the entire underparts were rich ferruginous without a trace of any other color; the back was very dark, and the long tail was full and bushy. One specimen was pure glossy black with faint shadowy indications of vermiculation on the back. Among the specimens in the National Museum from Fort Yukon are several in this melanistic condition, showing that it is not uncommon. The range of this species on the Yukon begins about 20 miles above Circle and extends at least to Fort Yukon and probably to the mouth of the Tanana.

Arctomys caligatus Eschscholtz. Hoary Marmot.

Six specimens of the hoary marmot were secured in the White Pass region and about Lake Bennett, where it was common. It is confined

[1] This valuable series was unfortunately destroyed.

to rocky, mountainous parts of the Hudsonian zone, and consequently we did not meet with it during the latter part of our trip, and only heard of it through reports from the mountains at the headwaters of the White and Tanana rivers. As elsewhere, it is familiarly known as the 'whistler,' although occasionally rather inappropriately called 'ground hog.' Its long drawn whistle is peculiarly mournful, particularly when it breaks the deathly silence of some rocky canyon. It loves to stretch at full length on top of a large rock and bask in the sun. I frequently found it quietly enjoying itself in this manner.

Castor canadensis Kuhl. American Beaver.

It hardly seems possible that half a million or more beaver skins have been secured in the Territory of Alaska. The animal is now almost as rare there as it is in the United States, the inevitable result of continued pursuit by both whites and natives, which has so many parallels that it is useless to emphasize it here. At Fort Selkirk I saw several beaver skins taken on a small tributary of Stewart River, and at St. Michael I found a very few in the warehouses of the trading companies. Beyond this I saw or heard nothing of them.

Mus decumanus Pallas. Norway Rat.

Large rats are exceedingly abundant at St. Michael. Their introduction must have been effected very recently, as they were unknown there at the time of Nelson's work. Unalaska has long been their northern limit on the Pacific coast. They find shelter about the wharves and lumber piles at St. Michael and also infest the buildings, particularly food warehouses. Their distribution will undoubtedly soon be extended all along the Yukon by means of the many steamers now plying between St. Michael and Dawson.

Peromyscus oreas Bangs. Bangs White-footed Mouse.

Peromyscus oreas Bangs, Proc. Biol. Soc. Wash., XII, 84, March 24, 1898.

Long-tailed mice were taken at Skagway, Glacier, Summit, Bennett, Caribou Crossing, Fifty-Mile River, and Rink Rapids. In general they seemed to be more woodland loving than the short-tailed species, though at Bennett a number were taken among bare rocks at the very water's edge. I first noticed them here while walking along the shore at night. They were darting in and out among the rocks, chasing each other as if playing a game of tag, and often four or five were in sight at once. *P. oreas* from the type locality is somewhat intermediate between my specimens and those which come from the coast of Puget Sound and southern British Columbia. Northern specimens are slightly larger, paler, and less ruddy brown than typical *oreas*. They are very similar in color to *canadensis* and increase the probability that the latter has a transcontinental range. Their skulls are larger and have fuller braincases than those of either *oreas* or *canadensis*.

Peromyscus maniculatus arcticus (Mearns). Arctic White-footed Mouse.

Hesperomys leucopus arcticus Mearns, Bull. Am. Mus. Nat. Hist., N. Y., II, 285, Feb., 1890.

A short-tailed white-footed mouse was found to be very common from Lake Marsh to Lake Lebarge. Thirty specimens were secured, most of them about the crevices of low ledges of rock along the lake shores. The name *arcticus* is only tentatively used for these specimens, as its applicability can not be positively known until a series of Labrador specimens is obtained. My specimens do not differ from topotypes of *arcticus*, and these in turn, as stated by Bangs,[1] do not differ in color and size from typical *maniculatus*. The description of the skull of the Great Whale River specimen examined by Bangs, however, does not agree well with the characters of the skulls of *arcticus*, so it seems advisable to recognize *arcticus* as a subspecies of *maniculatus*. It is probable that more material will amply justify this treatment of the western form.

Neotoma saxamans sp. nov. Northern Bushy-tailed Rat.

Type from Bennett City, head of Lake Bennett, British Columbia. No. 98923, U. S. Nat. Mus., Biological Survey Collection, ♂ ad. Collected June 19, 1899, by W. H. Osgood. Original No., 462. (See Plate V, fig. 4.)

Characters.—Similar to *Neotoma cinerea drummondi*, but somewhat darker; underparts pure white; skull strongly characterized.

Color.—(Type:) Above, grayish fawn mixed with black, becoming brighter on sides, where the quantity of black is much diminished; underparts and feet pure white; eyelids intense black with a limited sooty area about them; nose and anterior cheeks ashy; tail slaty above, white below.

Skull.—Similar to that of *N. drummondi* (Plate V, fig. 3[3]) but with interorbital space narrower; nasals narrower and more attenuate posteriorly; maxillary arm of zygoma lighter; sphenopalatine vacuities open; ventral surface of occipital with a high trenchant median ridge; front of incisors very pale.

Measurements.—Type (from dry skin): Total length 452; tail vertebræ 192; hind foot 46. Skull of type: Basal length 52; zygomatic breadth 29; interorbital width 5; length of nasals 23.

Remarks.—*Neotoma saxamans* differs from *N. cinerea*,[2] *N. occiden-*

---

[1] Am. Naturalist, XXXII, 496, July, 1898.

[2] *Neotoma c. columbiana* Elliot does not differ cranially from *N. cinerea*, and therefore need not be considered in this connection.

[3] No. 75907, U. S. Nat. Mus., from Jasper House, Alberta.

talis, and *N. drummondi* chiefly in its long attenuate nasals, open sphenopalatine vacuities, and pale incisors. The only specimens secured were caught in a slide of large granite bowlders at the head of Lake Bennett. It was ascertained to occur, however, from White Pass to the Semenow Hills. In the cliffs above Glacier on the coast side of White Pass I found signs of *Neotoma*, and once one peeped out of a crevice at me while I was busily engaged stalking a hoary marmot. It also occurs sparingly in the cliffs about Lake Lebarge and in the Semenow Hills, where the last evidences of its presence were seen. This distribution makes it the northernmost species of the genus.

One night about 10.30, as I was returning to camp at Bennett, I saw one of these rats frisking about in the rocks. It was still quite light, and I immediately stopped and stood motionless while he darted in and out of the rocks. His movements were utterly noiseless and so quick that my eye could scarcely follow them. For some time his little whiskered nose appeared and disappeared at various openings in the rocks about ten feet away. Each time he would look steadily at me for a moment or two and then silently vanish. Gradually his curiosity overcame his caution, and in decreasing circles he came nearer and nearer until he bobbed out right before me and then cautiously approached until he could sniff at the toe of my shoe. A slight grating of my gun barrel against a rock caused him to vanish like a flash, and this time he did not reappear.

Evotomys dawsoni Merriam. Dawson Red-backed Mouse.

Red-backed mice are by far the most abundant mammals in the Yukon region. Although but one specimen was taken at Bennett, and none between there and Fifty-Mile River, in spite of considerable trapping, aside from this they were found all along our route from Skagway to Fort Yukon. The following are the most important localities at which specimens were secured: Skagway, Glacier, Bennett, White Horse Rapids, Lake Lebarge, Rink Rapids, Fort Selkirk, Dawson, Charlie Creek, and Circle. From a study of this series, which numbers over 100 specimens, it appears that all belong to one species, *E. dawsoni*. Its range probably reaches northward almost if not quite to the limit of trees.

Specimens were trapped in all sorts of localities; along cold streams, under logs, in heavy moss, in *Microtus* runways, and among rocks. They abound on the large islands, where they were generally caught in dry, brushy places, in the dead leaves which cover the ground. We occasionally saw them during the day, and often heard them rustling the dead leaves on the ground about us as we lay in our blankets at night. They are the vermin of the miner's larder, and are always to be found about log cabins.

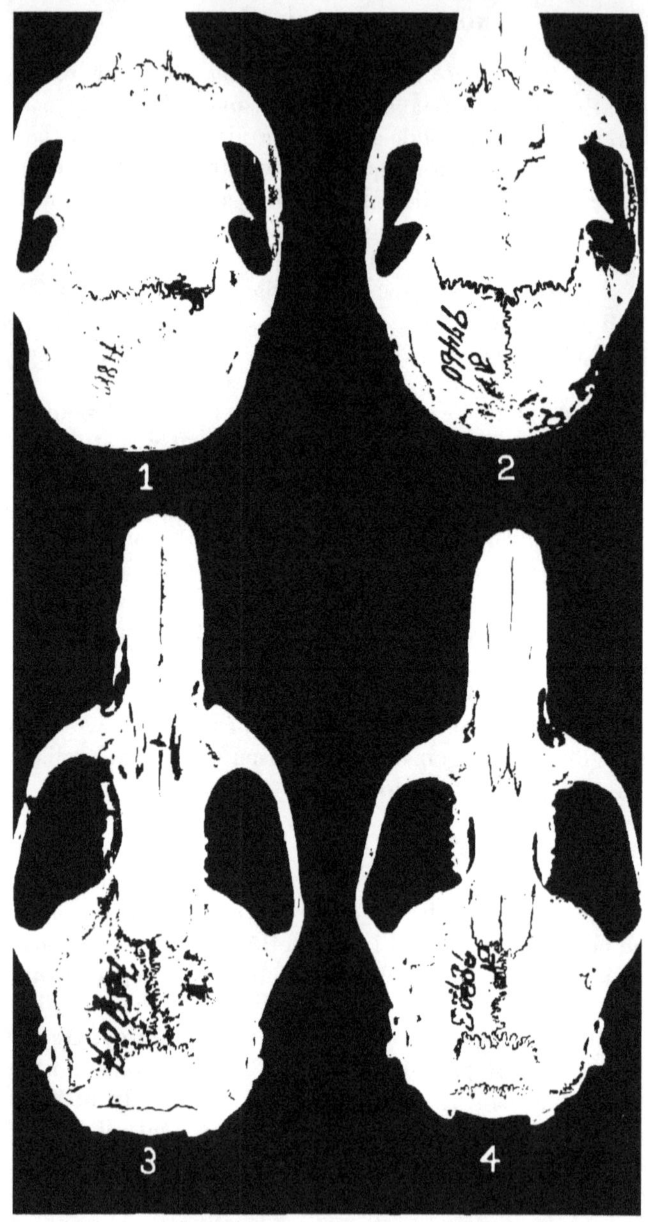

SKULLS OF SCIURUS AND NEOTOMA. (1½)

1. *Sciurus vancouverensis.*
2. *Sciurus hudsonicus petulans.*
3. *Neotoma cinerea drummondi.*
4. *Neotoma saxamans.*

Evotomys dawsoni alascensis (Miller). Tundra Red-backed Mouse.

Evotomys alascensis Miller, Proc. Acad. Nat. Sci. Phila., 1898, 364–367.

The *Evotomys* found at St. Michael has heretofore been compared only with the Asiatic *E. rutilus*. Its closest relationship is really with *E. dawsoni*, with which its range is doubtless continuous. On comparing the series secured at St. Michael with those in the same condition of pelage from Rink Rapids,[1] Northwest Territory, I am unable to find even the slightest difference in color or size. The skull of *alascensis* is slightly characterized by small, narrow molar series, and by nasals which have their posterior end truncate. The palate and audital bullae are not peculiar. The tail is often thick and bristly in winter pelage and in immature specimens of both *dawsoni* and *alascensis*. From this it appears that *alascensis* may be only a slightly marked subspecies.

The favorite habitat of these mice about St. Michael is in the heaps of broken lava scattered about over the tundra. They are very rarely taken in the *Microtus* runways. They are common in the warehouses, which they seem to enter more readily than other mice of the tundra.

Microtus mordax (Merriam). Long-tailed Vole.

Specimens of this vole were taken at Skagway, Glacier, Bennett, Lake Marsh, Lake Lebarge, Rink Rapids, and near Charlie Village. Specimens from near the coast are almost exactly like those of the interior and all are quite typical of the species. They were found in various environments, but the general habitat of the species was dry places rather than moist. At Glacier and Bennett they were secured on dry, rocky hillsides; at Lake Marsh two specimens were taken in the crevises of some granite rocks; at Lake Lebarge they were taken in the kitchen of a log cabin; at Rink Rapids, in an open, sandy place; and near Charlie Village, on the side of a cut bank, where they had made burrows and runways among the exposed roots of trees. Charlie Village is by far the northernmost locality from which the species has been recorded.

Microtus drummondi (Aud. and Bach.). Drummond Vole.

This is the most common meadow vole of the Yukon region. At Caribou Crossing and Lake Marsh its runways form interminable labyrinths in the level, open stretches of sedge at the margin of the water. It occurs in nearly all moist, grassy places from Caribou Crossing to Fort Yukon. From there it undoubtedly ranges farther on, at least to Nulato, where Dall took several specimens. It is most active during the day, as I easily learned by visiting traps night and

---

[1] No good series of specimens is available from any point nearer Finlayson River, the type locality of *E. dawsoni*, than Rink Rapids. This series is therefore used to represent the species.

morning. Near Fort Yukon I found its runways on recently deposited silt sparsely grown up to *Equisetum*. Its burrows in this soft material were very numerous, and at the entrance to each a little heap of earth in small globular lumps, as if carried in the mouth, was always to be seen.

Microtus xanthognathus (Leach). Yellow-cheeked Vole.

This fine species was met with only once. A small colony was found on a little stream near Charlie Village, occupying an old log jam, part of which had become embedded in a matrix of sand and mud and overgrown with weeds. Burrows perforated this structure in numerous places, and well-beaten, open runways connected various openings about the protruding logs. The little animals were quite active during the daytime, and as I walked over the logs I occasionally saw one flash from one opening under a log to another and heard sharp little squeaks sounding all about beneath me. A liberal number of traps placed about yielded nine specimens, chiefly immature. The colony was apparently confined to the log jam, as traps set in suitable places but a few yards away secured only *M. drummondi*. Four specimens of this vole collected by Robert Kennicott are in the National Museum, one from the mouth of the Porcupine and three from the Yukon, 200 miles southwest of that point.

Microtus operarius (Nelson). Nelson Vole.

This vole was taken on a small stream about 40 miles above Circle, and a few more were secured between that point and Fort Yukon. It doubtless ranges from there to the coast. Forty-nine specimens were taken at St. Michael. These represent all stages of growth and several phases of color and seem to offer pretty conclusive proof that but one species of *Microtus* occurs at St. Michael. It was found in all moist parts of the tundra, being particularly numerous along the banks of the small ponds in the tall grass and rank, weedy growths.

Fiber spatulatus[1] sp. nov. Northwest Muskrat.

Type from Lake Marsh, Northwest Territory, Canada. No. 98567, U. S. Nat. Mus., Biological Survey Collection, ♀ yg. ad. Collected July 3, 1899, by W. H. Osgood. Original No., 552. (See Plate VI, fig. 4.)

Characters.—Similar in general to *Fiber zibethicus;* size small; color rather dark; skull small; molar teeth very small; nasals short and much expanded anteriorly.

Color.—Similar to *Fiber zibethicus*, but apparently less suffused with fulvous.

Skull.—Similar to that of *Fiber zibethicus* (Plate VI, fig. 3[2]), but smaller; jugals more slender, and but slightly produced dorsally; audital bullæ smaller; molar teeth decidedly smaller; nasals much shortened and

---

[1] *Spatulatus*, spatulate, in allusion to the shape of the nasals.

[2] No. 76259, U. S. Nat. Mus., from Wilmington, Mass.

widely expanded anteriorly, rapidly becoming compressed posteriorly; angular process of mandible short, blunt, and upturned; condyle narrow and somewhat rounded.

Measurements.—Type (from dry skin): Total length 495; tail vertebræ 170; hind foot 73. Skull of type: Basal length 57; zygomatic breadth 38; length of nasals 21; alveolar length of molar series 14.

Remarks.—Specimens of this species from Ugashik, Fort Kenai, Nushagak, and Nulato, in Alaska, have been examined. Besides these, I find two specimens from Alberta which seem to be referable to it, one from South Edmonton and one from Henry House. These all agree in having very small molar teeth and short, spatulate nasals, characters which are amply sufficient to distinguish the species from all other forms in the genus. The specimens secured by Nelson at St. Michael can not now be found, but they doubtless show the same characters. *Fiber osoyoosensis* has larger teeth and a much longer rostrum than *spatulatus*, so need not be further compared with it. From these facts it appears that *Fiber spatulatus* is the form occupying all of northwest America, and is derived from a form east of the Rocky Mountains rather than from a western one.

Muskrats occur sparingly all along the Yukon, where they find particularly favorable environment about the many small swamp-invested ponds a short distance from the river banks. At St. Michael a few are still found about the open ponds on the tundra.

Synaptomys dalli Merriam. Dall Lemming Mouse.

Lemming mice were taken at the foot of Lake Lebarge, at Rink Rapids, and near the mouth of the Chandindu River. At Lake Lebarge they were found in the long grass at the edge of a small pond; at other localities in cold, boggy places near small streams. The external characters of *S. dalli* have been unknown up to the present time, but, as was to be expected, they are in accordance with the general type so uniform throughout this genus. The color of the upperparts is chiefly raw umber mixed with black; the lower parts are uniform bluish white, and the feet and tail are dusky. The ears are of medium size and partially hidden by long hairs growing from the anterior base; a conspicuous bluish-white side gland is present in the males. The skull of the type of *dalli* is not fully mature and does not agree in all particulars with my specimens from the Upper Yukon. In these the skull is somewhat larger and heavier and the nasals are a trifle longer and more noticeably constricted posteriorly.

Lemmus yukonensis Merriam. Yukon Lemming.

Lemmus yukonensis Merriam, Proc. Wash. Acad. Sci., II, 27, March 14, 1900.

This lemming was found at only two localities—Rink Rapids, where five specimens were secured, and Charlie Creek, where five more were taken. Considerable careful trapping was done at various points between these two places, but no other specimens were secured. At Rink Rapids they were caught about old logs and among dry leaves in places usually frequented by red-backed mice. At Charlie Creek one was caught in a *Microtus* runway and several were secured on the side of a cut bank. On one occasion one was seen running about under a brush heap in midday.

Lemmus alascensis Merriam. Alaska Lemming.

Lemmus alascensis Merriam, Proc. Wash. Acad. Sci., II, 26–27, March 14, 1900.

All efforts to secure this species at St. Michael proved fruitless. I kept large numbers of traps out for more than two weeks and set them in all conceivable locations about the tundra, but failed to catch any lemmings.

Dicrostonyx nelsoni Merriam. Nelson Pied Lemming.

Dicrostonyx nelsoni Merriam, Proc. Wash. Acad. Sci., II, 25–26, March 14, 1900.
Dicrostonyx hudsonius alascensis Stone, Proc. Acad. Nat. Sci. Phila., March 24, 1900, 37–38.

No specimens of this species were taken. Nelson says of it:

Specimens were brought me by the fur traders from above Fort Yukon and from Nulato, Anvik, and Kotlik, along the course of the Yukon, and also from the Kaviak Peninsula and about Kotzebue Sound. A few were taken near St. Michael, but they were not numerous there. They are more plentiful about Bering Straits than any other district visited by me, if the number of their skins among the native children can be taken as a guide.

Zapus hudsonius alascensis Merriam. Alaska Jumping Mouse.

Three typical specimens of this jumping mouse were taken in a sedgy swamp near the foot of Lake Lebarge. Similar swamps exist near the Yukon, at least as far as Fort Yukon, but I was unable to do any trapping in them. No specimens were taken elsewhere.

Erethizon epixanthus myops Merriam. Alaska Porcupine.

Erethizon epixanthus myops Merriam, Proc. Wash. Acad. Sci., II, 27–28, March 14, 1900.

Porcupines are quite common in all the forest region of Alaska. I noticed signs of them at many places along the Yukon. They were abundant about Glacier, in the White Pass region, and I shot one there one evening as it swayed back and forth in the top of a slender alder. It was eating the leaf buds which were just bursting.

Ochotona collaris (Nelson). Alaska Pika.

Two specimens of an ashy gray *Ochotona* were taken, one at the summit of White Pass, another at the head of Lake Bennett. The species was apparently quite rare at these localities and it was only with considerable difficulty that these individuals were secured. Both are very pale, ashy gray, with pure white underparts, no traces of fulvous, and very indistinct collars. They are in the early spring or left-over winter pelage, and agree quite well with specimens in the same pelage collected in the Chigmit Mountains, near Bristol Bay, by C. L. McKay. The type and topotypes of *O. collaris* are in the summer or post-breeding pelage and present quite a different appearance.

The species apparently occurs in the high mountains throughout Alaska. It was reported to me from the MacMillan Mountains, the Upper Stewart River, the Upper White, and the Upper Tanana. Fragments of a skull were found in an owl pellet picked up by Dr. Bishop near Windy Arm, Lake Tagish. The present record from White Pass is the most southern one. There is suitable country for it farther south, and it will be interesting to trace its range in this direction.

Lepus saliens sp. nov.

Type from Caribou Crossing, between Lake Bennett and Lake Tagish, Northwest Territory, Canada. No. 98956, U. S. Nat. Mus., Biological Survey Collection, ♂ ad. Collected June 26, 1899, by W. H. Osgood. Original No., 504.

Characters.—Similar to *Lepus bairdi*, but more yellowish and less ruddy; dorsal hairs with plumbeous roots; feet nearly white in summer; similar to *L. columbiensis*,[1] but with greater amount of black in dorsal region; feet much lighter; skull similar in general to that of *Lepus a. dalli;* audital bullæ very large.

Color.—Type in *worn spring pelage:* Upperparts mixed black and yellowish buff, with patches of plumbeous under-fur exposed in places; black hairs predominating on rump and middle of back, forming an ill-defined dorsal stripe; outer edge of thighs, outer side of forelegs and pectoral band buff; ears and head, except sides of nose, buff with black hairs sprinkled through; sides of nose gray; ears margined with white; hairs of fore and hind feet plumbeous at base, rufous in central part, and broadly white at tips; general appearance of feet white, lightly mixed with rufous; underparts, except pectoral band, white.

Skull.—Similar to that of *dalli* but somewhat larger; teeth heavier; nasals long, heavy, and very broad anteriorly; audital bullæ very large; palate short; malars rather wide, deeply channeled anteriorly; postorbital and antorbital processes of frontals well developed.

Measurements.—Type (measured from dry skin): Total length 395; hind foot 134; ear from crown 74. Skull of type: Occipitonasal length 77; greatest zygomatic breadth 38; length of nasals 33; greatest width of nasals 17; alveolar length of molar series 15.

Remarks.—The exact relation in which this species stands to *americanus, bairdi,* and *columbiensis* is difficult to determine at present. Its light feet point to relationship with *bairdi*, while its dark under color

---

[1] Rhoads, Proc. Acad. Nat. Sci. Phila., 1895, 242–243.

and general buffy appearance are more like *columbiensis*. Its skull is quite distinctive, the large audital bullæ and broad nasals being unequaled in the group. It seems probable that it is a northern form of *bairdi* not related to *columbiensis*, which is nearer to *washingtoni*. There are no specimens available to show whether or not it has any connection with *dalli*, which is the form found on the Lower Yukon. But two specimens were secured—the type, which I shot in a *Lepargyræa* thicket at Caribou Crossing, and one very young female which Dr. Bishop took in a willow bog near Bennett City. It seems to have been a decidedly 'off year' for rabbits, for these two were the only ones we saw on our entire trip, though numerous signs of their former abundance were seen daily.

Lepus americanus dalli Merriam. Dall Varying Hare.

Lepus americanus dalli Merriam, Proc. Wash. Acad. Sci., II, 29–30, March 14, 1900.

This rabbit is doubtless abundant at certain times all along the Lower Yukon, but we heard very little of it. It is subject to epidemics and frequently becomes locally extinct, which probably accounts for its scarcity last year.

Lepus othus Merriam. Alaska Arctic Hare.

Lepus othus Merriam, Proc. Wash. Acad. Sci., II, 28, March 14, 1900.

Signs of Arctic hares were occasionally noticed about St. Michael, but we did not see any of the animals. The Eskimos were hunting continually, and brought numbers of ducks and geese to the village to sell, but they brought no rabbits during our stay.

Lynx canadensis mollipilosus Stone. Arctic Lynx.

Lynx canadensis mollipilosus Stone, Proc. Acad. Nat. Sci. Phila., March 24, 1900, 48–49.

The Canada lynx is not as common in the interior of Alaska as might be expected. I saw no signs of it and could obtain only very scanty information as to its occurrence. The police sergeant in charge of the station at the foot of Lake Lebarge told me that the tracks of but one had been seen in that vicinity during the previous winter. Lynx-skin robes are in common use in the country, but the majority of them are imported. This I learned from a trader at Circle, who had several for sale that came from eastern and southern Canada.

Lynx skulls from the following localities are in the National Museum: Tanana River, Russian Mission, Nulato, Andraefski, and mountains near Unalakleet.

Canis occidentalis Richardson. Wolf.

The country along the Yukon is not well suited for wolves, and they are seldom seen there. A prospector showed me the skin of a large gray one from the upper waters of the MacMillan river—the only one I saw on the trip.

Vulpes fulvus (Desmarest)? Red Fox.

Occasional reports of foxes were received all along our route, but no specimens were secured. Owing to their natural sagacity, foxes are doubtless able to hold their own against trappers better than most other fur-bearing animals. Their skins are quite common among traders and natives.

Vulpes hallensis Merriam. Hall Island Fox.

Vulpes hallensis Merriam, Proc. Wash. Acad. Sci., II, 15–16, March 14, 1900.

White fox skins are common among the natives and traders at St. Michael, and could be bought at from $1 to $4 each, according to quality. During our stay there one of the animals was seen on the island, which indicates that they are still far from extermination.

Ursus americanus Pallas. Black Bear.

Black and brown bears are common all along the Yukon. We found them common on the upper river, and Nelson records them as far down as Anvik. We saw tracks very frequently, but owing to the thick forest and underbrush, and the fact that we made no special hunts for them, the animals themselves were rarely observed. A young adult female in glossy black pelage was killed at Glacier by A. G. Maddren, and several others were seen during our stay there. I was told at Lake Lebarge and at White Horse Rapids that brown bears were seen very frequently. At Fort Selkirk I saw skins brought from the Pelly River. Near Charlie Village I saw the skin of a large brown bear that had been killed there shortly before our arrival. One afternoon while sitting in the boat preparing specimens, about 20 miles above Circle, I saw a good-sized black bear walking deliberately across an open space on a hillside a short distance away. We gave chase, but did not see it again. At the mouth of the Tatondu River I saw numerous tracks, and on the border of a stagnant pool found evidences that bruin had been enjoying a mud bath. Moss uprooted by bears in digging for roots was noticed at several places.

Ursus horribilis alascensis Merriam. Alaska Grizzly Bear.

Very little accurate information is obtainable in regard to the grizzly in the Yukon region. It doubtless occurs sparingly all along the river, but miners and prospectors report any large bear as a grizzly, and without doubt often mistake the brown bear for it. There are a number of its skulls from Norton Sound in the Biological Survey collection.

Lutra canadensis (Schreber). American Otter.

The fate of the otter in Alaska is much the same as that of the beaver. There are doubtless a few on some of the smaller streams of the interior and about the Yukon delta, but they are now quite rare in comparison with their former abundance.

Lutreola vison ingens subsp. nov. Alaska Mink.

Type (skull) from Fort Yukon, Alaska. No. 6530, U. S. Nat. Mus., ♂ ad., old. Collected by Robert Kennicott. (See Plate VI, fig. 2.)

Characters.—Size largest of North American mink; similar to *L. v. energumenos*, but lighter in color and very much larger; skull and teeth very large and heavy.

Color.—Similar in general to *Lutreola v. energumenos*, but paler.

Skull.—Very large, angular, and ridged; rostrum very wide; braincase relatively shallow and very wide; zygomata heavy; audital bullæ large and relatively wide; dentition heavy. (Compare with skull of *Lutreola v. energumenos*, Plate VI, fig. 1.[1])

Measurements.—No. 13880, U. S. National Museum, St. Michael, Alaska (from dry skin): Total length 720; tail vertebræ 180; hind foot 75. Skull of type: Occipitonasal length 69; zygomatic breadth 47; mastoid breadth 41; breadth across postorbital processes 23; length of audital bulla 17. Average of five adults: Occipitonasal length 44.5; mastoid breadth 39.5; breadth across postorbital processes 21; length of audital bulla 17.5.

Remarks.—The large size of the Alaska mink has been noted by various authors,[2] but each has dismissed the subject by concluding that it is the natural result of the animal's northern range, and the form has remained unnamed, while less marked forms from other localities have been recognized. The largest mink previously described is *L. v. energumenos*, which is very much smaller than *ingens* and also averages much darker.

The minks of the Yukon region are caught mostly on the tributary streams, and, as stated by Nelson, are very abundant in the area between the deltas of the Yukon and the Kuskokwim. Along the Yukon itself our party did not see any, and very few signs of them were observed. Their skins were seldom seen among the Indians and Eskimos. They were reported, however, from the Porcupine, Koyukuk, Tanana, and various other streams tributary to the Yukon, and without doubt occur in suitable places all over Alaska.

Putorius arcticus Merriam. Tundra Weasel.

Putorius arcticus Merriam, N. Am. Fauna No. 11, 15, June, 1896.
Putorius cicognani richardsoni Merriam, *l. c.*, 11–12 (part).

Three immature specimens of this weasel were taken at St. Michael. They were caught in traps baited with sandpipers and set among the lava rocks along the shore. Several specimens which were also secured at St. Michael by Nelson and Turner are in the National Museum. Besides these I find specimens from Nulato, Fort Yukon, and Fort Reliance, which gives the species a more extensive range in the interior than it has been supposed to have. Most of these specimens are

---

[1] No. 5537, Bangs collection, from Sumas, B. C.

[2] See Allen, Bull. U. S. Geol. and Geog. Survey Terr., II, 327–328, 1876.

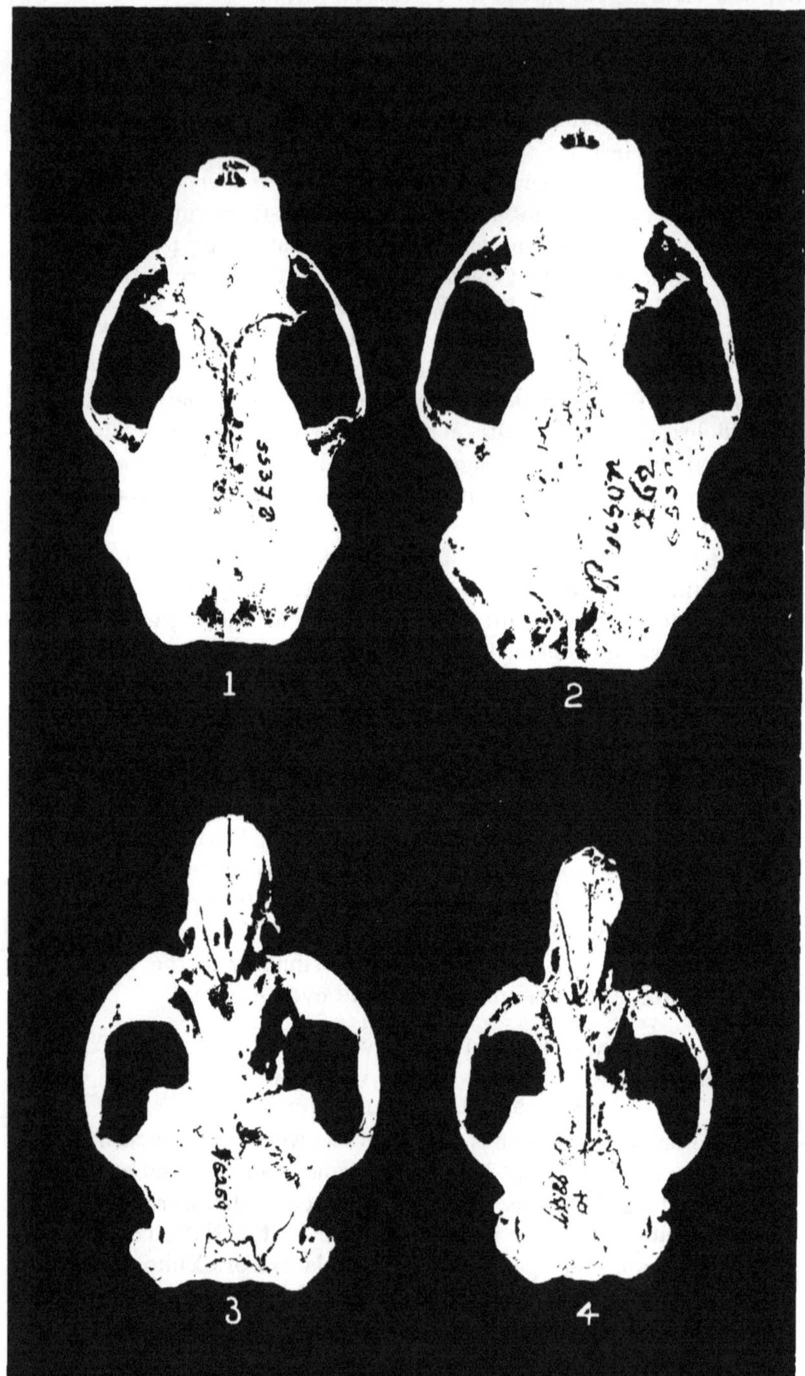

SKULLS OF LUTREOLA AND FIBER. (Natural size.)

1. [illegible]
2. [illegible]

imperfect, but enough skulls are now at hand to show conclusively that all the Yukon specimens heretofore identified as *richardsoni* are really practically identical with *P. arcticus* from Point Barrow.

Putorius cicognani alascensis (Merriam). Juneau Weasel.

A single immature specimen taken 20 miles below Fort Selkirk is referred to this form. Its skull is rather large and indicates a possible intergradation with *P. arcticus;* otherwise it agrees with *alascensis.*

Putorius rixosus eskimo Stone. Alaska Least Weasel.

Putorius rixosus eskimo Stone, Proc. Acad. Nat. Sci. Phila., March 24, 1900, 44–45.

No specimens of this rare weasel were obtained. There are three imperfect specimens in the National Museum, two from St. Michael and one from Fort Reliance. Besides these the only ones recorded are the type and four topotypes from Point Barrow, Alaska, and the specimen mentioned by Stone (loc. cit.) from Bethel, Kuskokwim River, Alaska.

Mustela americana actuosa subsp. nov. Alaska Marten.

Type (skull) from Fort Yukon, Alaska. No. 6043, U. S. Nat. Mus., ♂ ad., old. Collected by Robert Kennicott. (See Plate VII, fig. 2.)

Characters.—Similar to *M. brumalis*, but larger; cranial and dental characters distinctive.

Color.—(Topotype, No. 6416, U.S.N.M., ♂ ad.): Posterior half of upperparts pale ochraceous buff, shoulders and anterior part of upperparts gradually becoming grayish; entire upperparts, except head, overlaid with coarse brown hairs; head, including cheeks and throat, pale grayish-white lightly mixed with brown, especially on nose and chin; inside and edges of ears whitish, outside and bases of ears brown; underparts similar to upperparts, but darker and more brownish on chest; an irregular patch of creamy buff mixed with white on chest; legs and feet dark brown, front of legs with mixture of gray hairs; tail brown, somewhat darker at tip, and with a slight mixture of gray hairs.

Skull.—Similar to that of *M. brumalis* (Plate VII, fig. 1[1]), but somewhat larger; relatively longer and narrower; interorbital space slightly narrower; audital bullæ very much larger and longer; dentition relatively much weaker; last upper molar decidedly smaller.

Measurements.—Average of four adult male topotypes measured in the flesh by the collector: Total length 26.22 inches (665 mm.); tail vertebræ 8.08 inches (223 mm.); hind foot 4.36 inches (109 mm.). Skull of type: Occipitonasal length 85; greatest zygomatic breadth 55; breadth across postorbital processes 24; palatal length 44; length of audital bullæ 19.

[1] Type No. 7417, Bangs collection, from Okak, Labrador.

Remarks.—This form is the largest of the subspecies of *Mustela americana*. *M. brumalis* is also large, but does not equal *actuosa*, and notwithstanding its smaller size has heavier dentition. The enormous audital bullæ of *actuosa* are not equaled by those of any other member of the group. The skulls of *americana* (Plate VII, fig. 3[1]) and *caurina* are so very much smaller than those of *brumalis* and *actuosa* that they do not need to be closely compared. In a good series of *actuosa* from Fort Yukon and Fort McPherson the characters are very constant. A large number of skins from these localities present very little variation, and nearly all are quite light colored like the one described above. The marten is still the commonest fur-bearing animal of Alaska, notwithstanding the hundreds of thousands that have already been taken. Trappers are always confident of a harvest of martens whether other animals are abundant or not.

Mustela pennanti Erxleben. Fisher.

Dr. Elliott Coues[2] states that he has examined specimens of the fisher from Alaska, but does not give the exact locality. At present no specimens are at hand to corroborate this record, but there is little doubt that the animal occurs along the Upper Yukon, as it is known from similar latitudes to the eastward. It was not met with by our party, and I received no reliable information in regard to it.

Gulo luscus (Linnæus). Wolverine.

Wolverines seem to be quite common in the Yukon region. They were often reported, and I saw a number of skins among the natives on the lower river. One was said to have been trapped at Tagish in the winter of 1898, and others were seen in the vicinity. They are seen frequently about Lake Lebarge in winter, and trappers from the MacMillan River say they are abundant in that region.

Sorex personatus streatori Merriam. Streator Shrew.

Specimens of this shrew were secured as follows: Haines 1, Skagway 6, Glacier 1, Bennett 3, Caribou Crossing 2, Lake Lebarge 1, 50 miles below Fort Selkirk 1, mouth of Chandindu River 1, and 40 miles above Circle 1. Although the conditions along the Yukon seem to be ideal for shrews, I was unable to secure many specimens, and could only conclude that they were not common there, for the same methods of trapping were much more successful in the coast regions.

Sorex personatus arcticus Merriam. Arctic Shrew.

Sorex personatus arcticus Merriam, Proc. Wash. Acad. Sci., II, 17, Mar. 14, 1900.

Twenty specimens were taken at St. Michael. They occur throughout the tundra in much the same situations as *S. tundrensis*, but were also found in the lava heaps and along high banks near the coast.

[1] No. 4934, Merriam collection, from the Adirondacks, New York.

[2] Fur-bearing Animals, 69, 1877.

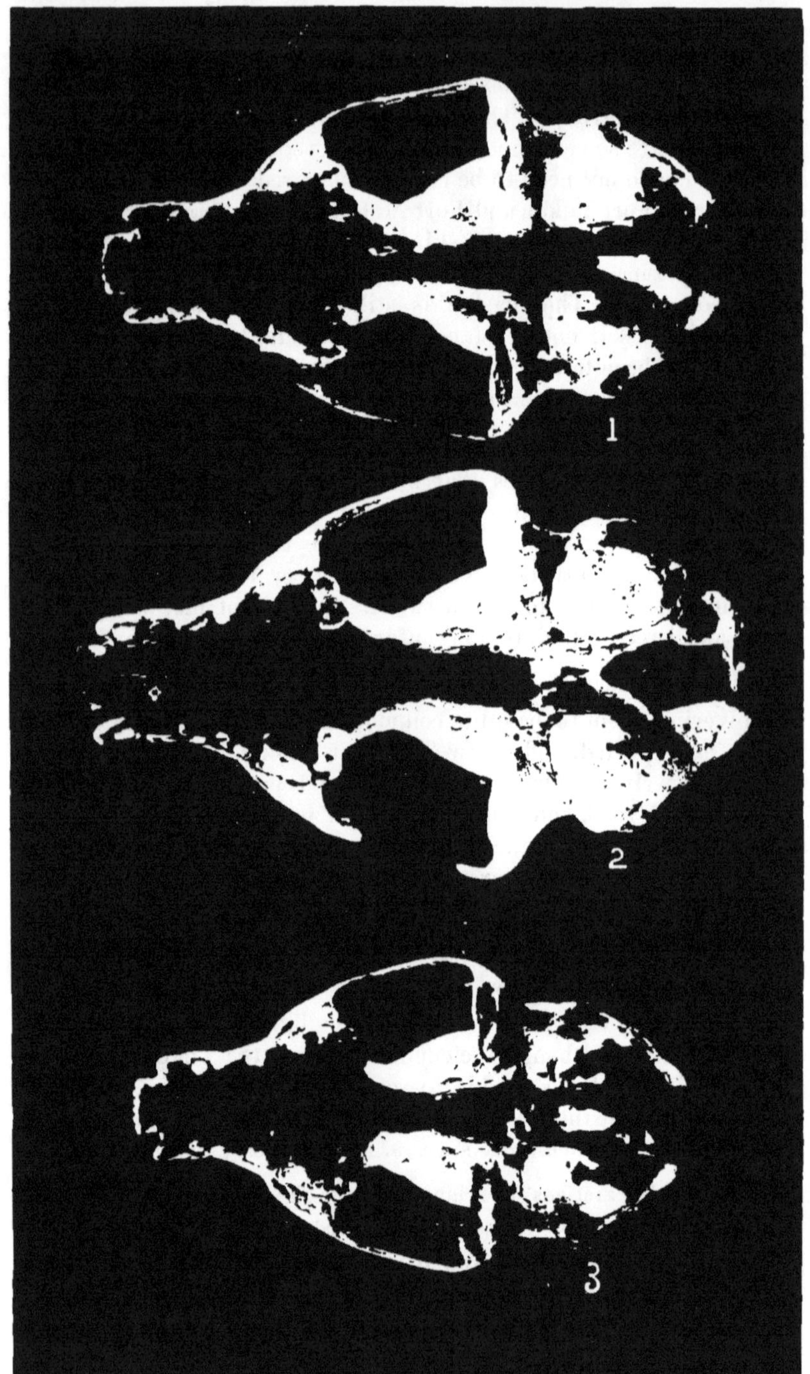

SKULLS OF MUSTELA. Natural size.

1. *Mustela americana brumalis.* 2. *Mustela americana actuosa.* *Mustela americana.*

Sorex obscurus Merriam. Mountain Shrew.

Two specimens were caught under tufts of grass on a rocky hillside at Bennett. This locality is much farther north than any from which this species has been previously recorded.

Sorex tundrensis Merriam. Tundra Shrew.

Sorex tundrensis Merriam. Proc. Wash. Acad. Sci., II, 16-17, March 14, 1900.

Eighteen specimens of this pretty shrew were taken at St. Michael. They were found in various parts of the tundra, but seemed to be in small localized colonies. About certain small ponds nearly all the shrews caught were of this species, while but a short distance away all were *arcticus*. A single imperfect specimen collected by Kennicott near Fort Yukon is in the National Museum. In size it does not differ from typical *tundrensis*, but in color it is somewhat darker, thus indicating a possible intergradation with *richardsoni*.

Myotis lucifugus (Le Conte). Little Brown Bat.

Bats were first seen at Caribou Crossing, and from that point were occasionally noticed at various places to our camp, 50 miles below Fort Selkirk, where they were last seen. Turner mentions their reported occurrence as far down as Fort Yukon and Nulato. In June and July we generally found them flying from 10 to 11.30 p. m., and sometimes even later. Two specimens only were secured. These are somewhat grayer and less glossy than specimens from the eastern United States.

BIRDS OF THE YUKON REGION, WITH NOTES ON OTHER SPECIES.

By Louis B. Bishop, M. D.

INTRODUCTION.

In preparing the ornithological part of this report I have thought it advisable to note as far as possible all species met with from the time we passed Dixon Entrance, northward bound, May 28, until we reached Cape Scott on the return trip, October 12, for the reason that articles on Alaska birds are not yet so numerous as to make such notes worthless. It was of course impossible to obtain specimens of water-fowl seen from the decks of steamers; therefore when specific identification was not positive I have referred genera seen to the species which previous observers—especially E. W. Nelson and William Palmer—have found most common in the waters visited.

Nowhere did we see the vast colonies of water birds which others have met with in Alaskan waters, probably because most of these birds had left their summer homes in Bering Sea when we passed in October; but various migrants were common in the Inside Passage in May, geese and ducks on the Lower Yukon in August, and waterfowl of many species in Akutan Pass in October.

The region from Skagway, at the head of Lynn Canal, to Circle, on the Yukon, was the scene of most of our work; and as very little was known of it ornithologically I have mentioned in my annotated list every occasion of our observation of all except the commonest species. Ornithologists, in referring to the Upper Yukon, include, as a rule, only that part of the river which lies between Dawson and Nulato; hence the avifauna of its head waters was with us largely a matter of conjecture. George G. Cantwell[1] mentions species he saw about the lakes; but his experience was in many ways so different from ours that, while crediting him with the first records for species which we also found, I have omitted others which we did not find and for which he may have mistaken closely allied birds.

The country we traversed between Skagway and Circle divides itself into three quite distinct faunal districts. The coast of Southeast Alaska belongs to the 'Sitkan district' of Nelson, White Pass Summit

[1] Birds of the Yukon Trail <Osprey, III, 25, Oct., 1898.

and the heights above Glacier belong to the Arctic-Alpine zone, and the Yukon Valley belongs to the Canadian and Hudsonian zones. In the last the Canadian element is most pronounced in the lake region, with a very slight infusion of Sitkan forms, the strictly Hudsonian species increasing and the others decreasing as the Yukon winds north toward Fort Yukon. Beyond this point Hudsonian forms predominate, giving place to Arctic where the Yukon loses its identity in the tundra of the delta. The Upper Yukon Valley may be divided faunally at Fort Selkirk, where the Pelly from the Rocky Mountains and the Lewes from the Coast Range unite to form the Yukon proper, 15 species of land and shore birds occurring above this point which have not been found between there and Fort Yukon, and 12 having been recorded between the Pelly and Fort Yukon which have not been taken above. Of the 128 species and subspecies found between Dixon Entrance and Fort Yukon, 22 per cent were common to the coast of southeast Alaska and the Yukon Valley, 19 per cent confined to the coast, 55 per cent to the Yukon Valley, and 4 per cent found only on White Pass Summit and at similar altitudes.

The avifauna of southeastern Alaska is already fairly well known, and the twelve days spent at Haines, Skagway, and Glacier resulted chiefly in extending the ranges of a few species, though the barn swallow proved to be the subspecies recently reinstated by Mr. Palmer, the myrtle warbler that lately described by Mr. McGregor, and the wood pewee an unrecognized form. Of the 52 species found between Dixon Entrance and Glacier, 2—*Colaptes auratus luteus* and *Merula migratoria*—were eastern, 8 Alaskan, 25 Pacific coast, and 17 common to northern North America. At Haines, which is situated on a narrow and for the most part heavily wooded peninsula, birds, although not common, were more numerous than they were either at Skagway, which is in a narrow cliff-bordered valley at the head of Lynn Canal, or at Glacier, 14 miles from Skagway, 1,870 feet higher, and surrounded by deep spruce woods and alder thickets. We found in the avifauna of Glacier a slight but decided difference from that of the tide-water level of Lynn Canal, *Junco hyemalis connectens* replacing *J. h. oregonus*, and *Wilsonia pusilla pileolata* replacing *Helminthophila celata lutescens*, while *Melospiza melodia rufina* and *Merula migratoria* were absent.

Among the thickets of alpine hemlock growing with moss and heather between the granite rocks of White Pass Summit and the heights above Glacier we found *Zonotrichia coronata* and *Anthus pensilvanicus* common, and *Lagopus rupestris*, *L. leucurus*, *Leucosticte tephrocotis littoralis* and *Sayornis saya yukonensis* in smaller numbers. *Sayornis s. yukonensis* reached the Yukon level at Fort Selkirk, and *Anthus pensilvanicus* at Circle, but the others were not seen again.

To one accustomed to the orchards, fields, and forests of Connecticut, the duck marshes of North Dakota, or even the balsam thickets of northern New England, the Yukon Valley seems wanting in bird life—not the center of abundance of its avifauna, but rather a deposit for the overflow from more favored regions. There are exceptions to this rule, notably wandering flocks of crossbills, the colonies of bank swallows of Fifty-Mile and Thirty-Mile rivers and the Yukon proper, the spotted sandpipers that continually flitted across our bow, the intermediate sparrows and juncos that seldom failed to greet us as we stepped ashore, and the Alma thrushes, whose songs sounded all night, wherever we happened to camp. Bird life is fairly abundant, too, in certain favored places such as Log Cabin, Caribou Crossing, the swampy shores of Lake Marsh, and the ponds and level country at the lower end of Lake Lebarge. Near Miles Canyon I noticed 23 species on July 11, but individuals of each, with the exception of bank swallows, were few. In the entire Upper Yukon Valley breeding colonies of shore and water birds were conspicuously absent. The precipitous shores of the lakes, the comparative absence of islands, the swift current of the Yukon, and its high banks cut by narrow, wooded valleys, are a sufficient explanation of this; and I can not believe that either geese, ducks, or shore birds ever bred abundantly in most of the region visited, though their number has doubtless been reduced in recent years.

In the Yukon flats the condition changes, and no doubt many of these birds find a summer home in the ponds a few miles back from the river, as they do at the foot of Lake Lebarge; but these we had no opportunity of visiting. Our study of the bird life of the Yukon was chiefly confined to what could be seen or heard from our boat or on the banks in the immediate vicinity of camping places. From the lakes to the Alaska boundary snow-capped peaks were absent, and no species were found that did not also occur upon the banks of the river, although we climbed hills, visited deep woods, and ascended small streams for some distance. As we proceeded north, however, several birds were found at lower altitudes than those at which they had been already noted. Away from the river, birds were rarer than immediately upon its banks.

We learned little regarding the Upper Yukon as a migratory highway for species breeding farther north, though we heard that thousands of geese and ducks passed Lower Lebarge in the spring. It was too late for the spring migration, and the southward movement of ducks and geese had hardly begun on August 20, when we left Circle. The fall migration of the Limicolæ should have been well under way at this date, but very few of these birds were observed. If they do pass in large numbers they must frequent the ponds back from the river. Several times at Circle, I walked a long distance over the sand flats left bare by the falling Yukon without seeing any shore birds, or anything on which they could feed. This was very different from the constantly passing flocks I saw on the Yukon Delta August 27–28, and the abundance of Limicolæ at St. Michael in September. The smaller land birds we often saw late in July and in August. They were usually in family parties, and most of them seemed to be traveling up the river. At Circle the intermediate sparrow, western tree sparrow, and western savanna sparrow were abundant, and were evidently migrating August 19–20.

Forty-two species of migratory birds, exclusive of those possessing a continental range, certainly occur as summer residents in the Yukon Basin above Fort Yukon. Of these, 13 (31 per cent) have their center of distribution in eastern North America, 14 (33 per cent) near the Pacific coast, and 15 (36 per cent) in western North America not far from the Rocky Mountains. The eastern birds reach the Yukon through the Rocky Mountains. Some of these, such as *Chordeiles virginianus*, were found only above the Tatchun River; others, as *Empidonax t. alnorum*, were absent above the Pelly and common from there to Fort Yukon; others, as *Wilsonia pusilla*, were not found above the Chandindu River; others, as *Helminthophila peregrina* were each found at a single place, while still others, as *Junco hyemalis* and *Merula migratoria*, were regularly distributed along the river. The Pacific coast forms probably all reach the Yukon over the Alaska coast range. These disappear as one goes north, *Hylocichla aonalaschkæ* extending through Lake Bennett, *Wilsonia p. pileolata* to Lake Marsh, *Dendroica townsendi* to Lake Lebarge, *Myadestes townsendi* to Dawson, and *Tachycineta thalassina* to Circle. Last and most important in number of species, abundance of individuals, and regularity of distribution are birds which breed in the Yukon Valley and spend the winter in the western United States, as *Zonotrichia l. gambeli*, *Spizella s. arizonæ*, and the small *Ammodramus s. alaudinus* of the Yukon lakes, and those which probably enter by the mouth of the Yukon, as the large *Ammodramus s. alaudinus*, found below Alaska boundary, and *Seiurus n. notabilis*, first met near Dawson.

In coloring, Yukon birds, especially in juvenile plumage, show a strong tendency to replace the buff-ochraceous markings of Eastern forms by white, cream color, and gray. *Canachites c. osgoodi*, *Parus h. evura*, and *Hylocichla u. alma* are good examples of this characteristic.

I take this opportunity to express my hearty thanks to Dr. Merriam for the privilege of visiting Alaska as a member of the Biological Survey party, of writing this report, and of using the collection of the Biological Survey in its preparation; also to Mr. Osgood and Mr. Oberholser of the Biological Survey for aid in determining species. I am also greatly indebted to Mr. Robert Ridgway and Dr. Charles W. Richmond for the opportunity of studying the collection of the

United States National Museum and for much valuable assistance; to Dr. J. A. Allen and Mr. F. M. Chapman for the hours which I spent with the birds in the American Museum of Natural History; to Mr. William Brewster for the courtesy of allowing me to compare my specimens with those in his valuable collection, and to Mr. Walter Deane for much help in this study.

CLASSIFIED LISTS OF SPECIES.

NEW SUBSPECIES.

Canachites canadensis osgoodi.
Sayornis saya yukonensis.
Contopus richardsoni saturatus.

SPECIES NOT HITHERTO RECORDED FROM WESTERN NORTH AMERICA.

Haliaeetus albicilla.

SPECIES NOT HITHERTO RECORDED FROM SOUTHEASTERN ALASKA.

Æchmophorus occidentalis.
Xema sabinii.
Lagopus leucurus.
Picoides americanus alascensis.
Contopus richardsoni saturatus.
Empidonax hammondi.
Junco hyemalis connectens.
Sitta canadensis.
Merula migratoria.

SPECIES NOT HITHERTO RECORDED FROM UNALASKA.

Larus philadelphia.
Tringa maculata.
Tringa acuminata.
Loxia curvirostra minor.

SPECIES NOT HITHERTO RECORDED FROM THE PRIBILOF ISLANDS.

Larus philadelphia.
Tringa acuminata.
? Arenaria melanocephala.

SPECIES NOT HITHERTO RECORDED FROM ST. MICHAEL.

Calidris arenaria.

SPECIES NOT HITHERTO RECORDED FROM THE YUKON ABOVE FORT YUKON.

Tringa bairdi.
Symphemia semipalmata inornata.
Buteo borealis calurus.
Falco sparverius.
? Megascops asio kennicotti.
? Dryobates villosus hyloscopus.
Contopus borealis.
Contopus richardsoni saturatus.
Empidonax trailli alnorum.
Empidonax hammondi.
Spinus pinus.
Spizella socialis arizonæ.
Passerella iliaca.
Helminthophila peregrina.
Dendroica townsendi.
Wilsonia pusilla pileolata.
Sitta canadensis.
Hylocichla aonalaschkæ.
Hylocichla aonalaschkæ pallasi.
Saxicola œnanthe.

LIST OF SPECIES KNOWN FROM THE YUKON BASIN.

Colymbus holbœlli.
Colymbus auritus.
Gavia imber.
Gavia arctica.
Gavia lumme.
Stercorarius pomarinus.[1]
Stercorarius parasiticus.[1]
Stercorarius longicaudus.[1]

[1] Known only from Fort Yukon or below.

Rissa tridactyla pollicaris.[1]
Larus barrovianus.[1]
Larus argentatus smithsonianus.
Larus vegæ.[1]
Larus brachyrhynchus.
Larus philadelphia.
Xema sabinii.[1]
Sterna caspia.[1]
Sterna paradisæa.
Sterna aleutica.[1]
Hydrochelidon nigra surinamensis.[1]
Phalacrocorax pelagicus robustus.[1]
Merganser americanus.
Merganser serrator.[1]
Anas boschas.
Mareca americana.
Nettion carolinensis.
Querquedula discors.[1]
Spatula clypeata.
Dafila acuta.
Aythya vallisneria.[1]
Aythya marila.
Aythya affinis.
Clangula clangula americana.
Clangula islandica.
Charitonetta albeola.
Harelda hyemalis.
Histrionicus histrionicus.
Arctonetta fischeri.[1]
Somateria v-nigra.[1]
Somateria spectabilis.[1]
Oidemia americana.[1]
Oidemia deglandi.
Oidemia perspicillata.
Chen hyperborea.[1]
Anser albifrons gambeli.[1]
Branta canadensis hutchinsi.
Branta canadensis minima.
Branta nigricans.[1]
Philacte canagica.[1]
Olor columbianus.[1]
Olor buccinator.[1]
Grus canadensis.
Fulica americana.[1]
Crymophilus fulicarius.[1]
Phalaropus lobatus.
Gallinago delicata.
Macrorhamphus scolopaceus.[1]
Tringa canutus.[1]
Tringa couesi.[1]
Tringa maculata.
Tringa bairdi.
Tringa minutilla.
Tringa alpina pacifica.
Ereunetes occidentalis.[1]
Calidris arenaria.[1]
Limosa lapponica baueri.[1]
Limosa hæmastica.[1]
Totanus flavipes.
Helodromas solitarius cinnamomeus.
Heteractitis incanus.[1]
Bartramia longicauda.[1]
Symphemia semipalmata inornata.[2]
Tryngites subruficollis [1]
Actitis macularia.
Numenius hudsonicus.
Numenius borealis.[1]
Squatarola squatarola.
Charadrius dominicus.
Charadrius dominicus fulvus.[1]
Ægialitis semipalmata.
Arenaria interpres.[1]
Arenaria melanocephala.[1]
Canachites canadensis osgoodi.
Bonasa umbellus umbelloides.
Lagopus lagopus.
Lagopus rupestris.
Pediœcetes phasianellus columbianus.[1]
Circus hudsonius.
Accipiter velox.
Accipiter atricapillus.
Buteo borealis calurus.[2]
Buteo swainsoni.[1]
Archibuteo lagopus.[1]
Haliæetus leucocephalus alascanus.
Falco rusticolus gyrfalco.
Falco peregrinus anatum.
Falco columbarius.
Falco columbarius richardsoni.[2]
Falco sparverius.[2]
Pandion haliaëtus carolinensis.[1]
Asio accipitrinus.[1]
Scotiaptex cinerea.
Scotiaptex cinerea lapponica.[1]
Nyctala tengmalmi richardsoni.
?Megascops asio kennicotti.[2]
Bubo virginianus pallescens.
Nyctea nyctea.[1]
Surnia ulula caparoch.
Ceryle alcyon.
Dryobates villosus leucomelas.
?Dryobates villosus hyloscopus.[2]

---

[1] Known only from Fort Yukon or below.
[2] Known only above Fort Yukon.

Dryobates pubescens nelsoni.
Picoides arcticus.
Picoides americanus alascensis.
Colaptes auratus luteus.
Chordeiles virginianus.[2]
Selasphorus rufus.[2]
Sayornis saya yukonensis.[2]
Contopus borealis.
Contopus richardsoni saturatus.[2]
Empidonax trailli.[1]
Empidonax trailli alnorum.[2]
Empidonax hammondi.[2]
Otocoris alpestris leucolæma.
Pica pica hudsonica.
Perisoreus canadensis fumifrons.
Corvus corax principalis.
Scolecophagus carolinus.
Pinicola enucleator alascensis.
Pyrrhula cassini.[1]
Loxia leucoptera.
Acanthis hornemanni exilipes.
Acanthis linaria.
Spinus pinus.[2]
Passerina nivalis.
Calcarius lapponicus alascensis.
Calcarius pictus.[1]
Ammodramus sandwichensis alaudinus.
Zonotrichia leucophrys gambeli.
Zonotrichia coronata.
Spizella monticola ochracea.
Spizella socialis arizonæ.[2]
Junco hyemalis.
Melospiza lincolni.
Passerella iliaca.
Petrochelidon lunifrons.
Hirundo erythrogastra unalaschkensis.
Tachycineta bicolor.
Tachycineta thalassina.[2]
Clivicola riparia.
Ampelis garrulus.
Lanius borealis.
Helminthophila celata.
Helminthophila peregrina.[2]
Dendroica æstiva rubiginosa.
Dendroica coronata hooveri.
Dendroica striata.
Dendroica townsendi.[2]
Seiurus aurocapillus.[1]
Seiurus noveboracensis notabilis.
Wilsonia pusilla.
Wilsonia pusilla pileolata.[2]
Budytes flavus leucostriatus.[1]
Anthus pensilvanicus.
Cinclus mexicanus.
Sitta canadensis.[2]
Parus atricapillus septentrionalis.
Parus cinctus alascensis.[1]
Parus hudsonicus evura.
Phyllopseustes borealis.[1]
Regulus calendula.
Myadestes townsendi.[2]
Hylocichla aliciæ.
Hylocichla ustulata almæ.
Hylocichla aonalaschkæ.[2]
Hylocichla aonalaschkæ pallasi.[2]
Merula migratoria.
Hesperocichla nævia.
Saxicola œnanthe.
Sialia arctica.[2]

SPECIES WHOSE OCCURRENCE ON THE YUKON IS DOUBTFUL.

Chaulelasmus streperus.
Eniconetta stelleri.
Branta canadensis.
Macrorhamphus griseus.
Aquila chrysaëtos.
Nucifraga columbiana.
Loxia curvirostra minor.
Motacilla ocularis.

---

[1] Known only from Fort Yukon or below.
[2] Known only above Fort Yukon.

SPECIES AND SUBSPECIES OCCURRING IN THE YUKON BASIN AND HAVING THEIR CENTER OF ABUNDANCE DURING THE BREEDING SEASON IN ALASKA AND BERING SEA.

Rissa tridactyla pollicaris.[1]
Larus barrovianus.[1]
Larus vegæ.[1]
Larus brachyrhynchus.
Sterna aleutica.[1]
Phalacrocorax pelagicus robustus.[1]
Arctonetta fischeri.[1]
Somateria v-nigra.[1]
Chen hyperborea.[2]
Branta canadensis minima.
Branta nigricans.[2]
Philacte canagica.[1]
Grus canadensis.[2]
Macrorhampus scolopaceus.
Tringa couesi.[2]
Ereunetes occidentalis.
Heteractitis incanus.
Arenaria melanocephala.
Canachites canadensis osgoodi.
Haliaeetus leucocephalus alascanus.
? Megascops asio kennicotti.
Picoides americanus alascensis.
Sayornis saya yukonensis.
Contopus richardsoni saturatus.
Perisoreus canadensis fumifrons.
Pinicola enucleator alascensis.
Leucosticte tephrocotis littoralis.
Calcarius lapponicus alascensis.
Hirundo erythrogastra unalaschkensis.
Dendroica æstiva rubiginosa.
Dendroica coronata hooveri.
Parus cinctus alascensis.
Parus hudsonicus evura.
Hylocichla ustulata almæ.
Hesperocichla nævia.

Of these 35 forms, 1 is a subspecies of an Asiatic bird, 5 are chiefly confined to Bering Sea, 2 range in winter to the western Pacific, 7 are resident subspecies of northern North American birds, and the remaining 20 pass in winter to the western United States or beyond.

EASTERN NORTH AMERICAN SPECIES FOUND IN THE YUKON BASIN.

Limosa hæmastica.
Numenius borealis.
Accipiter atricapillus.
Falco sparverius.
Colaptes auratus luteus.
Chordeiles virginianus.
Empidonax trailli alnorum.
Junco hyemalis.
Passerella iliaca.
Helminthophila celata.
Helminthophila peregrina.
Dendroica striata.
Seiurus aurocapillus.
Wilsonia pusilla.
Hylocichla aliciæ.
Hylocichla aonalaschkæ pallasi.
Merula migratoria.

WESTERN NORTH AMERICAN SPECIES FOUND IN THE YUKON BASIN.

Anser albifrons gambeli.
Branta canadensis hutchinsi.
Olor buccinator.
Symphemia semipalmata inornata.
Bonasa umbellus umbelloides.
Pedioecetes phasianellus columbianus.
Buteo borealis calurus.
Buteo swainsoni.
Falco columbarius richardsoni.
Bubo virginianus pallescens.
? Dryobates villosus hyloscopus.
Picoides americanus alascensis.
Selasphorus rufus.
Empidonax trailli.
Empidonax hammondi.
Otocoris alpestris leucolæma.
Pica pica hudsonica.
Calcarius pictus.
Ammodramus sandwichensis alaudinus.
Zonotrichia leucophrys gambeli.
Spizella monticola ochracea.
Spizella socialis arizonæ.
Seiurus noveboracensis notabilis.
Cinclus mexicanus.
Parus atricapillus septentrionalis.
Myadestes townsendi.
Sialia arctica.

[1] Reported only from the Yukon Delta.
[2] Known only as migrants.

AMERICAN PACIFIC COAST SPECIES FOUND IN THE YUKON BASIN.

Helodromas solitarius cinnamomeus.
Tachycineta thalassina.
Zonotrichia coronata.
Dendroica townsendi.
Wilsonia pusilla pileolata.
Hylocichla aonalaschkæ.

ASIATIC AND PACIFIC SPECIES FOUND IN THE YUKON BASIN.

Limosa lapponica baueri.[1]
Charadrius dominicus fulvus.[1]
Archibuteo lagopus.
Scotiaptex cinerea lapponica.[1]
Pyrrhula cassini.
Budytes flavus leucostriatus.[1]
Phyllopseustes borealis.[1]

MIGRATORY SPECIES NOT COMMON TO NORTHERN NORTH AMERICA FOUND DURING THE BREEDING SEASON IN THE YUKON BASIN ABOVE FORT YUKON.

Eastern species.

Accipiter atricapillus.
Falco sparverius.
Colaptes auratus luteus.
Chordeiles virginianus.
Empidonax trailli alnorum.
Junco hyemalis.
? Passerella iliaca.
Helminthophila celata.
Helminthophila peregrina.
Dendroica striata.
Wilsonia pusilla.
Hylocichla aliciæ.
Hylocichla aonalaschkæ pallasi.
Merula migratoria.

Western species.

Branta canadensis hutchinsi.
? Grus canadensis.
Symphemia semipalmata inornata.
Buteo borealis calurus.
? Otocoris alpestris leucolæma.
Pica pica hudsonica.
? Calcarius lapponicus alascensis.
Ammodramus sandwichensis alaudinus.
Zonotrichia leucophrys gambeli.
Falco columbarius richardsoni.
Selasphorus rufus.
Sayornis saya yukonensis.
Empidonax hammondi.
Spizella monticola ochracea.
Spizella socialis arizonæ.
Seiurus noveboracensis notabilis.
Myadestes townsendi.
Hylocichla ustulata almæ.
Sialia arctica.

Pacific coast species.

Larus brachyrhynchus.
Helodromas solitarius cinnamomeus.
Leucosticte tephrocotis littoralis.
Zonotrichia coronata.
Hirundo erythrogastra unalaschkensis.
Tachycineta thalassina.
Contopus richardsoni saturatus.
Dendroica æstiva rubiginosa.
Dendroica coronata hooveri.
Dendroica townsendi.
Wilsonia pusilla pileolata.
Hylocichla aonalaschkæ.
Hesperocichla nævia.

SPECIES OCCURRING ON THE COAST OF SOUTHEAST ALASKA AND IN THE YUKON VALLEY.

Gavia imber.
Larus philadelphia.
Anas boschas.
Histrionicus histrionicus.
Oidemia deglandi.
Oidemia perspicillata.
Phalaropus lobatus.
Actitis macularia.
Haliæetus leucocephalus alascanus.
Picoides americanus alascensis.

---

[1] Known only from the Yukon Delta.

Colaptes auratus luteus.
Selasphorus rufus.
Contopus richardsoni saturatus.
Empidonax hammondi.
Sterna paradisæa.
Corvus corax principalis.
Ammodramus sandwichensis alaudinus.
Spizella monticola ochracea.
Hirundo erythrogastra unalaschkensis.
Tachycineta bicolor.
Dendroica coronata hooveri.
Dendroica townsendi.
Wilsonia pusilla pileolata.
Anthus pensilvanicus.
Sitta canadensis.
Hylocichla aonalaschkæ.
Merula migratoria.
Hesperocichla nævia.

SPECIES OCCURRING ON WHITE PASS SUMMIT AND IN THE YUKON VALLEY.

Sayornis saya yukonensis.
Zonotrichia coronata.
? Wilsonia pusilla pileolata.
Anthus pensilvanicus.
Hirundo erythrogastra unalaschkensis.

SPECIES FOUND BY US ONLY ON WHITE PASS SUMMIT.

Lagopus rupestris.
Lagopus leucurus.
Zonotrichia coronata.
Leucosticte tephrocotis littoralis.

SPECIES FOUND BY US ONLY ON THE COAST OF SOUTHEAST ALASKA.

? Dendragapus obscurus fuliginosus.
Sphyrapicus ruber.
Cyanocitta stelleri.
Corvus caurinus.
Junco hyemalis oregonus.
Junco hyemalis connectens.
Melospiza melodia rufina.
Melospiza lincolni striata.
Passerella iliaca townsendi.
Helminthophila celata lutescens.
Anorthura hiemalis pacifica.
Parus rufescens.
Regulus satrapa olivaceus.
Regulus calendula grinnelli.

LAND BIRDS FOUND IN LYNN CANAL DISTRICT ONLY NEAR TIDE WATER.

Contopus richardsoni saturatus.
Sphyrapicus ruber.
Cyanocitta stelleri.
Corvus caurinus.
Spizella monticola ochracea.
Junco hyemalis oregonus.
Melospiza melodia rufina.
Melospiza lincolni striata.
Tachycineta bicolor.
Helminthophila celata lutescens.
Anthus pensilvanicus.
Sitta canadensis.
Merula migratoria.

LAND BIRDS FOUND IN LYNN CANAL DISTRICT ONLY NEAR THE LEVEL OF GLACIER.

Colaptes auratus luteus.
? Dendragapus obscurus fuliginosus.
Junco hyemalis connectens.
Wilsonia pusilla pileolata.
Cinclus mexicanus.
Anorthura hiemalis pacifica.
Regulus satrapa olivaceus.

SPECIES RECORDED FROM THE UPPER YUKON ONLY ABOVE THE PELLY RIVER.

Tringa minutilla.
Symphemia semipalmata inornata.
? Megascops asio kennicotti.
? Dryobates villosus hyloscopus.
Chordeiles virginianus.
Selasphorus rufus.
Contopus borealis.
Hirundo erythrogastra unalaschkensis.
Tachycineta bicolor.
Helminthophila peregrina.
Dendroica townsendi.
Wilsonia pusilla pileolata.
Hylocichla aonalaschkæ.
Hylocichla aonalaschkæ pallasi.
Sialia arctica.

SPECIES RECORDED FROM THE YUKON VALLEY BETWEEN FORT YUKON AND THE PELLY RIVER.

Falco peregrinus anatum.
?Falco columbarius.
?Falco columbarius richardsoni.
Empidonax trailli alnorum.
Otocoris alpestris leucolæma.
Calcarius lapponicus alascensis.
Acanthis hornemanni exilipes.
Passerella iliaca.
Seiurus noveboracensis notabilis.
Wilsonia pusilla.
Hylocichla aliciæ.
Saxicola œnanthe.

ANNOTATED LIST OF SPECIES.

1. **Æchmophorus occidentalis.** Western Grebe.

Several seen at Bocadequadra, near Dixon Entrance, May 28.

2. **Colymbus holbœlli.** Holbœll Grebe.

A young male was taken on the 'Canal' at St. Michael September 22. The irides were primrose yellow; basal two-thirds of the culmen, outside tarsi, and lobes, seal brown; rest of bill, ocher yellow; inside of the tarsi and lobes, maize yellow; nails, yellowish olive buff.

3. **Gavia imber.** Loon.

Seen at Bocadequadra May 28 and in the Inside Passage May 29. Several seen on Lake Bennett and a pair at Caribou Crossing between June 17 and 28. On Lake Marsh they were common and were frequently heard, especially at night. The last loon certainly referable to this species was seen there July 6.

4. **Gavia arctica.** Black-throated Loon.

A loon that flew over our boat on Thirty-Mile River July 18, and another seen near Big Salmon River July 20, I believe were *Gavia arctica*. I saw several loons at the Aphoon mouth of the Yukon August 27 and one at St. Michael on September 5 and 16. We obtained none of them, but the experience of others makes it probable that all were the black-throated. Dr. Romig, of the Moravian Mission on the Kuskokwim River, told me that his party killed two on August 27 on the portage from Bethel on the Kuskokwim to Hendricks Station on the Yukon Delta.

5. **Lunda cirrhata.** Tufted Puffin.

Osgood saw one at Whale Island, near St. Michael, September 8.

6. **Fratercula corniculata.** Horned Puffin.

We took two and saw about a dozen puffins near Whale Island September 8. Irides, drab gray; ring on eyelid and tip of bill, flame scarlet; rest of bill dull straw yellow; bare skin at gape and line along base of maxilla, cadmium yellow; line below lower eyelid and horns, black; palmations, cadmium orange; tarsi and toes, cadmium orange above, chrome yellow below; nails varying from drab gray to slate color.

7. **Simorhynchus pusillus.** Least Auklet.

Auklets were seen several times while we were crossing Bering Sea in the *Corwin* October 1–2 and increased in numbers as we approached the Pribilofs. They were common with various other (unidentified) species of water birds off Unalaska October 4 and abundant in Akutan Pass October 6. I refer them to this species, as Nelson found it the most abundant in these waters.

8. **Brachyramphus marmoratus.** Marbled Murrelet.

This bird was fairly common in the Inside Passage May 28–29, and one was killed at Bocadequadra. We saw a few on Lynn Canal May 30, and I shot one near Skagway May 31. Doubtless some of the many murrelets seen with auklets near the Pribilof and Aleutian islands in October were this species.

9. **Cepphus columba.** Pigeon Guillemot.

Seen at Bocadequadra and along the Inside Passage May 28–29. Guillemots which I saw near Unalaska October 4 were probably this species.

10. **Uria lomvia arra.** Pallas Murre.

The murres seen near St. Michael August 29 and about St. George Island and Unalaska in October were probably chiefly this species, though some may have been *Uria troile californica*.

11. **Stercorarius parasiticus.** Parasitic Jaeger.

Common at the Aphoon mouth of the Yukon August 27–28, and about St. Michael until September 10. About this time their numbers decreased, and the last one was seen September 16. All appeared to be adults (as were the four collected), and only one was in the black plumage.

12. **Stercorarius longicaudus.** Long-tailed Jaeger.

I saw one at the Aphoon mouth August 28, and both Osgood and I occasionally saw the species at St. Michael until September 12.

13. **Rissa tridactyla pollicaris.** Pacific Kittiwake.

Adult kittiwakes were tolerably common at St. Michael from September 19 to the end of our stay, but no young were seen. As we crossed Bering Sea October 1–5, and at Unalaska October 5–6, young kittiwakes were common, and we saw no adults except at St. George and Unalaska. The irides of the adult are vandyke brown; ring on eyelid orange rufous; bill sulphur yellow, whitish at tip; gape rufous; tarsi, toes, palmations, and nails slate black.

14. **Rissa brevirostris.** Red-legged Kittiwake.

One was seen by Osgood at Unalaska (Dutch Harbor) October 5.

15. **Larus barrovianus.** Point Barrow Gull.

Abundant on the Lower Yukon, at the Aphoon mouth, and during September at St. Michael, though most of the adults had gone by the middle of the month. While crossing Bering Sea we saw several young October 2 and others near Unalaska October 4. A young bird shot near St. Michael September 19 had the head of a recently killed ptarmigan in its throat. The irides of the young are Prout's brown; tip of bill and sides of nails black; rest of bill, toes, and palmations vinaceous buff; rest of nails drab gray.

16. **Larus glaucescens.** Glaucous-winged Gull.

Large gulls, which doubtless were chiefly this species, were common from Dixon Entrance to Lynn Canal May 28–30, and we saw a few near Skagway June 1–2. At Unalaska, where I collected two, they were abundant October 4–6. A few gulls that followed the *Corwin* in the North Pacific I think also belonged to this species.

17. **Larus argentatus smithsonianus.** American Herring Gull.

The only large gulls I took on the Yukon—a female which had finished laying, collected at Lake Tagish June 30, and another taken near Charlie Creek August 8—were this species, and no others came close enough to make identification positive; hence I must refer all the large gulls seen to *Larus a. smithsonianus*, although on several we could see no black on the primaries. I saw one flying over White Pass Summit June 12 and another at Bennett City June 19. We saw eight or ten at Caribou Crossing and a few on Lake Tagish. No more were observed until we reached Lake Lebarge July 13; but from this point to the mouth of the river large gulls slowly became more numerous, one or two being noted every few days. Three fully grown young, with their parents, were seen on a sand bar about 15 miles above Circle August 12.

18. **Larus brachyrhynchus.** Short-billed Gull.

Our acquaintance with this bird dates from our arrival at Lake Marsh, July 1, where we found it common, and took downy young the next day. From this time, until we reached the Tatchun River, July 23, hardly a day passed that we did not see several; on July 20 we counted fourteen on a sand bar near Little Salmon River. After July 23 we saw no more until September 6, when young of the year appeared at St. Michael, and were common there until the 23d. The only adult seen at St. Michael was noted on September 14.

The adult has the irides Prout's brown; ring on eyelids and skin at commissural angle reddish orange; gape orange; bill, tarsi, and toes olive yellow; nails black, french gray at base.

Natal plumage: Creamy white, becoming pale cream color on forehead, chin, and anterior breast, mottled with different shades of brown, except the center of chest and abdomen. Head markings slate-black, distinctly defined and numerous, the most characteristic being one that covers the entire nasal region, a V on the pileum, a W on the occiput, and a somewhat interrupted U on each side of the throat. On the upperparts the markings become pale seal brown, and with lighter tips render the lower neck, sides of breast, flanks, and anal region grayish. Bill brownish black; tip of bill, tarsi, toes, and palmations whitish; nails and edges of scutellæ of tarsi and toes hair brown.

19. **Larus philadelphia.** Bonaparte Gull.

I saw several small black-headed gulls, probably this species, in the Inside Passage May 29. I took a Bonaparte gull at Caribou Crossing on June 24 and saw several others. We saw one on Lake Marsh July 1, a few young at St. George Island October 3, and found them common at Unalaska October 4–5.

20. **Xema sabinii.** Sabine Gull.

Osgood found a dead bird of this species on the shores of Chilkat Inlet June 1. The specimen, unfortunately, was not in a condition to permit its preservation, but it was carefully identified at the time and showed no apparent variance from the description and figure in Ridgway's Manual.

21. **Sterna paradisæa.** Arctic Tern.

We saw a large flock of terns in the Inside Passage May 29, and two days later at Skagway saw a few more, securing two, which proved to be of this species. At Bennett, between June 15 and 20, we frequently saw two or three, and I was informed that arctic terns bred on a small lake near Log Cabin, British Columbia. We found a breeding colony of about twenty on a small rocky island lying in the entrance to Windy Arm, Lake Tagish, July 1. I found four single eggs (three fresh and one well advanced in incubation), one set of two (one fresh and the other at point of hatching), and also a young bird which had just left the shell. There were no nests; the young bird and eggs were in the short grass on the top of the island. Except a single bird, seen at Lake Marsh and probably belonging to this colony, we did not meet with terns again until August 27, when I found this species common at the Aphoon mouth. A single tern with injured primaries was seen frequently at St. Michael up to September 21. The downy young differs from the description given in Baird, Brewer and Ridgway's 'Water Birds,' in having the forehead plain dusky, the chin whitish, the basal half of bill, tarsi, and toes salmon pink, and the rest of bill and nails black.

22. **Diomedea albatrus.** Short-tailed Albatross.

A dark-brown albatross, probably the young of *D. albatrus*, joined the *Corwin* October 1, about 150 miles from St. Michael. It was soon accompanied by others, and until we reached Cape Scott, October 12, a glance astern would seldom fail to show two or three following the vessel.

23. **Fulmarus glacialis glupischa.** Pacific Fulmar.

A single dark-colored fulmar, possibly this form, was seen October 4, between St. George and Unalaska.

24. **Oceanodroma furcata.** Forked-tailed Petrel.

To this species I refer a few light-colored petrels seen October 3, on Bering Sea north of the Pribilof Islands.

25. **Phalacrocorax pelagicus.** Pelagic Cormorant.

Cormorants were seen October 4 near Unalaska, where this species is reported as common.

26. **Phalacrocorax pelagicus robustus.** Violet-green Cormorant.

We saw a single cormorant at Whale Island September 8; and one—possibly the same bird—was seen by Osgood several times at St. Michael.

27. **Phalacrocorax urile.** Red-faced Cormorant.

This is the only cormorant reported by William Palmer from St. George, where we saw several October 3.

28. **Merganser americanus.** American Merganser.

A pair of mergansers was breeding on a small, rocky island in Lake Tagish, at the entrance to Windy Arm, June 30–July 1. The nest was found by Osgood in a crevice in the cliffs about 15 feet above the water. It was made of down, and contained seven eggs about one week advanced in incubation. Retrieving would have been impossible had we shot the bird, but as I succeeded in watching the female on the nest from a distance of less than 6 feet I feel positive of the species.

A few other mergansers, usually in pairs, were seen on Lake Tagish July 1, on Lake Marsh July 8, at Fifty-Mile River July 9 and 12 (a flock of a dozen males flying up the river in the evening of the latter date), near Little Salmon River July 20, and about 25 miles above Circle August 12. Near Charlie Creek we found the dried wing of an adult male of this species August 8.

29. **Anas boschas.** Mallard.

On the flats of Chilkat Inlet I saw seven June 2. In no part of the Yukon Valley above Circle did we find ducks abundant, except surf scoters, but the mallard undoubtedly occurs at all suitable places throughout the region. It must breed very early, as on June 24, only three weeks after the lakes were open to steamer navigation, I found a female with two young at Caribou Crossing, and on June 28 I shot another female there and caught two of her half dozen downy young. Two ducks, probably mallards, were seen on Lake Marsh July 6, and at Miles Canyon Maddren was informed they had been common there earlier in the season. We saw several females with young in the marshy ponds at the foot of Lake Lebarge July 17, a few adults near the Little Salmon July 20, and a good-sized flock near Charlie Village August 10. Osgood shot one near Fort Yukon August 21.

In the large flocks of geese and ducks disturbed by the steamer on the Lower Yukon were two young mallards, secured at Hendricks Station August 25. Mallards were common at the Aphoon mouth August 27, and we saw a few at St. Michael September 2.

30. **Mareca americana.** Baldpate.

Five ducks that I took to be baldpates were seen a short distance above Fort Selkirk July 25.

31. **Nettion carolinensis.** Green-winged Teal.

Three teal that I saw in the creek at Circle, August 19, were probably this species. Green-winged teal were common in the tundra ponds about St. Michael during the first half of September, but apparently did not occur after September 16. All that were taken were young birds.

32. **Dafila acuta.** Pintail.

Maddren was told at Miles Canyon, July 11, that pintails were common, but we saw none near enough for identification until August 27, when I found them abundant at the Aphoon mouth. Seven were here killed by a passenger on the steamer. During September young pintails far outnumbered all other ducks on the marshes and tundra ponds about St. Michael. Large numbers were killed by the Eskimos, but no adults were seen. Their numbers had greatly decreased by September 20.

33. **Aythya marila.** Scaup Duck.

We saw a flock of about a dozen adult males at Caribou Crossing June 24, and another of about twenty on the Yukon, a short distance above Fort Selkirk, July 25.

34. **Aythya affinis.** Lesser Scaup Duck.

We found a pair with young on a small pond at Lower Lebarge July 17. Osgood secured the female.

35. **Clangula clangula americana.** American Golden-eye.

I am confident that a flock of ducks seen about 25 miles above Circle August 12 were males of this species or of *C. islandica.*

36. **Charitonetta albeola.** Buffle-head.

I shot a female on a small pond near Lake Marsh July 8, and saw a male near Little Salmon River July 20. Maddren was informed that buffle-heads were common near Miles Canyon, and a boy at Lower Lebarge said they bred commonly on the ponds near there, and that he took two young July 16.

37. **Harelda hyemalis.** Old Squaw.

Single young birds were found frequently during September in the small ponds about St. Michael, and a flock of about a dozen was seen in the harbor September 11. No adults were observed. One young bird, taken early in September, still retained natal down on the hind-neck.

38. **Histrionicus histrionicus.** Harlequin Duck.

We saw a male and two females in Wrangell Narrows May 29. A flock of twelve males came close to the shore at Bennett June 18; and on June 23 a single male swam so near that men sitting on the beach threw stones at it. One other harlequin was seen a few miles above Fort Selkirk July 25. Dr Romig told me he saw a number on the portage from the Kuskokwim to the Yukon August 24-25.

39. **Somateria v-nigra.** Pacific Eider.

We saw the head of a male of this species lying in the window of the hotel at St. Michael, and the soldiers at the barracks had a mounted bird, shot near St. Michael in the spring, but we saw no living eiders of any species during our trip.

40. **Oidemia americana.** American Scoter.

We noticed a few in Wrangell Narrows May 29, and I saw a number off Unalaska October 5.

41. **Oidemia deglandi.** White-winged Scoter.

This species was fairly numerous at Bocadequadra, Wrangell Narrows, and Lynn Canal May 28–30. We saw two on Lake Marsh July 6, two on Lake Lebarge July 14, and a flock of about twenty-five flying up Fifty-Mile River from Lake Lebarge on the evening of July 12.

42. **Oidemia perspicillata.** Surf Scoter.

In Lynn Canal, near Haines, June 1 we noted a large flock of surf scoters, most of which had disappeared the next day. They were abundant on all the Yukon lakes except Bennett, which was almost destitute of bird life. On Lake Tagish we saw fourteen June 30, and at Lake Marsh thirty to forty males almost every day between July 1 and 8. We saw no more, except a pair on July 11 on Fifty-Mile River, which connects Lake Marsh with Lake Lebarge, until we entered Lake Lebarge on the evening of July 12, when a flock of at least a hundred flew high overhead from the direction of the lake. About 8 p. m. and at 10 p. m. of the same evening, and on the next morning, we saw what we took to be the same flock. The birds were probably taking a morning and evening flight, such as E. S. Bryant has noticed in the case of the white-winged scoters breeding at Devils Lake, North Dakota; and I believe that with both species these flights are taken chiefly to exercise the wing muscles. We saw no females on any of the lakes, nor could we find them on the shore, though they were undoubtedly nesting in the vicinity. We observed several on Thirty-Mile River July 18 and two near the Little Salmon July 20. Near Whale Island, at St. Michael, we saw a number September 8, and two scoters, probably young of this species, September 21. I think there were a few with the American scoters I saw at Unalaska October 5.

43. **Chen hyperborea.** Lesser Snow Goose.

I saw five snow geese at the Aphoon mouth August 28, and a large flock at St. Michael September 11.

44. **Anser albifrons gambeli.** American White-fronted Goose.

A single white-fronted goose was seen by Osgood among a number of other birds killed by natives about the Yukon Delta August 29.

45. **Branta canadensis hutchinsi.** Hutchins Goose.

Although Maddren was informed that a goose with four young was seen near White Horse Rapids about July 11, and although the sergeant in charge of the police station of Lower Lebarge told us that thousands of geese and ducks passed there in the spring, and that he had counted twenty-four distinct species, and had killed both Hutchins and cackling geese, we did not see a goose of any species until we were in the neighborhood of Charlie Village, August 10. There we saw a flock of about twenty of the *Branta canadensis* group, and Osgood shot two *hutchinsi* and saw many more near Fort Yukon August 21. Brown geese, apparently chiefly this subspecies, were common on the Yukon flats and on the lower river, especially the Yukon Delta. A Hutchins goose was brought to the steamer *Robert Kerr* by an Eskimo August 26, and I found the bird common at the Aphoon mouth August 27–28. Prospectors on the *Kerr* told me that geese bred abundantly at the head waters of the Porcupine and the marshes at the source of Birch Creek.

During September this species was common about St. Michael in small flocks, but very shy; Osgood took one September 23.

[*Philacte canagica.* Emperor Goose. Dr. Romig told me they were common on the tundra along the Kuskokwim.]

[*Olor columbianus*. Whistling Swan. We were told that a swan—probably this species—was killed at Circle during the spring.]

46. **Grus canadensis.** Little Brown Crane.

Along the Yukon we did not see any cranes, although I thought I heard one near the Little Salmon July 21, and a man who had spent the summer at Circle told me he had heard and seen the 'sand-hill crane' there frequently during the past two months. I was also informed by prospectors that these cranes were found in small numbers at the head waters of Birch Creek and Porcupine River.

Near St. Michael we saw flocks of from two to six individuals each almost daily during the first half of September, but none later than September 15. On the night of September 13 and all the following day there was a hard southwest gale. On the 14th we saw large numbers—Osgood counted ninety-six—flying south, high in the air.

47. **Crymophilus fulicarius.** Red Phalarope.

We saw a small flock near Skagway in Lynn Canal June 2, and others I believed to be this species near Wrangell Narrows and in Prince Frederick Sound May 29. Osgood took one at St. Michael September 17 during a heavy storm.

48. **Phalaropus lobatus.** Northern Phalarope.

Large flocks were seen near Dixon Entrance May 28, and smaller ones on the Inside Passage May 29. From a flock of about twenty on Lake Lebarge July 13 I shot a female that was changing to winter plumage, and on a small pond at Lower Lebarge July 17 I took a male that was in worn breeding plumage. At St. Michael September 2 I caught a young bird that had but one wing, and on St. George Island October 3 I shot one that was swimming alone in a pool. Phalaropes, probably this species, were seen on Bering Sea October 1 and 4.

49. **Gallinago delicata.** Wilson Snipe.

At Haines May 31 I was told that several Wilson snipe had been seen that day, but was unable to find them. We saw one on Fifty-Mile River not far below Lake Marsh July 10, and another in the marsh at Lower Lebarge July 17. Osgood saw one at Circle August 18, and I killed two from a small flock at Hendricks Station August 25. At St. Michael we saw eight or ten single birds between September 12 and 22.

50. **Tringa couesi.** Aleutian Sandpiper.

Common about the lava rocks that line the shore at St. Michael, where flocks of five to fifty were observed, but only small flocks after September 15. A few were occasionally seen on the tidal mud flats, but none about the ponds in the interior of the island or on the salt meadow behind the town. Out of eighty specimens taken only eight were adults, and five of these were taken before September 9. On the rocky shores of a point opposite Dutch Harbor, Unalaska, I found them common October 5. Those taken at St. Michael were molting into first winter plumage, which is practically complete in the Unalaska birds. In this plumage there is considerable individual difference in the width and shade of the pale edgings of the feathers of the upperparts.

The irides were vandyke brown; bill, black changing to olive buff in basal half; tarsi and toes, yellowish olive buff washed with black; nails black.

I find great sexual variations in size in this species, the females, as in many other species of Limicolæ, averaging considerably larger, especially in length of bill. Measurements of twenty-nine males: Length 8.06 to 8.94 (average 8.57) inches; wing 4.37 to 5.12 (average 4.89) inches; exposed culmen 0.96 to 1.13 (average 1.06) inches; tarsus 0.91 to 1.03 (average 0.96) inches. Measurements of thirty-four females: Length 8.56 to 9.56 (average 9.03) inches; wing 4.47 to 5.31 (average 4.98) inches; exposed culmen 1.16 to 1.42 (average 1.24) inches; tarsus 0.96 to 1.05 (average 0.99) inches.

51. **Tringa ptilocnemis.** Pribilof Sandpiper.

We saw a number on St. George October 3, but too close to the rookery of fur seals to be obtained.

52. **Tringa acuminata.** Sharp-tailed Sandpiper.

First found September 18, when six were seen with a large flock of *T. a. pacifica* at St Michael. We did not see more than a dozen of this species during the rest of September. Although the species has not been hitherto recorded from St. George Island, we took three and saw about a dozen during the short time we were there October 3. At Unalaska, October 5, I secured one which was with *T. couesi* on the rocky beach. The irides were vandyke brown; maxilla and distal half of mandible, dark seal brown, mandible changing to dull olive buff at base; gape ecru drab; tarsi and toes, greenish maize yellow; nails black.

53. **Tringa maculata.** Pectoral Sandpiper.

This species was present throughout our stay at St. Michael, usually associating with flocks of *T. a. pacifica*, but in very small numbers, not more than twenty being seen. All the specimens taken were young birds. Osgood took one at St. George October 3, and I one at Unalaska October 5.

54. **Tringa bairdi.** Baird Sandpiper.

Two sandpipers, probably of this species, flew by us on Lake Marsh, and we saw four more in the marshes of Lower Lebarge, but failed to secure any of them. I shot one young bird near the Tahkandik River August 7. Osgood shot one from a flock of four at Circle August 15, and another near Fort Yukon August 21.

55. **Tringa minutilla.** Least Sandpiper.

At the southern end of Lake Marsh, not far from where Six-Mile River enters, the surrounding country is level, and at high water the lake stretches far back through a dense growth of willows. At the time of our visit the retreating water of the lake had left a belt of grass between these willows and its margin. Here on the evening of July 2 I found three pairs of least sandpipers, and after a long search, somewhat interrupted by hordes of mosquitoes, I came upon a female surrounded by four downy young. Both parents tried time and again the well-known 'wounded-bird' tactics to lure me from the spot where the young were hidden in the bunches of grass, and, finding this a failure, would circle around me only a few yards off, uttering a plaintive twitter. I saw two other least sandpipers on the west shore of Lake Marsh July 7.

Natal plumage: Lower parts, forehead and orbital region, brownish white. Upperparts bright cinnamon rufous mottled with black; many feathers, especially on head, rump, and tail, tipped with white. Post-orbital line and loral line blackish, and spot of bright cinnamon rufous on sides of chest. Irides dark brown; bill and nails, slate black; tarsi and toes, pale slate.

56. **Tringa alpina pacifica.** Red-backed Sandpiper.

Young red-backed sandpipers were very abundant at St. Michael during our stay, many times outnumbering all other Limicolæ. Early in September they frequented chiefly the mud flats on the coast, but after the middle of the month large numbers were found only about the pools of the salt marsh. September 24, when the tundra was quite thoroughly frozen, with snow in every hollow and a skimming of ice on the pools, I saw at least one hundred in this latter place.

In several taken early in September the back of the neck was still covered with down, but the majority were in full juvenile plumage. Some still retained this plumage at the time we left St. Michael, but the larger portion had molted into winter plumage. Only two adults were taken, September 1 and 5. A few were seen at St. George Island October 3.

57. **Calidris arenaria.** Sanderling.

I saw three at St. Michael September 11 and collected one, which proved to be a young female.

58. **Totanus flavipes.** Yellow-legs.

On July 1, while floating down Six-Mile River close to its entrance into Lake Marsh, we were attracted by the anxious cries of a pair of yellow-legs. Osgood shot both birds, and we found two downy young in the grass on the shore of the river. Entering Lake Marsh we heard a yellow-legs' whistle, and on July 2 I saw a yellow-legs near where I found the least sandpiper. I collected a female on the west shore of Lake Marsh July 8, and a male, the last bird of this species seen, near a small pond at Lower Lebarge July 17. Both these birds undoubtedly had eggs or young close by, for they alighted exclusively in trees, scolded vociferously, tilting the body with each cry, and refused to leave. Bare spaces on the breast show that both sexes assist in incubation.

Natal plumage: Upperparts and thighs, dark seal brown, many of the feathers tipped with cream buff and whitish; longitudinal lines on rump, cream color, inclosing central, seal-brown space. Forehead, buffy white, extending in narrow lines on sides of crown to occiput, and in broader lines above eye to nape, the latter crossed by transverse dark lines extending from eye to occiput. Line beginning at base of culmen enlarged to dark space on crown and occiput, extending down neck to back, seal brown; other dark lines extending from crown above eye to occiput, and from nostrils through eye to nape. Throat and center of abdomen silvery white; rest of lower parts and sides of neck, buffy white; each feather of lower parts becoming brownish black at base. Irides, vandyke brown; bill, black at tip, changing to greenish olive at base; tarsi and toes, yellow, paler than in adult, and mottled with brown; nails, brown. The juvenile plumage is appearing, in this specimen, on wings, wing coverts, chest, and sides.

59. **Helodromas solitarius cinnamomeus.** Western Solitary Sandpiper.

At Log Cabin, British Columbia, on the evening of June 14, we noticed a sandpiper wheeling through the air, like the woodcock at its breeding place, occasionally uttering a rather musical whistle. The next morning I found it feeding in a small swamp. It proved to be a solitary sandpiper, as I had suspected on the previous evening. Osgood saw another near Lake Marsh July 5, and I saw two near Little Salmon River July 21. On July 8, after rowing a few miles down Lake Marsh, we stopped for lunch on the west shore, where a forest fire had killed most of the trees, and fallen trunks piled in endless confusion, brush, small pools, and hordes of mosquitoes rendered the place anything but a paradise. Here I startled a solitary sandpiper and a yellow-legs at the same instant. They lighted on the half-fallen trees and scolded me, tilting their bodies at each cry. The solitary sandpiper, which doubtless had a nest there, differed chiefly from eastern specimens of *solitarius* in having dark, wavy markings on inner webs of outer primaries. Osgood took a typical young of *cinnamomeus* and saw another on an island near Sixty-Mile Creek August 1.

60. **Symphemia semipalmata inornata.** Western Willet.

While in a meadow a short distance back from the southeast end of Lake Marsh July 2 I heard a willet whistle several times its unmistakable 'pill-willet,' but failed to see the bird.

61. **Heteractitis incanus.** Wandering Tattler.

Osgood took an adult at Skagway May 31. I shot a young bird from a flock of three at St. Michael September 1, saw one on Whale Island September 8, and secured two at Unalaska October 5.

The irides of the adult were vandyke brown; bill, black, base of mandible brownish; tarsi and toes, brownish ocher; nails, black. In the young, the bill changed from black to sage green in basal third of mandible, and to greenish olive at base of maxilla; tarsi and toes, dull gallstone yellow, greenish at joints.

62. **Actitis macularia.** Spotted Sandpiper.

I saw one at Skagway June 3, and Osgood one at Glacier June 8. This is preeminently the shore bird of the Yukon Basin; we saw two at Bennett June 18, and until we reached Circle, August 15, hardly a day passed without our seeing many running along the shore, or skimming over the river. They were especially abundant between White Horse Rapids and Lake Lebarge. After the 1st of August most of the spotted sandpipers seen seemed to be traveling upstream in small flocks. We saw no adults after August 4.

The first set of eggs was found at Caribou Crossing June 27; the last at the Tatchun River July 23. The first young noticed were in a nest containing three young and one pipped egg found on Lake Marsh July 7. Both sexes were incubating. Nests were close to the shore, and also on small rocky islands in the lakes.

63. **Numenius hudsonicus.** Hudsonian Curlew.

I secured one from a flock of four curlews on the marshes of Chilkat Inlet, and Osgood found a dead bird in the woods at Haines, June 1. Three young were brought to the steamer by an Eskimo at the Aphoon mouth August 28. I saw one at St. Michael September 2, and, I think, another September 14.

Adult: Irides, vandyke brown; bill and nails; black; tarsi and toes cinereous. Young (Massachusetts specimen): Irides, raw umber; maxilla, black; mandible, clove brown, blackish at tip, vinaceous toward base; tarsi and toes, olive gray; nails, black.

64. **Squatarola squatarola.** Black-bellied Plover.

At the Aphoon mouth of the Yukon I saw a flock August 28. Osgood saw three young which had been shot on the mainland near St. Michael September 10, and from this date to the end of our stay we saw occasionally one or two birds of the year, one of which was taken September 16.

65. **Charadrius dominicus fulvus.** Pacific Golden Plover.

None were seen until September 16, after which young birds became fairly common on the boggy tundra about St. Michael and the mud flats along the shore. The only adult seen was taken by Osgood September 25. We saw a number of young birds on St. George Island October 3, and Osgood secured one. Crossing Bering Sea we saw some near Unalaska October 4, and I saw one on October 8, when we were several hundred miles south of the Aleutian Islands. This bird flew several times around the *Corwin*, answering my every whistle, and seemed anxious to alight. The specimens collected differ greatly in the amount of the golden coloring, but all are far more golden than Massachusetts skins of *dominicus*, and all have the shorter wings of *fulvus*. Irides, vandyke brown; bill and nails, black; tarsi and toes, slate gray.

66. **Ægialitis semipalmata.** Semipalmated Plover.

Osgood collected a male at Caribou Crossing June 24, and a pair of adults and one pipped egg at the southern end of Lake Marsh July 2. I removed the young bird from the shell, and within half an hour the down was almost dry, the eyes were open, and it could hop about on its 'knees.' Maddren took another adult at this place July 6, and I a female and four eggs nearly hatched, on the west shore of Lake Marsh on the same day. The nest was a hollow, lined with a few grasses and dead leaves, and was situated about 8 feet from the water in the drift débris among the stones of the beach. We saw three or four on a sand flat near Charlie Village August 10; a few about 15 miles above Circle August 12, and the last at Circle August 15.

Bare pectoral spaces showed that both sexes assist in incubation. Natal plumage: Lower parts, white, separated by broad bare space on neck, changing to cream color on lower tail coverts. Above, cream color, mottled with black, changing to buff on wings and tail. Forehead and infraorbital patches, cream color; broad band on neck encircling head, white, bordered above by narrow band of black extending from bill around occiput, and connecting in malar region with black line leading to inner canthus of eye. Spot on forehead, on sides of chest at lower border of bare space, on sides and on flanks, black. Irides, dark; bill and nails, black; tarsi and toes, slate color, whitish posteriorly.

67. **Arenaria melanocephala.** Black Turnstone.

We found a small flock on the rocky shore at St. Michael August 31: I took three young there the next day, and on September 5 I saw a single turnstone flying across the marsh. On St. George Island, October 5, we saw a number of birds that we had no doubt were black turnstones, but I do not find this species recorded from the Pribilofs, and we were unable to obtain specimens. Irides, vandyke brown; bill, olive black; tarsi and toes varying from clay color to vinaceous cinnamon, and washed with black; nails, black.

68. **Dendragapus obscurus fuliginosus.** Sooty Grouse.

We were told that grouse were common on the heights above Skagway, but although we often found droppings we saw no birds, and the spring 'calling' of the male had ceased. Maddren and I heard a bird that must have been this species 'booming' far up on the hillside from the ravine above Glacier June 8.

69. **Canachites canadensis osgoodi.** Alaska Grouse.

Canachites canadensis osgoodi Bishop, Auk, XVII, 114, April, 1900.

We first met the Alaska grouse at Bennett City, where Osgood shot a laying female June 22. At Caribou Crossing he found feathers of this grouse in a magpie's nest and in one of his mammal traps. At Lake Marsh he shot four females and four young July 4–5, at Lake Lebarge a female July 14, at Lower Lebarge a female and one young July 17, and on Thirty-Mile River an adult male July 19. He found the birds frequenting the thickets of poplars and young spruces and remarkably easy to approach. I saw a male at Lake Lebarge July 16, and shot a well-grown young near the Tatchun River July 16, but did not meet with the species elsewhere. This bird was reported as common at Lower Lebarge by the police sergeant stationed there; at Rampart City by Mr. Burkman, and along the Kuskokwim by Dr. Romig.

70. **Bonasa umbellus umbelloides.** Gray Ruffed Grouse.

I secured a female and one young bird on the west shore of Lake Lebarge July 14, and another female that had a brood of young, two-thirds grown, at Lower Lebarge July 17. Osgood took a young bird from a covey near Rink Rapids July 22. The sergeant at Lower Lebarge called this species rare, but I was told it was common near Rampart City.

71. **Lagopus lagopus.** Willow Ptarmigan.

Two flocks were found on the tundra at the Aphoon mouth August 28, one alighting close to the steamer. Not seen at St. Michael until September 11, when about one hundred appeared. These were seen frequently after this date, but were exceedingly shy. Most of those taken were young birds, and all were in full molt. The irides of a young male taken September 19 were vandyke brown; skin above eye, rufous; bill, slate black, whitish at tip and salmon buff at base of mandible; nails, white.

We were told that ptarmigan were very abundant near Atlin, British Columbia, at the head waters of the Porcupine River and Birch Creek, near Rampart City, along the Kuskokwim, and in winter at Glacier and Lower Lebarge. Doubtless some of these statements refer to the following species.

72. **Lagopus rupestris.** Rock Ptarmigan.

At White Pass Summit, June 11 and 13, we took three males still in white plumage (excepting a few dark feathers on head and lower neck), and saw a few others. Osgood found two eggs there, probably of the previous year, lying on the moss under an alpine hemlock. Dr. Romig told me that this species was more common than *L. lagopus* along the Kuskokwim.

[*Lagopus rupestris nelsoni.* Nelson Ptarmigan. We were told at Unalaska that this species had been abundant during the summer on Unalaska Island, but that the birds had been almost exterminated by the officers of an English man-of-war. We saw none during the day and a half we were there.]

73. **Lagopus leucurus.** White-tailed Ptarmigan.

Osgood took a white-tailed ptarmigan June 8 on the summit of the cliffs above Glacier, and saw several other ptarmigan, probably of this species. On June 8 he found at the same place, on the moss under an alpine hemlock, fragments of two ptarmigan eggs, sparingly dotted with brown as in *leucurus*.

74. **Circus hudsonius.** Marsh Hawk.

We saw one at Lake Marsh July 8, one at Lake Lebarge July 12, a young bird on which duck hawks were feeding near the Tahkandik River August 7, one about 20 miles above Circle August 12, and two at Circle August 15 and 20. At the Aphoon mouth I saw several August 28. At St. Michael we secured a young bird September 2, and saw single marsh hawks on September 6, 7, and 11. The young bird taken is noticeably darker than young from Dakota and New England.

75. **Accipiter velox.** Sharp-shinned Hawk.

I saw one at Lower Lebarge July 17, and two near White River July 30; Osgood found one feeding on a thrush near Charlie Creek August 9; at Circle I saw one August 17 and shot an adult female August 19. Osgood found a nest of this species, about 15 feet from the ground, in a small spruce in the center of an island near the Nordenskiold River July 22, and I secured the female, whose crop held the tibia, tarsus, and toes of a flicker. The nest contained three downy, but very pugnacious young, one infertile egg, and the remains of a young intermediate sparrow. I kept two of the young alive until July 31, when both were well feathered and trying to fly and were as irascible as ever. The last survivor succeeded in getting out of his box while we were moored at Dawson, flew into the Yukon, and was carried rapidly along by the current, though struggling valiantly to reach the shore. I suspect that it succeeded, as I heard a man who hurried after it say later that he would have 'fricasseed chicken for dinner.'

76. **Accipiter atricapillus.** American Goshawk.

I saw an adult flying high above the shore of Lake Marsh July 8 with a mammal, probably a ground squirrel, in its talons.

77. **Buteo borealis calurus.** Western Red-tail.

This is presumably the common hawk of the Upper Yukon; for the two large hawks taken are this species, and the numerous others seen resembled these in appearance, flight, and cry. About half were in the melanistic plumage.

Passing down Six-Mile River July 1 we saw three large buteos circling, and we noticed others frequently, usually in pairs, until we left Circle. Osgood and Maddren found a nest near Lake Marsh July 5 regarding which a pair of these birds were very solicitous. It was high in a spruce, and was empty except for a dead ground squirrel. On Fifty-Mile River July 10 I found a nest that was about 55 feet up in a spruce and contained two downy young. Osgood shot the female, which was in light plumage; the male, a melanistic bird, escaped. Osgood shot a melanistic female at Lower Lebarge July 17, and I found a pair—one light, the other dark—near Fort Selkirk July 25. These had a nest that was 60 feet up in a spruce and contained two young able to fly. I saw another nest with the birds about it on an island near the White River July 31.

78. **Archibuteo lagopus.** Rough-legged Hawk.

On September 1, 6, 7, and 9 we saw at St. Michael large hawks which from their proportions and flight were either buteos or archibuteos. Osgood shot one on Whale Island September 8, but could not retrieve it. Mr. Nelson's experience with the hawk family at St. Michael leads me to refer these birds to this species.

79. **Haliæetus albicilla.** Gray Sea Eagle.

Lieutenant Satterlee, of the *Corwin*, found a dead bird of this species at Unalaska October 5, which proved to be a young female.

The wings had been removed at the carpal joint, but the unruffled plumage--the down yet persisting on the ends of the secondaries—removes all probability that it had been a caged bird. This is the first record of the occurrence of this species in western North America, although it is common in Japan and occurs in Kamchatka and occasionally on the Commander Islands.

80. **Haliæetus leucocephalus alascanus.** Northern Bald Eagle.

We found this bird common along the Inside Passage, especially near Wrangell Narrows, and from the steamer I noticed three occupied nests. We visited one which was high in a gigantic dead cedar on a small island near Bocadequadra. Broken shells at the foot of the tree made it probable that the nest contained young. The female parent was secured by Maddren. On the flats of Chilkat Inlet June 1 I saw 28 eagles feeding. Here I found another occupied nest at least 100 feet up in a living spruce (it was so high that heavy charges of No. 4 shot did the bird no harm). A man passing by shot the male with a rifle. The next day I saw the female again on the nest. In the interior this bird is much rarer, though I saw one at Log Cabin June 20, and another at Bennett June 19. We saw the birds occasionally about the lakes (I found a deserted nest on Lake Marsh), and once or twice along the river, the last being observed near the White River July 31.

81. **Falco rusticolus.** Gray Gyrfalcon.

A female was caught in a steel trap set on a post at St. Michael September 21. Its stomach contained feathers. The irides were vandyke brown; tip of bill and nails, black; tarsi, toes, cere, gape, and rest of bill, pearl gray, the bill changing to pearl blue on maxilla near commissure.

82. **Falco peregrinus anatum.** Duck Hawk.

At Fort Selkirk the character of the Yukon Valley changes, and the high, sandy bluffs which have been constantly visible on one bank or the other are frequently replaced by rocky cliffs of varying height. Flying about one of these cliffs near Stewart River July 31 was the first duck hawk we noted. From that point to the Yukon Flats, a few miles above Circle, a day seldom passed without our seeing or hearing them, and from Camp Davidson to Circle I think there was at least one breeding pair every 10 miles. We saw a number of their nests on shelves on the cliffs, but at this time, the first half of August, the young had flown.

Osgood secured a young female August 5 on the cliff known as 'Old Woman,' and an adult female August 7 near the Tahkandik River, and shot several others which he failed to bag. I took a young male from a family on 'Castle Rock' August 5. We found that those taken had been feeding on marsh hawks, Alaska jays, white-winged crossbills, intermediate sparrows, and varied thrushes.

I saw two duck hawks near Nulato August 24, and a tame young bird spent part of the rainy evening of August 30 perched on the back of a chair in the hotel at St. Michael. The cere and bill of the young male were french gray, changing to black on tip of bill and along culmen and cere above; tarsi and toes, pale, grayish green; soles, tarsi behind, and edges of scutellæ in front, yellow; nails, black.

83. **Falco peregrinus pealei.** Peale Falcon.

One flew around the *Corwin* when we were some distance south of the Aleutian Islands and out of sight of land October 7.

84. **Falco columbarius.** Pigeon Hawk.

We saw a pigeon hawk feeding on a large vole near Charlie Creek August 8. Osgood took a young male at a point 12 miles above Circle August 13, and I saw one at the Aphoon mouth August 28.

85. **Falco columbarius richardsoni.** Richardson Merlin.

At Circle August 18 I shot a young female merlin which is intermediate between *columbarius* and *richardsoni*. In general coloring both above and below, it is between typical examples of the two forms and approaches very closely a specimen of *richardsoni* taken by Captain Bendire at Walla Walla, Washington, December 3, 1880, and now in the American Museum of Natural History. My bird has light spots on outer webs of primaries and six light bars on tail similar to those of *richardsoni*, but the bars are narrower and more interrupted. The crop and stomach contained the remains of a red-backed mouse. The irides were vandyke brown; cere, greenish-yellow; maxilla, slate black at tip, changing to greenish-white toward cere and pale french gray at commissure; mandible, pale dull greenish, changing to pale french gray toward tip and commissure; tarsi and toes, straw yellow, the latter inclining toward sulphur yellow; nails, black. Mr. Cantwell writes in the 'Osprey'[1] of having *seen* Richardson's merlin, but does not state that he took specimens. These are the only records for this bird in the Yukon Valley.

86. **Falco sparverius.** American Sparrow Hawk.

We saw this species at Log Cabin June 14, Semenow Hills July 19 and 20, near the Tatchun River July 23, near Fort Selkirk July 26, about 30 miles below Fort Selkirk July 28, and, I think, at Circle August 15. We took specimens on July 19 and 28. This species has not previously been reported from along the Yukon.

---

[1] Osprey, III, 25, Oct., 1898.

[*Pandion haliaëtus carolinensis.* American Osprey. While the steamer was anchored near Holy Cross Mission August 25, one of the passengers, Mr. J. F. Burkman, fired at, but failed to get, a large hawk which he was positive was this species. As Mr. Nelson records it from the Lower Yukon, I see no reason to doubt Mr. Burkman's identification.]

87. **Asio accipitrinus.** Short-eared Owl.

We saw a short-eared owl flying overhead at St. Michael on the evening of September 7, and I flushed one from some bushes on Whale Island the next day. September 9 I set three steel traps near St. Michael on poles in the tundra. One of them failed to catch anything, but before September 25 the others yielded 6 short-eared owls and the nails of another. These birds had been feeding on mice and shrews. Osgood shot a short-eared owl at Unalaska October 5. These specimens average slightly *darker*, with the white of the face purer, than fall birds from New England.

88. **Scotiaptex cinerea.** Great Gray Owl.

From some low growth on a steep hillside at Miles Canyon July 11 we flushed a large gray owl that I am confident was this species. We saw a mounted specimen in Dawson August 2 and I was told at Circle that an owl answering the description of this species had been killed there recently.

89. **Nyctala tengmalmi richardsoni.** Richardson Owl.

While lying awake under my mosquito netting in a clearing at the base of the Semenow Hills on the night of July 19 I saw a small, round-headed owl alight on the limb of a dead tree only a few feet away. It flew before I could bring my gun to bear, but I have no doubt it was this species. Osgood took a young bird near Rink Rapids July 22. I was told at Circle that a small owl was common there, and that one had been caught recently.

90. **? Megascops asio kennicotti.** Kennicott Screech Owl.

A reddish-brown owl, of the size and appearance of a screech owl, was seen by Maddren and myself at Caribou Crossing on the afternoon of June 27. We were drawn from camp by its peculiar notes, and saw it fly from a poplar across an opening to a spruce thicket. Later that day Osgood caught a glimpse of another, or perhaps the same bird, as it flew from the top of a small poplar.

91. **Bubo virginianus pallescens.** Western Horned Owl.

Owl pellets, some of them remarkably large, containing chiefly bones of rabbits, ground squirrels, and red squirrels, were found in great abundance, especially at Caribou Crossing and on Windy Island, Lake Tagish, but the most careful hunting failed to disclose the owls.

On Fifty-Mile River, near Lake Marsh July 8, we heard the hooting of a horned owl; and at our camp at Lower Lebarge one flew over, about midnight July 16, and lit in the top of a spruce just out of gunshot. I hurried after it but merely succeeded in seeing the bird swoop into the surrounding gloom.

At our camp near the Tatchun River July 22 one flew by and settled for an instant not far off; and the next day Osgood saw three extremely light-colored horned owls near by. We also heard the hooting of this species near the Yukon at the following places: Near Little Salmon River July 21, 20 miles below Fort Selkirk July 27, 20 miles below the Selwyn River July 29, near the Tatondu River August 6, about 15 miles above Circle August 12, and opposite Circle August 14. In the last case the identification is not without doubt, but the notes of the others were unmistakable.

[*Surnia ulula caparoch.* American Hawk Owl. At Bennett, June 18–22, a bird with a peculiarly weird cry flew about the cliffs above our camp every night. By a process of elimination I have attributed the serenade to this species.]

92. **Ceryle alcyon.** Belted Kingfisher.

This bird occurs about the Yukon lakes, but in small numbers. Osgood saw one at Bennett June 20; I heard one at Caribou Crossing June 29, and saw another on Lake Lebarge July 13.

We found kingfishers fairly common on Fifty-Mile River, and still more common on Thirty-Mile River. As the cliffs replaced the high banks below Fort Selkirk kingfishers became fewer, and none were seen after August 4, when we were about 40 miles below Dawson. Young able to fly were seen near Five-Finger Rapids July 22.

93. **Dryobates villosus leucomelas.** Northern Hairy Woodpecker.

Osgood took a single hairy woodpecker on Fifty-Mile River a few miles above Miles Canyon July 10.

94. ? **Dryobates villosus hyloscopus.** Cabanis Woodpecker.

Near the Little Salmon River July 21 I took a young female that corresponded in size and plumage with some young of this subspecies. It was seated in the entrance to a cavity in a burnt spruce. This is the first record of the occurrence of this bird in the Yukon Basin.

95. **Picoides arcticus.** Arctic Three-toed Woodpecker.

On July 1 I was attracted by the loud cries of a young bird, and traced the sound about 100 yards through a spruce grove on the bank of Six-Mile River. The noise proceeded from a full-fledged young woodpecker of this species that had thrust its head out of the opening to its house and kept up a continual screaming. With Osgood's assistance, the nest was opened but only this young bird was found. The entrance was 5 feet 8 inches from the ground, on the lower side of a living, slightly leaning spruce, and the cavity was 10 inches deep. Osgood shot what we both supposed was one of the parents, for it certainly came in answer to the cries of the young; yet this bird proved to be a typical adult male of *P. americanus alascensis*. We saw no other woodpeckers there, except flickers.

96. **Picoides americanus alascensis.** Alaska Three-toed Woodpecker.

Osgood found the remains of an Alaska three-toed woodpecker at Haines June 1, and I shot a laying female near Glacier June 10. In the Yukon Valley we secured one on Six-Mile River; three on Fifty-Mile River above Miles Canyon July 10–11, two of them young adults; two on the Lewes River between Big Salmon and Little Salmon rivers July 20–21, and two at Circle, August 19–20. The young have whiter backs than the adults.

97. **Sphyrapicus ruber.** Red-breasted Sapsucker.

I took an adult male at Skagway May 31, and heard what I suppose was its mate.

98. **Colaptes auratus luteus.** Northern Flicker.

We saw and heard flickers several times at Glacier. One, which Osgood flushed from a hole high in a dead pine June 8, had yellow quills. In the Yukon Valley this is by far the most common woodpecker. We found it quite regularly from Log Cabin to Circle, but, like most Yukon birds, it was shy. At Caribou Crossing June 27 Osgood secured a female and found her nest, containing 8 young and 3 eggs, in a cavity 3 feet from the ground in a partly dead poplar. At Six-Mile River we found a nest about 6 feet from the ground, and at Lower Lebarge July 17 I found 7 well-fledged young in a cavity about 5 feet from the ground in a small dead tree in a burnt tract. July 25 I took a full-grown young near Selkirk.

Adult flickers from Alaska average slightly darker than *luteus* from Canada and farther south, the wings, tail, and bars of upperparts being somewhat blacker, and the light parts more olive and less buffy. Three young—one from near Fort Selkirk, the others nestlings from Lower Lebarge—show this difference in a marked degree, having the wings, tail, and bars of upperparts deep black, and the ground color above smoky olive, instead of buffy olive as in *luteus;* they are even darker than the young of *auratus* from Florida. But the slightness of the difference shown by the adults, the small number of specimens from Alaska, and the possibility that the plumage of the three young may have been discolored by the burnt trees where they were found—though microscopic examination shows no sign of this—make their separation as a subspecies inadvisable at present.

99. **Chordeiles virginianus.** Nighthawk.

From Caribou Crossing, where I shot two females June 27, until after passing the mouth of the Tatchun River July 24, we met with nighthawks on numerous occasions. I took an adult male at White Horse Rapids July 11. These birds were very fat, as might be expected from the abundance of flying insects. They are slightly darker than *virginianus* from the East.

100. **Selasphorus rufus.** Rufous Hummingbird.

We saw a rufous hummingbird on 'Eagle Island' at Bocadequadra May 28. At Glacier Osgood saw one June 6, and on June 10 I found a nest with two slightly incubated eggs 3½ feet from the ground on the branch of a small conifer near the falls of the river. I secured the female, and also one of two males which I saw the same day in the open country below Glacier. On Lake Bennett we saw one opposite West Arm June 24. Mr. George G. Cantwell has already added both this species and *Chordeiles virginianus* to the Yukon avifauna.[1]

101. **Sayornis saya yukonensis.** Yukon Phœbe.

Sayornis saya yukonensis Bishop, Auk, XVII, 115, April, 1900.

Osgood took the type specimen of this phœbe on the heights above Glacier June 8, and I saw one on the mountainside at Bennett June 17. We next met the bird about some cliffs below Fort Selkirk July 26, and after this saw family parties almost daily. Near Stewart River July 31, we saw a pair about their nest on the face of a cliff a few feet above the water. After passing Charlie Creek August 10, we saw no more until we reached Circle, where I killed a young one August 19. Full-grown young were taken July 30. The note is harsh, somewhat resembling that of *Contopus richardsoni*, but louder and shriller. We found the birds only about the cliffs, or the steep, grass-grown banks of the Yukon, a favorite perch being rocks along the shore. Those we met in August seemed to be migrating up the river.

102. **Contopus borealis.** Olive-sided Flycatcher.

At Six-Mile River I took a pair July 1, the female of which had finished laying. A bird which I heard near Bennett June 20, and a large flycatcher which I shot, but could not find, at Caribou Crossing June 25, I believe were this species.

103. **Contopus richardsoni saturatus.** Alaska Wood Pewee.

Contopus richardsoni saturatus Bishop, Auk, XVII, 116, April, 1900.

Osgood took a wood pewee at Skagway May 30, and I two males at Haines June 2. In the Yukon Valley, from Windy Island, Lake Tagish, where I took a male June 30, until we passed Little Salmon River July 21, we often heard this bird's 'pee-ah' coming from the

[1] Osprey, III, 25, Oct., 1898.

wooded banks. We next saw the bird about 12 miles above Circle, where I took a pair August 14. It was more common at Miles Canyon than elsewhere on the Yukon, and here on July 11 I found an unfinished nest (which resembled that of *C. richardsoni*) in the fork of a half-dead poplar about 10 feet from the ground. No form of wood pewee has previously been recorded from the Yukon.

104. **Empidonax trailli alnorum.** Alder Flycatcher.

We first found this species July 26 at Fort Selkirk, where the Pelly River, from the Rocky Mountains, joining the Lewes, forms the Yukon, and hardly lost it again until we reached Circle; later I heard one 15 miles below Fort Yukon on August 21. Wherever we landed we found this or the Hammond flycatcher in the alders and willows. Full-grown young in juvenile plumage were taken on August 5. The adults are apparently typical *alnorum*, having the greener upperparts, more conspicuous wing bars, and shorter bill of this form.

105. **Empidonax hammondi.** Hammond Flycatcher.

We saw several Hammond flycatchers at Skagway, and collected three. I took one at Glacier June 8, and another on a hill above Caribou Crossing June 26. After this we did not again meet with the bird until about 15 miles below Selwyn River, where Osgood shot a young one July 29. From that point to Charlie Creek it was almost equally common with *Empidonax t. alnorum*, frequenting the same localities; but after passing Charlie Creek, August 9, we saw no more of it. The young secured were molting. The male collected at Caribou Crossing is unusually pale for *hammondi*, but this is doubtless the result of wear, as the same thing is shown in *Contopus saturatus* and *Hylocichla almæ*.

106. **Pica pica hudsonica.** American Magpie.

Maddren saw a pair at Caribou Crossing June 26, and Osgood found their deserted nest. At Fort Selkirk July 26 I took two young—male and female—which had just assumed first winter plumage. They were feeding about the houses of the town. I was told that another young bird had been seen there recently.

107. **Cyanocitta stelleri.** Steller Jay.

Osgood found the remains of a Steller jay in the woods at Haines June 1.

108. **Perisoreus canadensis fumifrons.** Alaska Jay.

We first met this bird at Log Cabin, noted it also at Bennett and Caribou Crossing, and found it common from Lake Marsh to Circle, generally in families. Between White River and Circle it was less common than farther up the Yukon. I saw one 15 miles above Fort

Yukon, heard several at Hendricks Station August 25, and saw one at St. Michael September 18.

Adults had completed the summer molt by July 20; the young were in full juvenile plumage on June 20, and in first winter plumage on August 20. The molt is complete in the adults, while in the young the wings and tail remain unchanged.

The adults collected are all intermediate between *capitalis* and *fumifrons;* each has a black orbital ring, but this is broader in those from Circle. All those in juvenile plumage have the head dull plumbeous, like the back, as in *fumifrons*.

109. **Corvus corax principalis.** Northern Raven.

Of all the birds we met the raven occurred most regularly. On our entire trip down the Yukon hardly a day passed without our seeing the birds in twos and threes. We saw a few at Wrangell, found them more common at Wrangell Narrows, saw several at Skagway, and noticed the wing of one at Glacier. A few were noted across White Pass at Middle Lake and they were abundant at Log Cabin. A flock of at least 200 was observed at the latter place June 20, and another of 50 at Bennett two days later. During September at St. Michael we saw them frequently, but never in large numbers. At Unalaska they were abundant and remarkably tame.

An adult taken on June 20 is in full molt; a young taken July 22 is in juvenile plumage; on one taken August 23 the body feathers of the first winter plumage have replaced most of the juvenile, and the change is complete in one taken September 9.

[*Corvus americanus.* American Crow. I was told by one of the prospectors whom I met on a Yukon steamer that the crow, as well as the northern raven, occurs at the head waters of the Porcupine.]

110. **Corvus caurinus.** Northwest Crow.

Common on 'Eagle Island' at Bocadequadra, where Osgood found a finished but empty nest May 28. Crows were very common near Vancouver June 26, but we saw none after leaving Bocadequadra.

111. **Scolecophagus carolinus.** Rusty Blackbird.

Two blackbirds which I saw at Log Cabin June 15 were probably this species, and I was told that rusty blackbirds had been abundant there a few days before our visit.

Osgood took a specimen near Fort Yukon August 21, and I saw a small flock at the Aphoon mouth August 28. I was informed that these birds breed in large numbers on the tundra by the Kuskokwim and at the head of the Porcupine.

[*Coccothraustes vespertinus montanus.* Western Evening Grosbeak. A prospector told me that a grosbeak, whose description answered that of this species, was common on the Copper River. He assured me it was not the pine grosbeak, which he knew well.]

[*Pinicola enucleator alascensis.* Alaska Pine Grosbeak. A red bird with dark wings—certainly not a crossbill—which I saw at Lake Marsh July 8 was probably a pine grosbeak, but we did not meet with any others during our trip. I was told this bird occurs along the Porcupine.]

112. **Loxia curvirostra minor.** Red Crossbill.

Osgood took a red crossbill and saw another at Unalaska October 5. We did not take any along the Yukon, but I feel positive that a red male crossbill which I shot at Lake Lebarge July 16, but could not find, belonged to this species.

113. **Loxia leucoptera.** White-winged Crossbill.

Crossbills in flocks of from half a dozen to one hundred individuals were often seen from Lower Lebarge to Charlie Village July 16 to August 11. Most of these flocks were probably *leucoptera*, and some certainly were. They were exceedingly restless, and the only ones taken (besides those found in the crop of a duck hawk), were three young at Camp Davidson August 5–6.

114. **Leucosticte griseonucha.** Aleutian Leucosticte.

We saw a number of Aleutian leucostictes on St. George October 3. At Unalaska I saw a flock of about twenty and another of two young October 5, and secured an adult and one of the young. The latter is in juvenile plumage, feathers of the first winter appearing only on the sides of the chest.

115. **Leucosticte tephrocotis littoralis.** Hepburn Leucosticte.

We found this bird only at White Pass Summit, where Osgood took two males and I one female June 13. It is doubtless this species of *Leucosticte* to which Cantwell refers in his paper on the 'Birds of the Yukon Trail.'[1]

116. **Acanthis hornemanni exilipes.** Hoary Redpoll.

I secured two young from a flock about 15 miles above Circle August 13, and Osgood one from a flock at Circle August 19. I saw several at the Aphoon mouth August 27, and we found them rather common in small flocks at St. Michael during September. All taken were young and were molting from juvenile to first winter plumage.

117. **Acanthis linaria.** Redpoll.

We saw several, usually in pairs and very shy, at Bennett June 17. One stopped for an instant on a bush close to our tent. Near Charlie Village I saw a male in high plumage August 11.

[1] Osprey, III, 25, Oct., 1898.

118. **Spinus pinus.** Pine Siskin.

A very restless family of this species was seen on Windy Island, Lake Tagish, June 30, and Osgood secured one. I saw one at Lower Lebarge July 18, and took one from a small flock near the Selwyn River July 29, and Osgood one from a large flock near Sixty-Mile Creek July 31. We saw a large flock near Dawson August 1, a few near Forty-Mile Creek August 4, and Osgood saw one 15 miles above Circle August 12. Flocks of either this bird or redpolls were heard near the Tatondu River and Charlie Creek August 7–8. I find no former record of this species for the Yukon Valley.

119. **Passerina nivalis.** Snowflake.

At White Pass Summit I shot a female June 12 that had an old fracture of the wing, which had healed in such a manner as to make long flight impossible. I was informed snowflakes had been very abundant there earlier in the year. At St. Michael I saw two September 16, and a flock of about twenty September 19. Osgood took one from a small flock September 25, and I three on September 28.

Snowflakes were common on St. George October 3, but the two young taken (♂ and ♀) are indistinguishable from those from St. Michael, and have bills smaller than the young of *townsendi*.

120. **Calcarius lapponicus alascensis.** Alaska Longspur.

I saw several small flocks at the Aphoon mouth August 27, and secured one specimen. A few were found at St. Michael the last of August, and large flocks there September 1–2. After that several were seen almost every day until September 22, when the last were taken. Osgood saw several at St. George October 3, and I saw one at Unalaska October 5.

121. **Ammodramus sandwichensis.** Sandwich Sparrow.

A few were seen at Unalaska October 5–6, and two young secured.

122. **Ammodramus sandwichensis alaudinus.** Western Savanna Sparrow.

I saw several savanna sparrows on the marshes of Chilkat Inlet June 1, and we took one at Haines, one at Skagway, and two at Glacier. Several pairs were found on the marshes near Log Cabin, a few at Caribou Crossing, and one pair on an island in Lake Tagish. They were fairly common about Lake Marsh, and Osgood found a set of four eggs there, securing the female July 5. After leaving Lake Marsh these sparrows were not seen again until we reached the Alaska boundary, when I took a young August 5. Osgood took a young specimen from a flock near Charlie Village August 10, and young were common at Circle August 14–19. I saw a number at the Aphoon mouth August 27–28, and we found a few at St. Michael up to September 11.

Breeding specimens from the Yukon lake region are indistinguishable in size and color from *alaudinus* from North Dakota. Those from Haines and Glacier are larger in bill and other measurements, slightly darker, and more buffy, but evidently belong to the same form. A male taken at Skagway May 31 is identical in color with a female *savanna* taken in Connecticut about the same date, but in measurements intermediate between *sandwichensis* and *alaudinus*. Young birds from St. Michael and the Yukon below Camp Davidson resemble closely the young of *savanna* in coloring, and have bills slightly shorter and deeper than adult *alaudinus* from the Yukon lakes, but are larger and have longer wings and tails than the latter.

123. **Zonotrichia leucophrys gambeli.**[1] Intermediate Sparrow.

Descending from the bleak, snow-covered rocks of White Pass, we reached at Portage June 14 a country of a more luxuriant vegetation. Here the intermediate sparrow appeared, and it stayed with us constantly until we left Circle, August 20. At Fort Gibbon August 23 I saw one adult and one young. With the exception of the bank swallow, this is the most abundant species inhabiting the Yukon Basin.

At Log Cabin June 20 I took a laying female; on Windy Island June 30 Osgood took a young, able to fly, and at Lower Lebarge I shot one molting into first winter plumage. We found young abundant in this plumage at Circle August 15–20, but saw no adults. This species has two distinct songs. That most often heard is a very mediocre performance, but the other, which I heard in its full perfection only on a hill at Caribou Crossing June 26, and about 2 a. m. on Fifty-Mile River July 9, possesses all the sweetness and clearness of the song of *Z. albicollis*. By July 15 the song season was practically over, but I heard one bird singing as late as August 10.

124. **Zonotrichia coronata.** Golden-crowned Sparrow.

Osgood found the golden-crowned sparrow on the heights above Glacier June 5. It was common at White Pass summit June 11–14, and was the only bird we saw along the trail to Portage June 14. We thought we heard it singing at Log Cabin. The song does not equal those of others of the genus. Osgood found an almost finished nest in a conifer at Summit Lake June 12. It was composed of sticks and moss, lined with grass, and placed about 2½ feet from the ground. The next day I shot a female that contained an egg ready for the shell.

125. **Spizella monticola ochracea.** Western Tree Sparrow.

At Haines I took a female June 2. At Caribou Crossing we took two pairs June 29, one of them with a nest containing three fresh eggs. The nest was buried in the moss at the base of a clump of willows in

---

[1] This is the species formerly known as *Zonotrichia leucophrys intermedia* Ridgway. See Ridgway, Auk, XVI, 36–37, 1899.

a willow swamp near the lake, and it was composed of fine, dry grasses. lined with feathers, covered externally with a thick coating of living moss. The eggs, which average 0.80 by 0.57 inches, are pale pea green, heavily mottled over the entire surface with reddish fawn color. At Lake Marsh July 8 I took an adult female, and 15 miles above Circle August 8 a young bird molting from the striped juvenile into the fall plumage. The species was abundant at Circle, and a number were seen on an island 15 miles above Fort Yukon August 21. I saw one at the Aphoon mouth August 27, and noticed seven during September at St. Michael, taking the last September 21.

126. **Spizella socialis arizonæ.** Western Chipping Sparrow.

We found this species almost daily from Log Cabin to Dawson, or between June 15 and August 1. In point of numbers it follows the intermediate sparrow and the slate-colored junco. It was last observed about 10 miles below Dawson August 3, but the range of the species may extend much farther north, as a large flock seen near the Selwyn River July 29 showed that the fall migration had begun.

We found a nest with four eggs at Lake Bennett June 24, large young in a nest on Lake Tagish June 30. Young able to fly were met with at Lake Marsh July 5, and a set of three eggs on Thirty-Mile River July 18. The nests were in small spruces, one 4 inches and another about 3 feet from the ground.

Yukon chipping sparrows, females especially, average darker than typical *arizonæ*, but coincide in measurements. Turner reports this species from Fort Yukon.[1]

127. **Junco hyemalis.** Slate-colored Junco.

From Log Cabin to Circle this bird occurs everywhere, contesting with the intermediate sparrow for supremacy in numbers. Two broods are, I think, regularly reared. Females taken at Log Cabin and Bennett had finished laying. On Windy Island June 30 I shot a young bird able to fly, and on the east shore of Lake Tagish the following day saw one pair building a nest and another feeding young. Maddren found a nest with four fresh eggs at Lake Marsh July 4, Osgood one with three fresh eggs at Lower Lebarge July 16, and I one with five young on Thirty-Mile River July 18, and another with four just hatched young near the Tatchun River July 23. By July 20 young in striped plumage were common, and August 2 I took one near Dawson molting into first winter plumage. The slate-colored junco, the intermediate sparrow, and the western chipping sparrow were most common about brush heaps left by lumbermen, weed-grown clearings resulting from forest fires, and cabins of the towns. Every nest found was sunk in the ground to the rim in an open place

[1] Contrib. Nat. Hist. Alaska, 174, 1886.

under a weed or a tussock of grass. One contained a few dark hairs besides the usual fine grass lining. Twenty adults differ from eastern summer specimens of *hyemalis* only in having in both sexes bills averaging 0.02 inch longer (measured from the nostril).

128. **Junco hyemalis oregonus.** Oregon Junco.

Tolerably common at Skagway and more so at Haines. At Skagway I took a female and four fresh eggs May 31. The nest, of dried grass lined with short, white hairs, was sunk in the ground and concealed by dead weeds under a birch only about 30 feet above the water of Lynn Canal.

129. **Junco hyemalis connectens.** Shufeldt Junco.

Maddren took a female at Glacier June 7, a male was taken near White Pass City June 9, and Maddren saw several near there that day. I took a male below Glacier June 10, and saw and heard a number singing a few hundred feet above White Pass City, where the spruce woods gave place to more open country. Their song is quite distinct from that of *oregonus*. This is a new record for Alaska.

130. **Melospiza melodia rufina.** Sooty Song Sparrow.

We heard several singing at Skagway May 31, and Osgood saw some at Haines June 2. At Haines I took a male June 1, and a pair, the female of which had finished laying, June 2.

131. **Melospiza cinerea.** Aleutian Song Sparrow.

Abundant at Unalaska, October 5–7, frequenting the roofs of buildings, lumber piles, wharves, beaches, and weeds of the level country and hillsides. The males were singing constantly, their song having the usual song sparrow character, but not the usual strength or beauty.

132. **Melospiza lincolni.** Lincoln Sparrow.

At Log Cabin June 15 we saw what was apparently a Lincoln sparrow. Osgood took a female and a set of five fresh eggs near Lake Marsh July 5, I another female on the west shore July 8, and we heard several singing near the lake. The nest found was composed of coarse grass lined with fine, and was in a tuft of grass in a swamp, about 4 inches above the water. We again met this species at Lower Lebarge, near Fort Selkirk, near the White River, at Camp Davidson, at Charlie Village, 15 miles above Circle, and at Circle, where one was taken August 19. July 27 a full-grown young was taken, and August 12 one that had almost finished molting into winter plumage.

133. **Melospiza lincolni striata.** Forbush Sparrow.

A Lincoln sparrow which Osgood saw at Haines June 1 should be referred to the northwestern subspecies.

134. **Passerella iliaca.** Fox Sparrow.

A wave of sparrows occured at Circle August 19 just after a frosty night, and among other species I saw a single fox sparrow. The bird was too close to leave identification doubtful.

135. **Passerella iliaca townsendi.**[1] Townsend Fox Sparrow.

Osgood saw one at Skagway, and we noticed several at Glacier which were exceedingly shy. Osgood collected two at Glacier June 8–9, one of which was too badly shot to preserve: the other Mr. Ridgway pronounces somewhat nearer this form than *annectens*.

136. **Petrochelidon lunifrons.** Cliff Swallow.

This species was common at Log Cabin June 15 and 20. At Caribou Crossing we saw a few June 29, probably members of the small colony breeding on the cliffs of an island in Lake Tagish July 1. We next saw cliff swallows near the Hootalinqua River July 19, and from this point to a few miles above Dawson, August 1, we frequently met with colonies of varying size, the largest being near White River. Their nests were attached to cliffs bordering the river, except at Fort Selkirk, where they were breeding under the eaves of houses. Full-fledged young were taken July 25, and I think the absence of this species below Dawson was due to their having already migrated. I was told that both cliff and bank swallows were exceedingly abundant along the Porcupine.

137. **Hirundo erythrogastra unalaschkensis.**[2] Alaska Swallow.

A few were flying over the marshes of Chilkat Inlet June 1; I heard that they were common at White Pass City June 9, and we saw two about the buildings of White Pass Summit June 10. At Log Cabin they were common on June 14, 15, and 20, and on the last date I took a male. A few were noticed at Bennett June 19–21. I refer all seen to this subspecies, for all had remarkably long tails. The single specimen taken had a length of 7.96 inches, wing 4.68 inches, tail 4.10 inches, fork of tail 2.33 inches. The forehead, lower wing-coverts, and abdomen are more highly colored than eastern skins of *H. erythrogastra*, and the shafts of the long tail feathers are whitish.

138. **Tachycineta bicolor.** Tree Swallow.

I saw several at Skagway May 31 and June 3, and over the Chilkat marshes June 1. We saw others near Caribou Crossing June 29; one July 6 and a pair July 7 at Lake Marsh; and several at Miles Canyon July 11. A few miles above Fort Selkirk July 25 I saw several entering and leaving an old flicker hole in a dead spruce.

---

[1] Auk, XVII, 30, Jan., 1900.

[2] Rept. Fur-Seal Investigations, 1896–97, pt. 3, 422, 1899.

139. **Tachycineta thalassina.** Violet-green Swallow.

Mr. Cantwell[1] has already added this species to the list of birds known to inhabit the Yukon Valley. We saw a single male among flocks of bank swallows flying over Fifty-Mile River above Miles Canyon July 11 and another between White Horse Rapids and Lake Lebarge. On July 18 I took a male from several that we saw near Hootalinqua, and at the Semenow Hills July 20 Osgood secured a female, finding her nest with four young in a crevice in the cliffs. Maddren shot a young July 28. After this we frequently saw colonies of from six to ten birds of this species, and one near White River that must have contained over fifty.

They were nesting about the cliffs as a rule, but several times we saw them enter holes in banks similar to those of *Clivicola riparia*, while at Fort Selkirk they were nesting in the interstices between the logs of the cabins. We often met with small colonies until within 15 miles of Circle, but after August 5 they kept so high about the cliffs that identification was possible only by their characteristic twitter. The two adult males have green rumps.

140. **Clivicola riparia.** Bank swallow.

We found a small colony nesting at the northern end of Lake Tagish July 1, and a larger one on the west shore of Lake Marsh July 7, but we were entirely unprepared for the great abundance of this species on Fifty-Mile River above Miles Canyon. There almost every bank was honeycombed with their holes. Along the rest of the Yukon as far as Circle bank swallows were common and often abundant, but after August 1 their former presence was generally manifested only by the deserted holes. At Circle I saw about thirty August 17, and a single bird on the following day. Eggs advanced in incubation were found July 7, and by the 22d the young were flying, and all acting as if preparing to migrate. As it grew dusk on the evening of August 5 we watched a large flock which circled over the Yukon, rising higher with each revolution, and at last disappeared toward some mountains due south.

141. **Ampelis garrulus.** Bohemian Waxwing.

We saw several on Six-Mile River July 1, two at Lake Marsh July 7, one on Fifty-Mile River July 10, two pairs (one of which was secured) at Miles Canyon July 11, one at Lower Lebarge July 16, two about a mile apart on Thirty-Mile River July 18, and four near the Selwyn River July 28. We took two adults and three young from a flock of twenty about 15 miles below the Selwyn July 29, and four from a similiar flock near Sixty-Mile Creek July 31. We saw them again at the Chandindu River August 4, Camp Davidson August 5,

[1]Osprey, III, 25, Oct., 1898.

50 miles above Circle August 11, and 15 miles lower August 12. The female taken July 11, which lacked the wax tips on the secondaries, contained an egg ready for the shell. The young resemble those of *A. cedrorum*, but are grayer, have less white on the abdomen, no pale streaking above, and have the wings, tail, and lower tail-coverts like adult *garrulus*. They lack the cinnamon suffusion of the head of the adult, have only a few black feathers on the throat, a much shorter crest, the wax-like tips of the secondaries peach-blossom pink instead of scarlet, and the lower tail-coverts paler. A still younger bird than the two described is obscurely streaked with whitish both on back and lower parts. On one of the young the wax tips are very small.

In habits and notes the Bohemian waxwing closely resembles the common cedar waxwing. Two males that we noticed while descending Thirty-Mile River were perched on the topmost sprays of tall spruces, uttering a lisping whistle at frequent intervals. One of them flew after a passing insect in the manner of a flycatcher. Flocks were easily approached, and when one bird was shot the rest would scatter, and each would alight on the top branch of some spruce and utter a characteristic call note. This note, which we often heard from passing flocks, was similar to the whistle just mentioned. The birds that we collected had been feeding on the purple berries of some unidentified plant.

142. **Helminthophila celata.** Orange-crowned Warbler.

Osgood took an adult male at Caribou Crossing June 26; I a female and two young 20 miles below Fort Selkirk July 27, and a young near Dawson August 2. Osgood secured an adult and one young at Camp Davidson August 5 and 6, and I saw one young 15 miles above Fort Yukon August 21. All taken were in alders or willows close to the water.

143. **Helminthophila celata lutescens.** Lutescent Warbler.

Common at Haines, where we took five June 1 and 2.

144. **Helminthophila peregrina.** Tennessee Warbler.

Found only at Caribou Crossing, where I heard four males singing and secured three of them June 25 and 27. They were in comparatively open swamps of willows and low spruces.

145. **Dendroica æstiva rubiginosa.**[1] Alaska Yellow Warbler.

I am positive I often heard the song of this species at Bennett June 17–22. I took an adult male at Caribou Crossing June 27, and think I heard the song about Lake Marsh. An adult female was taken by

[1] Auk, XIV, 76, 123, 1897.

Osgood near the Nordenskiold River July 22, and family parties were often found in the alders and willow thickets between the Pelly River and Circle. I took a young from a small flock 15 miles above Fort Yukon August 21, saw one at the Aphoon mouth August 28, and a few I thought this species at Hendricks Station August 25. Birds from the Yukon Valley do not differ from those of the Alaska coast. A young female is duller above and more buffy below than the young female of *D. æstiva*.

146. **Dendroica coronata hooveri.**[1] Hoover Warbler.

We found Hoover warblers common at Skagway, Glacier, Log Cabin, and Caribou Crossing, and also noted them at Haines, Bennett, Lake Tagish, Miles Canyon, White River, Sixty-Mile Creek, and 12 miles above Circle. At Skagway May 31 they were still in flocks, but at Glacier June 4–10 they seemed to be mated and settled for the summer. At Log Cabin we found a flock June 15, but five days later those still remaining there were beginning to nest. A small flock seen on an island near Sixty-Mile Creek August 1 showed that the return migration had begun. I took a young in striped plumage August 1. Adult males average paler below than typical *D. coronata*, the black markings being narrower, thus giving an effect of broad longitudinal markings rather than black clouding on the chest. Eight specimens of both sexes average slightly larger in length of wing and tail than the corresponding sexes from eastern and central United States. In six males, the exposed culmen averages 0.02 inch longer than in males from Connecticut, but the bill from nostril averages the same, as do both measurements in females. In juvenile plumage *hooveri* is darker than *coronata*, the black markings are broader and blacker, both above and below, and the brownish edgings to the feathers greatly restricted—entirely wanting on the lower parts and middle back.

147. **Dendroica striata.** Black-poll Warbler.

At Log Cabin June 15 this species was common, but on my return June 20 I saw only one pair—which I secured—and one other male. July 5 I took a male at Lake Marsh. Two birds taken at Caribou Crossing are somewhat smaller than average specimens from Dakota and Connecticut.

148. **Dendroica townsendi.** Townsend Warbler.

Osgood took a male at Skagway May 31. At Glacier it was tolerably common in the dense woods of spruce and fir, and unquestionably nesting; altogether we noticed about twenty individuals during our stay. Osgood took an adult at the southern end of Lake Marsh July 1, and I an adult female and young female on the west shore of Lake

[1] Bull. Cooper Ornith. Club, I, 32, 1899.

Lebarge July 14. The juvenile plumage differs from that of *D. virens* only in being slightly less brown on crown and back. This is a new species for the Yukon Valley.

149. **Seiurus noveboracensis notabilis.** Grinnell Water Thrush.

The first sound that I heard on the morning of August 1, when we were on a small island about 10 miles below Sixty-Mile Creek, was the unmistakable alarm note of the water thrush. This was the first time we had met with this species, and before starting that morning on our daily Yukon drift, Osgood and I each secured a young bird. Near Forty-Mile Creek, Tatondu River, and Charlie Creek water thrushes were again met with. At Circle I saw several August 16–20, took one 15 miles above Fort Yukon August 21, and saw two in a thicket at the Aphoon mouth August 28. The young in fall plumage taken on the Yukon are clove-brown above, including wings and tail—far darker than is usual in *notabilis*—and have darker streaks below.

150. **Wilsonia pusilla.** Wilson Warbler.

Osgood took an adult female near the Chandindu River August 4, and I a young female near Charlie Village August 11 and a young male 25 miles above Circle August 12. I also saw one 30 miles below Circle August 20. These birds, while not typical *pusilla*, are, like those of the Lower Yukon, nearer it than *pileolata*.

151. **Wilsonia pusilla pileolata.** Pileolated Warbler.

We found this the most abundant bird at Glacier June 5–10, frequenting the alder thickets from the valley as far as they extended up the hills. I saw a yellow warbler I thought this species on White Pass Summit June 12. Pileolated warblers were common at Log Cabin, Bennett, and Caribou Crossing, and I am confident I heard them singing at Lake Marsh. Adult males from Glacier resemble normal *pileolata* closely, but have the back rather more green; those from the Yukon Valley, while having the orange forehead and lower parts of this form, have the duller green back of *pusilla*.

[*Motacilla ocularis.* Swinhoe Wagtail. On the morning of August 28 the *Robert Kerr*, on which I was a passenger, was hindered from proceeding by a gale and low water on the bar, and was made fast to the bank at the Aphoon mouth of the Yukon. As I came on deck I saw half a dozen white wagtails fly about the vessel and settle in the grass close by. While I returned for my gun they left, but a thorough acquaintance with *Motacilla alba* in Egypt, where it is abundant during the winter, leaves me no doubt that these birds were wagtails.]

152. **Anthus pensilvanicus.** American Pipit.

A male taken at Skagway June 3 was probably a belated migrant. On the heights above Glacier Osgood saw several June 5, and we found them common at Summit June 11–13. A female taken June 13 was laying, and a fresh but empty nest I found the same day I attributed to this species, no other being near. This nest was loosely formed of fine dry grass in a hollow in the deep moss which covered the almost perpendicular side of a bowlder lying on a hill high above Summit, only a small hole for entrance showing in the moss. We often saw the song-flight at Summit. Launching himself with a sharp preliminary 'chip' from one of the granite bowlders that abound there, the male would rise rapidly to a height of a hundred feet or more, uttering a sweet, clear song. After poising high in air and repeating this song for several minutes the singer would slowly float toward earth and alight 100 yards from where he started, soon to repeat the same performance. We found a pair on the heights above Bennett June 17, and a few, possibly members of one family, at Circle August 15–20.

153. **Cinclus mexicanus.** American Dipper.

We collected a female and set of four fresh eggs at the falls at Glacier June 8. A single ouzel seen farther down the river June 10 was probably the mate of the one taken. Osgood also took one at Unalaska October 5.

154. **Anorthura hiemalis pacifica.** Western Winter Wren.

We noticed a few at Glacier June 4–10, and I took a male there June 6.

155. **Anorthura alascensis.** Alaska Wren.

I saw one at St. George October 3, and we collected five at Unalaska October 5. The young were then molting.

156. **Sitta canadensis.** Red-breasted Nuthatch.

I took a male at Skagway May 31 and another near Log Cabin June 20, and heard one on an island at the junction of the Lewes and Pelly rivers July 26. This species has not heretofore been noted in the Yukon Valley.

157. **Parus atricapillus septentrionalis.** Long-tailed Chickadee.

We took this species at Bennett June 19, west shore of Lake Bennett June 24, Caribou Crossing June 26, Lake Marsh July 7, and Lake Lebarge July 15, but did not notice it again until we reached the Lower Yukon, although chickadees were heard several times whose specific identity was not determined. Thirty miles below Holy Cross Mission I took two August 25, and at the Aphoon mouth I saw a small flock August 28. Young able to fly were taken July 7. One taken August 25 had completed the molt into first winter plumage, while an adult taken the same day was in fresh plumage.

158. **Parus hudsonicus evura.**[1] Yukon Chickadee.

We took the Yukon chickadee at Caribou Crossing June 27, Lake Tagish June 30, Lake Marsh July 5, and Lake Lebarge July 14, and after reaching Thirty-Mile River July 19, found it regularly distributed in families or large flocks all the way to Fort Yukon, 15 miles above which I saw a flock August 21. At St. Michael I took a young female in first winter plumage September 20. Young able to fly were first taken July 5 and molting birds August 13. We took adults in full molt June 27, and one in which the molt was almost completed July 24.

159. **Parus rufescens.** Chestnut-backed Chickadee.

We found a few at Haines and Skagway, and I took one and heard another at Glacier June 5. A female taken at Skagway June 3 had finished laying.

160. **Regulus satrapa olivaceus.** Western Golden-crowned Kinglet.

Tolerably common at Glacier; often heard but seldom seen, and difficult to procure. A female that I took June 10 had the last egg ready for the shell.

161. **Regulus calendula.** Ruby-crowned Kinglet.

I took a male at Log Cabin, and on June 20, between that point and Bennett, heard another singing. Osgood took two young specimens, one at Lower Lebarge July 17, and the other 20 miles below Dawson August 4.

162. **Regulus calendula grinnelli.**[2] Sitka Kinglet.

At Skagway I heard a Sitka kinglet singing May 31, and at Haines took a male and heard another singing June 1. At Glacier I took a male June 6, and during our stay heard two or three others singing. While the Log Cabin bird is normal *calendula*, the Haines and Glacier birds have the more olive back and darker sides of crown of *grinnelli*.

163. **Myadestes townsendi.** Townsend Solitaire.

On the heights above Bennett I took an adult male June 17. On the hot noon of June 26, while seated on the summit of a hill some 1,500 feet above Caribou Crossing, I heard the most beautiful bird song that has ever delighted my ear. It seemed to combine the strength of the robin, the joyousness and soaring quality of the bobolink, and the sweetness and purity of the wood thrush. Starting low and apparently far away, it gained in intensity and volume until it filled the air, and I looked for the singer just above my head. I finally traced the song to a Townsend solitaire that was seated on

---

[1] Auk, XVII, 118, April, 1900. [2] Auk, XIV, 399, 1897.

a dead tree about 150 yards away, pouring forth this volume of melody without leaving its perch. The singer came close enough later to make identification certain.

Osgood and Maddren saw one at Lake Lebarge July 14. Osgood took an adult at Miles Canyon July 11, another at the Semenow Hills July 20, a young in the spotted plumage 20 miles below the Selwyn River July 29, and another young 30 miles above the White River July 30. I saw an adult near the Selwyn River July 29, and took a molting adult near Sixty-Mile Creek August 1. Mr. Cantwell found this species in the Yukon Valley.

164. **Hylocichla aliciæ.** Gray-cheeked Thrush.

Several thrushes which we heard singing on the west shore of Lake Marsh July 8 were, I think, this species, as their song differed from that of the dwarf, hermit, and Alma thrushes. I saw two, but they were so shy that I could not secure either. Near Sixty-Mile Creek, July 31, I took a young in spotted plumage, which was with the young of *almæ* which Osgood shot. At Circle I took a young in first winter plumage, also with *almæ*.

165. **Hylocichla ustulata almæ.**[1] Alma Thrush.

This is the common thrush of the Yukon basin, occurring everywhere from Log Cabin to Circle, perhaps in largest numbers at Caribou Crossing and Lake Marsh. Fifteen miles above Fort Yukon I took one, and saw others August 21. We found many nests, usually 6 to 10 feet from the ground in thick growths of young spruces, but none contained eggs. A nest containing four young just hatched, which I found at Caribou Crossing June 25, was about 8 feet from the ground in a thicket of small spruces. The nest resembled that of *H. u. swainsoni.* At Miles Canyon July 11 we saw young able to fly. Osgood took young in spotted plumage July 31, but those taken August 20 had assumed first winter plumage.

They were usually silent by day, but sang frequently during the short nights. At Caribou Crossing, the last of June, their song could be heard constantly from 8 p. m. to 8 a. m., one taking up the strain as another stopped. The song is much superior to that of *Hylocichla aonalaschkæ* and almost equal to that of *H. fuscescens.* It has whispered notes like that of *H. mustelinus.* By the middle of July the song season was practically over, though we heard one of the birds singing July 23. When the nights became really dark in August, I often heard the call-note of this bird near our camp between 2 and 3 a. m.

[1] Auk, XVII, 119, April, 1900.

166. **Hylocichla aonalaschkæ.** Dwarf Hermit Thrush.

We heard several singing at Skagway, and Osgood took one at Haines June 2. At Glacier they were tolerably common, and we secured several, but they were very shy, keeping in the thickets during the day and singing for several hours in the evening from the topmost spray of some spruce well up the mountainside. Several thrushes' nests in small spruces 6 to 8 feet from the ground were empty, for which condition the abundant red squirrels were probably responsible. At Log Cabin and Bennett we heard a few singing, and at Caribou Crossing Osgood took one June 27.

167. **Hylocichla aonalaschkæ pallasi.** Hermit Thrush.

About 15 miles below Little Salmon River July 22 we secured a pair, whose nest, containing four well-grown young, Osgood had found the evening before. Far from selecting the secluded nesting site usual with this species, this pair had placed their nest between two small bunches of flowers on an open southern hillside, just above a small piece of burnt poplar woodland, and exposed to the full glare of the sun.

168. **Merula migratoria.** American Robin.

Tolerably common at Haines and Skagway, but not found at Glacier. At Haines I took a female and four well-incubated eggs June 2. Robins were common at Log Cabin June 15, and were found regularly, but in gradually decreasing numbers, until August 1, when the last was noted near Sixty-Mile Creek. A flock seen July 29 showed that the southern migration had commenced. We found an empty nest 30 miles below Dawson and heard that the birds bred near Fort Yukon.

Although robins were by no means common at Caribou Crossing, I found, on June 25, 13 empty nests, most of them evidently built that year, and 4 empty nests of the Alma thrush, in a small patch of spruces. The red squirrels which lived in a hollow tree near by probably knew the location of most of these nests. Osgood took a well-grown young robin here on June 26.

169. **Hesperocichla nævia.** Varied Thrush.

At Haines I saw several June 1, and Osgood took one June 2. At Glacier varied thrushes were rather common, but exceedingly shy. About an hour before sunset they would fly to the top of some tall tree and repeatedly utter a long-drawn, plaintive whistle until darkness fell. Sometimes on cloudy days we would hear their song, but it was infrequent and had about stopped when we left Glacier, June 11. We next saw this species near the Tatchun River, where I took a young bird July 23. Thirty miles below Dawson we took young, and met with the birds several times until August 21, when large flocks were seen near Fort Yukon.

At Glacier I found on June 7 a nest containing four eggs, varying from fresh to several days incubated. It was very large, built of sticks and moss and lined with dry grass, and was situated 15 feet from the ground, near the top of a small spruce growing in dense woods a short distance from the river. When I put my hand on the tree, the female flew from the nest with a hoarse, cackling cry and settled a few feet away; the male did not appear. The eggs average 1.25 by 0.84 inches and are nile blue sparingly spotted with écru drab and seal brown.

170. **Saxicola œnanthe.** Wheatear.

Osgood saw two young wheatears at Circle August 19, and secured one. At the Aphoon mouth I shot one on August 27, which fell into the river and was carried away by the rapid current, but I saw the white rump plainly.

171. **Sialia arctica.** Mountain Bluebird.

We found a pair on Fifty-Mile River a short distance above Miles Canyon July 10. The next day I secured the female and found the nest with four well-incubated eggs in a hole about 8 feet from the ground in a dead spruce in the midst of a burnt tract. July 22 I shot a male near the point where Fifty-Mile River empties into Lake Lebarge. Mr. Cantwell also found this species on Fifty-Mile River.

INDEX.

[Names of new species in **black-face** type.]

O

U. S. DEPARTMENT OF AGRICULTURE
DIVISION OF BIOLOGICAL SURVEY

NORTH AMERICAN FAUNA

No. 20

[Actual date of publication, August 31, 1901]

REVISION OF THE SKUNKS OF THE GENUS CHINCHA

BY
ARTHUR H. HOWELL
ASSISTANT, BIOLOGICAL SURVEY

Prepared under the direction of
Dr. C. HART MERRIAM
CHIEF OF DIVISION OF BIOLOGICAL SURVEY

WASHINGTON
GOVERNMENT PRINTING OFFICE
1901

STATISTICAL DEPARTMENT
SEP 21 1901
BOSTON PUBLIC LIBRARY

LETTER OF TRANSMITTAL.

U. S. DEPARTMENT OF AGRICULTURE,
DIVISION OF BIOLOGICAL SURVEY,
Washington, D. C., July 5, 1901.

SIR: I have the honor to transmit herewith for publication as No. 20 of North American Fauna a report entitled "Revision of the Skunks of the Genus *Chincha*," by Arthur H. Howell, assistant in the Biological Survey.

Respectfully,

T. S. PALMER,
Acting Chief, Biological Survey.

Hon. JAMES WILSON,
Secretary of Agriculture.

CONTENTS.

ILLUSTRATIONS.

REVISION OF THE SKUNKS OF THE GENUS CHINCHA.[1]

By Arthur H. Howell.
Assistant, Biological Survey.

INTRODUCTION.

HISTORY.

Skunks have figured in literature since the early part of the seventeenth century. Their peculiar means of defense has served to make them conspicuous, but the repugnance in which they are commonly held has doubtless prevented as thorough study of their habits and characteristics as has been accorded to other common mammals.

Apparently the first account of them is that given by Gabriel Sagard-Théodat in his history of Canada, published in 1636, in which he refers to the ill-smelling qualities of these '*enfans du diable*.'[2] In 1651 Hernandez gave an account of the Mexican skunks, which he confused somewhat with the nasua.[3] It was on this description that Linnæus primarily based his *Viverra memphitis*. Buffon, in 1765, gave a description and figure of a skunk, which he called '*le chinche*'[4]—a name taken from Feuillée, who in 1714, in an account of his travels in South America, had recorded this as the native name of the skunk occurring near Buenos Ayres.[5] Buffon's account has served as the basis for many of the later names applied to the skunks. He ascribes '*le chinche*' to South America, but his figure evidently represents one of the skunks of the genus *Chincha*, which is strictly North American. The confusion of the skunks of the two continents, for which this error is partly responsible, has continued, though in a lessening degree, to the present day.

[1] *Chincha* of this paper is the equivalent of *Mephitis* of recent authors. For a discussion of the reasons for this change, see p. 14.

[2] Histoire du Canada, p. 748, Paris, 1636.

[3] Thesaurus Rer. Med. Novæ Hisp., p. 332, Rome, 1651.

[4] Histoire Naturelle, XIII, p. 294, pl. 39, Paris, 1765.

[5] Journal du P. Feuillée, p. 272; Paris, 1714.

Schreber, in 1776, gave a good description of the large North American skunk, and his specific name *mephitis* is the first one available for any member of the genus. Cuvier, in 1800, applied the same name in a generic sense to the '*mouffettes*,' his group including both the large and small skunks of North America, since separated into different genera.[1] During the early years of the nineteenth century the large skunks, which form the basis of this paper, were treated by many authors, and some new names were proposed; but for the most part the descriptions are hopelessly confused and the names are unidentifiable. In 1829 Richardson described *hudsonica* from central Canada; in 1832 Lichtenstein named *mesomelas* from Louisiana and *macroura* and *vittata* from Mexico; and in 1837 Gray described *varians* from Texas and *mexicana* (the same as *macroura*) from Mexico.

Up to 1838 the North American skunks had all been included in a single undivided genus, *Mephitis*, but in this year Lichtenstein published an extended revision of the group, in which he classed the white-backed skunks as a subgenus under the name *Thiosmus*, but did not separate the little spotted skunks from their larger relatives.[2] In 1842 Lesson published a brief but important synopsis of the genus, in which he arranged the species in three groups, restricting *Mephitis* in a subgeneric sense to the little spotted skunks, naming the large striped skunks *Chincha* and retaining *Thiosmus* for the white-backed skunks.[3] Lesson's *Mephitis* and *Chincha*, proposed as subgenera, are perfectly tenable as genera, and are here used in that sense, *Mephitis* for the little spotted skunks, commonly known as *Spilogale*, and *Chincha* for the large striped skunks, commonly known as *Mephitis*.

Boitard, in 1842, proposed three new names for the large skunks, all of which probably refer to the same species, though only one, *putida*, is certainly identifiable. This is available for the eastern skunk. In 1857 Baird named the California skunk *occidentalis*. For several decades after this the group received very little attention from systematists, but during the last few years there has been considerable activity in the naming of species. In 1890 *estor* was proposed by Merriam; in 1895, *elongata* by Bangs; in 1896, *scrutator* by Bangs, and *fossidens* and *orthostichus* (fossils) by Cope; in 1897, *holzneri* and *milleri* by Mearns; in 1898, *avia* and *spissigrada* by Bangs, and in 1899, *fœtulenta* by Elliot, and *leptops* and *obtusatus* (fossils) by Cope.

DISTRIBUTION.

The skunks of the genus *Chincha* range over the greater part of North America, from the Hudsonian zone in Canada to Guatemala. The northern limits of their distribution are not definitely known.

[1] Leçons d'Anat. Comp., I, tabl. 1, 1800.

[2] Abh. Akad. Wiss. Berlin for 1836, pp. 249–312, 1838.

[3] Nouv. Tabl. de Règne Anim., p. 67, 1842.

In the interior they occur as far north as Great Slave Lake, as shown by a specimen from that locality in the U. S. National Museum, and by the statement of B. R. Ross, who says that while he has never seen a living specimen on the Mackenzie River, he has found the bones and part of the skin a short distance from the shores of Great Slave Lake.[1] On the Atlantic coast they have not been recorded north of Nova Scotia. On the Pacific slope the most northerly record is Stuart Lake, British Columbia, though on the immediate coast of the Pacific they are not reported farther north than Howe Sound, in southern British Columbia, and probably do not occur much beyond this point. In the United States they are generally distributed, except in the higher mountains, and in many places are extremely numerous. They are absent or very rare in eastern North Carolina, according to Mr. Outram Bangs,[2] and perhaps the same is true of other small areas within their range.

The subgenus *Chincha* ranges but a short distance into Mexico: *C. estor* reaches southern Chihuahua in the Sierra Madre, *holzneri* enters Lower California, and *varians* occurs in eastern Mexico in the Rio Grande Valley. The subgenus *Leuconitra*[3] occupies nearly the whole of Mexico, and its range overlaps that of the subgenus *Chincha* in southern Arizona. It is chiefly confined to the Sonoran and Transition zones, not occurring in the tropical lowlands except on the coast of Oaxaca. The southern limit of its range is not known, but it probably does not extend beyond the highlands of Guatemala.

HABITS.

The large skunks are wholly terrestrial, living in caves, in the deserted burrows of other animals, or in burrows of their own excavation. They do not avoid the habitations of man, but seem rather to prefer the clearings and pastures about the farm, and frequently make their abode under a house or other building. They are sluggish in movement, and usually show little fear of human beings, depending for safety more on the efficacy of their malodorous discharge than on attempts to escape. Although chiefly nocturnal, they are often seen moving about in the daytime, especially in the morning and evening. They hibernate only during the severest part of the winter, and probably only in northern latitudes.

Their food consists largely of small mammals, reptiles, batrachians, insects, and birds' eggs; and they apparently have no difficulty in securing an abundance, for they frequently become excessively fat. They are particularly fond of insects, and during the seasons when grasshoppers are abundant feed extensively on these pests. In many parts of

[1] Canadian Nat. and Geol., VII, p. 139, 1862.

[2] Proc. Biol. Soc. Wash., X, p. 139, 1896.

[3] See p. 39.

the United States they have been found useful in destroying the 'white grub,' a great pest in lawns and meadows. Fish and crustaceans, when available, and even carrion or other refuse, are readily eaten. They occasionally rob the poultry yard, but such depredations as they commit are more than offset by their destruction of noxious mammals and insects. Vegetable matter is sometimes eaten. Mr. E. A. Goldman reports that he has found roots and wild fruits in the stomach of a hooded skunk.

Skunks have been extensively trapped for furs ever since the settlement of the country by white men,[1] and within the last few years attempts have been made to breed them in confinement; but although 'skunk farms' have been started in several States, the industry seems not yet well established.

EXTERNAL CHARACTERS.

Skunks are too well known to need more than a passing reference to their external appearance. They are stocky animals, particularly heavy behind, with slender nose, small ears, short legs, long fore claws, and long bushy tail. The pelage is long, loose, and silky; the under-fur is quite dense, especially in northern latitudes; and the tail hairs are coarse and flaccid. The colors are black and white, the white usually arranged in a narrow frontal stripe and two na row dorsal bands or one broad band covering the whole back, but the pattern of coloration is subject to much variation. During spring and summer the pelage often becomes much worn and faded, and the glossy black of the winter pelage is frequently replaced by a dull brown. There is apparently but one molt in a year, and this usually occurs in late summer or in autumn.

The females are always smaller than the males and less specialized in their cranial characters, which are often of so little prominence that they do not serve to readily distinguish closely related species. The difference in size is most marked in the larger species. Thus in *Chincha notata* the average length of the skull is 10 percent greater in males than in females, and the average zygomatic breadth 14 percent greater.

The teats usually number 10 to 14. The young are born in litters of from 4 to 10, and reach maturity at a very early age, the bones of the skull completely coalescing before the teeth show any appreciable signs of wear.

That which particularly distinguishes skunks from other animals is their means of defense, consisting of a characteristic malodorous fluid, which, when ejected, speedily disarms the boldest aggressor. The

[1] During the forty years from 1850 to 1890 the Hudson Bay Company alone shipped over 250,000 skunk skins to England.

fumes from the fresh liquid are overpowering in their pungency, and are possessed of remarkably penetrating and lasting properties.[1] The fluid is secreted by two anal glands, similar in character to those possessed by other members of the Mustelidæ, but larger and more muscular. Dr. C. Hart Merriam has given a concise description of these glands, as follows:

> The glands lie on either side of the rectum, and are imbedded in a dense, gizzard-like mass of muscle which serves to compress them so forcibly that the contained fluid may be ejected to the distance of four or five metres (approximately 13 to 16½ feet). Each sac is furnished with a single duct that leads into a prominent, nipple-like papilla that is capable of being protruded from the anus, and by means of which the direction of the jet is governed.[2]

MATERIAL.

The present revision is based chiefly on a study of the specimens in the collections of the U. S. Biological Survey and the U. S. National Museum, supplemented by much additional material from the collection of the American Museum of Natural History, New York, and the private collections of Dr. C. Hart Merriam and Mr. Outram Bangs. The total number of specimens examined exceeds 950, among which are all the existing types. The Biological Survey collection contains good series of skins and skulls of most of the species, and is particularly rich in large series of extra skulls, which have proved of great value in showing the range of individual variation.

My thanks are cordially extended to Dr. C. Hart Merriam for the privilege of using the collections under his charge. Thanks are also due Dr. J. A. Allen and Mr. Frank M. Chapman, of the American Museum of Natural History, Mr. Gerrit S. Miller, jr., of the U. S. National Museum, Mr. D. G. Elliot, of the Field Columbian Museum, Mr. Witmer Stone, of the Academy of Natural Sciences, Philadelphia, and Mr. William Brewster, of the Museum of Comparative Zoology, Cambridge, for the loan of material under their charge. I am especially indebted to Mr. Outram Bangs for the use of a fine series of skunks from his collection, including the types of all his species.

The illustrations in the present paper are from photographs taken by Dr. A. K. Fisher, of the Biological Survey.

One new subgenus (*Leucomitra*) and three new species and subspecies (*Chincha occidentalis major*, *C. occidentalis notata*, and *C. platyrhina*) are here described.

NOMENCLATURE.

Three generic and 35 specific names have been applied to the skunks of this group. The general similarity in external characters among the members of the genus has led to the inclusion of widely different

---

[1] For a valuable account of the chemical properties of this fluid see T. B. Aldrich, Journ. Exp. Med., I, pp. 323–340, 1896; II, pp. 439–452, 1897.

[2] Mammals of the Adirondack Region, Trans. Linn. Soc., N. Y., I, p. 76, 1882.

forms under one name, while the extreme variability in some species has multiplied synonyms. The confused state of the nomenclature makes necessary a brief statement of the status of each name.

GENERIC NAMES.

Viverra Linnæus, **1758.** Syst. Nat., ed. X, pp. 43–44,

The genus comprehended five species: *V. ichneumon*, *V. memphitis*, *V. putorius*, *V. zibetha* (usually considered the type), and *V. genetta*. The American skunks were referred to it by authors prior to Cuvier, 1800, but it is now restricted to the Old World civets.

Mephitis Cuvier, **1800.** Leçons d'Anat. Comp., I, tabl. I.

Cuvier proposed this genus to include the '*mouffettes*,' but as no species are mentioned, it is necessary, in order to determine its specific constituents, to refer to his 'Tableau Elementaire,' published two years previously.[1] In this work he places the *mouffettes* as a subgroup of *Mustela*, and mentions two species, *Viverra putorius* Linn. and *V. mephitis* Linn.[2] The latter was removed in 1842 to become the type of the genus *Chincha* Lesson, which leaves *V. putorius*, one of the little spotted skunks, as the type of *Mephitis*. The name *Spilogale*, proposed in 1865 by Gray for the little spotted skunks, will therefore have to be abandoned, becoming a synonym of *Mephitis*, which thus unfortunately proves to be the name of this group of skunks instead of the group for which it has so long been used.

Chincha Lesson, **1842.** Nouv. Tableau Règne Anim., Mamm., p. 67.

Lesson proposed this as the name of a subgenus of *Mephitis*, with *Chincha americana* as the type species; *hudsonica* Richardson is given as a variety, but no other species are placed in the group. The references show that his type species is based on *Viverra mephitis* Erxleben, which in turn is based on *V. mephitis* Schreber—a plainly recognizable species. If we assume (as we can with all propriety) that Cuvier, in placing '*Viverra mephitis* L.' as one of the types of his genus *Mephitis* referred to *V. mephitis* of Gmelin's edition, we then have for the type of *Chincha* a species which is one of the two originally composing the Cuvierian genus *Mephitis*, and one that is likewise identifiable, for *V. mephitis* Gmelin is based on *V. mephitis* Schreber.

---

[1] Tabl. Element. de l'Hist. Nat. des Anim., p. 116, 1798.

[2] This name does not appear in either the 10th or 12th editions of Linnæus, so we must assume that the reference is to Gmelin, 1788; *memphitis* of the 10th edition is a different name (see p. 18) and Cuvier's description shows that the animal he had in mind was a skunk, and not the composite species which Linnæus described under the name *memphitis*.

It is perfectly clear, therefore, that Lesson intended to apply the name *Chincha* to the large two-striped North American skunks, and it is used for these in a generic sense in the present paper.

[NOTE.—In a list of names of Mexican animals by A. L. Herrera,[1] '*Mammephitisus macrura*' is used for the '*zorillo*' (the vernacular name of the hooded skunk in Mexico). All of Herrera's names are modified forms of accepted generic names, and until such time as his system shall have been adopted, his names require no consideration.]

SPECIFIC AND SUBSPECIFIC NAMES.

americana (*Mephitis*) Desmarest, **1818**. Nouv. Dict. d'Hist. Nat., Paris, XXI, p. 514.

This name was applied to a composite species, including as varieties all the skunks of North and South America. The first author to use it in a restricted sense was Sabine, who in 1823 applied it to the skunks of Canada.[2] He says that the animals "under examination are the particular sort designated as the *Viverra mephitis* of Gmelin," which is variety No. 7 of Desmarest. The name *americana* therefore becomes a synonym of *mephitis*.

avia (*Mephitis*) Bangs, **1898**. Proc. Biol. Soc. Wash., XII, p. 32.

Under this name Bangs described a form from San Jose, Illinois. It proves to be closely related to *mesomelas*, of which it is considered a subspecies.

bivirgata (*Mephitis americana*) Ham. Smith, **1839**. Jardine's Nat. Libr., Mamm., I, p. 196.

The description under this name probably refers to the eastern skunk (*Chincha putida*), but no type locality is assigned, and the species can not be identified with certainty.

cinche (*Viverra*) Müller, **1776**. Natursyst. Suppl., p. 32.

The species thus named is based on the *chinche* of Buffon, which probably belongs to this genus, but is not certainly identifiable. Müller follows Buffon in ascribing the animal to South America, where *Chincha* does not occur.

concolor (*Mephitis*) Gray (Verreaux MS.?), **1865**. Proc. Zool. Soc. London, 1865, p. 149.

Published by Gray under his 'var. *c. vittata*', with 'Verreaux MS.?' as authority. The form thus named is evidently only one of the many variations to which *vittata* is subject, and although Gray afterwards raised it to subspecific rank[3], the name must be regarded as a synonym.

---

[1] Sinonimia Vulgar y Cientifica de los Principales Vertebrados Mexicanos, p. 30, 1899.

[2] Franklin's Narrative of a Journey to the Polar Sea (1819–1822), App., p. 653, 1823.

[3] Cat. Carnivora, Brit. Mus., p. 138, 1869.

edulis (*Memphitis*) Coues (Berlandier MSS.), **1877**. Fur-Bearing Animals, p. 236.

Berlandier (as quoted by Coues) gives a brief description of a skunk from Mexico which is probably *macroura*. The name will stand as a synonym of the latter.

elongata (*Mephitis mephitica*) Bangs, **1895**. Proc. Boston Soc. Nat. Hist., XXVI, p. 531.

Under this name Bangs described the Florida skunk. It proves to be a distinct species, most nearly related to *putida*.

estor (*Mephitis*) Merriam, **1890**. N. Am. Fauna No. 3, p. 81.

The species described under this name by Merriam is a well-marked form inhabiting Arizona and northern Mexico.

fetidissima (*Mephitis*) Boitard, **1842**. Jardin des Plantes, Mamm., p. 147.

Boitard proposed this name on the same page with *olida* and *putida*. The form thus named is probably one of the many variations of the common eastern skunk, but as no more definite type locality is assigned than 'United States,' the name can not be specifically applied.

fœda (*Viverra*) Boddaert, **1785**. Elenchus Animalium, I, p. 84.

This is one of the many early names which can not be specifically identified. Boddaert quotes Buffon, Schreber, and other early authors, but assigns his species to Mexico.

fœtulenta (*Mephitis*) Elliot, **1899**. Field Columbian Museum, Pub. 32, Zool. Ser., I, no. 13, p. 269.

The form to which this name was applied proves to be practically identical with that to which Bangs had previously given the name *spissigrada*, of which *fœtulenta* is, therefore, a synonym.

fossidens (*Mephitis*) Cope, **1896**. Proc. Acad. Nat. Sci. Phila., p. 386.

This name was applied by Cope to a well-marked fossil species from the Pleistocene bone caves at Port Kennedy, Pennsylvania.

frontata (*Mephitis*) Coues, **1875**. Bull. U. S. Geol. & Geog. Surv. Terr., 2d series, no. 1, p. 7.

The form described by Coues under this name was based on a skull found in the Post-pliocene deposits of Pennsylvania. While the characters assigned to it by him are of slight weight, it differs from the living species in dental characters, and seems worthy of specific recognition.

holzneri (*Mephitis occidentalis*) Mearns, **1897**. Proc. U. S. Nat. Mus., XX, p. 461.

Mearns proposed this name for the form of *occidentalis* occupying southern California and northern Lower California (type locality, San Isidro Ranch). It is a rather poorly marked subspecies that ranges north to the vicinity of Monterey Bay.

hudsonica (*Mephitis americana*) Richardson, **1829**. Fauna Boreali-Americana, I, Mamm., p. 55.

Richardson's name, applied to the skunk of the Northern plains, proves available for the species ranging from Colorado north to the interior of Canada.

intermedia (*Mephitis vittata*, var. *b.*) Gray, **1869**. Cat. Carnivora Brit. Mus., p. 138.

This name is applied to one of the numerous varieties of the hooded skunks due to individual variation. It is a synonym of *vittata*.

laticaudata (*Mephitis*) E. Geoffroy, **1803**. Cat. Mamm. Mus. National d'Hist. Nat., Paris, p. 109.

This name can not be satisfactorily identified. The references indicate that the type specimen belonged to the genus *Chincha*, but the description applies more nearly to one of the South American skunks.

leptops (*Mephitis*) Cope, **1899**. Journ. Acad. Nat. Sci. Phila., 2d series, XI, pt. 2, p. 235.

This name was applied to a fossil species from the Port Kennedy bone caves, Montgomery County, Pennsylvania. I have not had an opportunity to compare it with specimens of recent species.

longicaudata (*Mephitis*) Tomes, **1861**. Proc. Zool. Soc. London, 1861, p. 280.

Under this name Tomes described the form from Dueñas, Guatemala. Specimens from near the type locality have been compared with those of *macroura* and are found to be practically identical.

macroura (*Mephitis*) Lichtenstein, **1832**.[1] Darst. Säugeth., pl. 46, with accompanying text.

Lichtenstein's plate and description are sufficient to identify the species thus named by him, which came from the 'mountains northwest of the City of Mexico.' This is the first name applied to any of the hooded skunks.

memphitis (*Viverra*) Linnaeus, **1758**. Syst. Nat., p. 44.

A name applied by Linnaeus to an unrecognizable species, evidently part skunk and part nasua There is nothing in the description to indicate even the genus to which it refers, and it seems best to reject the name as indeterminable. It is quite distinct from *mephitis*, as shown below.

mephitica (*Viverra*) Shaw, **1792**. Museum Leverianum, p. 171.

This name has been generally adopted by recent authors for the eastern skunk, and was restricted by Bangs in 1895 to the northeastern

[1] The date of publication of that portion of the 'Darstellung' in which the skunks are described is fixed by Lichtenstein himself in a later paper (Abh. Akad. Wiss. Berlin for 1836, p. 303, 1838).

species. Schreber's name *mephitis*, however, is perfectly tenable for the same species, and being of earlier date is adopted instead of *mephitica*.

mephitis (*Viverra*) Schreber, **1776**. Säugth., III, p. 44' tab. 121.

Recent authors have rejected this name on the ground that it is preoccupied by *Viverra memphitis* Linn., 1758, supposed to be a misprint for *mephitis*. Through the kindness of Mr. J. E. Harting, of London, who has examined for me a copy of the tenth edition of Linnæus's 'Systema Naturæ' in the Linnæan Society Library, which contains numerous corrections and annotations in the author's own handwriting, sufficient evidence has been brought to light to show that Linnæus intended to write *memphitis*, which must therefore be considered entirely distinct from *mephitis*. In this copy of Linnæus Mr. Harting finds that although certain alterations[1] are made in the diagnosis, the spelling of *memphitis* is not corrected. Still other evidence of the validity of the name is adduced by Mr. Harting, who writes: "That he [Linnæus] meant *memphitis* to stand is clear, not only from his leaving the spelling uncorrected in his annotated tenth edition, but by his rewriting it in a marginal note to his copy of Ray's 'Synopsis Animalium,' 1693, wherein, on p. 181, he identifies it with Ray's '*Yzquiépatl seu Vulpecula*.' Opposite these words he has written '*Viverra memphitis*,' distinctly."

Schreber's name *mephitis*, then, is not preoccupied, and being accompanied by a recognizable description, is adopted as the first name for the Canada skunk.

mesomelas (*Mephitis*) Lichtenstein, **1832**. Darst. Säugeth., pl. 45, with accompanying text.

Lichtenstein's specimen, on which the species thus named is based, was secured from a natural history dealer, and was said to have come from Louisiana. The measurements show it to be the small species inhabiting the southern Mississippi Valley, subsequently named *scrutator* by Bangs.

mexicana (*Mephitis*) Gray, **1837**. Charlesworth's Mag. Nat. Hist., I, p. 581.

Gray's brief description under this name probably refers to one of the forms of *macroura*, of which *mexicana* is therefore a synonym.

---

[1] The alterations referred to consist in the substitution for the original diagnosis, of the following:

"*V. alba subtus nigro maculata;*" and the entry in the margin of the query, "*an nasua?*"

The diagnosis as altered agrees with the later portion of the description, and the nasua element is largely removed, although the references point principally to that animal (cf. Bangs, Proc. Boston Soc. Nat. Hist., XXVI, p. 529, 1895).

That Linnæus himself was in doubt as to the validity of the species is shown both by the annotations above mentioned, and by the omission of the name from his twelfth edition.

milleri (*Mephitis*) Mearns, **1897**. Proc. U. S. Nat. Mus., XX, p. 467.

The form thus named was described from Fort Lowell, Arizona. It proves to be a northern race of the hooded skunk, for which this is the only name.

obtusatus (*Mephitis*) Cope, **1899**. Journ. Acad. Nat. Sci. Phila., 2d series, XI, pt. 2, p. 236.

Cope gave this name to an extinct species based on a single jaw discovered in the Port Kennedy bone deposit (Pennsylvania), but his type can not now be found. It is described as a very small species, the size of a weasel.

occidentalis (*Mephitis*) Baird, **1857**. Mamm. N. Am., p. 194.

Under this name Baird described the western skunk. It is a wide-ranging species for which there is no other name.

olida (*Mephitis*) Boitard, **1842**. Jardin des Plantes, Mamm. p. 147.

As in the case of *fetidissima*, described on the same page, this name was probably intended to apply to one of the forms of the eastern skunk, but in the absence of a definite type locality, no specific application can be made.

orthostichus (*Mephitis*) Cope, **1896**. Proc. Acad. Nat. Sci. Phila., p. 389.

Cope described the species thus named from remains found in the Port Kennedy bone caves, Pennsylvania. It appears to be quite distinct from any living species.

putida (*Mephitis*) Boitard, **1842**. Jardin des Plantes, Mamm. p. 147.

This is the first name applicable to the eastern skunk in connection with which the type locality is definitely fixed. Boitard refers to it as '*La Moufette de New Jersey*' and gives a brief description of the animal. The name is adopted for the species long known as *mephitica*.

scrutator (*Mephitis mephitica*) Bangs, **1896**. Proc. Biol. Soc. Wash., X, p. 141.

Under this name Bangs described the form from Louisiana as a subspecies of the Canada skunk. It proves to be quite distinct from the latter, and is a well-marked form, for which, however, Lichtenstein's name *mesomelas*, of much earlier date, must be used.

spissigrada (*Mephitis*) Bangs, **1898**. Proc. Biol. Soc. Wash., XII, p. 31.

The form thus named was described by Bangs from Sumas, British Columbia. It proves to be a northern race of *occidentalis*.

varians (*Mephitis*) Gray, **1837**. Charlesworth's Mag. Nat. Hist., I, p. 581.

The description of the form thus named is inadequate, and the type locality ('Texas') indefinite, but Gray's statement in a later paper that the tail is as long as the body fixes the name to the large skunk of southern and western Texas. The type specimen can not be found.

vittata (*Mephitis*) Lichtenstein, **1832.** Darst. Säugeth. plate 47, with accompanying text.

Lichtenstein described under this name the species found at San Mateo del Mar, Oaxaca, Mexico. It proves to be a well-marked form of the hooded skunk group.

Genus CHINCHA Lesson.

Generic characters.—Skull highly arched, highest in frontal region; rostrum truncated with slight obliquity; posterior margin of palate nearly on a line with posterior border of last molars; periotic region not inflated; mastoid and paroccipital processes prominent; post-orbital processes not prominent; coronoid process of mandible conical, erect. Dental formula: $i.\ \frac{3-3}{3-3},\ c.\ \frac{1-1}{1-1},\ pm.\ \frac{3-3}{3-3},\ m.\ \frac{1-1}{2-2}=34$. Snout not greatly produced; nostrils lateral; tail long and bushy.

The genus *Chincha* is a member of the subfamily Mephitinæ, which also includes three other genera, *Mephitis*, *Thiosmus*,[1] and *Conepatus*, of which only the first two occur in North America. *Mephitis* may be readily distinguished by the color pattern, which consists of numerous white stripes (always more than two) broken into many patches or spots. *Thiosmus* differs from the other two North America genera in having the snout produced, and bare for a considerable distance from the tip, with the nostrils inferior; tail short and sparsely haired; and usual color pattern black, with a solid white band covering the entire back and part or all of the tail.

Key to species and subspecies.

[Based on adult males.]

Audital bullæ greatly inflated; back usually either wholly black or wholly white ....(Subgenus *Leucomitra*, p. 39)
- Smaller; bullæ much inflated ....*vittata* (p. 43)
- Larger; bullæ less inflated.
  - Tail much longer than body ....*milleri* (p. 42)
  - Tail not longer than body ....*macroura* (p. 41)

Audital bullæ not greatly inflated; back usually with a white stripe, divided posteriorly ....(Subgenus *Chincha*, p. 22)
- Palate with prominent spine.
  - Tail longer than body ....*elongata* (p. 27)
  - Tail shorter than body ....*putida* (p. 25)

---

[1] *Thiosmus* Lichtenstein, Abh. Akad. Wiss. Berlin for 1836, p. 270, 1838. Material in the collection of the Biological Survey indicates that *Conepatus* Gray, described from Patagonia, will have to be separated from the white-backed skunks usually known under this name. The white-backed skunks will, in such event, require another name; and since *Thiosmus* Lichtenstein seems to be the earliest one that is available, it is here provisionally adopted for this group.

Palate without prominent spine.

Tail less than half the length of body .... *mephitis* (p. 22)

Tail more than half the length of body.

Tail usually more than 350 mm. (Texas) .... *varians* (p. 31)

Tail usually less than 350 mm.

Skull small (basilar length[1] less than 66 mm.).

Body stripes very broad .... *estor* (p. 32)

Body stripes narrower.

Tail more than 250 mm .... *holzneri* (p. 38)

Tail less than 250 mm.

Skull smaller (basilar length usually under 60 mm.) . *mesomelas* (p. 29)

Skull larger (basilar length usually over 60 mm.) .... *aria* (p. 30)

Skull large (basilar length more than 66 mm.).

Body stripes narrow; frequently not continuous .... *notata* (p. 36)

Body stripes broad; always continuous.

Tail usually more than 285 mm.

Skull very broad (mastoid breadth more than 45 mm.) .. *major* (p. 37)

Skull narrower (mastoid breadth less than 45 mm.).

Rostrum very broad (breadth across post-orbital processes more than 24 mm.) .... *platyrhina* (p. 39)

Rostrum narrower (breadth across post-orbital processes less than 24 mm.) .... *occidentalis* (p. 34)

Tail usually less than 285 mm.

Palate extending back of last molars .... *hudsonica* (p. 24)

Palate ending on a line with last molars .... *spissigrada* (p. 35)

List of Species and Subspecies, with Type Localities.

Subgenus Chincha.

Species and subspecies.	Type localities.
Chincha mephitis (Schreber)	America (restricted to eastern Canada).
hudsonica (Richardson)	Plains of the Saskatchewan.
putida (Boitard)	New Jersey.
elongata (Bangs)	Micco, Brevard County, Florida.
mesomelas (Lichtenstein)	Louisiana.
mesomelas aria (Bangs)	San Jose, Illinois.
mesomelas varians (Gray)	Texas.
estor (Merriam)	San Francisco Mountain, Arizona.
occidentalis (Baird)	Petaluma, California.
occidentalis spissigrada (Bangs)	Sumas, British Columbia.
occidentalis notata nobis	Trout Lake, Mount Adams, Washington.
occidentalis major nobis	Fort Klamath, Oregon.
occidentalis holzneri (Mearns)	San Isidro Ranch, Lower California.
platyrhina nobis	South Fork Kern River, California.

Subgenus Leucomitra.

Chincha macroura (Lichtenstein)	Mountains northwest of City of Mexico.
macroura milleri (Mearns)	Fort Lowell, Arizona.
macroura vittata (Lichtenstein)	San Mateo del Mar, Oaxaca, Mexico.

[1] Basilar length of Hensel, measured from inferior lip of foramen magnum to posterior rim of alveolus of middle incisor.

Subgeneric characters.—Skull (Pls. V, VI, and VII) long and relatively narrow interorbitally; zygomata usually spreading broadly; interpterygoid fossa broad; palate ending either squarely or with a median notch or spine; audital bullæ not greatly inflated; anterior palatine foramina usually small and narrow; mastoids, and sagittal and supraorbital crests well developed. Size medium to large; build heavy; soles broad; ears not prominent; fur dense.

The usual color pattern is as follows: A narrow median white stripe extends from nose to nape; a white dorsal band beginning with a broad nuchal patch, and narrowing between the shoulders, divides into two lateral stripes, which continue to the tail and sometimes down its sides; the rest of the body is black; the tail is of mixed black and white hairs, the white ones much the longer; all the tail hairs are white at the base. (See Pl. II, fig. 1.)

Great variability is exhibited by the species inhabiting the eastern United States—*putida*, *elongata*, and *mesomelas*—which grade in color pattern from specimens wholly white above, including the tail, to specimens in which the only white areas are the frontal stripe and a dash on the nape. In the western species and in *C. mephitis* from Canada the variability is slight, and consists chiefly in the breadth of the body stripes and the amount of white on the tail. (See Pls. I, II, and III.) The white areas are often of a creamy hue, and are never mixed with black hairs, as they are in the hooded skunks (subgenus *Leucomitra*), and specimens in the black phase never show any trace of the white on the sides that is usual in the black phase of *Leucomitra*.

Descriptions of Species and Subspecies.

CHINCHA MEPHITIS (Schreber). CANADA SKUNK.

Viverra mephitis Schreber, Säugth., III, p. 444, tab. 121, 1776.
Viverra mephitica Shaw, Mus. Leverianum, p. 171, 1792 (part).
Viverra mephitis Oken, Lehrbuch der Naturg., p. 994, 1816.
Chincha americana Lesson, Nouv. Tabl. Règne Anim., Mamm., p. 67, 1842.
Mephitis mephitica Bangs, Proc. Boston Soc. Nat. Hist., XXVI, p. 533, 1895; Elliot, Synop. Mamm. N. A., Field Columbian Museum, Zool. Ser., II, p. 322, 1901 (part).

Type locality.—America.

Geographic distribution.—Eastern Canada—Nova Scotia, Quebec, and northern Ontario; west and north at least to Oxford House, Keewatin.

General characters.—Size large; *tail short and slender*, mixed black and white; markings constant; skull large and massive; palate ending in an even curve, *without notch or spine*.

Color.—The color pattern, which is quite constant, agrees with that described in the subgeneric diagnosis (p. 22). The white stripes are never very broad and are sometimes separated for their entire length, as in *C. notata*. They extend down the sides of the tail and project beyond its tip. (See Pl. II, fig. 1.)

Cranial characters.—Skull large and massive; *rostrum broad;* zygomata heavy, *not greatly expanded; mastoid processes not prominent;* braincase very broad; palate ending in an even curve, *without distinct notch or spine;* posterior border of palate usually slightly back of posterior alveoli of last molars. (See Pl. V, figs. 1 and 2.)

Measurements.[1]—Average of 7 adult males from Canada:[2] Total length, 613; tail vertebræ, 188; hind foot, 78. Average of 2 adult females from Canada:[2] 578; 165; 70. *Skull:* (See table, p. 44).

General remarks.—Until quite recently the specific name *mephitica* has been applied indiscriminately to all the skunks of eastern North America. In 1895, Mr. Outram Bangs[3] restricted the name to the form inhabiting the Hudsonian and Canadian zones,[4] and pointed out the characters which distinguish it from its congeners. Schreber's name *mephitis*, which has sixteen years' priority over Shaw's name *mephitica*, is strictly applicable to the Canada skunk (as shown by Bangs), since the tail is said to be half the length of the body, a proportion not found in any other of the eastern species. The specific name *mephitis* has been rejected by all recent authors on the ground that it is preoccupied by *Viverra memphitis*, Linn.; but as shown in the remarks on nomenclature (p. 18), the two names are quite distinct.

The Canada skunk is a large, stocky animal with a heavy skull and a very short tail that narrows gradually to the tip. It is most nearly related to *hudsonica* of the western plains, from which it differs in the slender tail, as well as in skull characters. From *putida* of the Atlantic States it differs in larger size, shorter tail, and constancy of markings. Skulls of females are considerably smaller than those of males, and are equaled in size by skulls of males in *putida*, but in other respects they maintain the characters assigned to the species.

Specimens examined.—Total number, 13, from the following localities:

Nova Scotia: Annapolis, 1; Digby, 1; Halifax, 1.
Quebec: Lake Edward, 4.
Ontario: Moose Factory, 1; North Bay, Lake Nipissing, 2; Little Pic River, Lake Superior, 1.
Keewatin: Oxford House, 1; Pine Lake, 1.

---

[1] All measurements are in millimeters.

[2] *Fide* Bangs.

[3] Proc. Boston Soc. Nat. Hist., XXVI, p. 533, 1895.

[4] Examination of a large number of specimens from northern New York and New England shows that *putida* is the form occupying the greater part of the Canadian zone.

CHINCHA HUDSONICA (Richardson). Northern Plains Skunk.

Mephitis americana, var. *hudsonica* Richardson, Fauna Boreali-Americana, I, Mamm., p. 55, 1829.

Mephitis mephitica, var. *occidentalis* Merriam, Ann. Rep. U. S. Geol. Surv. for 1872, p. 662, 1873.

Mephitis mephitica hudsonia Elliot, Synop. Mamm. N. A., Field Columbian Museum, Zool. Ser., II, p. 322, 1901 (part).

Type locality.—Plains of the Saskatchewan.

Geographic distribution. Western Canada, from Manitoba to British Columbia (east of the Cascades); south in the United States to Colorado, Nebraska, and Minnesota.

General characters.—Size very large; *tail heavy* and of medium length, *ending in a blunt brush without a white pencil;* skull heavy, with a long palate; *zygomata broadly spreading.*

Color.—This species exhibits the usual color pattern of skunks of this group. The white stripes are of medium width, bifurcate just behind the shoulders, and extend nearly to the tip of the tail. The tail, which is very full and bushy, usually ends in a blunt, black brush, and has an indistinct band of white about two-thirds of the distance from the root to the tip.

Cranial characters. Compared with *mephitis:* Skull both longer and broader; *zygomata very widely expanded;* mastoid processes prominent; *palate long*, ending behind plane of posterior molars, without notch or spine; interorbital constriction marked; nasals and ascending branches of premaxillæ short; dentition heavy, the lower carnassial relatively large. (See Pl. VI, figs. 1 and 2.)

Measurements.—Average of three adult males from Saskatchewan, Montana, and Wyoming: Total length, 726; tail vertebræ, 268; hind foot, 82. Average of 3 adult females from Montana and Idaho, 602; 250; 71. *Skull:* (See table, p. 44).

General remarks.—This skunk may be readily distinguished from the skunks of the eastern States by its large size and bushy tail without a distinct white pencil. Compared with *mephitis*, its nearest ally, it has a longer and heavier tail, broad, heavy soles, and a skull of quite different proportions. The most noticeable difference between the skulls of the two species is in the widely spreading zygomata of *hudsonica* and the contracted mastoids of *mephitis*. The skull of *hudsonica*, viewed from above, appears narrower than that of *mephitis* in the frontal and parietal regions. It resembles that of *occidentalis* in general shape, but has more widely expanded zygomata. Typical *hudsonica* may be distinguished from both *varians* and *occidentalis* by the short ascending branches of the premaxillæ.

The species has an extensive range on the northern plains, and spreads over most of the northern portions of the Rocky Mountains from Colorado to British Columbia. Its range meets that of *mephitis* in Manitoba, and the two species may possibly intergrade, though present material does not fully show that such is the case. Specimens from British Columbia have slightly shorter tails, but in other respects are typical.

A large series from Arkins, Colorado, consisting of skulls and a few skins, show that two species are present in that region. Of 31 skulls of adult males, 16 are fairly typical *hudsonica*, 12 are just as typical *varians*, and 3 are indeterminate. It is not difficult to separate them, the *hudsonica* series having broad, heavy skulls, with spreading zygomata and long palates, the *varians* series much slenderer and narrower skulls, with shorter palates. No skins of the *hudsonica* form are available, but those of the *varians* form are typical, except that they have the shorter tails usual in specimens of *varians* from the northern part of its range. Under these rather unusual conditions it seems hardly possible to consider that the two species intergrade, but rather that their ranges overlap at this point, each remaining distinct, save for an occasional hybrid.

Specimens examined.—Total number, 80, from the following localities:

Mackenzie: Great Slave Lake, 1.
Saskatchewan: Wingard, 1.
Alberta: Jasper House, 1.
British Columbia: Shuswap, 6; Sicamous, 4; Ashcroft, 1; Okanagan, 4; Kamloops, 1; Stuart Lake, 2; Ducks, 1.
Washington: Fort Spokane, 1.
Idaho: Bear Lake (east side), 1; Cœur d'Alène, 1.
Montana: Stanford, 1; Great Falls, 1; St. Marys Lake, 1; Nyack, Teton Mountains, 1; Bear Paw Mountains, 3; Prospect Creek, near Thompson, 1; Tobacco Plains, 2; Yellowstone River, (26 miles from mouth), 1.
Wyoming: Rona, Sheridan County, 1; Lower Geyser Basin, 1; Shoshone Lake, 1; Fort Bridger, 1; Big Horn Mountains, 1; Bull Lake Creek, Fremont County, 1; Fort Laramie, 1.
Colorado: Arkins, Larimer County, 19 (skulls).
Nebraska: Johnstown, 1.
South Dakota: Custer, 1; Fort Pierre, 1; Fort Randall, 1; Fort Sisseton, Marshall County, 1; Rapid City, 1.
Minnesota: Elk River, 9; Fort Snelling, 1; Roseau River, at Point d'Orme, 1; Bois de Sioux, 1.

CHINCHA PUTIDA (Boitard). EASTERN SKUNK.

Mephitis putida Boitard, Jardin des Plantes, Mamm., p. 147, 1842.
Mephitis olida Boitard, Jardin des Plantes, Mamm., p. 147, 1842.
Mephitis fetidissima Boitard, Jardin des Plantes, Mamm., p. 147, 1842.
Mephitis mephitica Baird, Mamm. N. Am., p. 195, 1857 (part); Elliot, Synop. Mamm. N. A., Field Columbian Museum, Zool. Ser., II, p. 322, 1901 (part).
Mephitis varians, var. *b. mephitica*, Gray, Cat. Carnivora Brit. Mus., p. 137, 1869.
Mephitis varians, var. *c. chinga*, Gray, Cat. Carnivora Brit. Mus., p. 137, 1869.

Type locality.—New Jersey.

Geographic distribution.—New England and Middle Atlantic States; south to Virginia; west to Indiana.

General characters.—Size medium; tail longer than that of *mephitis*, black, *with a distinct white pencil;* skull short and relatively broad zygomatically, *with a prominent spine on the palate.*

Color.—This species exhibits the usual color pattern of the skunks of this group, with considerable variation in the breadth and extent of the white stripes. The tail is usually wholly black, excepting a white pencil, which extends from 100 to 150 mm. beyond the end of the vertebræ. The white stripes are usually broader than in *mephitis*, and specimens frequently occur in which the back is almost wholly white. In many individuals, however, the stripes are much reduced both in length and breadth, and occasionally are entirely wanting, the white being confined to a small patch on the nape and the usual frontal stripe. Only two individuals in the large series examined have the white stripes continued down the sides of the tail. (See Pl. I, fig. 1.)

Cranial characters.—Skull of medium size and *relatively broad across zygomata;* interorbital constriction marked; *posterior border of palate with a prominent spine;* palate usually ending about on a line with last molars; *mastoid processes prominent.* (See Pl. V, figs. 3 and 4).

Measurements.—Average of 6 adult males from Hastings, New York: Total length, 575; tail vertebræ 229, hind foot 60. Average of 6 adult females from same locality: 603; 223; 62. *Skull:* (See table, p. 44).

General remarks.—This species, the common skunk of the Eastern States, is generally distributed from Maine to Virginia, but reaches the Mississippi Valley only, so far as known, in Ohio and eastern Indiana. It has long borne the specific name *mephitica*, which until recently it shared with all its congeners in eastern North America; but since this is a synonym of *C. mephitis* (Schreber), which is here adopted for the Canada skunk, it becomes necessary to select the next available name. While a great many names were proposed in the early years of the present century for the American skunks, the earliest used in a sufficiently restricted sense to be available for the present species is *putida* of Boitard, proposed in 1842 for the skunk of New Jersey. There is but one species of skunk in New Jersey: hence, although Boitard's description is not in itself sufficient for identification, no doubt can exist as to the applicability of the name. The two other names (*olida* and *fetidissima*) proposed by him on the same page probably apply also to this form, but these are also inadequately described and lack the specific mention of a type locality.

In 1896 Bangs described a new form from Louisiana under the name *Mephitis mephitica scrutator* (=*mesomelas*), and expressed the opinion that the skunks inhabiting the New England and Middle Atlantic States are intermediates between it and *mephitica* (=*mephitis*).[1] A careful study of a large number of specimens from eastern North America, including Bangs's types, shows that the skunks of the Atlantic coast States are very distinct from those inhabiting the

[1] Proc. Biol. Soc. Wash., X, pp. 139–144, 1896.

Mississippi Valley, and that the form from New England and the Middle Atlantic States is also quite distinct from *mephitis*. Hence it is entitled to separation as a full species.

It may be at once distinguished from *mephitis* by its longer tail and very different skull. The skull is not only much smaller and weaker than that of *mephitis*, but the relatively great breadth across the zygomata and the marked interorbital constriction give it a very different appearance. The skulls of some of the largest males are almost as broad zygomatically as those of *mephitis*, though very much shorter. The spine on the posterior border of the palate, which is a fairly constant character (absent in only a very few individuals in the large series examined), distinguishes *putida* alike from *mephitis* and from *mesomelas* and its subspecies *avia*. The presence of a palatal spine, the great mastoid breadth, and other differences in the skulls of males of this species, as compared with those of *avia*, whose range meets that of *putida* in Indiana, show that the two forms are quite distinct. Skulls of females, however, have the spine on the palate less pronounced, the zygomata less abruptly spreading, and the mastoid processes reduced, and thus resemble rather closely the females of *avia*.

As in all the skunks, the skulls of the females of *putida* are very much smaller than those of the males, although occasionally the skull of a very large female will equal that of a small male. The skulls show a large amount of individual variation, particularly in size but also in other respects. Specimens from New Hampshire and northern New York average larger than the typical form, which might be regarded as due to the influence of *mephitis* but for the complete absence of other signs of intergradation with that species.

Specimens examined.—Total number, 182, from the following localities:

New York: Adirondack Mountains, 51 (skulls); Lake George, 10; Tomhannock, 1; Locust Grove, 11; Severance, 13 (skulls); Catskill Mountains, 1; Hastings-on-Hudson, 16; Highland Falls, 3; Big Moose Lake, Hamilton County, 1; Mayville, 2; Westchester County, 1; Sing Sing, 3; Montauk Point, 2; Shelter Island, 19 (skulls); Miller Place, Suffolk County, 2.

New Hampshire: Ossipee, 8.

Maine: Bucksport, 1; Brooklin, 2.

Massachusetts: Wilmington, 15; Burlington, 2; Taunton, 1; Ipswich, 1; Woods Hole, 2.

Connecticut: East Hartford, 3.

Pennsylvania: Carlisle, 1.

Ohio: Garrettsville, 1.

Indiana: Marion County, 3; Boone County, 1; Denver, 1.

Maryland: Jefferson, 1.

District of Columbia: Washington, 3.

CHINCHA ELONGATA (Bangs). FLORIDA SKUNK.

Mephitis mephitica elongata Bangs, Proc. Boston Soc. Nat. Hist., XXVI, p. 531, 1895.
Mephitis elongata Bangs, Proc. Biol. Soc. Wash., X, p. 142, 1896.

Type locality.—Micco, Brevard County, Florida.

Geographic distribution.—Florida (from vicinity of Lake Worth) to North Carolina, and in the mountains to West Virginia; west on the Gulf coast to the Mississippi River.

General characters. Size medium; *tail very long*, marked with white on the sides, and with a long white pencil; markings variable, but white stripes usually very broad; skull peculiar—larger than that of *putida*.

Color. Markings similar to those of *putida*, but white stripes averaging broader (about 45 mm. in width) and bifurcating about the middle of the back; tail mostly white above and on the sides, this color reaching around and almost meeting beneath the root of the tail. Great variability in markings is exhibited, some specimens being wholly black, save for a few irregular white patches on the shoulders, others nearly all white above, including the tail, and mixed with white below. (See Pl. I, figs. 2 and 3.)

Cranial characters.—Skull larger and heavier than that of *putida; highly arched in frontal region; rostrum very broad;* anterior palatine foramina large; zygomata spreading less abruptly; dentition heavy, lower carnassial especially large; audital bullæ rather large; spine on palate prominent; interpterygoid fossa broad.

Measurements. Average of 2 adult males from type locality: Total length, 703; tail vertebræ, 317; hind foot, 74. Average of 2 adult females from St. Marys, Georgia: 710; 315; 73. A series from Lake Harney and Mullett Lake are somewhat smaller, 3 adult males averaging 689; 292; 64. *Skull:* (See table, p. 44.)

General remarks.—The Florida skunk is closely related to *putida*, but is very distinct from *mesomelas* of the lower Mississippi Valley, from which it may be distinguished by its long tail and heavy skull. It was described by Bangs as a subspecies of the eastern skunk, but in a later paper he accorded it specific rank, and stated that its range does not meet that of *mephitica* (=*putida*) on the Atlantic coast, since the coastal region of North Carolina is practically uninhabited by skunks. More recently Mr. Bangs has received a specimen of *elongata* from Raleigh, North Carolina, the only one, so far as known, ever taken in the eastern part of the State. Two specimens from West Virginia are typical except that the tail is not quite the usual length—a character found occasionally even in specimens from Florida.

Specimens examined.—Total number, 39, from the following localities:

Florida: Micco, 2; New Berlin, 2; Blitches Ferry, Citrus County, 3; Lake Harney, 5; Mullett Lake, 5; Gainesville, 1; Sebastian, 1; Fort Kissimmee, 3; Lake Worth, 1; Hernando County, 1.

Alabama: Baldwin County, 2.

Mississippi: Bay St. Louis, 2.

Georgia: St. Marys, 2; Pinetucky, 2; McIntosh County, 1; Nashville, 1.

North Carolina: Raleigh, 1; Weaverville, 2.

West Virginia: Green Bank, 1; Travellers Repose, 1.

CHINCHA MESOMELAS (Lichtenstein). LOUISIANA SKUNK.

Mephitis mesomelas Licht., Darst. Säugeth., pl. 45, fig. 2, with accompanying text, 1832.
Mephitis mephitica scrutator Bangs, Proc. Biol. Soc. Wash., X, p. 141, 1896; Elliot, Synop. Mamm. N. A., Field Columbian Museum, Zool. Ser., II, p. 324, 1901.

Type locality. Louisiana.

Geographic distribution.—West side of Mississippi Valley from southern Louisiana to Missouri; westward along the coast of Texas to Matagorda Island; and up the Red River Valley as far at least as Wichita Falls.

General characters.—*Size very small;* tail short, usually wholly black; skull small and relatively narrow.

Color.—More variable than *varians*, but apparently less so than *putida*. In specimens from the type locality the white stripes are narrow, and usually terminate about the middle of the back, though they occasionally extend to the root of the tail;[1] tail usually wholly black; the white pencil generally absent, but if present, shorter than in *putida*. (See Pl. II, fig. 2.) In specimens from Texas the stripes usually reach to the tail, and the coloration is more constant.

Cranial characters.—*Skull very small and relatively narrow; mastoids contracted;* palate ending squarely, *without distinct spine;* teeth small; audital bullæ usually more inflated than in *putida*. (See Pl. VI, figs. 3 and 4.)

Measurements.—Average of 4 adult males from Louisiana: Total length, 576; tail vertebræ, 223; hind foot, 63. Average of 3 adult females from same localities: 566; 224; 62. Average of 4 adults (both sexes) from Marble Cave, Missouri: 628; 232; 67. *Skull:* (See table, p. 44.)

General remarks.—In Lichtenstein's original description of this species he remarked that his type was secured from a natural history dealer, and was said to have come from Louisiana. In a later paper[2] he gave its range as '*Ludoviciana et ad Missouri fluvium.*' By reason of his assignment of a definite type locality, and by the aid of the measurements of his specimen,[3] we are able to apply the name with certainty to the small species inhabiting the lower Mississippi

---

[1] Lichtenstein figures a specimen in which the white stripes reach to the tail and down its sides. None of those examined from Louisiana have as much white as this specimen, but it is stated by Mr. Levi Spalding, of Iowa, Louisiana, that all gradations of color occur in the skunks of that section. When more specimens are obtained from the State, many individuals will probably be found that agree perfectly with the figure of the type.

[2] Abh. Akad. Wiss. Berlin for 1836, p. 277, 1838.

[3] His measurements are: Head and body, 1 ft. 7 in.; tail, 9 in.; hind foot, 2 in. Reduced to millimeters (assuming that he used the Rhineland foot) these measurements are as follows: Total length, 733; tail vertebræ, 235; hind foot, 52; which do not differ radically, except in length of body, from measurements of specimens recently taken in Louisiana. The great length of body was undoubtedly due to stretching of the skin.

Valley. The hairy soles ascribed by him to this species have not been observed in any species of the genus.

The Louisiana skunk was later described by Bangs as a subspecies of '*mephitica*' (= *mephitis* and *putida*). With *mephitis* it apparently has no connection; from *putida* it may be distinguished by its small size, short tail, and narrow skull with square palate. These characters also distinguish it from *elongata*. In most of the specimens examined the tail ends in a blunt black brush, as in *varians* and *hudsonica;* but a few from both extremes of its range have a slender white pencil, shorter than in *putida*. One specimen from Marble Cave, Missouri, is nearly all white above, including the tail. The shape and size of the anterior palatine foramina are variable in this species, some individuals having the large rounded foramina which appear in the subgenus *Leucomitra*.

Specimens examined.—Total number, 39, from the following localities:

> **Louisiana:** Cartville, Acadia Parish, 2; Point aux Loups Springs, 5; Calcasieu Parish, 11; Calcasieu Pass, 1.
> **Texas:** Matagorda Peninsula, 3; Virginia Point, 1; San Antonio, 3; Aransas County, 1; Gainesville, 2; Henrietta, 1; Wichita Falls, 1.
> **Oklahoma:** Fort Cobb, Washita River, 1.
> **Missouri:** Marble Cave, Stone County, 7.

CHINCHA MESOMELAS AVIA (Bangs). ILLINOIS SKUNK.

Mephitis avia Bangs, Proc. Biol. Soc. Wash., XII, p. 32, 1898.

Type locality.—San Jose, Illinois.

Geographic distribution.—Prairie region of Illinois, western Indiana, and eastern Iowa; boundaries of range imperfectly known.

General characters.—Resembles *mesomelas* very closely, but skull slightly larger.

Color.—The series from the type locality is variable; in some the white stripes terminate about the middle of the back, in others they reach to the root of the tail. Tail wholly black, with or without a white pencil.

Cranial characters.—Skull slightly larger than that of *mesomelas;* zygomata more widely expanded; upper carnassial large; palate variable in length, ending sometimes in front of and sometimes behind plane of last molars.

Measurements.—Average of two adult males from type locality:[1] Total length, 641; tail vertebræ, 184; hind foot, 65; one adult female from Freeport, Illinois: 610; 220; 68.[2] *Skull:* (See table, p. 44).

General remarks.—This form is very closely related to *mesomelas* from which it differs chiefly in greater size and perhaps shorter tail.

---

[1] *Fide* Bangs.

[2] These measurements, taken in the flesh from a specimen which died in the National Zoological Park, show that the species may have a somewhat longer tail than Bangs's measurements indicate, and not appreciably shorter than that of *mesomelas*.

While its range meets that of *putida* in Indiana intergradation seems not to take place. The differences between *avia* and *putida* have already been pointed out (see p. 27).

Specimens examined.—Total number, 10, from the following localities:

Illinois: San Jose, 6; Freeport, 1; 'Illinois,' 1.
Indiana: Fowler, Benton County, 1.
Iowa: Delaware County, 1.

CHINCHA MESOMELAS VARIANS (Gray). LONG-TAILED TEXAS SKUNK.

Mephitis varians Gray, Charlesworth's Mag. Nat. Hist., I, p. 581, 1837.

Mephitis macroura Aud. & Bach., Quad. N. Am., III, p. 11, 1854 [not *M. macroura* Licht.].

Mephitis mesomelas Allen, Bull. Am. Mus. Nat. Hist., VI, p. 188, 1894 [not *M. mesomelas* Licht.]; Elliot, Synop. Mamm. N. A., Field Columbian Museum, Zool. Ser., II, p. 325, 1901 [not *M. mesomelas* Licht.].

Type locality.—Texas (specimens from lower Rio Grande Valley considered typical).

Geographic distribution.—Southern and western Texas, eastern New Mexico, and adjacent parts of Mexico; north into Oklahoma, Colorado, Kansas, and Nebraska.

General characters.—Size large; *tail very long;* markings similar to those of *hudsonica*, constant; skull longer than that of *mesomelas*.

Color.—Similar to that of *hudsonica;* white stripes narrower than in *estor;* tail ending in a black brush *without* a pencil; white hairs intermixed in the tail, usually showing prominently in upper surface to about the middle of the tail, where they form an indistinct band.

Cranial characters.—Skull of medium size, smaller and narrower than that of *hudsonica;* longer than that of *mesomelas;* zygomata spreading less abruptly, and palate averaging shorter than in *hudsonica;* ascending branches of premaxillæ very long.

Measurements.—Average of 4 adult males from vicinity of Brownsville, Texas: Total length, 758; tail vertebræ, 393; hind foot, 71. Average of 4 adult females from lower Rio Grande Valley (Laredo and vicinity): 681; 376; 69. *Skull:* (See table, p. 44.)

General remarks.—In Gray's original description of this species he remarks that it inhabits Texas, and in a later paper mentions that the tail is as long as the body.[1] Two forms are found in Texas, either of which might be the subject of the original description, but only one of these, the larger, has a tail as long as the body. To this form, therefore, the name is restricted.[2]

[1] Cat. Carnivora Brit. Mus., p. 136, 1869.

[2] Dr. J. A. Allen has endeavored to fix Lichtenstein's name, *mesomelas*, to this form (Bull. Am. Mus. Nat. Hist., VI, p. 188, 1894), but that this can not be done is evident upon comparing the measurements of this species with those of *mesomelas* as given by Lichtenstein (see footnote p. 29). The specimens on which Dr. Allen based his views came from Oklahoma and belong to the large form—not the small one, of whose presence in Louisiana he was at the time unaware. The body and tail measurements made by him from dry skins, do not correctly represent the average measurements of *varians;* but the size and characters of the skull leave no doubt as to the identity of the specimens.

The species varies greatly in size: thus specimens from the lower Rio Grande Valley have much longer tails and rather smaller skulls than those inhabiting central Texas, Oklahoma, and Colorado. In southeastern Texas, in the vicinity of Matagorda Bay, it intergrades with *mesomelas*, whose range extends westward along the coast from Louisiana. In a series of 7 specimens from O'Connorport (opposite Matagorda Island), which are evidently intermediate, the markings are like those of *varians* while the skulls are small like those of *mesomelas*, and the tails average 312 mm.—much longer than the average tail measurement of *mesomelas*. Seven skins from Mason, Texas, show much more white than typical specimens, in this respect approaching *estor*, from which some of them can be distinguished only by greater size and longer tail. More than half of these Mason skins have the white hairs of the tail extending beyond the tip, and some have a distinct white pencil.

Specimens examined.—Total number, 139, from the following localities:

Tamaulipas: Matamoras, 2; Mier, 1.

Texas: Brownsville, 7; Corpus Christi, 1; Nueces Bay, 2; Santa Tomas, 1; Hidalgo, 2; Rio Grande City, 1; Padre Island, 1; Laredo, 2; Eagle Pass, 3; Mouth of Pecos River, 1; Presidio County, 2; El Paso, 2; East Painted Cave, Valverde County, 1; Indianola, 1; O'Connorport, 7; Rockport, 3; Fort Richardson, Jack County, 1; Berne, 1; Mason 7; Gail, 1; Colorado, 5; Fort Clark, Kinney County, 1; Sherwood, 2; Langtry, 2; Pecos High Bridge (Southern Pacific R. R.), 2; Chisos Mountains, 1; Davis Mountains, 1.

New Mexico: Hall Peak, 2; Eddy, 1.

Oklahoma: Beaver River, Beaver County, 9.

Colorado: Arkins, Larimer County, 20; Chivington, 1; Cañon City, 1; Boulder County, 1; Costilla County, 2; Conrow, 1; Loveland, 1.

Kansas: Cedarvale, 1; Neosho Falls, 1; Trego County, 5; Long Island, 2; Onaga, 3.

Nebraska: Johnstown, 21 (skulls); Valentine, 1; Cherry County, 1; Loup Fork River, 1.

CHINCHA ESTOR (Merriam) ARIZONA SKUNK.

Mephitis estor Merriam, N. Am. Fauna No. 3, p. 81, 1890.

Type locality.—San Francisco Mountain, Arizona.

Geographic distribution.—Arizona, western New Mexico, Sonora, Chihuahua, and northern Lower California; south in the Sierra Madre to southern Chihuahua; limits of range unknown.

General characters.—Size rather small; tail shorter than that of *varians; much white on body and tail;* skull resembling that of *varians*, but smaller.

Color.—White stripes on back very broad—almost confluent; posterior back wholly white in some specimens; tail of black and white hairs, the white longer and chiefly on the upper surface, where they extend beyond and nearly conceal the black: white pencil at tip. (See Pl. II, fig. 3.)

Cranial characters.—Skull resembling that of *varians* in general shape but *smaller and slenderer;* palate ending about on a line with posterior molars, either square or with a very small notch: molars smaller than in either *varians* or *occidentalis;* anterior palatine foramina small and narrow. (See Pl. VII, figs. 3 and 4.)

Measurements.—Average of 7 adult males from Arizona and adjacent parts of Mexico: Total length, 639; tail vertebræ, 285; hind foot, 69. Average of 4 adult females from same localities: 580; 273; 63. *Skull:* (See table, p. 44.)

General remarks.—The Arizona skunk is a very distinct species inhabiting Arizona, New Mexico, and adjacent parts of Mexico. In southern Arizona and Chihuahua its range overlaps that of *milleri*, the two species being often found at the same place. By reason of this fact, and on account of the extreme variability of *milleri*, the two have been frequently confused by authors, and many references to *estor* really apply to *milleri*. There need be no difficulty, however, in distinguishing them, either by skin or skull. The hooded skunks (to which group *milleri* belongs), while extremely variable, are usually either wholly black or wholly white on the back, and never have the two white stripes of *Chincha;* the tail is longer than the head and body (about 52 per cent of the total length). *Estor* is rather constant in markings, and has the white stripes of the other United States species inclosing a small patch of black, while the tail is shorter than the head and body (about 44 per cent of the total length). In the few cases in which *estor* has the whole back white, the purity of the white will serve to distinguish it from *milleri*, in which the white is of a grayish hue, due to the intermixture of black hairs. The very pronounced skull characters distinguishing the two groups are pointed out under the description of the subgenus *Leucomitra* (p. 39).

Estor differs from both *varians* and *occidentalis* in smaller size and shorter tail, and in the much greater extent of white on its body and tail. Specimens from the Mexican boundary line at the west base of the San Luis Mountains are somewhat larger than the typical form of *estor*, and one of them is plainly referable to *varians*, so it is possible that intergradation takes place between the two.

Specimens examined.—Total number, 55, from the following localities:

Arizona: San Francisco Mountain, 3; Springerville, 3; Holbrook, 1; Calabasas, 2; Yuma, 1; Fort Mohave, 1; Fort Verde, 12; Fort Huachuca, 1; Huachuca Mountains, 2; Huachuca Station, 1; Prescott, 1; Pinal County, 4; Whipple Barracks, 1.

New Mexico: Fort Wingate, 1; Cloverdale, Grant County, 1, Hachita, 1.

Sonora (near Mexican boundary line): Santa Cruz River, 1; San Pedro River, 1; Patagonia Mountains, 2; west side San Luis Mountains, 1; San Luis Springs, 1; Animas Valley, 1; San Bernardino Ranch (monument 77, Mexican boundary line), 1; La Noria (monument 111, Mexican boundary line), 1.

Chihuahua: White Water, 1; Cajon Bonita Creek (near Mexican boundary), 1; Colonia Garcia, 6; Sierra Madre (near Guadalupe y Calvo), 1.

Lower California: Poso Vicente, 1.

CHINCHA OCCIDENTALIS (Baird). California Skunk.

Mephitis occidentalis Baird, Mamm. N. A., p. 194, 1857.

Type locality.—Petaluma, California.

Geographic distribution.—Northern and central California, from vicinity of Monterey Bay northward, west of the Sierra and Cascades, to the Willamette Valley, Oregon.

General characters.—Size rather large; resembling *hudsonica* quite closely, but *tail longer; skull relatively narrow across zygomata;* palate rather long, sometimes with a distinct median notch.

Color.—Closely resembling *hudsonica* in pattern of coloration; white stripes of medium width, and frequently extending down sides of tail, though the white hairs never reach beyond the tip. The markings show little variation.

Cranial characters.—Compared with that of *hudsonica*, the skull of *occidentalis* is much *narrower across the zygomata, which spread less abruptly* and are more nearly parallel to the axis of the skull. Palate rather long, usually with a distinct median notch, though this is not always present, even in the typical form. The largest skulls of *occidentalis* equal those of *hudsonica* in length and in mastoid breadth, but the majority are somewhat shorter and narrower. (See Pl. VII, figs. 1 and 2.)

Measurements.—Type:[1] Total length, 800; tail vertebræ, 312; hind foot, 77. Average of 5 adult males from vicinity of San Francisco Bay: 693; 303; 78. One adult female from Auburn, California: 700; 330; 75. *Skull:* (See table, p. 44.)

General remarks.—The California skunk was recognized as distinct by Baird in 1857. With its four subspecies it forms a well-marked group, quite distinct from any of the eastern members of the genus, and has an extensive range. It equals *hudsonica* in size, but has a longer tail, and differs materially in cranial characters. The colors are very constant, but the skulls show great individual variation. The length of the tail is also variable, ranging from 265 to 370 mm. A series of skulls from Cassel, Shasta County, California, average larger than the typical form, and two specimens in the series are clearly intermediate between *occidentalis* and *major*. A specimen from Lake Tahoe likewise shows characters intermediate between

[1] *Fide* Baird.

these two forms. On the north intergradation with *spissigrada* is shown by a large series from Lake Cushman, Washington; while to the southward the typical form merges gradually into *holzneri*. Its relationships with *notata* and *platyrhina* are not clear, but intergradation with these is probable.

Specimens examined.—Total number, 107, from the following localities:

> **California:** Petaluma, 1; Glen Ellen, 1; Novato, 1; Nicasio, 5; Point Reyes, 2; Mt. Tamalpais, 1; Fairfield, 2; Walnut Creek, 1; Santa Clara, 1; Salt Springs, Fresno River, 1; Wawona, 12 (skulls); Pine City, 7 (skulls); Yosemite Valley, 4; Mariposa County, 6 (skulls); South Fork Merced River, 1; Carbondale, 1; Hope Valley, Alpine County, 2; Markleeville, 1; Lake Tahoe, 2; Blue Canyon, 1; Auburn, 2; Tehama, 1; Red Bluff, 2; Sherwoods, 1; Cahto, 1; Cassel, Shasta County, 29 (skulls); Baird, Shasta County, 1; Shasta Valley, 2; Sisson, 2; Pitt River, Shasta County, 4; Fort Crook, Shasta County, 1; Big Valley Mountains, Lassen County, 1.
> **Oregon:** Roseburg, 1; Eugene, 1; Grant Pass, Rogue River Valley, 5.

CHINCHA OCCIDENTALIS SPISSIGRADA (Bangs). Puget Sound Skunk.

Mephitis spissigrada Bangs, Proc. Biol. Soc. Wash., XII, p. 31, 1898.
Mephitis fœtulenta Elliot, Field Columbian Museum, pub. 32, Zool. Ser., 1, no. 13, p. 269, 1899.

Type locality.—Sumas, British Columbia.

Geographic distribution.—Shores of Puget Sound and coast region of Washington and northern Oregon.

General characters.—Resembling *occidentalis*, but with *shorter tail*, and more white on body and tail; *palate short*, without notch.

Color.—As in *occidentalis*, but body stripes very broad, and much white in tail. The white stripes bifurcate about the middle of the back (instead of between the shoulders as in *occidentalis*) and extend down the sides of the tail, the long white hairs frequently reaching beyond the tip. In most of the specimens examined the body stripes have a distinctly yellowish cast, but this is not a constant character. (See Pl. III, fig. 3.)

Cranial characters.—Skull similar to that of *occidentalis*, but *shorter, and relatively broader across zygomata;* rostrum averaging broader; bullæ slightly larger; *palate ending squarely*, with no trace of a notch, *even with last molars;* lower carnassial smaller; nasals long; ascending branches of premaxillæ very long and narrow.

Measurements.—Average of 3 adult males from type locality: Total length, 653; tail vertebræ, 246; hind foot, 79. Average of 2 adult females from type locality: 625; 235; 75. Average of 3 adult males from Neah Bay, Washington: 630; 230; 84. *Skull:* (See table, p. 44.)

General remarks.—This is a handsome skunk and is said to be very abundant, feeding in large numbers on the ocean beaches. It occupies a comparatively small area and is a strongly marked subspecies. Its short tail and the great amount of white in the markings distinguish it from typical *occidentalis*. From *hudsonica*, it differs chiefly in its shorter palate, weaker and less abruptly spreading zygomata, and less mastoid breadth. Skulls of females of *spissigrada* and *hudsonica* bear a close resemblance to one another; those of the young may be distinguished by the longer nasals of *spissigrada*.

Since the original description of this form, additional material from the type locality shows clearly its relationship to *occidentalis* and establishes its identity with *fœtulenta* of Elliot. Specimens from the shores of the Olympic peninsula, on which *fœtulenta* was based, show the greatest extreme in the characters assigned, but the differences between them and specimens from Sumas are too slight to warrant even subspecific recognition. The slight notch in the palate exhibited by Elliot's specimens is due to the fact that they are all immature. In the comparisons on which the present results are based, a series of adult specimens from Neah Bay has been used. Specimens from Washington south of the Olympics are intermediate in characters between *spissigrada* and *occidentalis*, as is shown by the longer tail, longer skull, and notched palate.

Specimens examined.—Total number, 52, from the following localities:

British Columbia: Sumas, 6.
Washington: Neah Bay, 5; The Lagune, near Port Angeles, 3; Lapush, 1; Port Townsend, 1; Steilacoom, 5; Tenino, 2; Lake Cushman, 26 (skulls); Chehalis County, 2.
Oregon: McCoy, 1.

CHINCHA OCCIDENTALIS NOTATA subsp. nov. CASCADE SKUNK.

Type from Trout Lake, Mount Adams, Washington, ♂ adult, No. 87043, U. S. Nat. Mus., Biological Survey, Coll. Collected March 22, 1897, by Peter Schmid.

Geographic distribution.—Southern Washington and northern Oregon, east of the Cascades; exact limits of range unknown.

General characters.—Similar to *occidentalis*, but tail shorter; skull slightly larger and dentition heavier; *body stripes very narrow, and separated for their entire length*.

Color.—Similar to that of *occidentalis*, but body stripes much narrower (about 15 mm. broad in average specimens) and sometimes interrupted; usually, but not always, joined at nape, and not confluent anywhere else. Tail usually all black exteriorly, but sometimes with a little white on each side near the base. In the type and some other specimens, the body stripes terminate about the middle of the back. (See Pl. III, fig. 2.)

Cranial characters.—Skull slightly larger than that of *occidentalis*; palate ending nearly squarely, with no distinct notch; nasals short; *upper molars large*, often exceeding those of the largest specimens of *major*; *lower carnassial broad*.

Measurements.—Average of 3 adult males from type locality: Total length, 633; tail vertebræ, 249; hind foot, 76. Average of 5 adult females from The Dalles, Oregon: 659; 286; 69. *Skull:* (See table, p. 44.)

General remarks.—This form shows greater variability in markings than any other of the western skunks and is the only one in which the body stripes are ever interrupted. In skull characters it resembles *occidentalis* quite closely, but lacks the notch in the palate and has much larger molars. It probably intergrades with both *occidentalis* and *major.*

Specimens examined.—Total number, 41, from the following localities:

Washington: Trout Lake, Mount Adams, 31 (skins with skulls, 6; skulls only, 25); Rockland, Klickitat County, 1; Goldendale, 1.
Oregon: The Dalles, 8.

CHINCHA OCCIDENTALIS MAJOR subsp. nov. GREAT BASIN SKUNK.

Type from Fort Klamath, Oregon. ♂ adult, No. 92238, U. S. Nat. Mus., Biological Survey Coll. Collected Jan. 5, 1898, by B. L. Cunningham. Original number, 80.

Geographic distribution.—Eastern Oregon, northern California, and Nevada; east to the Wasatch Mountains in Utah.

General characters.—Similar to *occidentalis* but much larger; *hind foot longer; skull larger* and more heavily built.

Color.—Much as in *occidentalis:* white stripes broad, bifurcating near the middle of the back, and extending only a short distance on the tail, which is nearly all black exteriorly.

Cranial characters.—*Skull larger* and more heavily built than that of *occidentalis; rostrum broader and much flattened; braincase broader and not so deep,* thus giving a flattened appearance to the upper surface of the skull; dentition heavier; palate long, usually ending in a concave line, sometimes irregularly notched; ascending branches of premaxillæ short and broad.

Measurements.—Average of 5 adult males from type locality: Total length, 705; tail vertebræ, 306; hind foot, 84. *Skull:* (See table, p. 44.)

General remarks.—This subspecies seems to be the largest and heaviest skunk in the genus; the hind foot is both longer and broader than in any other member of the genus, and the front foot is correspondingly large. The large skull with its broad braincase readily distinguishes the form from its congeners. In the series from the type locality, the characters are constant, but the subspecies undoubtedly intergrades with *occidentalis* in northern California, and possibly with *notata* in northern Oregon. Specimens from western Nevada are provisionally included with *major*, although by reason of the fact that no males have been examined—female skunks are less readily separable—their exact relationships are uncertain. Specimens from Ogden, Utah, are clearly referable to this form. Immature skulls of *major* may be distinguished from those of both *occidentalis* and *hudsonica* by the broad premaxillæ.

Specimens examined.—Total number, 27, from the following localities:

> **Oregon:** Fort Klamath, 6; Tule Lake, 2; Plush, Lake County, 1; Shirk, Harney County, 1; Harney, 2; Elgin, 1.
> **Washington:** Touchet, 1.
> **California:** Lassen Creek, Shasta County, 1; Honey Lake, 1; Sierra Valley, 1.
> **Nevada:** Carson, 1; Reno, 1; Quinn River Crossing, Humboldt County, 1.
> **Utah:** Ogden, 6; Provo, 1.

CHINCHA OCCIDENTALIS HOLZNERI (Mearns). SOUTHERN CALIFORNIA SKUNK.

Mephitis occidentalis holzneri Mearns, Proc. U. S. Nat. Mus., XX, p. 461, 1897.

Type locality. San Isidro Ranch, Lower California.

Geographic distribution.—Southern California, from vicinity of Monterey Bay south into Lower California; east to the Sierra Nevada and San Bernardino Range; limits of southward range unknown.

General characters.—Similar to *occidentalis* but smaller.

Color. There are no appreciable color differences to distinguish this form from *occidentalis*. (See Pl. III, fig. 1.)

Cranial characters.—*Skull smaller* than that of *occidentalis* and *relatively narrow in mastoid region;* audital bullæ more circumscribed, but rather prominent; palatal notch usually absent.

Measurements.—Type (adult male):[1] Total length, 665; tail vertebræ, 273; hind foot, 72. Average of 3 adult males from Twin Oaks, San Diego County: 637; 292; 71. Average of 3 females (barely adult) from San Diego County: 605; 291; 64. *Skull:* (See table, p. 44.)

General remarks.—This form does not differ greatly from *occidentalis*, but averages considerably smaller in cranial measurements. The body and tail measurements of *holzneri* are slightly smaller than those of *occidentalis*, and the hind foot is decidedly shorter. Specimens from Ventura and adjacent counties are intermediate between the two forms, having the long tail of *occidentalis*, but very small skulls.

Specimens examined.—Total number, 41, from the following localities:

> **Lower California:** San Isidro Ranch, 3.
> **California:** Dulzura, 3; Twin Oaks, 5; Witch Creek, 2; Unlucky Lake, San Diego County, 1; Pacific coast at Mexican boundary, 1; Santa Ysabel, San Diego County, 1; Santa Paula, 2; San Fernando, 1; Ventura River, 1; Santa Ynez Mission, 3; Gaviota Pass, Santa Barbara County, 2; San Emigdio, Kern County, 1; Morro, 1; San Luis Obispo, 3; San Simeon, 2; Monterey, 3; South Fork Kern River (25 miles east of Kernville), 1; Three Rivers, 5.

[1] *Fide* Mearns.

CHINCHA PLATYRHINA sp. nov. BROAD-NOSED SKUNK.

Type from South Fork of Kern River (25 miles east of Kernville), California, ♂ adult, No. $\frac{29384}{41351}$, U. S. Nat. Mus., Biological Survey Coll. Collected July 5, 1891, by Vernon Bailey. Original number, 2998.

Geographic distribution. Known only from the type locality and from Owens Valley.

General characters. Externally much like *occidentalis;* skull with peculiarly shaped zygomata and *very broad rostrum.*

Color. Resembling *occidentalis:* white stripes of medium width, produced only a short distance on the sides of the tail; tail black externally, except for an indistinct band of white on the upper surface about ⅗ the distance from base to tip. Specimens from Owens Valley have most of the tail hairs *chestnut* instead of black, exteriorly; and in one case most of the body is chestnut. This is probably due to fading.

Cranial characters.—Skull resembling that of *occidentalis,* but shorter and relatively broader; much flattened in frontal region; *rostrum very broad*—actually and relatively broader than in largest specimens of *occidentalis;* zygomata spreading less abruptly and in an even curve nearly parallel to the axis of the skull; palate nearly square, with only a very slight notch; audital bullæ small, and slightly inflated; tube of auditory meatus short; nasals short and broad.

Measurements. Type (adult male): Total length, 750; tail vertebræ, 320; hind foot, 90. Average of 3 adult females from Owens Valley: 679; 332; 79. *Skull:* (See table, p. 44).

General remarks.—This species shows no marked characters of pelage to distinguish it from *occidentalis,* which it equals or exceeds in size, but its peculiar skull at once serves to separate it. As typical *holzneri* occurs at the type locality of *platyrhina,* it is evident that intergradation does not take place between the two species at this point. It is quite likely, however, that *platyrhina* intergrades with *major;* but until there is material available to show such intergradation, a binomial designation seems preferable. The skull resembles that of *major* in general shape, but is very much smaller, and relatively broader across the postorbital processes.

Specimens examined.—Total number, 9, from the following localities:

California: South Fork Kern River, 5; Owens Valley, 2; Owens Lake, 2.

Subgenus LEUCOMITRA.

LEUCOMITRA, *subgen. nov.* HOODED SKUNKS.

Type.—*Chincha macroura* (Lichtenstein). From mountains northwest of the City of Mexico.

Subgeneric characters.—Skull (Pl. VIII) short and broad; interpterygoid fossa narrowly U-shaped; palate without notch or spine;

anterior palatine foramina large and rounded; audital bullæ large and greatly inflated; tube of auditory meatus short; periotic region slightly inflated; mastoid processes and sagittal crest never greatly developed; zygomata never spreading abruptly, and often nearly parallel to the axis of the skull; interorbital constriction not marked; paroccipital processes directed outward and not sharply pointed; posterior margin of coronoid distinctly concave. Size medium to small; build much slenderer than that of *Chincha;* feet slender; tail very long; ears prominent; fur long and silky but not dense; hairs on the nape elongated and spreading sidewise, forming a sort of hood.

In addition to the characters given above, the hooded skunks differ radically from the large striped skunks (subgenus *Chincha*) in the pattern of coloration (see Pl. IV), and although they exhibit great individual variation, one description will answer for the three forms comprised in the subgenus.

Two patterns occur, one in which the upperparts are chiefly white, the underparts black; the other in which the upperparts are nearly all black, with narrow lateral stripes, and under surface of tail white. Between these two extremes are many intermediate phases. The frontal stripe is narrow and often absent.

In the *white-backed* phase a broad band of white begins between the ears and covers the whole back and upper surface of the tail, the long white hairs drooping gracefully from the sides and over the tip of the tail. This band is of varying width, but is never bifurcated as in true *Chincha*. There may or may not be a white lateral line separated from the dorsal band by a black area. The dorsal band is never of the deep creamy hue so frequent in the stripes of *Chincha*, but is composed of nearly pure white hairs with numerous black ones intermixed, and so is more or less distinctly grayish in effect. The lateral stripes, when present, are always without this mixture of black hairs.

The *black-backed* phase usually has the narrow frontal stripe, and may have the white hood, which, however, is often absent. The white lateral stripe is almost always present, though varying in width from an inch or more to a mere trace; it is frequently replaced by two narrow stripes on each side. The lower surface of the tail is usually white, though sometimes the whole tail is black externally.

As in the typical subgenus, the bases of the tail hairs in both phases are invariably white. Irregular white spots on the ventral surface of the body are frequent.

In some of its characters, this subgenus shows an approach toward the genus *Mephitis* (*Spilogale* of authors), particularly in the shape of the audital bullæ and the anterior palatine foramina. In the concave border of the coronoid it resembles the white-backed skunks (*Thiosmus*); but this is a character that sometimes appears in true *Chincha*. The parasites found occasionally in the skulls of all skunks are especially frequent in this subgenus.

Descriptions of Species and Subspecies.

CHINCHA MACROURA (Lichtenstein). Hooded Skunk.

Mephitis macroura Lichtenstein, Darst. Säugeth., pl. 46, with accompanying text, 1832.—Baird, Mamm. N. Am., 1857, p. 200. [Not *M. macroura* Aud. and Bach.]
Mephitis mexicana Gray, Charlesworth's Mag. Nat. Hist., I, p. 581, 1837.
Mephitis longicaudata Tomes, Proc. Zool. Soc. London, 1861, p. 280.

Type locality. Mountains northwest of the city of Mexico.

Geographic distribution.—Highlands of central and southern Mexico; south to Guatemala.

General characters.—Size medium; coloration as in other members of the subgenus extremely variable; skull with sagittal crest and mastoid processes well developed.

Color.—The description of the color given in the remarks on the subgenus will apply to this species, with the added note that quite a large proportion of the specimens examined are in the black phase, and that the white pencil is frequently absent. (See Pl. IV, fig. 1.)

Cranial characters.—Skull of medium size (for the subgenus): sagittal crest and mastoid processes (in adult males) well developed; molars small; anterior palatine foramina large and rounded; posterior margin of palate an even curve, ending on a line with last molars.

Measurements.—Average of 4 adult males from near the type locality (Querendaro, Nahuatzin, and Salazar): Total length, 623; tail vertebræ, 299; hind foot, 67. Average of 4 adult females from same localities: 594; 297; 60.5. *Skull:* (See table, p. 44).

General remarks.—This species was described by Lichtenstein in 1832, and his name has been very generally adopted by subsequent authors. His type came from the mountains northwest of the city of Mexico, and specimens from Salazar are considered typical. The Guatemala form has been described by Tomes under the name *longicaudata*, but specimens from the vicinity of Dueñas show that it is not separable from the typical form. The southern limit of range is unknown, but it will probably be found little south of the highlands of Guatemala. To the northward *macroura* grades imperceptibly into *milleri*.

Specimens examined.—Total number, 60, from the following localities:

Mexico: Tlalpam, 4; Amecameca, 1; Salazar, 2.
Hidalgo: Marques, 1; El Chico, 1; Irolo, 1; Encarnacion, 1; Zimapan, 1; Real del Monte, 1.
Michoacan: Querendaro, 4; Patzcuaro, 1; Nahuatzin, 2.
Colima: Colima, 4; Hacienda Magdalena, 1.
Jalisco: Ameca, 2; San Sebastian, 1.
Tepic: Santa Teresa, 1.
Zacatecas: Valparaiso, 1.
San Luis Potosi: Hacienda La Parada, 6.
Tamaulipas: Jaumave, 1.
Morelos: Cuernavaca, 5.

Puebla: Chalchicomula, 1; Tehuacan, 1; Piaxtla, 2.
Vera Cruz: Perote, 1; Las Vigas, 2; Jico, 4; Orizaba, 1.
Oaxaca: Oaxaca (mountains 15 miles west), 1.
Guatemala: Dueñas (vicinity), 5.

CHINCHA MACROURA MILLERI (Mearns). NORTHERN HOODED SKUNK.

Mephitis milleri Mearns, Proc. U. S. Nat. Mus., XX, p. 467, 1897.

Type locality.—Fort Lowell (near Tucson), Arizona.

Geographic distribution.—Southern Arizona, Sonora, and parts of Chihuahua, Sinaloa, Durango, and Coahuila.

General characters.—Very similar to *macroura* but averaging *larger, with heavier skull.*

Color.—Much as in *macroura*, with probably a larger proportion of the white-backed phase. In a series of 15 from Camoa, Sonora, 6 have black backs and 9 white backs. Two of the black-backed ones are with and 4 without the white hood; 1 is wholly black save for a trace of the side stripe and a very narrow frontal stripe. (See Pl. IV, figs. 2, 3, and 4.)

Cranial characters.—*Skull averaging larger* than that of *macroura*, the greatest difference being in the length. Lower carnassial considerably larger, both in length and breadth. (See Pl. VIII, figs. 1 and 2.)

Measurements.—*Type* (an abnormally large specimen): Total length, 790; tail vertebræ, 435; hind foot, 73. Average of 7 adult males from Arizona and adjacent parts of Mexico: 672; 359; 65. Average of 7 adult females from same localities: 668; 357; 61. *Skull:* (See table, p. 44.)

General remarks.—The northern hooded skunk is a rather poorly marked subspecies of *macroura*, the two forms intergrading in central Mexico. The type is a greatly overgrown specimen, as is shown by comparison with a series of adults from near the type locality. Average specimens are somewhat larger than *macroura*, the tail usually exceeding 350 mm. in length and the skulls averaging larger, though it is possible to select specimens from both extremes of their combined range that are almost identical in cranial characters.

The most northern point at which the subspecies has been taken is Fort Grant, Graham County, Arizona, whence it spreads southward over northwestern Mexico, passing into the typical form in the central states of Mexico. Apparently it does not occupy northeastern Mexico, since it has not been recorded from the Rio Grande Valley or from the northern parts of Coahuila, Nuevo Leon, or Tamaulipas.

Specimens examined.—Total number, 55, from the following localities:

Arizona: Fort Lowell, 1; Nogales, 1; Fort Huachuca, 7; Fort Grant, Graham County, 2; Tucson, 1; Calabasas, 1; Fairbank, 2; Santa Catalina Mountains, Pinal County, 5.

Sonora: Patagonia Mountains, 1; Santa Cruz River, 3; Santa Cruz, 1; Hermosillo, 1; Magdalena, 1; Camoa, 15; Alamos, 1.

Chihuahua: Chihuahua, 3; Casas Grandes, 2; Guadalupe y Calvo (mountains near), 3.

Sinaloa: Sierra de Choix, 1.

Coahuila: Jimulco, 2; La Ventura, 1.

CHINCHA MACROURA VITTATA (Lichtenstein). LEAST HOODED SKUNK.

Mephitis vittata Lichtenstein, Darst. Säugeth., pl. 47, with accompanying text, 1832.
Mephitis concolor Gray (Verreaux MS.?), Proc. Zool. Soc. London, 1865, p. 149.
Mephitis vittata var. *b. intermedia* Gray, Cat. Carnivora Brit. Mus., p. 138, 1869.
Mephitis vittata var *c. concolor* Gray, Cat. Carnivora Brit. Mus., p. 138, 1869.

Type locality.—'San Matteo el Mar,' Oaxaca, Mexico.

Geographic distribution.—Known only from the type locality: probably ranges over the coast region of Oaxaca and Chiapas.

General characters.—*Smaller than macroura*, with very small skull, and slightly developed mastoids and sagittal crest.

Color.—As in *macroura;* pencil not distinct. In the series of 18 topotypes examined, 5 are in the black phase.

Cranial characters.—*Skull decidedly smaller* than that of *macroura;* relatively narrow across zygomata, and *mastoids much reduced;* sagittal crest very slightly developed; *bullæ disproportionately large.* (See Pl. VIII, figs. 3 and 4.)

Measurements.—Average of 6 adult males from the type locality: Total length, 558; tail vertebræ, 275; hind foot, 60.4. Average of 12 adult females from the type locality: 585; 300; 59.5. *Skull:* (See table, p. 44.)

General remarks.—This form was described by Lichtenstein at the same time as *macroura*, and his description was accompanied by a good figure of a specimen in the black phase. The description is too meager in details, in the absence of material from the type locality, to clearly establish the validity of the subspecies, but all uncertainty has been removed by the fine series of specimens collected at San Mateo del Mar in 1895 by E. W. Nelson and E. A. Goldman. The form is well marked, differing more from *macroura* than does *milleri*. It is the smallest of the genus, and may be recognized by this fact as well as by its peculiar skull characters. The pelage is rather thin and coarse. The tail is relatively longer than that of *macroura*. Although, as usual, the males have the greater average skull measurements, yet in total length the average of the females is greater. A very large proportion of the skulls examined had been infested with parasites, and the distortion of the cranium through this cause is greater than in any other species examined. One specimen in particular has the swelling produced fully 7 mm. above the normal top of the cranium and spread to a width of 28 mm., although the mastoid breadth of the same specimen is but 31 mm.

Specimens examined.—Eighteen, all from the type locality.

Average cranial measurements of Chincha.

Species.	Sex.	Localities.	Basal length.	Basilar length of Hensel.[1]	Greatest zygomatic breadth.	Greatest mastoid breadth.	Breadth across postorbital processes.	Least interorbital breadth.	Palatal length.	Post-palatal length.	Foramen magnum to plane of last molars.	Number of specimens averaged.
Chincha mephitis	♂	Lake Edward, Quebec	71.5	70	52	44	25.5	23	29.5	40.5	43	1
	♀	....do	65	62.5	47	40.5	24	20.5	27	35.5	38.5	1
C. hudsonica	♂	Saskatchewan, Montana, Wyoming, and Nebraska	75.5	73.8	54.4	45	24.3	19	31	42.8	44.1	4
	♂	Sicamous and Shuswap, British Columbia	72.1	70.1	52.6	43.4	23.5	19.4	29.6	40.5	42.1	4
	♀	....do	66.6	65	46.9	39.1	22.3	18.8	27.8	37.3	38.4	4
C. putida	♂	Westchester and Orange counties, New York	63.5	61.6	45.6	38.8	20.5	17.9	26.3	35.3	36.7	5
	♂	Adirondack Mountains, New York	66	63.6	47.7	40.8	21.5	19.2	26.3	37.3	38.5	5
	♀	Westchester County, New York	58.6	56.7	43.1	36.7	20.8	18	21	32.7	33.5	5
C. elongata	♂	Florida (peninsula)	65	62.9	47.4	38.9	22.3	20.3	27.2	35.7	37.1	5
	♀	....do	60.8	58.9	44.6	37	21.5	19.7	25	33.9	35.1	5
C. mesomelas	♂	Louisiana	60.5	58.5	42.5	34.5	20.9	18.8	24.8	33.6	34.1	6
	♀	....do	57.8	56	41.1	34.1	20.2	17.6	24.2	31.8	32.9	5
C. m. avia	♂	San Jose, Illinois	64.1	62.4	44.7	36.6	21.1	17.4	26.6	36	36	4
	♀	....do	60.8	59	42	35.8	19.5	17.5	24.5	34.5	34.8	2
C. m. varians	♂	Southern Texas and Tamaulipas	64.6	62.5	45.2	37.5	22	19.1	26.6	36	37	7
	♀	Southern Texas	62.8	60.5	44.2	36.4	21.3	19.2	26	34.5	35.8	5
C. estor	♂	Arizona	62.2	60.2	44.3	36.2	20.7	18.2	25.3	34.8	35	3
	♀	Arizona and New Mexico	59.3	57.3	41.2	34.3	20.7	18.5	24.2	33.2	33	3
C. occidentalis	♂	Vicinity of San Francisco Bay, California	72.3	70	49.8	42.4	23.3	20.1	28.8	41.2	42.6	5
	♀	Nicasio and Monterey, California	65.3	62.7	44.7	37.8	21.3	18.5	27.7	35.3	36.8	3
C. o. spissigrada	♂	Sumas, British Columbia	68.8	66.8	50	41.5	23.5	18.9	28	38.8	39.8	2
	♂	Neah Bay, Washington	69.8	68.5	50.4	41	24	20	28	40.5	40.5	4
	♀	Sumas, British Columbia, and Neah Bay, Washington	67.5	66	48	40	22.5	20	27.5	38.5	39.5	

C. o. notata	♂	Trout Lake, Washington	73.1	71.2	52	44.9	23.7	19.7	30.2	41	42.9[1]	7
	♀	.....do	66.1	64.5	45.4	39.4	20.9	18.6	27.8	36.7	37.8	5
C. o. major	♂	Fort Klamath, Oregon	75.3	73.5	52.6	45.8	24.1	20.8	30.5	43	44.6	4
	♀	Plush, Warner Lake, Oregon	67	65	47	41	23	21	27.5	38.5	39	1
C. o. holzneri	♂	Southern California and northern Lower California	66.9	64.5	47.4	38.7	22.2	19.4	27.3	37.2	38.9	7
	♀	Southern California	61.5	59.5	42.5	35.4	19.7	18.3	25.5	34	31.8	5
C. platyrhina	♂	Kern River and Owens Lake, California	69.2	67.7	48.7	40.7	21.7	19.8	29	38.7	40.7	3
	♀	Owens Valley, California	67	65	46.8	39	23.5	19.3	28.5	36.5	38.8	2
C. macroura	♂	Michoacan and Puebla, Mexico	56.3	54.5	42.2	34.1	21.2	(²)	22.9	31.6	31.8	8
	♀	Michoacan and Distrito Federal, Mexico	51.5	53.3	39.2	32.7	19.6	(²)	22.8	30.5	30.5	5
C. m. milleri	♂	Arizona and Sonora	60.3	58.4	43.9	36	22.3	19.4	24.1	34.2	33.8	7
	♀	Arizona	56.4	54.4	39.6	32.6	20.4	18.3	23.5	31	31.3	5
C. m. vittata	♂	San Mateo del Mar, Oaxaca	54.6	52.6	38.3	32.3	20.1	(²)	22	30.6	30.7	5
	♀	.....do	52.3	50.5	38.3	31.8	20.6	(²)	21.2	29.3	29.3	7

[1] See footnote, p. 21.

[2] On account of the diseased condition of these skulls this measurement could not be taken.

INDEX.

[New names in **black-face type**; synonyms in *italics*.]

PLATE I.

[Greatly reduced, and relative sizes not accurately shown, owing to differences in preparation of skins.]

Fig. 1. *Chincha putida* (Boitard). Burlington, Massachusetts. (No. 77878, U. S. Nat. Mus.)

2. *Chincha elongata* (Bangs). Fort Kissimmee, Florida. (No. 64017, U. S. Nat. Mus.)

3. *Chincha elongata* (Bangs). Fort Kissimmee, Florida. (No. 64016, U. S. Nat. Mus.)

SKINS OF CHINCHA.

1. *Chincha putida.* 2, 3. *Chincha elongata.*

PLATE II.

[Greatly reduced, and relative sizes not accurately shown, owing to differences in preparation of skins.]

Fig. 1. *Chincha mephitis* (Schreber). Pine Lake, Keewatin. (No. 107226, U. S. Nat. Mus.)

2. *Chincha mesomelas* (Licht.). Calcasieu Parish, Louisiana. (No. 99831, U. S. Nat. Mus.)

3. *Chincha estor* (Merriam). Holbrook, Arizona. (No. 53209, U. S. Nat. Mus.)

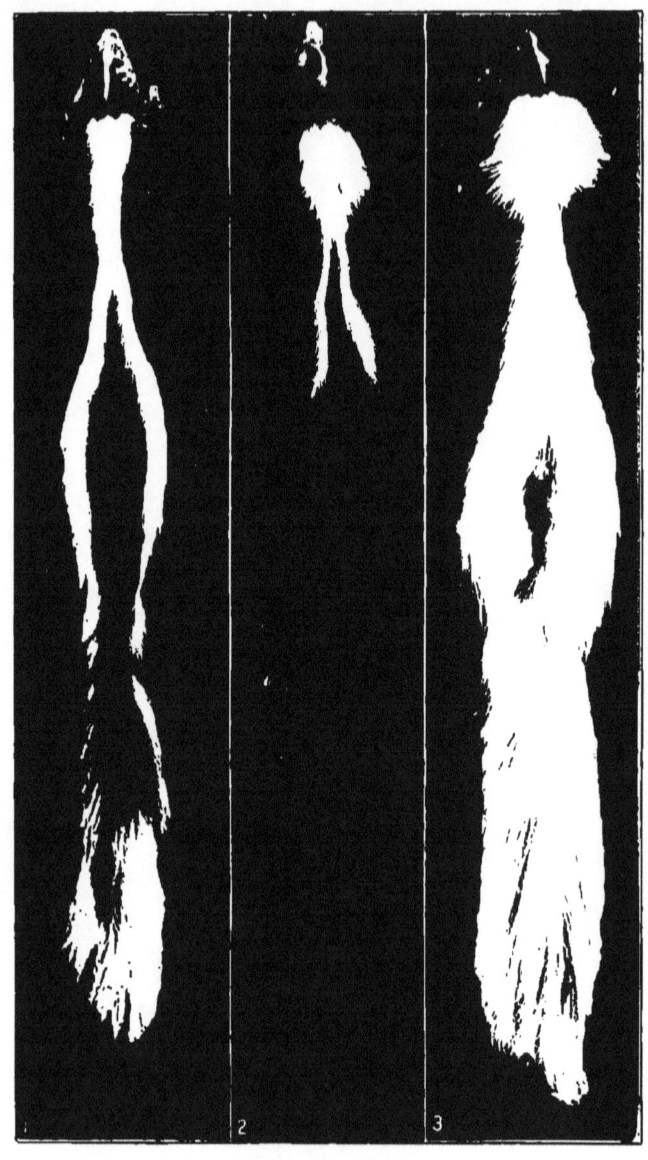

SKINS OF CHINCHA.

1. *Chincha mephitis*. 2. *Chincha mesomelas*. 3. *Chincha estor*.

PLATE III.

[Greatly reduced, and relative sizes not accurately shown, owing to differences in preparation of skins.]

FIG. 1. *Chincha occidentalis holzneri* (Mearns). Three Rivers, California. (No. 31244, U. S. Nat. Mus.)

2. *Chincha occidentalis notata* Howell. Trout Lake, Mt. Adams, Washington. (No. 87042, U. S. Nat. Mus.)

3. *Chincha occidentalis spissigrada* (Bangs). Neah Bay, Washington. (No. 88650, U. S. Nat. Mus.)

PLATE IV.

[Greatly reduced, and relative sizes not accurately shown, owing to differences in preparation of skins.]

FIG. 1. *Chincha* (*Leucomitra*) *macroura* (Licht.). Perote, Vera Cruz. (No. 54225, U.S. Nat. Mus.)

2. *Chincha* (*Leucomitra*) *macroura milleri* (Mearns). Camoa, Sonora. (No. 95927, U. S. Nat. Mus.)

3. *Chincha* (*Leucomitra*) *macroura milleri* (Mearns). Camoa, Sonora. (No. 95923, U. S. Nat. Mus.)

4. *Chincha* (*Leucomitra*) *macroura milleri* (Mearns). Camoa, Sonora. (No. 95931, U. S. Nat. Mus.)

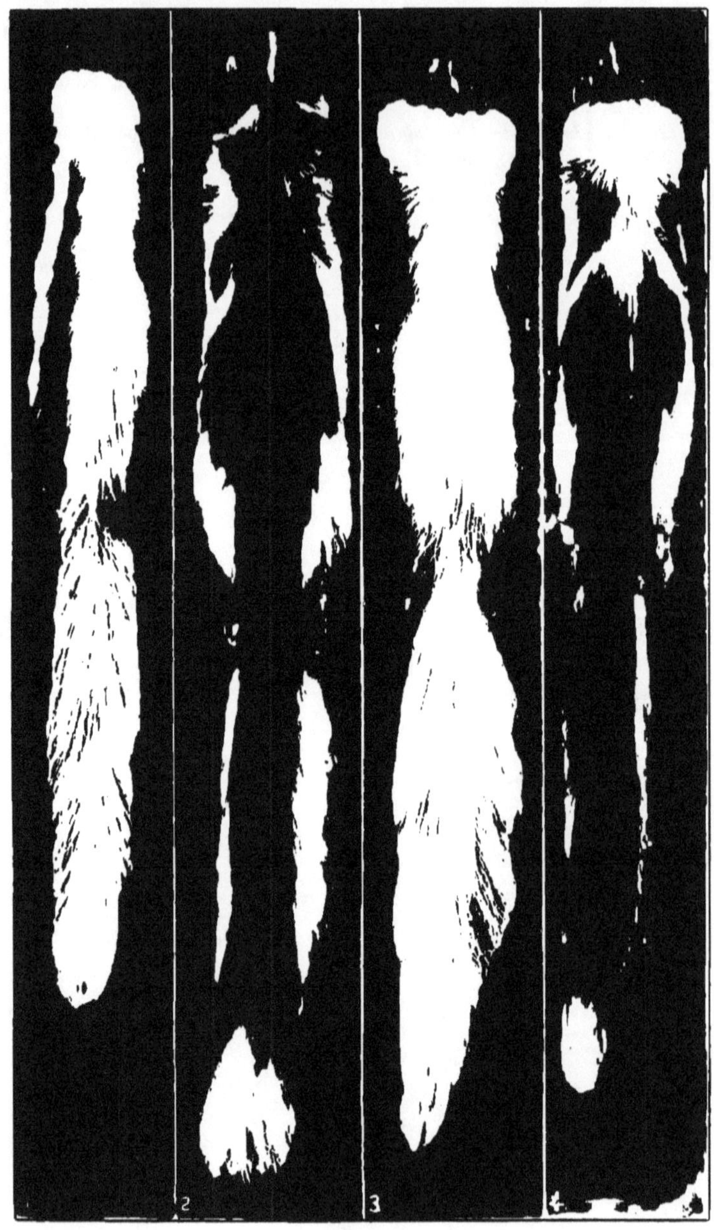

SKINS OF CHINCHA (LEUCOMITRA).

1. *Chincha macroura.* 2, 3, 4. *Chincha macroura milleri.*

PLATE V.

[Natural size.]

FIGS. 1 and 2. *Chincha mephitis* (Schreber). ♂, Lake Edward, Quebec. (No. 3805, Coll. E. A. & O. Bangs.)

3 and 4. *Chincha putida* (Boitard). ♂, Highland Falls, New York. (No. 2020, Am. Mus. Nat. Hist.)

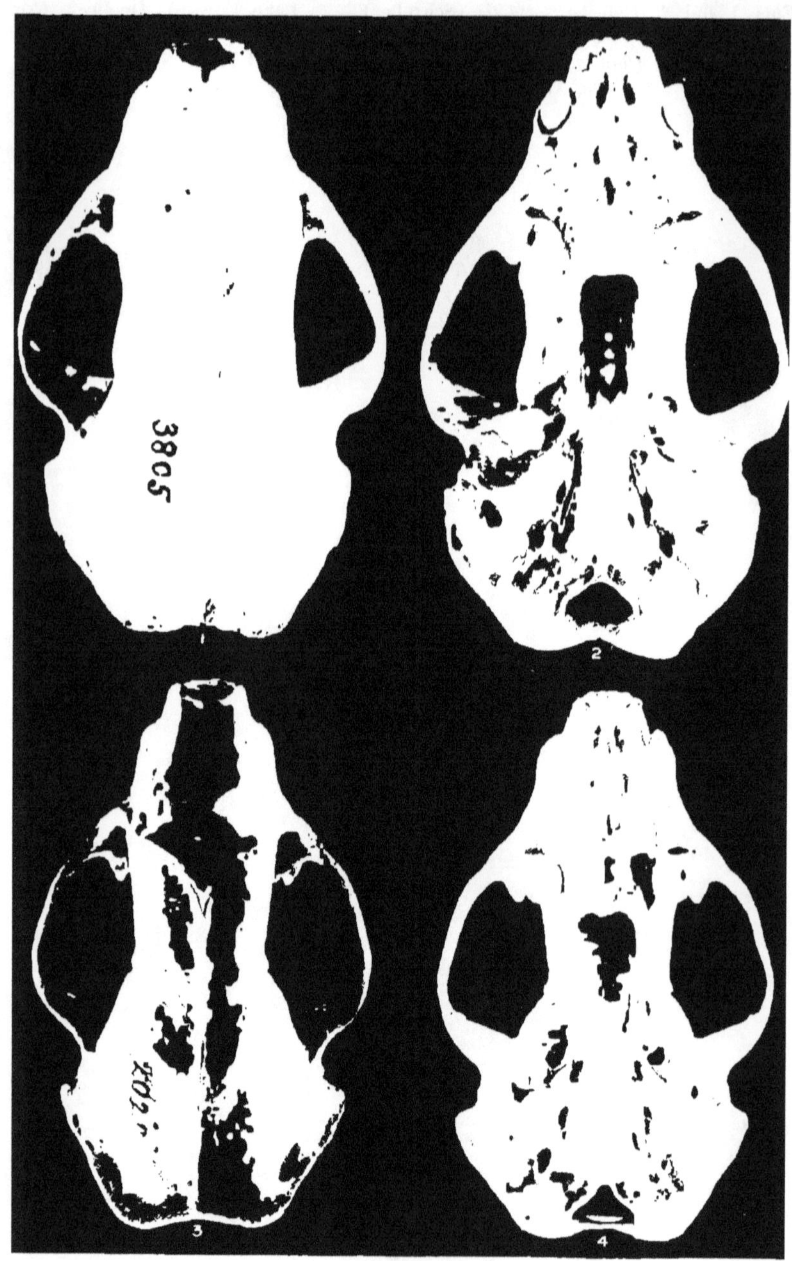

SKULLS OF CHINCHA

PLATE VI.

[Natural size.]

FIGS. 1 and 2. *Chincha hudsonica* (Richardson). ♂, Sicamous, British Columbia. (No. 69957, U. S. Nat. Mus.)

3 and 4. *Chincha mesomelas* (Licht.). ♂, Calcasieu Parish, Louisiana. (No. 99969, U. S. Nat. Mus.)

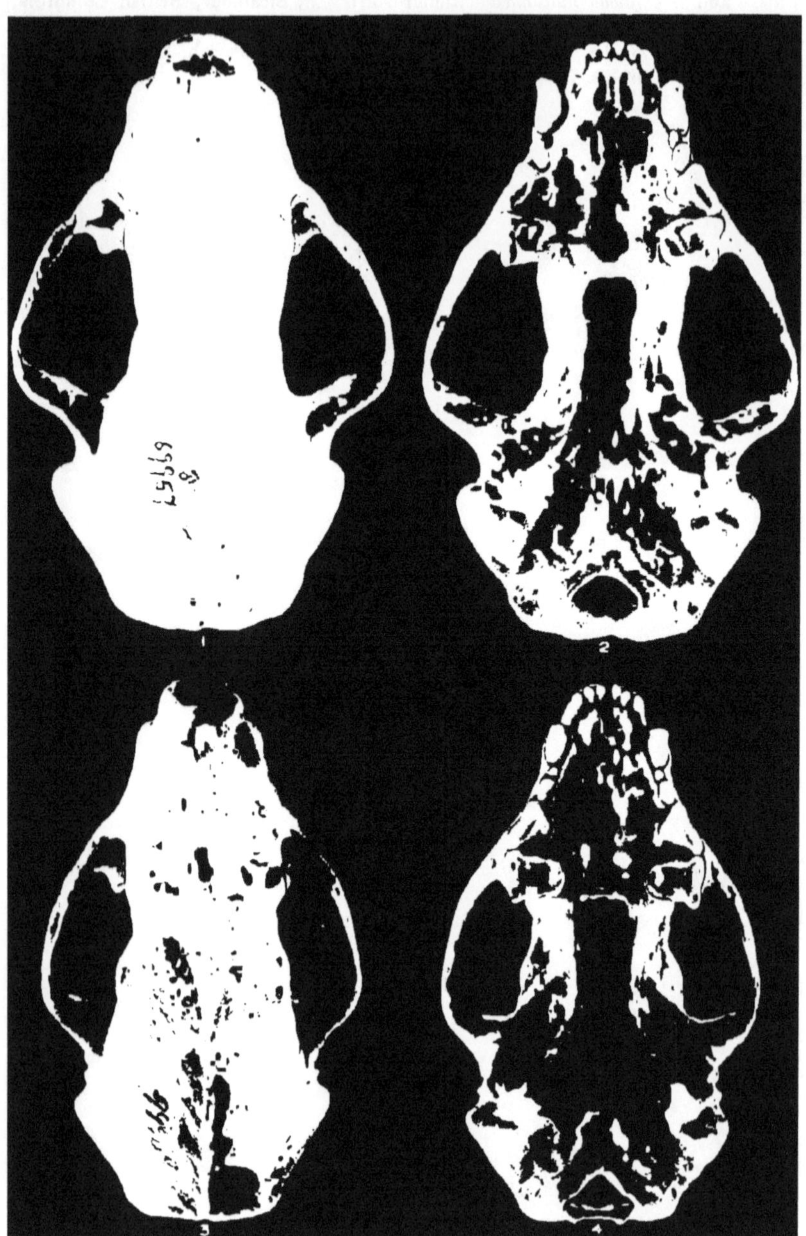

SKULLS OF CHINCHA

1, 2. ... sandles.

PLATE VII.

[Natural size.]

FIGS. 1 and 2. *Chincha occidentalis* (Baird). ♂, Type. Petaluma, California. (No. 2617, U. S. Nat. Mus.)

3 and 4. *Chincha estor* (Merriam). ♂, Type. San Francisco Mountain, Arizona. (No. 24645, U. S. Nat. Mus.)

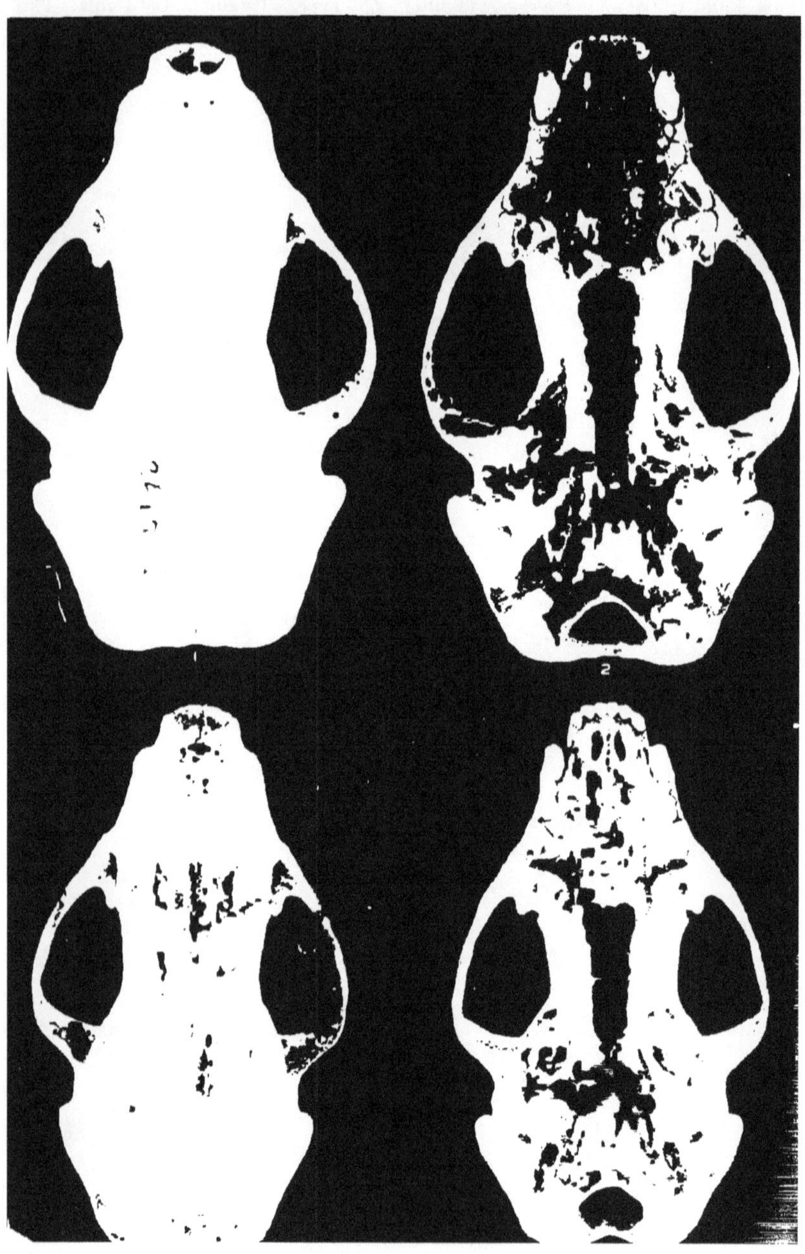

PLATE VIII.

[Natural size.]

FIGS. 1 and 2. *Chincha (Leucomitra) macroura milleri* (Mearns). ♂, Fort Grant, Arizona. (No. 96129, U. S. Nat. Mus.)

3 and 4. *Chincha (Leucomitra) macroura vittata* (Licht.) ♂, San Mateo del Mar, Oaxaca. (No. 73478, U. S. Nat. Mus.)

O

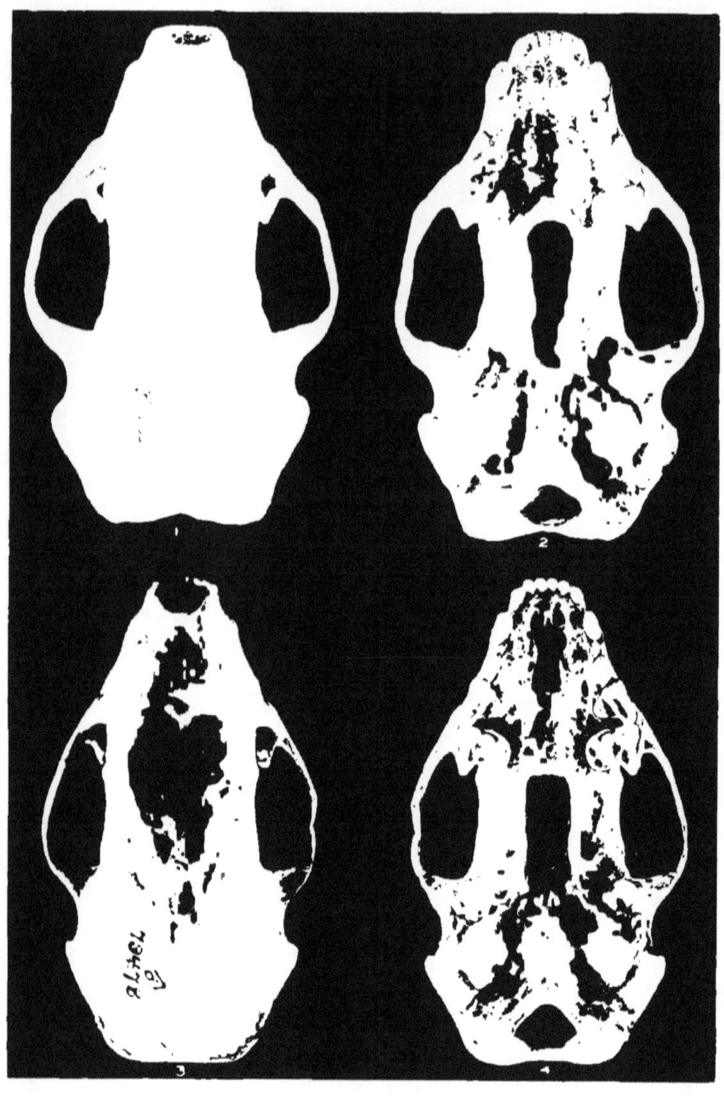

SKULLS OF CHINCHA (LEUCOMITRA).

1, 2. *Chincha macroura milleri.* 3, 4. *Chincha macroura vittata.*

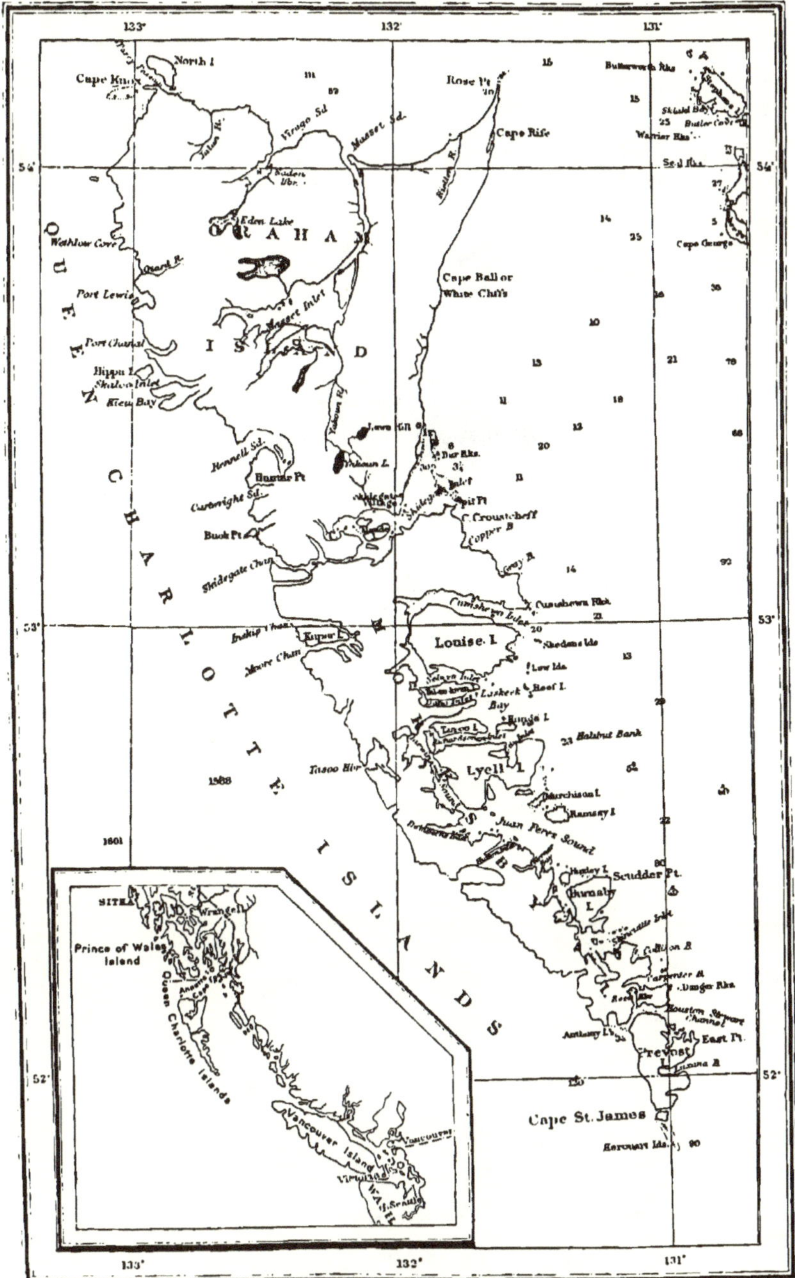

MAP OF THE QUEEN CHARLOTTE ISLANDS.

From United States Coast and Geodetic Survey chart No. 3089.

U. S. DEPARTMENT OF AGRICULTURE
DIVISION OF BIOLOGICAL SURVEY

NORTH AMERICAN FAUNA

No. 21

[Actual date of publication, September 26, 1901.]

NATURAL HISTORY OF THE QUEEN CHARLOTTE ISLANDS, BRITISH COLUMBIA
NATURAL HISTORY OF THE COOK INLET REGION, ALASKA

BY

WILFRED H. OSGOOD
ASSISTANT, BIOLOGICAL SURVEY

Prepared under the direction of

Dr. C. HART MERRIAM
CHIEF OF DIVISION OF BIOLOGICAL SURVEY

WASHINGTON
GOVERNMENT PRINTING OFFICE
1901

Sup^t of Documents
a c/o
STATISTICAL DEPARTMENT
OCT 8 1901
BOSTON PUBLIC LIBRARY

LETTER OF TRANSMITTAL.

U. S. Department of Agriculture,
Washington, D. C., July 5, 1901.

Sir: I have the honor to transmit herewith for publication, as No. 21 of North American Fauna, two special reports on the natural history of little-known parts of the northwest coast of North America, the Queen Charlotte Islands, British Columbia, and Cook Inlet, Alaska, both by my assistant, Wilfred H. Osgood.

Owing to the absence of definite information concerning the faunas of these areas, Mr. Osgood was sent there to conduct biological explorations during the field season of 1900. The results of his trip form an important contribution to the natural history of the northwest coast region.

Respectfully,

C. Hart Merriam,
Chief, Biological Survey.

Hon. James Wilson,
Secretary of Agriculture.

CONTENTS.

ILLUSTRATIONS.

PLATES.

TEXT FIGURE.

NATURAL HISTORY OF THE QUEEN CHARLOTTE ISLANDS.

By Wilfred H. Osgood,

Assistant Biologist, Biological Survey.

INTRODUCTION AND ITINERARY.

The Queen Charlotte Islands lie off the coast of British Columbia, just south of the Alaskan boundary, between latitude 51° 55′ N. and 54° 15′ N. They are slightly farther from the mainland than any of the islands of the Alexander Archipelago, to the north of them, and are not in the track of regular coasting steamers. They were visited by several of the early navigators of the northwest coast, but until 1787 no name was given them and no account of them had been published. In this year Capt. George Dixon cruised about the islands from July 1 to August 3, trading with the natives and roughly charting the coast. He named the group after Queen Charlotte, the consort of George III of England, and in the report of his voyage which appeared later included a very interesting account of his visit, together with maps and illustrations. In the early part of the nineteenth century various fur-trading vessels stopped frequently at the Queen Charlottes, and later the discovery of gold and coal in small quantities has caused sporadic invasions by prospectors. No important attempt has been made, however, to develop the resources of the islands.

The interior has not been explored to any great extent, and probably will not be for some time to come, since the difficulties of travel are insurmountable to ordinary expeditions. The principal harbors and most of the east coast have been surveyed from time to time by officers of the royal navy, and the late Dr. George M. Dawson spent the summer of 1878 in studying the geology and littoral topography of the group. The report[1] that he published is exceedingly interesting and important. It contains accounts of the history, geology, and ethnology of the islands, with some notes on the natural history, more especially of invertebrates. The vertebrate fauna as a whole had never been studied until the present year, however, and the little that was known of it was entirely due to the zeal of Rev. J. H. Keen, who for

[1] Geol. Survey of Canada, Report of Progress for 1878–79, Pt. III, pp. 1–239, Montreal, 1880.

eight years was engaged in missionary work at Massett. The present white population consists of several missionaries and three or four other men, who are engaged in trading and fishing. To supply necessities to these and to carry mail to them and the educated natives, small steamers make irregular trips to the islands.

From one of these steamers I was landed June 13, 1900, with my assistant, Mr. Edmund Heller, at the fishing station in Cumshewa Inlet, known as Clew (also spelled Klue), on the north side of the inlet about 10 miles from its mouth. We were very kindly received by Mr. W. H. Dempster, who conducts a plant here for obtaining oil from the dogfish which abound in the waters about the islands. We made small collections at Clew, and then devoted a number of days to cruising about the inlet in a canoe, collecting and making observations at various points. The first and most important trip was to the head of West Arm of Cumshewa Inlet, where we camped from June 17 to 25. After the coast at this point had been worked an expedition was made to the top of the highest mountain near the head of West Arm.[1] This trip was exceedingly trying, as we were obliged to carry food, bedding, and traps on our backs and beat our way through the deep forest and heavy underbrush. We reached an altitude estimated at 4,500 to 5,000 feet. After working here in the Hudsonian zone as much as possible in the brief time available we returned to Clew, and from there made short trips to Louise Island and the mouth of the inlet. The next move was to Prevost Island, at the south end of the group, which was reached by means of a small fishing schooner. We dropped anchor July 3 in Houston Stewart channel, near Ellen Island, from which point both Prevost Island and the south end of Moresby Island were worked. In a few days we sailed north again and landed at Skidegate July 9. Here work was done about Skidegate Inlet and on the south end of Graham Island until July 18, when the trip was drawn to a close. Our entire time on the islands was thus about five weeks, and we were able to visit the three largest and most important of the group, namely Graham, Moresby, and Prevost. The weather during practically all this time was extremely disagreeable, rain being not only continuous for long periods, but often so severe as to interfere seriously with work. Most of the natives, unfortunately, were away at the time. They find employment in summer at the salmon canneries on the coast, the men being skilled in fishing and the women in packing. They are much reduced in numbers, and the entire population is divided between the two small villages of Massett and Skidegate, though ruins of former villages are abundant on other parts of the islands.

---

[1] This mountain is indicated, but not named, on the map published by Dawson. The only Indian I was able to interview said it was sometimes called Haida Mountain; but the white men had no name for it, though it is a very conspicuous peak.

ACKNOWLEDGMENTS.

The preparation of this report and the following one has been greatly facilitated by the free access which has been accorded me to the collections of the U. S. National Museum. Mr. Robert Ridgway, curator, and Dr. Charles W. Richmond, assistant curator of birds, have been exceedingly kind, not only in placing at my disposal the collections under their charge, but in numerous other ways as well. I am likewise indebted to Mr. Gerrit S. Miller, jr., assistant curator of mammals, for similar favors from his department. Mr. Joseph Grinnell, of California, has kindly forwarded me specimens for examination from his private collection.

Through the kindness of Rev. J. H. Keen it has been possible to include in the list of birds many migratory and winter resident species. Mr. Keen lived at Massett for eight years, and during that time collected and identified a large number of birds and other animals. He has very generously furnished a list of Massett birds, with notes for use in the present report, giving all the species positively identified by him. Special acknowledgment has been made elsewhere to Mr. Frederick V. Coville and other botanists who have so kindly assisted in the determination of plants (see p. 13).

PHYSIOGRAPHY.

The Queen Charlottes are part of a submerged mountain chain like most of the other large islands of the same coast, and were regarded by Dawson as a continuation northwestward of the ranges of Vancouver Island and the Olympic Peninsula. Their general trend is northwest and southeast, practically parallel with the mainland. The greatest length of the entire group is 156 miles,[1] and the greatest width 52 miles; the area is unknown. The main islands of the group are, consecutively from north to south: North, Graham, Moresby, and Prevost (see frontispiece). All are very closely connected, the width of each intervening channel being reduced, at least at some points, to less than a mile. The shortest distance between the Queen Charlottes and the islands very closely connected with the mainland is 27 miles, from Rose Point, Graham Island, to Stephens Island. The wide channel known as Hecate Strait, which lies between the Queen Charlottes and the mainland, is rather shallow; that part between Graham Island and the mainland seldom exceeds 20 fathoms in depth.

Graham Island is the largest of the group; its greatest length is 67 miles, and its width 52 miles. The coast on the north end is very deeply indented by Massett Inlet, and to a lesser degree by Naden Harbor; on the east side it is comparatively regular, and the west is characterized by deep, unsurveyed sounds. The east side of the island is low

[1] Distances are stated in nautical miles, and on the authority of Dawson

and comparatively level: the northwest part is slightly higher and somewhat rolling; and the southwestern corner is quite mountainous. This mountainous district may be roughly indicated as extending from Cartwright Sound south to Skidegate Inlet and east to Bearskin Bay. Some of the mountains are high enough to maintain perennial banks of snow, which feed numerous streams; these, however, are not very large, and are exceeded by the lowland streams, which drain greater areas, particularly those in the northern part of the island, emptying into Massett Inlet and Naden Harbor. Heavy forest covers almost the entire island and fringes the coast to the very water's edge. In the northern part, not far from Massett, a few open, swampy meadows are known, and near Rose Point there are grassy sand hills, but elsewhere all is dense forest.

Moresby Island is next in size to Graham; it is, in fact, about 5 miles longer, but is so dissected by sounds and inlets that its area is very indefinitely known, though certainly less than that of Graham Island. Its east coast bounds a succession of inlets, which make deep, transverse cuts into it and frequently unite with each other to detach small islets. The island is thus a mere skeleton; or perhaps it might be more properly called a mere backbone, since it is a nearly continuous mountain chain. Apparently the only part of it which is not mountainous is the peninsula lying between Skidegate and Cumshewa inlets. Cumshewa Inlet, the first deep indentation in the east coast south of Skidegate Inlet, is one of the largest of the island. Its south side, formed by Louise Island, and the region about its upper end are very mountainous. Among the peaks is one which rises near the head of West Arm to a height of 4,000 feet or more. From the inlet, its sharp cliffs and heavy snow banks present a rugged, imposing appearance, much heightened by contrast with the low, rounded, and somewhat undulating, forest-covered hills near the shore line. One very deep canyon cuts down its east side and through it a stream of moderate size runs into West Arm. Numerous smaller streams enter the inlet in the same vicinity. To the east, west, or south from the summit of this mountain one looks out over many other snow-laden peaks not so high, but of the same character and crowded together in tremendous masses. These mountains are practically continuous from the north end of Moresby Island south at least as far as Skincuttle Inlet. From a boat about 10 miles offshore in Hecate Strait one can look across Darwin and Juan Perez sounds and obtain an excellent view of the San Cristoval mountains, the best part of the chain.

Prevost Island is the southernmost of the Queen Charlottes. It is quite small, being but 11½ miles in length and about 8 miles in extreme width. It is low and rolling, and not so densely forested as Graham and Moresby. The hills on the north end are perhaps 600 feet in altitude and in other parts of the island they probably do not exceed 1,000 feet. A few small streams take their rise in the interior, several of which empty into Houston Stewart Channel.

FLORA.

The vegetation of the northwest coast region, which is so well known for its almost tropical luxuriance, is probably nowhere more highly developed than in the Queen Charlotte Islands. The magnificent dark forests are comparatively endless, the underbrush is omnipresent and well-nigh impenetrable, and mosses and lichens everywhere festoon the trees and shrubs and carpet the rocks and soil (see Plate II, fig. 1). The Indian, on the rare occasions when he can not travel by canoe, discreetly follows the beach; hence the interior wilderness remains almost as trackless as if human beings had never set foot on the islands. Relatively open forest is found on the higher slopes of the mountains, but can be reached only by hand to hand conflict with the tangle lying between it and tide water. From the tops of the mountains in the northern part of Moresby Island one can look out over vast stretches of forest to the northeast on Moresby and Graham islands as far as the eye can perceive. Of coniferous trees at least seven species are found, namely, the Sitka spruce (*Picea sitchensis*), the western hemlock (*Tsuga heterophylla*), the alpine hemlock (*Tsuga mertensiana*), the giant cedar (*Thuja plicata*), the yellow cedar (*Chamaecyparis nootkatensis*), the northwest coast pine (*Pinus contorta*), and the Pacific yew (*Taxus brevifolia*).

The Sitka spruce is the most important. It is well distributed and generally becomes large, being second in size only to the giant cedar. It was found to be the predominating tree about Cumshewa and Skidegate inlets, and, though smaller, equally abundant on Prevost Island and the south end of Moresby Island. Owing to exposed position and rocky soil, the trees on Prevost Island are rather small, and dead tops are so mixed with the live ones that from a little distance the dark green hillsides appear to be uniformly overcast with a light hoariness. Dawson reports spruce[1] as abundant in Skincuttle Inlet and about Darwin Sound on Moresby Island; he also found large forests of it in the eastern and northern parts of Graham Island, particularly about Naden River. Much of this timber is merchantable, though as yet no serious attempt to exploit it has been made. About Cumshewa Inlet the spruces stand in magnificent groves, the grandeur of which is appreciated only when one gets above the tangle of undergrowth and obtains an unobstructed view of the tall, straight, reddish-barked trunks, column after column extending far into the forest, until the dim light is finally entirely obscured and individual trees can no longer be distinguished.

[1] The spruce spoken of by Dawson was called in his report *Abies menziesii*, but from the context it is evident that *Picea sitchensis* was meant.

The giant cedar (*Thuja plicata*) is sparingly scattered through the forest in all parts of the islands except the higher mountains. Large individuals were rarely seen near the shore, owing, doubtless, to the fact that for many years the Indians have used the most accessible ones for making canoes.

The yellow cedar (*Chamæcyparis*) is rather rare, except at high elevations. A few individuals of small size were found at the head of West Arm, Cumshewa Inlet, and at the head of Rose Harbor. Dawson found the species in cold places about other inlets of Moresby Island, as well as on the more exposed west coast. And Rev. J. H. Keen reports it in small quantities near Massett. On the mountains of the northern part of Moresby Island it is very abundant, and ranges from an altitude of about 2,000 feet to the upper limit of timber.

The western hemlock (*Tsuga heterophylla*) is probably second in abundance to the spruce, and its distribution, below an altitude of 2,000 feet, is also general. It does not, however, occur in great numbers on the actual shores, like the other conifers, but becomes more abundant inland. The very deepest, darkest forests are largely composed of this hemlock.

The alpine hemlock (*Tsuga mertensiana*) was found only in the mountains near the head of Cumshewa Inlet. It appears with the yellow cedar at an elevation of about 2,000 feet, and soon becoming well established, persists to the highest limit of trees. It is slightly more abundant than the yellow cedar, and with it straggles in fantastic shapes up the ridges or flattens in thick mats on sunny slopes. Now and then it attains fair size and regularity of branching in cold canyons or about small seepage pools in little heather meadows.

The northwest coast pine (*Pinus contorta*) is rather rare. A few small individuals were noticed on rocky detached islets in Skidegate and Cumshewa Inlets and in Houston Stewart Channel; and, curiously, a few very depauperate individuals were found well toward the summit of the mountain near the head of Cumshewa Inlet.

The Pacific yew (*Taxus brevifolia*) was found on Cumshewa Inlet from Clew to the head of West Arm, being most abundant about West Arm. It is quite common around the shores of Prevost Island and the south end of Moresby Island. It is said to occur toward the west coast in Skidegate Inlet, and Rev. J. H. Keen reports that it is not uncommon in some places near Massett. It does not grow to large size and is always found near the shore or on the very edge of the water, which it overhangs at high tide.

The only deciduous trees of importance are alders, willows, and wild crab apples. All are abundant but are somewhat limited to the immediate vicinity of the coast or the borders of streams. The alder (*Alnus oregona*) grows to a relatively large size; individuals from 10 to 15 inches in diameter were frequently seen. The willow (*Salix*

FIG. 1.—MOSS-GROWN SPRUCE, CUMSHEWA INLET.

FIG. 2.—SALMONBERRY THICKET, CUMSHEWA INLET.

Habitat of *Melospiza f. rufina.*

scouleriana) is also of fair size and is generally distributed. The Oregon crab apple (*Pyrus rivularis*) is found about streams and along the borders of sandy beaches as well as in occasional open places in the forest, in company with elderberry (*Sambucus racemosus*), dogwood (*Cornus occidentalis*), honeysuckle (*Lonicera involucrata*), and wild currants (*Ribes bracteosum* and *R. laxiflorum*). A hawthorn (*Cratægus brevispina*) also occurs, but evidently is rare, as we found it at but one locality, near a small stream on Louise Island.

The underbrush is largely composed of several species of huckleberry (*Vaccinium*), the sallal (*Gaultheria shallon*), and the salmon berry (*Rubus spectabilis*—see Plate II, fig. 2). *Menziesia*, *Viburnum*, and *Amelanchier* are locally abundant. Throughout the damp forest are many ferns, mosses, liverworts, and numerous species of flowering plants that thrive in such an environment. The few and small open meadows that are occasionally to be found teem with grasses, sedges, buttercups, beach peas, vetches, monkey flowers, thistles, lilies, and large cow parsnips The flora of the higher parts of the mountains is much like that of similar altitudes on the mainland. As the forest becomes more open the character of the herbaceous plants changes, and numerous pretty glades are found carpeted with heathers (*Phyllodoce* and *Cassiope*) and sprinkled with dwarf laurel (*Kalmia glauca*), dwarf huckleberries (*Vaccinium cæspitosum*), and cowslips (*Caltha palustris*).

Most of our time was devoted to collecting mammals and birds, but a small collection of plants, representing nearly all the species observed was also made. From these specimens the following list has been prepared through the kindness of Mr. Frederick V. Coville, Botanist of the U. S. Department of Agriculture and Honorary Curator of the U. S. National Herbarium. Mr. Coville has not only generously given his own time and that of his assistants to the identification of specimens, but has enlisted the very necessary aid of several eminent specialists, who have authoritatively named specimens in special groups. The liverworts were very kindly determined by Prof. A. W. Evans, of Yale University; the mosses by Prof. J. M. Holzinger, of the Minnesota State Normal School, with the exception of the two species of Dicranaceæ, which are given on the authority of Dr. R. H. True; the ferns by Mr. William R. Maxon, of the U. S. National Museum, and the flowering plants by Mr. Frederick V. Coville and Mr. W. F. Wight, of the U. S. National Museum.

JUNGERMANNIACEÆ.

Porella navicularis (L. & L.) LINDB.
Radula bolanderi GOTTSCHE.
Scapania bolanderi AUST.
Diplophylleria albicans (L.) TREVIS.
Frullania nisquallensis SULLIV.
Herberta adunca (DICKS.) S. F. GRAY.

DICRANACEÆ.

Dicranum fuscescens TURN.
Dicranella heteromalla SCH.

BRYACEÆ.

Bartramia glaucoviridis C. M. & K.
Mnium glabrescens KINDB.

HYPNACEÆ.

Eurhynchium oregonum (SULLIV.) L. & J.
Isothecium cardoti KINDB.
Plagiothecium undulatum (L.) SCH.
Hypnum callichroum BRID.
Hylocomium squarrosum (L.) SCH.
Hylocomium loreum (L.) SCH.
Hylocomium splendens (L.) SCH.

POLYPODIACEÆ.

Filix fragilis (L.) UNDEW.
Dryopteris spinulosa dilatata (HOFF.) UNDEW.
Polystichum munitum (KAULF.) UNDEW.
Athyrium cyclosorum RUPR.
Struthiopteris spicant (L.) WEISS.
Adiantum pedatum L.
Polypodium falcatum KELLOGG.

PINACEÆ.

Pinus contorta LOUD.
Picea sitchensis (BONG.) TRAUTR. & MEY.
Tsuga mertensiana (BONG.) CARR.
Tsuga heterophylla (RAF.) SARGENT.
Thuja plicata DON.
Chamaecyparis nootkatensis (LAMB.) SPACH.
Taxus brevifolia NUTT.

SCHEUCHZERIACEÆ.

Triglochin maritima L.
Triglochin palustris L.

POACEÆ.

Agrostis exarata TRIN.
Deschampsea caespitosa (L.) BEAUV.
Dactylis glomerata L.

POACEÆ—continued.

Puccinellia distans (L.) PARL.
Festuca rubra L.
Bromus aleutensis TRIN.
Elymus mollis TRIN.
Elymus sp.

CYPERACEÆ.

Scirpus pauciflorus LIGHTF.
Carex sp.

JUNCACEÆ.

Juncus balticus WILLD.
Juncus bufonius L.
Juncoides parviflorum (EHRH.) COVILLE.

LILIACEÆ.

Fritillaria kamtschatcensis (L.) KER.
Unifolium bifolium (L.) GREENE.
Streptopus roseus MICHX.

IRIDACEÆ.

Sisyrinchium littorale GREENE.

ORCHIDACEÆ.

Habenaria hyperborea (L.) R. BR.
Peramium menziesii (LINDL.) MORONG.
Corallorhiza mertensiana BONG.

SALICACEÆ.

Salix scouleriana BARRATT.

BETULACEÆ.

Alnus sinuata (REGEL) RYDB.
Alnus oregona NUTT.

URTICACEÆ.

Urtica lyallii WATS.

POLYGONACEÆ.

Rumex sp.
Polygonum sp.

CHENOPODIACEÆ.

Atriplex gmelini C. A. MEYER.
Salicornia herbacea L.

PORTULACACEÆ.

Montia parvifolia (MOC.) GREENE.
Montia sibirica (L.) HOWELL.

ALSINACEÆ.

Cerastium sp.
Sagina crassicaulis WATS.
Ammodenia peploides (L.) RUPR.
Tissa marina (L.) BRITTON.

RANUNCULACEÆ.

Caltha palustris L.
Coptis asplenifolia SALISB.
Aquilegia formosa FISCHER.
Anemone narcissiflora L.
Ranunculus occidentalis NUTT.
Ranunculus sp.

BRASSICACEÆ.

Cochlearia oblongifolia DC.
Brassica campestris L.
Cardamine angulata HOOK.
Arabis hirsuta SCOP.
Arabis sp.

CRASSULACEÆ.

Sedum roseum (L.) SCOP.
Sedum spathulifolium HOOK.

SAXIFRAGACEÆ.

Saxifraga mertensiana BONG.
Saxifraga nutkana MOC.
Tiarella trifoliata L.
Heuchera cylindrica DOUGL.
Heuchera glabra WILLD.

RIBACEÆ.

Ribes bracteosum DOUGL.
Ribes lacustre (PERS.) POIR.
Ribes laxiflorum PURSH.

ROSACEÆ.

Lutkea pectinata (HOOK.) KUNTZE.
Aruncus aruncus (L.) KARST.
Pyrus rivularis DOUGL.
Crataegus brevispina DOUGL.
Sorbus sambucifolia (C. & S.) ROEM.
Amelanchier alnifolia NUTT.
Rubus pedatus SMITH.
Rubus spectabilis PURSH.
Rubus parviflorus NUTT.
Fragaria chiloensis (L.) DUCHESNE.
Potentilla anserina L.
Potentilla villosa PALL.
Geum calthifolium MENZIES.
Geum sp.
Sanguisorba sp.
Rosa sp.

VICIACEÆ.

Lupinus nootkatensis DONN.
Lupinus nootkatensis unalaskensis WATS.
Trifolium involucratum WILLD.
Lathyrus maritimus (L.) BIGEL.
Vicia gigantea HOOK.

VIOLACEÆ.

Viola glabella NUTT.

ONAGRACEÆ.

Epilobium glandulosum LEHM.
Epilobium hornemanni REICHENB.
Epilobium minutum LINDL.

ARALIACEÆ.

Echinopanax horridum (SMITH) D. & P.

APIACEÆ.

Washingtonia sp.
Conioselinum gmelini COULT. & ROSE
Heracleum lanatum MICHX.

CORNACEÆ.

Cornus canadensis L.
Cornus occidentalis (TORR. & GR.) COVILLE.

PYROLACEÆ.

Moneses uniflora (L.) GRAY.

VACCINIACEÆ.

Vaccinium cæspitosum MICHX.
Vaccinium ovalifolium SMITH.
Vaccinium parvifolium SMITH.
Vaccinium uliginosum L.

ERICACEÆ.

Menziesia ferruginea SMITH.
Chamæcistus procumbens (L.) KUNTZE.
Kalmia glauca AIT.
Phyllodoce glanduliflora (HOOK.) COVILLE.
Cassiope mertensiana (BONG.) DON.
Cassiope stelleriana DC.
Gaultheria shallon PURSH.
Arctostaphylos uva-ursi (L.) SPRENG.

PRIMULACEÆ.

Glaux maritima L.
Dodecatheon viviparum GREENE.

NEPETACEÆ.

Prunella vulgaris L.
Stachys sp.

SCROPHULARIACEÆ.

Collinsia parviflora DOUGL.
Mimulus langsdorfii DON.
Veronica americana SCHWEIN.
Castilleja pallida KUNTH.
Castilleja parviflora BONG.
Pedicularis lanata WILLD.
Pedicularis pedicellata BUNGE.

PLANTAGINACEÆ.

Plantago maritima L.
Plantago sp.

RUBIACEÆ.

Galium aparine L.

CAPRIFOLIACEÆ.

Lonicera involucrata (RICH.) BANKS.

VIBURNACEÆ.

Viburnum pauciflorum PYLAIE.
Symphoricarpos racemosus MICHX.
Linnaea borealis L.

CAMPANULACEÆ.

Campanula langsdorffiana FISCH.

CARDUACEÆ.

Grindelia sp.
Erigeron peregrinus (PURSH.) GREENE.
Achillea borealis BONG.
Matricaria discoidea DC.
Senecio resedifolius LESS.
Carduus edulis (NUTT.) GREENE.

FAUNA.

The mammal fauna of the Queen Charlottes, in view of the proximity of the group to the mainland, may be said to be rather meager. There are only 11 species of indigenous land mammals, and 4 of these are bats. The genera represented are *Peromyscus*, *Ursus*, *Lutra*, *Mustela*, *Putorius*, *Sorex*, *Lasionycteris*, and *Myotis*. Characteristic genera of the adjacent mainland, such as *Odocoileus*, *Lutreola*, *Sciurus*, *Evotomys*, and *Microtus*, are entirely wanting. The absence of these genera, which are common on the mainland and well adapted to all the conditions of the islands, indicates that the water barrier between the islands and the mainland is an effective one. The channel is certainly too wide to swim, and small mammals could not easily be carried on driftwood, as the strong tides would be apt to sweep everything of that nature out at either end of Hecate Strait rather than across it.[1] The presence of the marten, an animal which is terrestrial and arboreal, and the absence of the mink, which is semiaquatic, and the deer, which swims considerable distances, might be considered evidence that the islands must have been peopled with animals at a time of complete connection with the mainland. But if at that time the mainland fauna was approximately the same as at present, it is difficult to explain the present absence of the deer and mink. Whatever the case may have been, it is certain that the mammals have been isolated for a

[1] This means of communication is possible, however, as is shown by the fact that the dead bodies of deer have been washed up on the shores of the islands (see p. 25), but the course of the tides is such that they could not have been carried directly across the strait. It is therefore probable that the journey would be too long and perilous for any living animal to survive.

sufficient length of time to become differentiated into well-marked insular forms. All the land mammals that have been studied have proved distinct from the species of the same genera found on the mainland.[1] Moreover, the larger, less mutable forms (*Ursus*, *Mustela*, and *Putorius*), which are also the ones least likely to have been introduced through accident or human agency, are the most strongly characterized, thus affording additional evidence of isolation of rather long standing.

Still more interesting than the general peculiarity of the entire mammal fauna as contrasted with that of the mainland, is the individuality shown among closely adjacent islands when contrasted with one another. A mouse (*Peromyscus prevostensis*) and a shrew (*Sorex l. prevostensis*) are peculiar to Prevost Island. The island is quite small, possibly 50 square miles in area, yet these mammals are found in great abundance, and do not occur on other islands which lie less than a mile away. The channels between these islands are said by Dawson to be accidental, but at least they can not have been formed very recently or this distribution of animals would not be found. *Peromyscus prevostensis*, though from the *southernmost* island of the group, is most closely related to *P. sitkensis* which has been found only on islands some distance *north* of the Queen Charlottes, while *P. keeni*, of Graham and Moresby islands is not related to northern forms, but is scarcely distinguishable from the comparatively remote *P. akeleyi* of the Olympic Peninsula, Washington. The small mammals of the northwest coast are still so imperfectly known and they are all so interrelated, that it may be unprofitable to speculate at present on the derivation of various insular forms, and it is possible that the animals are so plastic that their present characters can not safely be taken as indicative of their true relationships.

Birds are comparatively abundant. The species are much the same as those found throughout the whole district from Puget Sound to Glacier Bay, but individuals seem to be more numerous than on the mainland. The land birds frequent the thickets of salmon berry, alder, and willow bushes found bordering sandy beeches or small streams (see Pl. II, fig. 2). In these restricted areas certain species are very common. Many such environments are afforded in Cumshewa and Skidegate inlets, and it was there that most of our collecting was done. The steamer which carried us to the islands dropped anchor in Cumshewa Inlet, about 50 yards from the shore at 5 o'clock in the morning of June 13, and through the little port of my stateroom sounded such an avian chorus as I have heard nowhere else on the northwest coast. The greatest volume of song came from song spar-

[1] The land otter, *Lutra*, is the only one known to occur on the islands of which no specimens have been critically examined.

rows and fox sparrows, but the rich tones of the russet-backed thrushes contributed much to strength and quality; winter wrens bubbled and spluttered on all sides, the clear notes of a robin were heard now and then, and from farther back in the forest came the weird call of the varied thrush, while all the time several huge ravens carried on a lively conversation with one another. The deep forest is for the most part dark and quiet, and save for an occasional creeper or winter wren contains no birds. Water birds are reported to breed in large colonies on islets near the west coast of Moresby Island. We were unable to visit these, but observed small rookeries at the mouth of Cumshewa Inlet and in Houston Stewart Channel.

Very few of the land birds are definitely separable from those of the Sitkan district, but the tendency to dark colors and heavy markings is extreme. Two forms are peculiar to the islands, a jay (*Cyanocitta stelleri carlottæ*) and a woodpecker (*Dryobates picoideus*), both of which belong to groups not known to be as variable as others, such as *Junco*, *Melospiza*, and *Passerella*, whose Queen Charlotte representatives are not strongly characterized. Ninety-six species are known to occur on the islands, 62 of which are found in the breeding season. For notes on the occurrence of many of the species, I am greatly indebted to Rev. J. H. Keen, whose observations at Massett covered a period of eight years.

Besides the mammals and birds but one land vertebrate has been found on the islands, a toad (*Bufo halophilus columbiensis*), which is common on the adjacent mainland. We noted no strictly fresh-water fishes, but trout are reported to inhabit some of the streams. A collection of several hundred littoral species of fish was made, chiefly by Mr. Edmund Heller, who has determined the species and found none of them peculiar. Conditions about the islands are exceedingly favorable for marine invertebrate life and it flourishes in profusion, but no attempt at collecting such forms was made.

The vertebrate fauna of the islands, as a whole, is very similar to that of the adjacent mainland, but is nevertheless peculiar in many respects. The vertebrates known to occur on or about the islands are as follows:

MAMMALS.

1. Globicephala scammoni.
2. Balænoptera velifera.
3. Lagenorhynchus obliquidens.
4. Phocæna phocæna.
5. Odocoileus columbianus sitkensis (introduced).
6. Peromyscus keeni.
7. Peromyscus prevostensis.
8. Ursus carlottæ.
9. Lutra canadensis subsp.
10. Putorius haidarum.
11. Mustela nesophila.
12. Eumetopias stelleri.
13. Otoes alascanus.
14. Phoca largha.
15. Sorex longicauda prevostensis.
16. Sorex longicauda elassodon.
17. Lasionycteris noctivagans.
18. Myotis yumanensis saturatus.
19. Myotis subulatus keeni.
20. Myotis californicus caurinus.

BIRDS.

1. Gavia imber.[1]
2. Gavia pacifica.[1]
3. Gavia lumme.[1]
4. Lunda cirrhata.[1]
5. Fratercula corniculata.
6. Synthliboramphus antiquus.
7. Brachyramphus marmoratus.[1]
8. Cepphus columba.[1]
9. Uria troile californica.[1]
10. Rissa tridactyla pollicaris.
11. Larus glaucescens.[1]
12. Larus brachyrhynchus.
13. Larus philadelphia.
14. Puffinus griseus.
15. Puffinus tenuirostris.
16. Oceanodroma furcata.
17. Phalacrocorax pelagicus.[1]
18. Merganser americanus.[1]
19. Merganser serrator.[1]
20. Anas boschas.[1]
21. Mareca americana.
22. Nettion carolinensis.
23. Dafila acuta.
24. Aythya sp.[1]
25. Clangula clangula americana.
26. Charitonetta albeola.
27. Harelda hyemalis.
28. Histrionicus histrionicus.
29. Oidemia deglandi.[1]
30. Oidemia perspicillata.[1]
31. Anser albifrons gambeli.
32. Branta canadensis occidentalis.[1]
33. Olor sp.
34. Ardea herodias fannini.
35. Porzana carolina.
36. Gallinago delicata.
37. Tringa acuminata.
38. Ereunetes occidentalis.
39. Calidris arenaria.
40. Totanus melanoleucus.
41. Actitis macularia.
42. Squatarola squatarola.
43. Charadrius dominicus.
44. Arenaria melanocephala.
45. Hæmatopus bachmani.[1]
46. Dendragapus obscurus fuliginosus.[1]
47. Zenaidura macroura.
48. Accipiter velox.[1]
49. Accipiter atricapillus striatulus.
50. Buteo borealis calurus.[1]
51. Haliaeetus leucocephalus alascanus.[1]
52. Falco peregrinus pealei.[1]
53. Falco columbarius suckleyi.[1]
54. Pandion haliaëtus carolinensis.[1]
55. Megascops asio kennicotti.[1]
56. Nyctala acadica scotæa.[1]
57. Nyctea nyctea.
58. Ceryle alcyon.[1]
59. Dryobates picoideus.[1]
60. Sphyrapicus ruber flaviventris.[1]
61. Colaptes cafer saturatior.[1]
62. Selasphorus rufus.[1]
63. Empidonax difficilis.[1]
64. Cyanocitta stelleri carlottæ.[1]
65. Corvus corax principalis.[1]
66. Corvus caurinus.[1]
67. Pinicola enucleator flammula.[1]
68. Loxia curvirostra minor.[1]
69. Loxia leucoptera.[1]
70. Spinus pinus.[1]
71. Passerina nivalis.
72. Ammodramus sandwichensis alaudinus.
73. Calcarius lapponicus alascensis.
74. Zonotrichia coronata.
75. Junco hyemalis oregonus.[1]
76. Melospiza melodia rufina.[1]
77. Passerella iliaca townsendi.[1]
78. Hirundo erythrogastra.[1]
79. Tachycineta bicolor.[1]
80. Tachycineta thalassina.[1]
81. Helminthophila celata lutescens.[1]
82. Dendroica æstiva rubiginosa.[1]
83. Dendroica townsendi.[1]
84. Wilsonia pusilla pileolata.[1]
85. Anthus pensilvanicus.[1]
86. Cinclus mexicanus.[1]
87. Anorthura hiemalis pacifica.[1]
88. Certhia familiaris occidentalis.[1]
89. Sitta canadensis.[1]
90. Parus rufescens.[1]
91. Regulus satrapa olivaceus.[1]
92. Regulus calendula grinnelli.[1]
93. Hylocichla ustulata.[1]
94. Hylocichla aonalaschkæ verecunda.[1]
95. Merula migratoria propinqua.[1]
96. Hesperocichla nævia.[1]

BATRACHIAN.

Bufo halophilus columbiensis.

[1] Known, or supposed, to breed.

FISHES.[1]

Squalus sucklii.
Hydrolagus colliei.
Clupea pallasi.
Oncorhynchus sp.
Salmo sp.
Gasterosteus cataphractus.
Aulorhynchus flavidus.
Cymatogaster aggregatus.
Sebastodes melanops.
Sebastodes caurinus.
Hexagrammus stelleri.
Ophiodon elongatus.
Artedius lateralis.
Hemilepidotus hemilepidotus.
Euophrys bison.
Leptocottus maculosus.
Oligocottus maculosus.
Blenicottus globiceps.
Ascelichthys rhodorus.
Caularchus maendricus.
Apodichthys flavidus.
Pholis ornatus.
Anoplarchus atropurpureus.
Xiphistes chirus.
Xiphidion mucosum.
Xiphidion rupestre.
Theragra fucensis.
Gadus macrocephalus.
Hippoglossus hippoglossus.
Psettichthys melanostictus.
Limander asper.

LIFE ZONES.

The zones of the Queen Charlottes are the Canadian and the Hudsonian. The greater part of the group, at least all that which lies below an altitude of 2,500 feet, is Canadian, and the remainder above that elevation is Hudsonian. This zonal definition is determined almost entirely by the plant life. The insular occurrence of various species of mammals, and to some extent of birds, may be due to accident and unnatural agency; therefore the absence of certain forms obviously can not be considered significant in correlating island and continental zones. Moderate insular isolation restricts vegetation much less than animal life; so that in determining the faunal position of the Queen Charlottes the fact that practically all the trees and shrubs are those usually found in the Canadian and Hudsonian zones is of much more importance than that no mammals occur other than those of the Canadian zone.

The latitude of the southern part of the group is about the same as that of the mainland where the Transition zone merges into the Canadian, and since the average temperature on the islands may be assumed to be slightly warmer than on the mainland, some Transition intrusions might be expected. These are comparatively few, however, and may safely be disregarded in a general definition of the zones of the group. The characteristic Transition tree, *Pseudotsuga mucronata*, which reaches its northern limit on the mainland in about this latitude, has not been found on the Queen Charlottes. None of the coniferous trees, with the possible exception of *Taxus brevifolia*, can be considered unequivocal Transition species. *Picea sitchensis*, *Tsuga heterophylla*, *Thuja plicata*, and *Pinus contorta* all range throughout the Sitkan district, which is Canadian, and, roughly speaking, extends

[1] Species determined by Mr. Edmund Heller.

FIG. 1.—MOUTH OF STREAM, CUMSHEWA INLET.

FIG. 2.—SHORE OF BARE ISLAND, SKIDEGATE INLET.

from River Inlet, British Columbia, to Cross Sound, Alaska. Among the small shrubs of the Queen Charlottes that also range through this Canadian district may be mentioned *Alnus sinuata*, *Ribes laxiflorum*, *Ribes bracteosum*, *Vaccinium ovalifolium*, *Menziesia ferruginea*, *Viburnum pauciflorum*, *Arctostaphylos uva-ursi*, and *Echinopanax horridum*.

The Hudsonian zone occupies those parts of the islands which are above an elevation of about 2,500 feet.[1] The mountains which exceed this height are distributed in four principal groups, namely, those of the southwestern part of Graham Island, those of the head of Cumshewa Inlet, Moresby Island, those of the central part of Louise Island, and those lying along Darwin Sound, and known as the San Cristoval Range[2] (see fig. 1). The only part of the Hudsonian zone actually traversed by us is that on the principal peak at the head of West Arm of Cumshewa Inlet. There the timber on the lower slopes of the mountain was found to be very heavy and of much the same character as that of the shore, the principal difference being the absence of *Taxus*, which seems to be confined to the immediate border of the inlets. At an altitude of about 2,000 feet a more decided difference in the flora began to be noticeable. This change progresses rapidly. The giant cedar (*Thuja*) disappears entirely, and the spruce (*Picea*) and western hemlock (*Tsuga heterophylla*) are much reduced in numbers. The alpine hemlock (*Tsuga mertensiana*) and the yellow cedar (*Chamæcyparis nootkatensis*) take a place in the forest and soon become well established as the predominating trees, the hemlock being slightly in the ascendancy. Many smaller Hudsonian plants appear simultaneously with these trees and continue with them nearly or quite to the summit. Among these may be mentioned *Cassiope stelleriana*, *Cassiope mertensiana*, *Phyllodoce glanduliflora*, *Kalmia glauca*, *Vaccinium cæspitosum*, *Vaccinium uliginosum*, *Chamæcistus procumbens*, *Luetkea pectinata*, *Caltha palustris*, *Pedicularis lanata*, and *Pedicularis pedicellata*. On the higher ridges a few individuals of *Tsuga heterophylla*, *Picea sitchensis*, and *Pinus contorta* still persist, but in an exceedingly depauperate condition. Between the ridges are characteristic glades and heather meadows, and in occasional suitable basins clear pools of seepage water. Thus the

FIG. 1.—Outline map of the Queen Charlotte Islands, showing extent of the Hudsonian zone.

[1] The elevations given are estimates only, as I was not equipped with a barometer.

[2] Very few of the mountains have been named, therefore it is necessary to make awkward descriptive reference to them.

general aspect of this belt between 2,500 feet altitude and the summit is that of a pure Hudsonian-Alpine zone, such as is found on the mountains of the mainland in the same latitude. Although this mountain and others near it carry large banks of snow the year round, a definite timberline does not exist on them, for a few trees straggle practically to the summits, and smaller plants flourish on the favored sides of the very highest pinnacles.

BIBLIOGRAPHY.

The following titles are merely those of such books as have been consulted in the preparation of this report. By far the most important is the general report by the late Dr. George M. Dawson. Other published references to the natural history of the islands are largely such as have resulted from the work of Rev. J. H. Keen.

1789. Dixon, Captain George. A Voyage Round the World, but more particularly to the North-West Coast of America performed in the *King George* and *Queen Charlotte*, Captains Portlock and Dixon. pp. 199–234. London, 1789.

General account of a visit to the islands with notes on fur trading with the natives.

1868. Brown, Robert. Synopsis of the Birds of Vancouver Island. <Ibis, IV, 424, 1868.

Mentions occurrence of *Hæmatopus niger* on Queen Charlotte Islands.

1869. Brown, Robert. Physical Geography of the Queen Charlotte Islands. <Rept. Brit. Ass. Adv. Sci. XXXVIII (for 1868), Notices and Abstr., pp. 133–134, 1869.

Brief notes on geography, geology, and ethnology.

1871. Brown, Robert. Notes on Arctic Zoology. <Ann. and Mag. Nat. Hist., ser. 4, VII, 64–66, Jan., 1871.

Reference to a skull of *Phocæna communis* (=*Phocæna phocæna*) brought from the islands and sent to the British Museum.

1872. Poole, Francis. Queen Charlotte Islands. A Narrative of Discovery and Adventure in the North Pacific. Edited by J. W. Lyndon. 8vo., pp. 1–347. London, Hurst and Brackett.

An account of two years' sojourn in the islands. The natural history notes which are summarized on pp. 305–309 are of questionable value. The following is the list of birds given as frequenting the neighborhood of Burnaby Island exactly as printed on p. 308:

Night-hawk—*falco nocturnus.*
Sparrow-hawk—*falco sparverius.*
Gos-hawk—*astur atricapillus.*
White-headed eagle—*haliaetus leucocephalus.*
Belted kingfisher—*alcedo accinctus.*
Western bluebird—*cyanurus occidentalis.*
North Western fish-crow—*corvus caurinus.*
Wilson's snipe—*gallinago wilsonii.*
Canadian goose—*bernacla canadensis.*
White-cheeked goose—*bernacla leucoparsia.*
Mallard (stock duck)—*anas boschas.*
Canvas-back duck—*aythia vallisneria.*
Golden-eye (whistle-wing duck)—*bucephala americana.*
Buffle-head duck—*bucephala albeola.*
Harlequin duck—*histrionicus torquatus.*
Velvet duck—*malanetta velvetina.*
Glaucous-winged duck—*larus glaucescens.*
Suckley's gull—*larus suckleyii.*
Great Northern diver—*colymbus torquatus.*
Red-necked grepe—*podicetus griseygeria.*

1880. DAWSON, GEO. M. Report on the Queen Charlotte Islands. Geological Survey of Canada. Report of Progress for 1878–79, III, 1–239, Montreal, 1880.

Consists of a general report subdivided as follows: (1) Position, Discovery, and Early History of the Islands; (2) General Description; (3) Geological Observations; and seven appendixes, as follows:

A. On the Haida Indians of the Queen Charlotte Islands, by Geo. M. Dawson.
B. Vocabulary of the Haida Indians, by Geo. M. Dawson.
C. On Some Marine Invertebrata from the Queen Charlotte Islands, by J. F. Whiteaves.
D. Notes on Crustacea from the Queen Charlotte and Vancouver Islands, by S. I. Smith.
E. List of Plants from the Queen Charlotte Islands, by John Macoun.
F. Meteorological Observations, by Geo. M. Dawson.
G. Notes on Latitudes and Longitudes, by Geo. M. Dawson.

1883–1892. MACOUN, JOHN. Catalogue of Canadian Plants, pts. i–iv, pp. 1–398; pt. vi, pp. 1–277, Montreal, 1883–1892.

Contains references to various plants collected by Dawson on the Queen Charlotte Islands.

1887. CHAMBERLAIN, MONTAGUE. Catalogue of Canadian Birds, St John, New Brunswick, 1887.

Mentions occurrence of *Fratercula corniculata*, *Uria troile californica*, *Larus glaucescens*, and *Larus brachyrhynchus*.

1890. DAWSON, GEO. M. Later Physiographical Geology of the Rocky Mountain Region in Canada. Trans. Royal Soc. Canada, VIII, Sec. IV, 51–52, 1890.

Contains reference to supposed occurrence of caribou on Queen Charlotte Islands.

1891. MACKENZIE, ALEX. Descriptive notes on certain implements, weapons, etc., from Graham Island, Queen Charlotte Islands, British Columbia. < Trans. Royal Soc. Canada, IX, Sec. II, 45–59, 1891.

Refers to tomahawk made of native (?) reindeer antler (p. 50).

1894. RHOADS, S. N. *Sitomys keeni* sp. nov. <Proc. Acad. Nat. Sci. Phila. (1894), 258–259, Oct. 23, 1894.

Type from Massett, Graham Island.

1895. KEEN, J. H. List of *Coleoptera* collected at Massett, Queen Charlotte Islands, British Columbia. < Canadian Entomologist, XXVII, 165–172, July, 1895; cont. pp. 217–220, Aug., 1895.

A list of 222 species, with brief notes by the collector.

1895. MERRIAM, C. HART. Bats of Queen Charlotte Islands, British Columbia. < American Naturalist, XXIX, 860–861, Sept. 1, 1895.

Vespertilio subulatus keenii subsp. nov. Type from Massett, Graham Island.

1897. KEEN, J. H. Three interesting Staphylinidæ from Queen Charlotte Islands. < Canadian Entomologist, XXIX, 285–287, Dec., 1897.

1898. PREBLE, E. A. Description of a New Weasel from the Queen Charlotte Islands, British Columbia. < Proc. Biol. Soc., Wash., XII, 169–170, Aug. 10, 1898.

Putorius haidarum sp. nov. Type from Massett, Graham Island.

1898. DORSEY, GEORGE A. A Cruise Among Haida and Tlingit Villages about Dixon's Entrance. <Popular Science Monthly, LIII, 160–174, 1898.

An account of an ethnological expedition which visited Massett and other native villages. No important notes on natural history.

1898. Fannin, John. A preliminary catalogue of the collections of natural history and ethnology in the Provincial Museum, Victoria, British Columbia, 1898.

Mentions several mammals from the Queen Charlotte Islands and the following birds: *Fratercula corniculata*, *Rissa tridactyla pollicaris*, *Puffinus griseus*, and *Tringa acuminata*.

1900. Seton-Thompson, Ernest. *Rangifer dawsoni*, preliminary description of a new Caribou from Queen Charlotte Islands. Ottawa Naturalist, XIII, 257–261, Pls. IV, V, Feb., 1900.

1900. Macoun, John. Catalogue of Canadian Birds, Part I, Geological Survey of Canada, Ottawa, 1900.

Mentions occurrence of *Puffinus griseus*, *Puffinus tenuirostris*, *Tringa acuminata*, and *Bonasa umbellus sabinei*.

1901. Osgood, Wilfred H. New subspecies of North American birds. Auk. XVIII, 179–185, April, 1901.

Contains description of *Hylocichla aonalaschkæ verecunda*. Type from Cumshewa Inlet. Moresby Island.

1901. Chapman, Frank M. A New Race of the Great Blue Heron, with Remarks on the Status and Range of *Ardea wardi*. Bull. Am. Mus. Nat. Hist., New York, XIV, 87–90, April 15, 1901.

Ardea herodias fannini subsp. nov. Type from Skidegate, Graham Island.

Globicephala scammoni Cope. Blackfish.

Common in Hecate Strait. While our little schooner was *en route* to and from Prevost Island small schools played about it every day, and often with a familiarity that was alarming.

Balænoptera velifera Cope. Finback Whale.

A party of about half a dozen whales was seen in Hecate Strait July 2.

? Lagenorhynchus obliquidens Gill. Striped Porpoise.

A porpoise supposed to be this species kept within a few feet alongside the schooner for some time while we were sailing in Hecate Strait July 7.

Phocæna phocæna Linn. Common Porpoise.

A school of porpoises numbering at least 100 individuals was seen in Hecate Strait July 7; others were frequently seen in the strait.

A skull, evidently of this species, was sent from the islands by Dr. Robert Brown in 1868 to the British Museum.[1]

[**Odocoileus columbianus sitkensis** Merriam. Sitka Deer.

Deer have been introduced on the islands, but have not yet thoroughly established themselves there. I was told by Rev. Mr. Collinson, who was formerly in charge of the missionary work at Massett, that he was instrumental in the introduction of a few deer on Graham Island some years ago. Nine individuals from the vicinity of Port Simpson were liberated at Massett, and within a year signs of them were seen near Skidegate. Mr. Tennant, of Skidegate, states that a deer was killed by Indians about two years ago on Moresby Island, near Skidegate, but that since that time no deer or tracks have been reported. According to Rev. Mr. Collinson the dead bodies of deer from the mainland are occasionally found washed up on the beaches of the islands.]

[1] Ann. and Mag. Nat. Hist., ser. 4, VII, 64, 1871. This specimen is probably the one referred to by Flower as having come from Vancouver Island. (List Cetacea Brit. Mus., p. 16, 1885.)

[**Rangifer dawsoni**[1] (=**Rangifer arcticus**).

The description of a caribou from the Queen Charlotte Islands was, to say the least, somewhat unexpected, so in visiting the islands I was particularly interested in obtaining information in regard to it. I could find no evidence, however, that native caribou ever existed on any of the islands. Rev. Mr. Keen, who lived at Massett for eight years, and who was specially interested in matters pertaining to natural history, says that from his own experience and that of the oldest Indian hunters, whom he questioned closely, he is decidedly of the opinion that no caribou are to be found in any part of the islands. Rev. Mr. Collinson, who was one of the earliest missionaries at Massett, has the same belief as Mr. Keen, though he did not express such definite conclusions. Besides the missionaries I also interviewed a Mr. Stevens, who has kept the general store at Massett for the past nine years, and obtained from him the same opinion. All these persons are familiar with the story of 'Mackenzie's caribou,' which is doubtless the cause of the mistaken idea that a peculiar species is native to Graham Island. According to this story, which was told me independently and without essential variation by Messrs. Keen, Collinson, and Stevens, some fifteen or twenty years ago Mr. Alexander Mackenzie, a trader for the Hudson Bay Company at Massett, conceived the idea that in such a favorable place as Graham Island there must be deer and caribou, though the Indians had never killed them or even seen their tracks. Accordingly he offered a reward to anyone who should kill one or bring him evidence of having done so. The offer remained open for a long time, but finally a claimant appeared with fragments of a caribou, including the head. This imperfect specimen passed through several hands and finally found its way to the Provincial Museum in Victoria, where it was unearthed to receive the name *Rangifer dawsoni*. If the reward was incident to such a statement the Indian who brought this specimen to Mackenzie no doubt solemnly averred that he killed it on Graham Island. An Indian's testimony in a case of this kind, however, would not hang very heavy in the balance, even against a small amount of circumstantial evidence. Mr. Mackenzie is not now living, but the testimony of Mr. W. Charles, who received the caribou head from him, indicates that for its absolute origin we have the word of the Indians only. In response to a letter to Mr. Charles I received an answer from Mr. J. R. Anderson, deputy minister of agriculture at Victoria, from which the following is extracted:

> Some time ago Mr. W. Charles, who is an invalid, handed me your letter of the 10th January last regarding the occurrence of caribou on Queen Charlotte Islands. Mr. Charles asked me to communicate with you and say that the head referred to, and which had deformed antlers, undoubtedly came from Queen Charlotte Islands, hav-

[1] Seton-Thompson, Ottawa Naturalist, XIII, 257-261, Feb., 1900.

ing been sent to him by the Hudson Bay Company agent there, and was equally that of a caribou. The animal, Mr. Charles has no reason to doubt, was actually killed by the Indians, and they being unacquainted with it, brought the skull to Mackenzie, and reported more of the same kind in the interior of the island.

From this it seems that all the information in regard to the Mackenzie specimen came from the Indians, and that no white man has given any direct first-hand testimony as to its absolute origin.

At the instance of Mr. Anderson a brief request for information was inserted in the 'Daily Colonist,' of Victoria, B. C., and several replies were received. One of these, from Mr. S. M. Harrison, of Massett, which is of considerable interest, was kindly forwarded to me. It is addressed to Mr. Anderson under date of April 30, 1901, and is as follows:

SIR: I noticed a paragraph in the Colonist under the heading of "Who knows" re the existence of caribou on Queen Charlotte Islands. I have lived here twenty years, and know the account given is quite correct. I have made diligent inquiries amongst the Indians, and have gained the following information:

(1) Three years ago an Indian named Shakwan saw a female caribou feeding near a lake up Virago Sound, but failed to kill it, although he fired twice. Yethgwonas, another Indian, was with him at the time.

(2) This March a man named Stlinga with his two sons saw the tracks of a big herd near the headwaters of Malon River, near Viragó Sound.

(3) Men who were with the man who killed the two referred to in the Colonist are ready to show me the place where he killed them. This is near Lthum, up Virago Sound.

(4) The Haidas refused to eat the flesh of the caribou and left their carcasses. Mr. Mackenzie then paid them to go and bring the meat in and kept it for his own use.

(5) As the Indians are not interested in the killing of caribou, they refusing to eat the meat and there being no market for the antlers, etc., they consequently do not hunt them. They say they are afraid to go up the mountains and into much danger for no recompense, there being, according to their traditions, one-eyed monsters, hobgoblins, spirits, etc., to be met with on the mountains which they frequent. * * *

This, though much more definite than any other report received, contains little which did not emanate from the Indians, and it is therefore difficult to be certain that it contains any element of reliability. Surely men who believe in "one-eyed monsters, hobgoblins," etc., could easily indulge themselves with an imaginary caribou. However, Mr. Harrison's statement that meat was brought to Mackenzie and used by him is much more worthy of consideration and might lead one to entertain a belief in the *possibility* that caribou were killed on Graham Island, but the *probability* that such was the case is still doubtful.

If the type specimen of *Rangifer dawsoni* originally came from the mainland, as seems probable, instead of from Graham Island, it may either have been deliberately bartered for with the intention of obtaining a reward, or it may have been innocently brought to the islands to be used in the native arts. More or less communication has always existed between the islands and the coast, and between the coast and the interior, both in times of peace and during hostilities. Hence either explanation is probable. The fact that the Haidas used caribou horn for making implements and ornaments is not particularly pertinent to the question, since articles made of mountain goat and mountain sheep horns are even more commonly used by them. If they could obtain horns of elk, deer, mountain goats, and mountain sheep from the mainland, which they undoubtedly did, it certainly must have been just as easy to get the useful parts of the caribou. The Haidas, it is true, are better fishermen than hunters; but this is probably more on account of lack of game than otherwise, for they are physically and mentally a very superior tribe.

In view of the conflicting nature of the reports it does not seem safe or advisable to recognize *Rangifer dawsoni* as a distinct form, particularly as the specimen itself furnishes no indication that it represents a peculiar species, even granting that it came from the islands, for it does not essentially differ from specimens of *Rangifer arcticus*, the only difference claimed being that of darker color, and this is merely an opinion, as the piece of skin was long since destroyed.]

[**Mus musculus** Linn. House Mouse.

Said to have been abundant at Clew and Skidegate, but recent liberal importations of cats seem to have cleared it out.]

[**Mus norvegicus** Erxl. Norway Rat.

A few have escaped from ships occasionally, but in most cases each individual was detected and hunted down immediately, so the pest has not yet obtained a foothold on the islands.]

Peromyscus keeni (Rhoads). Keen's Mouse.

This is the common white-footed mouse of Graham and Moresby islands and the small detached islets near them. We found it abundant about Cumshewa Inlet from sea level to timberline, and equally common in Skidegate Inlet. It infests the few inhabited houses in company with shrews, and elsewhere is found indiscriminately all over the islands. Most of our specimens were caught near the shore in rocky or relatively dry places in the underbrush, but a few were taken high up in the mountains. In all, 98 specimens were secured, as follows: Cumshewa Inlet, 40; Skidegate Inlet, 50; near Rose Harbor, south end Moresby Island, 8. I have not recently examined the type of the species which was taken at Massett by Rev. J. H. Keen, but several alcoholic specimens from that locality are at present in the Biological Survey collection. I have compared four good skulls of these Massett specimens with a series from Skidegate, at the other end of Graham Island, and found them identical. Those from Moresby Island average a trifle larger than those from Skidegate, but the difference is extremely slight. The mainland species most similar to *P. keeni* seems to be *P. akeleyi*[1] from the Olympic Mountains, Washington. Specimens from various parts of the Olympic Peninsula (Neah Bay, Lake Cushman, Queniult Lake, etc.) do not differ from *keeni* in color, or appreciably in cranial characters. The only distinctions of consequence are the rather smaller ears and shorter tail of *keeni*. Ten specimens from Neah Bay, Washington, assumed to be *P. akeleyi*, average as follows: Total length, 203.8; tail vertebræ, 114.7; hind foot, 23.3. Twenty males of *keeni* from Skidegate average 197; 102; 22.7; fifteen females of *keeni* from Skidegate average 199.8; 103.4; 22.4.

Peromyscus prevostensis sp. nov. Prevost Island Mouse.

Type from Prevost Island, Queen Charlotte Islands, British Columbia, ♀ adult (old), No. 100818, U. S. Nat. Museum, Biological Survey collection. Collected July 5, 1900, by W. H. Osgood and E. Heller. Original No. 1135.

Characters. Similar to *Peromyscus macrorhinus*, but larger and with shorter tail. Similar to *Peromyscus sitkensis*, but with slightly shorter tail and cranial differences.

Color. Similar to *P. sitkensis* and *P. macrorhinus*, but slightly darker. Upperparts with dusky concentration on middle of back, forming a wide, ill-defined dorsal stripe; space around and in front of eyes black; ears dusky, with faint pale edgings; under parts grayish-white, occasionally with a faint narrow stripe of pinkish buff down middle of breast; hind feet generally somewhat dusky; tail sharply bicolor.

Skull. Similar to that of *Peromyscus sitkensis*, but slightly heavier; nasals decidedly shorter and not so distinctly attenuate posteriorly; posterior palatine foramina nearly or fully twice as long as in *sitkensis*.

Measurements.--Average of forty-seven adults: Total length, 217; tail vertebræ, 104; hind foot, 26. (Average of ten adult topotypes of *P. sitkensis:* 224; 113.6; 26.5. Average of two adult topotypes of *P. macrorhinus:* 231; 128; 26.) Average of five skulls of *prevostensis* (adult males): Basilar length of Hensel, 23; zygomatic breadth, 15.2; nasals, 11.5. (Average of five skulls of *sitkensis* (adult males): 23.4; 15.2; 12.7.)

Remarks.—This large mouse is very abundant on Prevost Island, where forty-seven fine adults were easily trapped in the three nights of our stay. They were taken in the dark woods along the shore of the island, under old logs or roots and in damp, mossy places. They are stoutly built, and individuals often made their escape from the ordinary 'out o' sight' traps by beating about until they freed themselves. Occasionally a sprung trap and a dead mouse would be found lying on the ground side by side or a foot or more apart. *P. prevost-*

[1] Elliot, Field Columbian Mus. Zool. Ser., I, 226–227, Feb., 1899.

ensis appears to be entirely confined to Prevost Island, as it was taken nowhere else in the Queen Charlotte group. Exactly opposite Prevost Island, on the south end of Moresby Island, *P. keeni* only was caught, as was the case elsewhere on Moresby Island. The distance between the two islands at this point is less than a mile, but it is probable that the strong tides which sweep through the channel would carry logs or any floating débris out to sea rather than from shore to shore, and thus prevent small mammals from being transferred. *P. prevostensis* is so different from *P. keeni* that even if the islands were more closely connected it would be improbable that either species was derived from the other. Moreover, it is also questionable whether either is the immediate descendant of *P. macrorhinus*, the form of the adjacent mainland, as both are more similar to forms found at a greater distance. Notwithstanding its geographic position, *P. prevostensis* is related neither to the mainland species of the same latitude nor to any of those farther south, but to *P. sitkensis*, which is found much farther north. In fact, considering its distribution, its resemblance to *sitkensis* is remarkable; it is characterized only by a combination of slight peculiarities. It seems best, however, to treat it as a full species until the rather difficult subject of the distribution of the section of the genus to which it belongs is more clearly understood.

Ursus (Euarctos) carlottæ sp. nov. Queen Charlotte Black Bear.

Type from Massett, Graham Island, Queen Charlotte Islands, British Columbia. Skull only, No. 87620, U. S. National Museum, Biological Survey collection, Nov., 1896, J. H. Keen.

Characters. Size slightly greater than *Ursus americanus;* skull more elongate; rostrum relatively more slender; cranium less arched; teeth larger and heavier, particularly last molars; last upper molar with posterior 'heel' quite elongate. (See Plate IV.)

Measurements.—The following table gives measurements of 6 full-grown old skulls of *U. americanus* from western North America and one adult (the type) and 6 immature skulls of *carlottæ*. The measurements of the type of *carlottæ* are the only ones fairly comparable with those given of *americanus*, but those of the young skulls are introduced to show the relatively large size of the last upper molar. No. 87618, for example, is the skull of a mere cub, yet its last upper molar is much larger than any of those of the full-grown specimens of *americanus*.

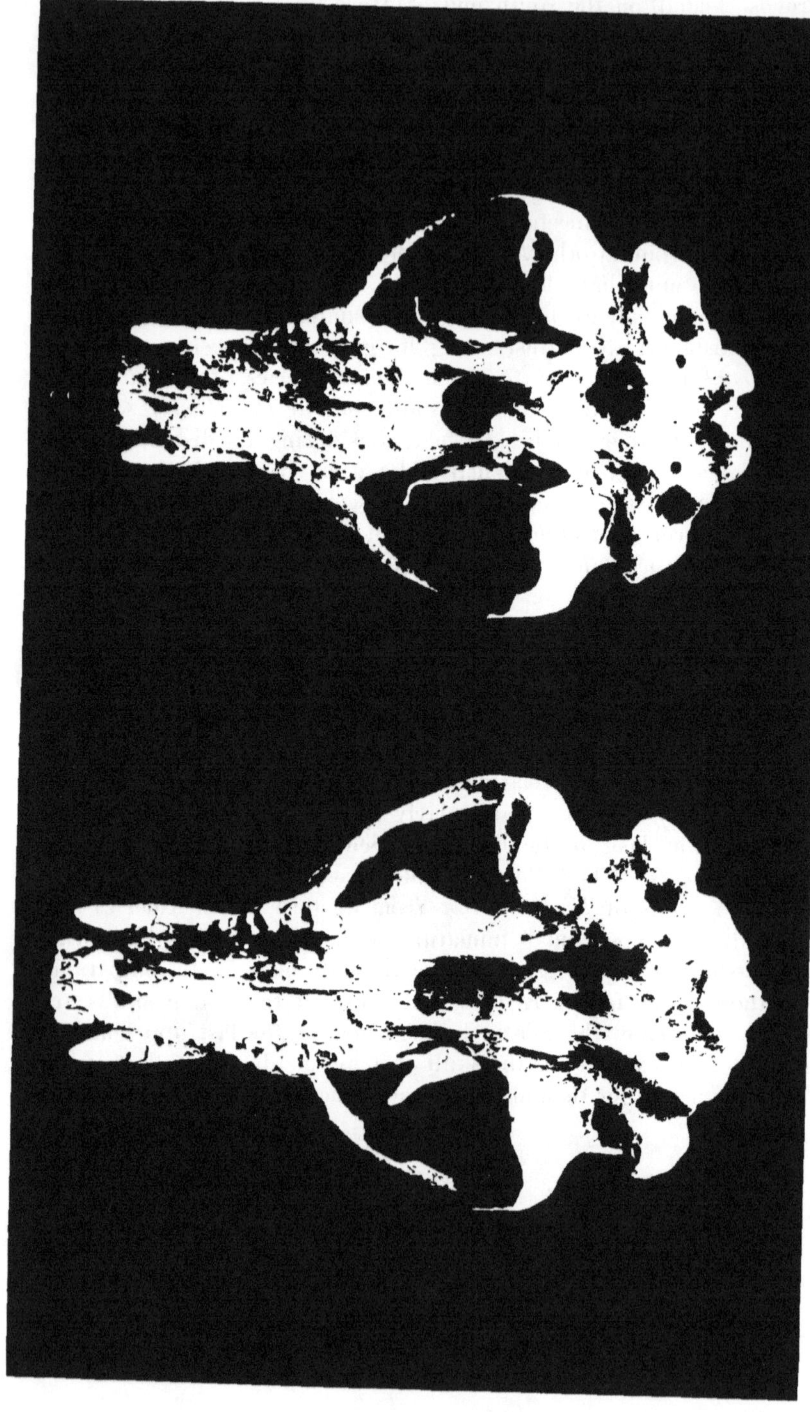

SKULLS OF URSUS · ABOUT ONE-THIRD NATURAL SIZE

URSUS CARLOTTÆ.

Number.	Locality.	Sex.	Basilar length of Hensel.	Palatal length.	Postpalatal length.	Zygomatic breadth.	Width across postorbital processes.	Length of nasals.	Posterior edge of alveolus of canine to posterior edge of alveolus of last molar.	Crown of last upper molar.
87620	Massett, British Columbia	? ♂ ad.	282	155	127	193	116	82	88	30
78065	.....do	? im.	260	146	114	162	97	70	81	31
87619	.....do	? im.	240	139	101	145	88	76	80	31.5
75052	.....do	? im.	230	130	100	137	81	66	78	30
87617	.....do	? im.	233	134	99	137	86	70	78	27
87621	.....do	? im.	223	127	96	132	78	59	76	27
87618	.....do	? im.	200	115	85	119	73	55	73	29

URSUS AMERICANUS.

Number.	Locality.	Sex.	Basilar length of Hensel.	Palatal length.	Postpalatal length.	Zygomatic breadth.	Width across postorbital processes.	Length of nasals.	Posterior edge of alveolus of canine to posterior edge of alveolus of last molar.	Crown of last upper molar.
72332	Shuswap, British Columbia	♀ old	260	144	116	168	94	78	81	26
75301	Jasper House, Alberta	♀ ad.	260	140	120	186	101	65	76	25
31277	Salmon River Mountains, Idaho.	? ad.	255	139	114	181	91	67	77	25.5
75053	Cook Inlet, Alaska	? ad.	254	138	116	172	109	73	81	27
48211	Stuart Lake, British Columbia.	? ad.	260	142	118	187	113	78	77	26.5
53577	.....do	? ♂ ad.	266	143	123	186	101	75	77	24

Remarks.—Seven perfect skulls of the Queen Charlotte black bear are in the Biological Survey collection, and although most of them are those of immature animals there is no difficulty in distinguishing any of them from mainland specimens. In comparisons specimens from western British Columbia and Alaska have been used to represent *americanus*, which, as far as known at present, ranges from the Atlantic to the Pacific. No specimens are available from the west side of the coast mountains on the mainland opposite the Queen Charlottes, but a few from the region immediately north (Cook Inlet and White Pass) and south (Olympic Peninsula) have been examined, and none of them show any approach to *carlottæ*. The skull of *Ursus luteolus* is equal in size to that of *carlottæ*, but the teeth, though as long as in *carlottæ*, are much wider and heavier. In *carlottæ* the brain case is fuller, the arch of the cranium much greater, and the interorbital region wider. At the suggestion of G. S. Miller, jr., the skull of *U. carlottæ* was compared with that of *Ursus procerus*, the fossil species from Ohio, and more or less superficial resemblance between them was found, though, of course, detailed diagnostic characters are numerous. Both agree in general form of skull, particularly in the elongate rostrum in contrast with that of *U. americanus*. The dentition of *procerus*, however, is much heavier and otherwise different, while other characters are abundant, indicating that the resemblance to *carlottæ* is merely coincidental and not indicative of close relationship. No skins of *carlottæ* have been examined, but they are said to be glossy black at all times, the cinnamon form being absolutely unknown on the islands. The skulls on which the foregoing description is based were secured from the natives at Massett, in 1895 and 1896, by Rev. J. H. Keen, to whom we are indebted for so many other specimens from the Queen Charlottes. I saw signs of bear only on the mountain at the head of West Arm of Cumshewa Inlet, and there the indications were not fresh enough to raise any hopes of securing a specimen. The Haidas hunt bears to some extent, and also secure them in dead-fall traps. I noticed several of these traps near the head of Cumshewa Inlet, but they had not been used for some time. Mr. Tennant, of Skidegate, secures from the Indians 10 to 30 skins annually.

Lutra canadensis subsp.? Land Otter.

Otters are rather rare on the Queen Charlottes, though perhaps no more so than on the mainland. Mr. Tennant, of Skidegate, says that his annual receipts from the Indians have seldom included more than a dozen otter skins. This is the only mammal known to occur on the islands of which I have not examined specimens.

Putorius haidarum Preble.[1] Haida Weasel.

Three specimens of this weasel were caught about Cumshewa Inlet and one at Skidegate. Traps baited with fish and set along the shore in the rank grass or in the rocks were the most successful. The type of this species is in winter pelage with slight traces of the summer coat, indicating, as Mr. Preble noted, a probable color difference from *Putorius kadiacensis*. This surmise is amply borne out by our specimens in full summer pelage. They are much darker than *kadiacensis*, displaying very nearly the rich chocolate of *P. streatori*, and having much less of the lemon-yellowish wash on the underparts than *kadiacensis*. The color of the upperparts does not encroach on that of the belly, and the black on the tail is extensive, occupying nearly half its length; in these respects it is more like *kadiacensis* than *streatori*. The white of the underparts extends on the under side of the tail for about one-third of its length. The fore feet are entirely white; the toes and one-third of the upper side of the hind feet are white. Its skull is quite distinct from that of any other of the *cicognani* group. In general terms, it is flat, short, and stocky. The most conspicuous point of difference from *kadiacensis*, *cicognani*, or *streatori* is the very broad rostrum and interorbital region. Young specimens entirely free from parasites show this to good advantage, and adult skulls which are infested are so much wider than those of

[1] Proc. Biol. Soc. Wash., XII, 169–170, Aug. 10, 1898.

kadiacensis in a similar condition that they are easily distinguishable. The flesh measurements of a fully adult male are as follows: Total length, 283; tail vertebræ 70; hind foot 39. Adult female: 252; 63; 31. Skull of adult male: Basal length 39; mastoid breadth 19.2; width across postorbital processes 14.5; palatal length 17.5; length of audital bullæ 13. Skull of adult female: 35.5; 17.5; 12; 16.5; 11.6.

Mustela nesophila sp. nov. Queen Charlotte Marten.

Type from Massett, Graham Island, Queen Charlotte Islands, British Columbia. Skull only. Male (?), No. 78066, U. S. National Museum, Biological Survey Collection. J. H. Keen.

Characters.—Similar to *Mustela caurina*, but larger; rostrum shorter and heavier; dentition heavier; premolars larger and more crowded. Last upper molar similar to that of *Mustela americana*, internal length being more nearly equal to external length than in *caurina*. (See Pl. V.)

Measurements.—Type: Basilar length of Hensel 73; palatal length 39; postpalatal length 34; zygomatic breadth 45.5; width across postorbital processes 23; interorbital constriction 19. Topotype No. 76429, female: Basilar length of Hensel 69; palatal length 37; postpalatal length 32; width across postorbital processes 23; interorbital constriction 15. (*Mustela caurina* No. 87075, female adult, Port Moody, British Columbia: Basilar length of Hensel 64; palatal length 33; postpalatal length 31; zygomatic breadth 41; width across postorbital processes 20; interorbital constriction 15.)

Remarks.—This form is represented by two skulls which Mr. Keen secured several years ago from natives at Massett. These are not sexed, but the difference in size and other slight characters make it evident that one is a young male and the other an adult female. The peculiarities shown by these skulls are so marked that there can be no doubt that they represent an insular species. In the Biological Survey series of nearly 500 skulls of *Mustela americana* and its allies I have been able to find no others showing the characters of these individuals from the Queen Charlotte Islands. The molar teeth of *nesophila* are heavier than in any other form of the group. The audital bullæ are actually about the same size as in *caurina* and thus relatively smaller. The maxillary region between the infraorbital foramen and the alveoli of the upper molars is wider and heavier than in *caurina*. The most obvious cranial character, however, and the one which distinguishes *nesophila* from all other members of the *americana* group is the thick, heavy rostrum. When skins are available for comparison they also will doubtless show some slight differences. The fur traders say the Queen Charlotte martens are always light colored and short haired and do not command as high a price as those from the mainland. The Haidas trap more or less for martens every winter, but the animal is evidently not abundant, for Mr. Tennant's annual receipts seldom exceed forty skins.

Latax lutris (Linn). Sea Otter.

Formerly very abundant, but quite rare at present. A few are occasionally taken on the west coast of the islands or off the southern end of Prevost Island. During his cruise about the islands in 1787 Dixon bartered with the Haidas for 1,821 sea-otter skins. He secured a great many in Cloak Bay, on North Island, and describes his experience as follows:[1]

> A scene now commenced which absolutely beggars all description, and with which we were so overjoyed that we could scarcely believe the evidence of our senses. There were 10 canoes about the ship, which contained, as nearly as I could estimate, 120 people. Many of these brought most beautiful beaver cloaks, others excellent skins, and, in short, none came empty-handed, and the rapidity with which they sold them was a circumstance additionally pleasing. They fairly quarreled with each other about which should sell his cloak first, and some actually threw their furs on board if nobody was at hand to receive them. * * * In less than half an hour we purchased near 300 beaver skins, of an excellent quality. * * * That thou mayest form some idea of the cloaks we purchased here I shall just observe that they generally contain three good sea-otter skins, one of which is cut in two pieces. Afterwards they are neatly sewed together so as to form a square, and are loosely tied about the shoulders with small leather strings, fastened on each side.

At another time, when near either Skidegate or Cumshewa Inlet, under date of July 29, he writes:

> Early in the afternooon we saw several canoes coming from shore, and by 3 o'clock we had no less than 18 alongside, containing more than 200 people, chiefly men. This was not only the greatest concourse of traders we had seen, but what rendered the circumstance additionally pleasing was the quantity of excellent furs they brought us, our trade now being equal, if not superior, to what we had met in Cloak Bay, both in the number of skins and the facility with which the natives traded. * * * Besides the large quantity of furs we got from this party (at least 350 skins) they brought several raccoon cloaks, each cloak consisting of 7 raccoon skins neatly sewed together.[2]

?**Eumetopias stelleri** (Lesson). Steller Sea Lion.

A sea lion, probably this species, is reported. It was not seen by us.

Otoes[3] **alascanus**[4] (Jordan and Clark) Alaska Fur Seal.

Fur seals still occasionally stop on or near the Queen Charlotte Islands. In former days the natives secured a great many in the region off the south end of the group.

---

[1] A Voyage Round the World in the *King George* and *Queen Charlotte*, pp. 199–234, London, 1789.

[2] Since raccoons are not found on the Queen Charlotte Islands, these skins were probably from Vancouver Island where the animals are common.

[3] Fide Palmer, Proc. Biol. Soc. Wash., XIV, 133–134, Aug. 9, 1901.

[4] Report Fur Seal Invest. 1896-1897, Pt. 3, pp. 2–3, 1899.

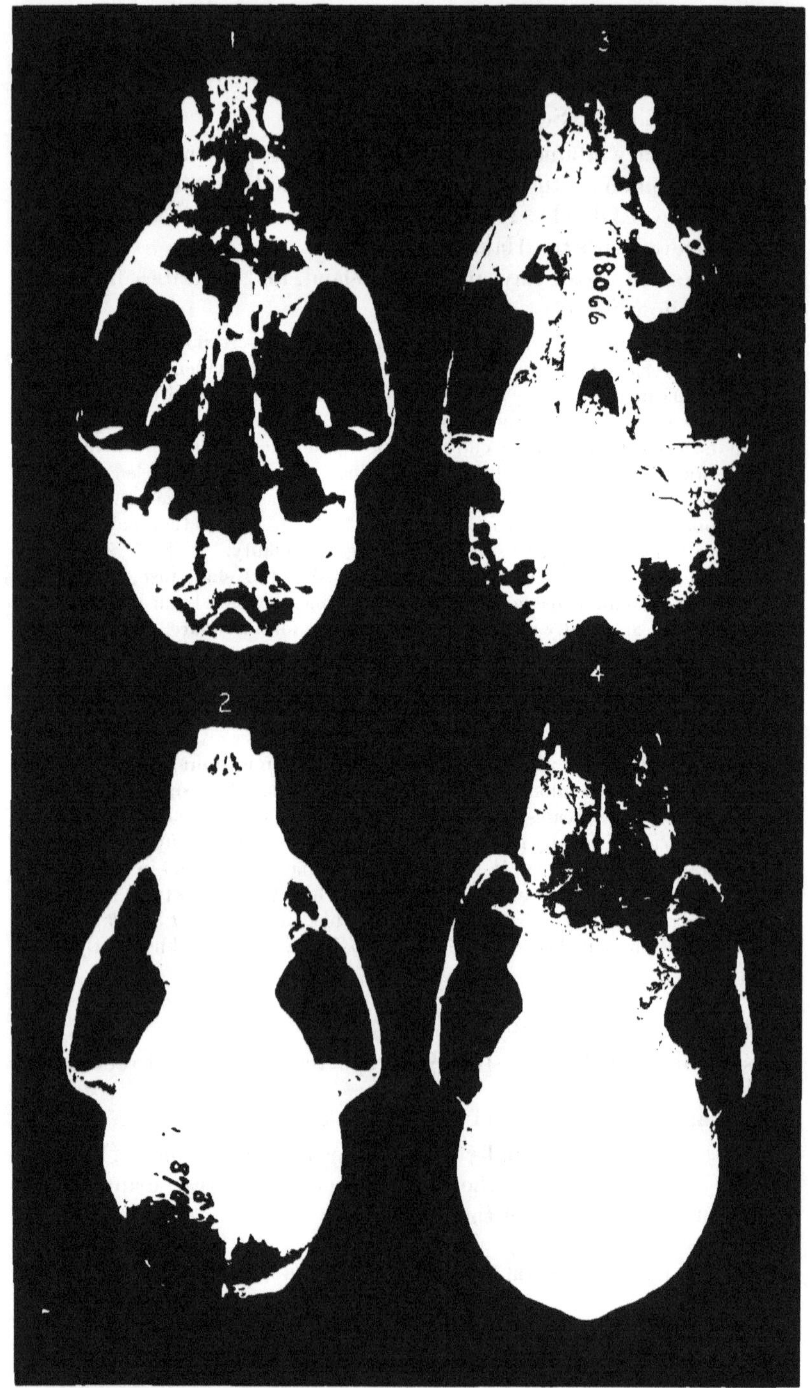

SKULLS OF MUSTELA NATURAL SIZE.

1, 2. Mustela caurina. 3, 4. Mustela nesophila.

Phoca largha Pallas. Pacific Harbor Seal.

Harbor seals are quite common. They bobbed up very often near our canoe as we were paddling about the inlets.

Sorex longicauda prevostensis subsp. nov. Prevost Island Shrew.

Type from Prevost Island, Queen Charlotte Islands, British Columbia. ♂ ad. No. 100618, U. S. Nat. Mus., Biological Survey Collection. Collected July 3, 1900, by W. H. Osgood and E. Heller. Orig. No. 1089.

Characters.—Similar to *Sorex longicauda;* tail, relative to head and body, shorter; dental characters distinctive.

Color.—Very slightly darker than *S. longicauda;* contrast between upper and lower parts less; otherwise similar.

Skull.—Size large, equal to *S. longicauda;* dentition slightly lighter; third unicuspid nearly equal to fourth, not about one-half smaller as in *longicauda.*

Measurements.—Type: Total length, 133; tail vertebræ, 58; hind foot, 15. Average of 7 topotypes: Total length, 135; tail vertebræ, 55; hind foot, 15; ratio of length of tail to total length, 40.7. (Average of 27 topotypes of *S. longicauda:* Total length, 129; tail vertebræ, 58; hind foot, 15.5; ratio of length of tail to total length, 45.)

Remarks.—The shrews of Prevost Island differ from those of Graham and Moresby islands to a greater degree than from the mainland species *S. longicauda.* In color they are very dark, even averaging darker than *longicauda;* occasional specimens are deep chocolate brown both above and below. On the whole, however, they are very similar to *longicauda,* but the combination of slight characters they show can hardly be explained, except by assuming that they were produced by insular isolation. The case is much like that of *Peromyscus prevostensis,* though not so striking. In both cases the Prevost Island form is distinct from its nearest geographical ally and very closely similar to forms found at a greater distance. The very close relationship of *prevostensis* to *longicauda* seems best indicated by a trinomial name.

This shrew was not particularly abundant on Prevost Island, and only 14 specimens were secured. They were caught in damp, mossy places such as shrews usually inhabit in the northwest coast region.

Sorex longicauda elassodon subsp. nov. Queen Charlotte Shrew.

Type from Cumshewa Inlet, Moresby Island, Queen Charlotte Islands, British Columbia. ♂ yg. ad. No. 100597. U. S. National Museum, Biological Survey Coll. Collected June 13, 1900, by W. H. Osgood and E. Heller. Orig. No. 1030.

Characters.—Similar to *Sorex longicauda,* but smaller and with relatively shorter tail; teeth actually about as in *Sorex obscurus,* relatively much smaller.

Color.—Almost exactly as in *S. longicauda;* lower parts paler than in *prevostensis.*

Skull. Similar in general to that of *Sorex longicauda*, but somewhat smaller; compared with those of *S. obscurus* the braincase is more elevated and the rostrum more slender and attenuate, the skull in general having a longer and narrower appearance; the teeth are decidedly smaller than in *longicauda*, but about equal to those of *obscurus*, thus being relatively smaller than those of *obscurus*. In the relative size of individual teeth I can find no departure from *longicauda* or *obscurus*.

Measurements.—Type: Total length, 123; tail vertebræ, 52; hind foot, 14. Average of 7 topotypes: Total length, 132; tail vertebræ, 55; hind foot, 14.

Remarks.—The shrews of Moresby and Graham Island are easily separable from *S. prevostensis*, which might be supposed to be their nearest relative. They are very closely related to *S. obscurus* and *S. longicauda*, however, and seem to be intermediate between them. As in the case of *prevostensis*, a trinomial name is proposed for them in order to group them with their very similar relatives.

They were found in abundance on Moresby Island, but for some reason were quite rare at Skidegate, on Graham Island. They have been taken at Massett by Mr. Keen, who reports that they are common there. In the vicinity of Cumshewa Inlet we took 25 specimens, but at Skidegate with equal effort, only one.

Lasionycteris noctivagans (Le Conte). Silver-haired Bat.

A single adult male was taken at Skidegate on the evening of July 10. Another was killed the same evening, but it fell in a dense thicket and could not be retrieved. Others, which by their large size were supposed to be this species, were occasionally seen. The specimen secured was in the deep-brown phase, with but slight silver tipping to the hairs.

I have seen no previous record of the occurrence of the silver-haired bat on the coast north of Puget Sound.

Myotis yumanensis saturatus Miller. Sooty Big-footed Bat.

Bats were very rarely seen in Cumshewa Inlet or in Houston Stewart Channel, near Prevost Island. At Skidegate, however, they were very abundant both about the village and along the edge of the forest. They were most easily secured about the village, and nearly every evening during our stay we spent several hours wandering among the deserted Indian cabins on the lookout for them. Early in the evening they were found flying quite high up along the edge of the forest and over the village; later they were lower down, darting in and out between the houses, never going much above them, and sometimes almost touching the ground. This made them very hard to shoot, but after considerable expenditure of ammunition we obtained representatives of several species. In some cases we secured specimens by striking them to the ground with long switches. Four adult males of

Myotis yumanensis saturatus were taken, and many other bats apparently of this subspecies were seen. It appears to be the most abundant bat at Skidegate. The specimens seem to be perfectly typical and can easily be matched by others from the Puget Sound region. The species has not been previously recorded from the Queen Charlotte Islands.

Myotis subulatus keeni (Merriam). Keen Bat.

This bat was originally described by Dr. C. Hart Merriam in 1895, from specimens collected at Massett by Rev. J. H. Keen. The type and three other alcoholic specimens are at present in the Biological Survey collection. Although no bat of the *subulatus* type has as yet been found elsewhere on the west coast, it does not seem probable that this form is peculiar to the Queen Charlotte Islands, and its capture on the adjacent mainland will probably occur before many years have passed. Strange to say, this bat was not taken by us at Skidegate.

Myotis californicus caurinus Miller. Northwest Bat.

This subspecies was described by Mr. G. S. Miller, jr., from specimens taken in 1895 by Rev. J. H. Keen. The type and eight topotypes in alcohol are now in the Biological Survey collection. Three specimens, one male and two females, taken at Skidegate July 10–12, are slightly darker than specimens from Mount Rainier and Ashford, Washington, and doubtless represent the extreme development of the form.

Gavia imber (Gunn). Loon.

Rather common about all the islands. Its cry when heard at night in one of the narrow, closely walled inlets is even more weird and mournful than usual.

Gavia pacifica (Lawr.). Pacific Loon.

Several were seen at Skidegate July 9.

Gavia lumme (Gunn). Red-throated Loon.

A pair of red-throated loons were seen flying down Cumshewa Inlet June 27.

Lunda cirrhata Pallas. Tufted Puffin.

Often seen flying in Hecate Strait. A moderate-sized breeding colony was found on an islet in Houston Stewart Channel. One specimen taken July 5.

Fratercula corniculata (Naum). Horned Puffin.

Recorded from Massett, Graham Island, by Mr. John Fannin, on the authority of Mr. Keen;[1] also mentioned by Chamberlain.[2]

Synthliboramphus antiquus (Gmelin). Ancient Murrelet.

Seen at Massett by Mr. Keen; not seen by us.

Brachyramphus marmoratus (Gmelin). Marbled Murrelet.

Occasionally seen in Cumshewa and Skidegate inlets.

Cepphus columba Pallas. Pigeon Guillemot.

This is the most abundant water bird about the islands. It breeds in crevices of the rocks along the shores of quiet inlets. In many of these places the branches of the trees overhang the rocks and almost touch the water at high tide, so that when the birds are startled from their nests it is possible to observe the strange circumstance of a guillemot flying out of a tree. In frequent instances nesting sites are chosen in and about clefts of the rocks under the roots of large trees.

Uria troile californica (Bryant). California Murre.

Several were seen in Hecate Strait a few miles off the mouth of Cumshewa Inlet July 1; they were apparently directing their course

[1] Preliminary Catalogue of Collections in the Provincial Museum, Victoria, British Columbia, p. 16, 1898.

[2] Catalogue of Canadian Birds, p. 4, 1887.

for the Skedans Islands at the mouth of the inlet, where they probably breed.

Rissa tridactyla pollicaris Ridgw. Pacific Kittiwake.

According to Mr. John Fannin, this bird was taken at the islands September, 1895, by Dr. C. F. Newcombe.[1] It was not seen by us.

Larus glaucescens Naum. Glaucous-winged Gull.

A breeding colony of about 100 of these gulls was found on an islet in Houston Stewart Channel. Fresh eggs, as well as young just hatched, were observed there July 3. Very few large gulls were seen elsewhere about the islands, but they are said to breed in numbers on the west coast of Moresby Island.

Larus brachyrhynchus Rich. Short-billed Gull.

Mentioned by Chamberlain.[2]

Larus philadelphia (Ord). Bonaparte Gull.

A few small gulls supposed to be this species were seen.

Puffinus griseus (Gmelin). Dark-bodied Shearwater.

Shearwaters supposed to be this species were seen in large flocks in Hecate Strait a few miles off Moresby Island, July 1–8. Reported in great numbers off the west coast of the islands in the fall of 1895, by Dr. C. F. Newcombe.[3]

Puffinus tenuirostris (Temm.). Slender-billed Shearwater.

Shot off the coast of Queen Charlotte Islands, by Dr. C. F. Newcombe, in August, 1894.[4]

Oceanodroma furcata (Gmelin). Forked-tailed Petrel.

A few small petrels supposed to be this species were seen in Hecate Strait, July 1–8.

Phalacrocorax pelagicus Pallas. Pelagic Cormorant.

Frequently seen. A few breed on the Skedans Islands off the mouth of Cumshewa Inlet, and on some of the islets off the west coast of Prevost Island.

Merganser americanus (Cassin). American Merganser.

A large merganser is abundant and evidently breeds. Mr. Keen reports both this species and the following:

Merganser serrator (Linn.). Red-breasted Merganser.

Reported by Mr. Keen. Not positively identified among those seen by us.

---

[1] Preliminary Catalogue Provincial Museum, Victoria, B. C., p. 17, 1898.

[2] Catalogue of Canadian Birds, p. 10, 1887.

[3] Fannin, Preliminary Catalogue Provincial Museum, Victoria, B. C., p. 17, 1898.

[4] Macoun, Catalogue of Canadian Birds, Part I, p. 61, 1900.

Anas boschas Linn. Mallard.

A flock of about a dozen mallards was seen frequently about the head of Cumshewa Inlet, June 17–26.

Mareca americana (Gmelin). Baldpate.

Reported by Mr. Keen.

Nettion carolinensis (Gmelin). Green-winged Teal.

Given in Mr. Keen's Massett notes. Not seen by us.

Dafila acuta (Linn.). Pintail.

Reported by Mr. Keen. Not seen by us.

Aythya sp.?

Scaup ducks were several times seen flying at a distance.

Clangula clangula americana (Bonap.). American Golden-eye.

Included in Mr. Keen's Massett list. Not seen by us.

Charitonetta albeola (Linn.). Butllehead.

Seen at Massett (Keen).

Harelda hyemalis (Linn.). Old-squaw.

A roughly stuffed skin of an adult male old-squaw was seen at an Indian village in Cumshewa Inlet.

Histrionicus histrionicus (Linn.). Harlequin Duck.

Said to occur at Massett (Keen).

Oidemia deglandi Bonap. White-winged Scoter.

Frequently seen in Cumshewa and Skidegate inlets.

Oidemia perspicillata (Linn.). Surf Scoter.

Common.

Anser albifrons gambeli (Hartl.). American White-fronted Goose.

Reported by Mr. Keen.

Branta canadensis occidentalis (Baird). White-cheeked Goose.

Six or seven were seen crossing Cumshewa Inlet June 16.

Olor sp.?

Swans are said to have been taken frequently.

Ardea herodias fannini Chapman.[1] Northwest Coast Heron.

Often seen feeding at low tide on the beaches and mud flats of Skidegate and Cumshewa inlets. No specimens taken.

Porzana carolina (Linn.). Sora Rail.

Included in Mr. Keen's manuscript list of birds seen at Massett.

---

[1] Bul. Am. Mus. Nat. Hist. XIV, 87–90, Apr. 15, 1901.

Gallinago delicata (Ord). Wilson Snipe.

Reported by Mr. Keen.

Tringa acuminata (Horsf.). Sharp-tailed Sandpiper.

Taken at Massett, Graham Island, December 27, 1897, by Rev. J. H. Keen.[1] Mr. Keen kindly forwarded me the specimen on which this record was made. I have compared it with others of the same species and found it typical.

Ereunetes occidentalis Lawr. Western Sandpiper.

A small sandpiper supposed to be this species was seen on a beach in Cumshewa Inlet, June 28. Mr. Keen reports its occurrence at Massett.

Calidris arenaria (Linn.). Sanderling.

Reported by Mr. Keen.

Totanus melanoleucus (Gmelin). Greater Yellowlegs.

Two were seen and one of them taken on the beach at Skidegate, July 17.

Actitis macularia (Linn.). Spotted Sandpiper.

One seen at Skidegate in July.

Squatarola squatarola (Linn.). Black-bellied Plover.

Reported by Mr. Keen.

Charadrius dominicus Müller. American Golden Plover.

Reported by Mr. Keen.

Arenaria melanocephala (Vigors). Black Turnstone.

A flock of 6 was seen near Lina Island, Skidegate Inlet, July 12.

Hæmatopus bachmani Aud. Black Oystercatcher.

Abundant. The shrill cries of the oystercatchers were heard about the inlets at all hours of day or night. Nearly every outlying rocky islet was occupied by a pair of 'sandpipers,' as they are locally called, and whenever a boat approached both birds would circle about it for some time, flying close to the water and crying shrilly. Dr. Robert Brown, writing in 1869,[2] says of this species:

> About Queen Charlotte Islands it is very plentiful. In March, 1866, while rowing along the narrow sounds among these islands we often saw it. It would sit on the rocks until we could almost touch it; then, uttering a low whistling cry, it would dart off to another skerry, repeating the same maneuver over and over again.

A nest found in Cumshewa Inlet, June 17, was merely a hollow about 2 inches deep and almost perfectly round, scooped out of a weedy turf a few feet above high-water mark. The bottom of the hollow was covered with bits of broken stone, evidently placed there by the old

[1] Fannin, Preliminary Catalogue Provincial Museum, Victoria, B. C., p. 28, 1898

[2] Ibis, IV, 424, 1868.

bird. A few feet from the nest a downy young bird was discovered squatting in the weeds and gravel. It may be described as follows: Upperparts chiefly mottled olive-gray and black, the gray predominating and the black distributed mainly in an ill-defined patch on the back of the head and two prominent parallel stripes that extend from the nape down the middle of the back to the rump; middle of back with a little buffy-tipped down; wings like back but with more buffy; flanks spotted with black; throat and breast slate gray, darker on sides of neck and indistinctly patched with paler on middle of throat; a conspicuous white spot far back on middle of breast with a white line extending forward from it on each side to the vicinity of the axillars; abdomen paler than breast and lightly washed with buffy, also having to some degree the vermiculated appearance of the upperparts.

Dendragapus obscurus fuliginosus Ridgw. Sooty Grouse.

Several were heard booming about Cumshewa Inlet early in June. A pair of adult birds was taken at an altitude of about 3,000 feet in the mountains at the head of Cumshewa Inlet June 23.

[**Bonasa umbellus sabini** (Dougl.). Oregon Ruffed Grouse.

In Macoun's Catalogue of Canadian Birds the following statement occurs under *Bonasa u. sabini:*

"One of the most abundant birds of the coast region of British Columbia including all the islands in the Gulf of Georgia, Vancouver Island, and *Queen Charlotte Islands.*"

We did not meet this bird anywhere on the Queen Charlottes and it is not mentioned in Mr. Keen's manuscript list, so it seems probable that this statement is erroneous.]

? Lagopus sp.

I was told by Mr. Tennant, of Skidegate, that eight 'white grouse' were killed several years ago by a party of prospectors in the mountains on Graham Island a few miles from Anchor Cove, Skidegate Inlet. We found conditions favorable for ptarmigan near the summit of the mountains about the head of Cumshewa Inlet, but did not see any during our short stay there.

Zenaidura macroura (Linn.). Mourning Dove.

Seen at Massett by Mr. Keen; not observed by us.

Accipiter velox (Wils.). Sharp-shinned Hawk.

Two small hawks supposed to be this species were seen at Skidegate July 12. Mr. Keen reports its occurrence at Massett.

Accipiter atricapillus striatulus Ridgw. Western Goshawk.

Seen at Massett (Keen).

Buteo borealis calurus (Cassin). Western Red-tailed Hawk.

A solitary red tail was seen flying near the head of Cumshewa Inlet June 22; no others were seen during our visit.

Haliæetus leucocephalus alascanus Towns. Northern Bald Eagle.

Very common; often seen in parties of from 2 to 10 individuals, the majority being birds of the year. They seem to feed largely on mollusks and crustaceans, which are very abundant. At one time I saw seven huge birds clumsily hopping over the rocks on the shore evidently looking for crabs. Eagles' nests were occasionally noted in the tops of tall, partially dead trees.

Falco peregrinus pealei Ridgw. Peale Falcon.

An immature female was taken July 2. It was shot as it circled around the small schooner in which we were drifting in Hecate Strait a few miles off Scudder Point, Burnaby Island. Several others were seen near Prevost Island, and while we were in Houston Stewart Channel a pair of them had daily altercations with a bald eagle in the tops of the trees on an islet near our anchorage.

Falco columbarius suckleyi Ridgw. Black Merlin.

A small dark hawk was indistinctly seen flitting out from the top of a tall spruce in Cumshewa Inlet June 14. Mr. Keen reports the black merlin from Massett.

Pandion haliaëtus carolinensis (Gmelin). American Osprey.

Ospreys were not seen by us, but they are evidently common in some parts of the islands, as we heard numerous reports of them. Mr. Keen has noted their spring arrival at Massett as follows: 1894, May 13; 1896, April 24; 1897, May 12; 1898, April 30.

Megascops asio kennicotti (Elliot). Kennicott Screech Owl.

Seen at Massett (Keen).

Nyctala[1] **acadica scotæa**[2] subsp. nov. Northwest Saw-whet Owl.

Type from Massett, Queen Charlotte Island, British Columbia, ♂ ad. No. 168171, U. S. National Museum, Biological Survey collection. Collected December 19, 1896, by J. H. Keen.

Characters.—Similar to *N. acadica*, but darker both above and below, dark markings everywhere heavier; flanks, legs, and feet more rufescent.

Color.—Upperparts, including head, neck, back, and upper tail-coverts, mummy brown;[3] head with light stripes on forehead and thence down sides of neck; neck and interscapulars irregularly marked with white; wings slightly lighter than back; five outer primaries with two to four white spots on outer and inner webs; inner prima-

---

[1] Dr. C. W. Richmond has proposed the name *Cryptoglaux* as a substitute for *Nyctala* on the ground that the latter is preoccupied by *Nyctalus* (cf. Auk, XVIII, 193, April, 1901). This disposition of *Nyctala* seems reasonable, but *Cryptoglaux* is not used here, since it has not been adopted by the American Ornithologists' Union.

[2] Scotæa=dark, dusky.

[3] The color names used are from Ridgway's Nomenclature of Colors.

ries and secondaries with white spots on inner webs only; tail pale clove brown, narrowly tipped with white, webs of rectrices crossed by three white bars; auriculars isabella color streaked with dusky; forehead and superciliary region white; orbital ring and outer feathers of lores sooty; chin, throat, and upper breast white, interrupted by a collar of mars brown; lower breast and abdomen white, heavily streaked with walnut brown; sides, flanks, legs, and feet clear ochraceous buff.

Measurements.—Type: Wing 85; tail 69; tarsus 26.

Remarks.—This dark-colored form of the Acadian owl doubtless ranges throughout the humid Pacific coast region. Its rarity probably accounts for its having been previously overlooked, for its characters are in general the same as those of the numerous other forms peculiar to the same region, which have long been recognized in nomenclature. The only specimens that I have examined beside the type are several imperfect ones from Puget Sound, which are in the National Museum collection. These agree with the type in richness of color and extent of dark markings. The type was collected by Rev. J. H. Keen, who very generously presented it to the Biological Survey collection. A small owl, apparently this species, flew over our vessel at 11 o'clock on the night of July 4, while we were at anchor in Houston Stewart Channel. This was the only owl seen at any time during our visit to the islands.

Nyctea nyctea (Linn.). Snowy Owl.

Mr. Tennant says he has killed large white owls at Skidegate. Mr. Keen reports them from Massett.

Ceryle alcyon (Linn.). Belted Kingfisher.

Generally found along the larger streams. One specimen was taken June 21.

Dryobates picoideus sp. nov. Queen Charlotte Woodpecker.

Type from Cumshewa Inlet, Moresby Island, Queen Charlotte Islands, British Columbia. ♀ ad., No. 166816, U. S. Nat. Mus., Biological Survey collection. Collected June 15, 1900, by W. H. Osgood and E. Heller. Orig. No. 386.

Characters.—Similar in general to *Dryobates v. harrisi;* bill slightly smaller; middle of back barred and spotted with black; flanks streaked with black.

Description.—Top of head, wings, and tail black; middle of back from nape to rump white, heavily barred, or spotted with black (streaked in immature specimens), primaries and secondaries more or less spotted with white; coverts usually with two to four elongate white spots; underparts smoky brownish, deepest on breast; flanks streaked, barred, or spotted with black and dusky (this sometimes extending forward on sides to axillars); three outer tail-feathers white, the innermost always partially black, the others sometimes more or less barred with black.

Measurements.—The Queen Charlotte specimens have rather smaller bills than *harrisi*, as appears from the following table:

DRYOBATES PICOIDEUS.

Number.	Sex.	Locality.	Wing.	Tail.	Exposed culmen.	Tarsus.
166816..	♀ ad....	Queen Charlotte Islands, British Columbia (type).	126	94	29	23
166821..	♀ ad....	.....do................................	120	96	28.5	22
166820..	♀ im	.....do................................	125	96	27.5	22
166818..	♀ im	.....do................................	124	95	24	22
166817..	♂ im	.....do................................	121	98	28.5	23
166819..	♂ im	.....do................................	122	99	29.5	21.5

DRYOBATES VILLOSUS HARRISI.

Number.	Sex.	Locality.	Wing.	Tail.	Exposed culmen.	Tarsus.
157586..	♂ ad....	Neah Bay, Washington........................	123	97	34	23
157619..	♂ ad....	.....do................................	126	99	36	24
157598..	♂ ad....	.....do................................	129	102	35	23
117689..	♀ ad....	Departure Bay, British Columbia..............	123	85	30	21
151483..	♀ ad....	Comox, British Columbia	123	95	31	22.5
132504..	♀ ad....	Seattle, Washington..........................	121	97	32	22

Remarks.—Adult specimens of *Dryobates picoideus* are easily distinguishable from all other members of the *villosus* group by the black markings on the back. Immature birds of *harrisi* and of other members of the group occasionally have a few median or lateral streaks of black on the back, but never the definite barring, as in *picoideus*. There is also an occasional tendency in young *harrisi* to show dusky on the flanks, but neither this nor the black in the back persists in the adult. This woodpecker is not abundant on the islands; during our stay we saw but six, all of which were secured. These consist of two adult females, two immature females, and two immature males. The only adult male examined was one brought to me by a boy at Skidegate, which was not preserved, as it was very much mutilated, but its characters, particularly the barred back, were noted.

Sphyrapicus ruber flaviventris (Vieill.). Northern Red-breasted Sapsucker.

Picus flaviventris Vieillot, Ois. Am. Sept., II, 67, 1807.
Sphyrapicus varius ruber Grinnell, Condor, III, 12, Jan. 15, 1901.

Common. Represented by ten specimens, which agree perfectly with birds from Vancouver Island and the mainland of British Columbia. This bird was discovered by Captain Cook in Nootka Sound on the west coast of Vancouver Island. In the narrative of his famous voyage he makes special mention of it and gives a very good description. Later, in 1807, Vieillot named it *Picus flaviventris*, and distinguished it from *Picus ruber* of Gmelin as follows:

'Diffère principalement par la teinte du ventre qui est d'un jaune olivatre.'

Colaptes cafer saturatior (Ridgw.). Northwestern Flicker.

Apparently quite rare, as we saw none. A few unmistakable tail feathers were found, however, by Heller in Cumshewa Inlet. Mr. Keen reports flickers abundant near Massett.

Selasphorus rufus (Gmelin). Rufous Hummingbird.

Common. During the month of June hummers were often seen visiting the abundant blossoms of *Lonicera involucrata*. Mr. Keen has observed their arrival at Massett for six years as follows: 1891, April 6; 1892, April 21; 1893, April 29; 1894, April 2; 1895, April 11; 1896, April 15.

Empidonax difficilis Baird. Western Flycatcher.

Rather common, but very shy and difficult to secure. Represented by two adult females from Cumshewa Inlet. Mr. Keen has noted its spring arrival at Massett as follows: 1892, May 15; 1894, May 20.

Cyanocitta stelleri carlottæ subsp. nov. Queen Charlotte Jay.

Type from Cumshewa Inlet, Moresby Island, Queen Charlotte Islands, British Columbia. ♂ ad., No. 166822, U. S. Nat. Mus., Biological Survey collection. Collected June 17, 1900, by W. H. Osgood and E. Heller. Orig. No. 400.

Characters.—Similar to *C. stelleri*, but larger and darker colored; abdomen and flanks deep Berlin blue instead of Antwerp or China blue as in *C. stelleri;* frontal spots much reduced; black of head extending on breast and merging into blue of abdomen without sharp demarcation.

Color.—Head, neck, and back deep dull black (bluish black in some specimens and very slightly brownish in others); forehead with very slight blue spots or immaculate; upper parts of wings and tail deep Berlin blue; inner secondaries and tip of tail with black bars of varying distinctness; rump and upper tail-coverts Paris blue; throat and neck black or brownish black; breast Berlin blue becoming slightly paler on flanks and crissum.

Measurements.—Type: Total length 350; extent 483; (length and extent measured in flesh) wing 153; tail 155; exposed culmen 32.5; tarsus 49. Average of three adult males from the type locality: Wing 155; tail 154; culmen 32; tarsus 49. (Average of six adult males of *C. stelleri* from Puget Sound: Wing 147; tail 147; culmen 30; tarsus 45.)

Remarks.—The large size and dark color of this jay were noticed in the field, and subsequent comparison of specimens in the museum showed these characters to be amply sufficient to distinguish it from the mainland form *C. stelleri*. It is accorded only subspecific rank because its derivation from the mainland form is scarcely to be doubted, and because individual variation in *C. stelleri* occasionally approaches the condition of *C. s. carlottæ*. It is represented in the collection by four adult and four immature birds, all but one of which were taken about Cumshewa Inlet. Jays are not very common on the islands. They were seen only occasionally and were generally in family parties of four to six adults and young.

Corvus corax principalis Ridgway. Northern Raven.

Very abundant. During June the majority of those seen were young birds of the year which were easily distinguishable by their juvenile manners and ludicrous colloquial attempts. I frequently watched them feeding on crabs. The general method of procedure seemed to be for one raven to catch a small crab on the shore and then retire to a log or the top of a stump, a few rods back in the forest, to eat it. While he was enjoying the tidbit several of his companions would perch in the trees near by preening themselves and making vigorous comments now and then until it was time to return to the beach for another morsel.

Corvus caurinus Baird. Northwest Crow.

Not common. A flock of about thirty was seen several times near the head of Cumshewa Inlet.

Pinicola enucleator flammula (Homeyer). Kadiak Pine Grosbeak.

A small flock of pine grosbeaks was seen in Cumshewa Inlet June 16, and one immature male was taken. No others were seen during our stay.

Loxia curvirostra minor (Brehm.). American Crossbill.

Large flocks of crossbills were seen frequently, but as none came within range of our guns, no specimens were secured. Mr. Keen reports this species from Massett.

Loxia leucoptera Gmelin. White-winged Crossbill.

Mr. Keen found this species at Massett. We were unable to distinguish species among the many crossbills that we saw in other parts of the islands.

Spinus pinus (Wilson). Pine Siskin.

Heard occasionally; no specimens taken. Seen at Massett by Mr. Keen.

Passerina nivalis (Linn.). Snowflake.

Seen at Massett by Mr. Keen.

Ammodramus sandwichensis alaudinus (Bonap.). Western Savanna Sparrow.

Reported from Massett by Mr. Keen; not seen by us.

Calcarius lapponicus alascensis Ridgw. Alaska Longspur.

Seen at Massett by Mr. Keen.

Zonotrichia coronata (Pallas). Golden-crowned Sparrow

Said to occur at Massett (Keen).

Junco hyemalis oregonus (Towns.). Oregon Junco.

Not common. It was very seldom seen near the coast and but few were noticed on the mountains. Although great pains were taken to secure every specimen seen, our total was but seven, and three of these were immature birds. If the junco that breeds at Sitka be considered typical *oregonus*, the Queen Charlotte birds are easily referable to this form. They seem to be identical in color, and the measurements differ too slightly to be of consequence.

Melospiza melodia rufina (Bonap.). Sooty Song Sparrow.

Very abundant. Their favorite haunts are the dense *Rubus* thickets along the shore, whence they occasionally wander out on the rocks and sandy beaches in search of insects and sand fleas. Hour after hour they sit swinging on the slender topmost twigs of the salmonberry bushes and look out over the water while they pour forth a jubilant ringing song. In some of the few open grassy places they were particularly numerous, and in skulking through the weeds frequently came to grief by encountering our mouse traps.

A nest which Heller found in Cumshewa Inlet June 24, was situated on the ground in a bunch of weeds near the water's edge. It contained two fresh eggs, which dissection of the female bird showed to be a complete clutch, though undoubtedly a second laying, as fledged young were abundant at that time. Another nest, which I stumbled upon near Skidegate July 14, was placed in much the same kind of situation and contained three fresh eggs. These eggs are slightly smaller than those of *Melospiza m. insignis* in the National Museum, but otherwise very similar. They measure as follows: 22.6 x 16.7, 22.7 x 16.8, 23 x 16.7, 22.8 x 16, 22 x 16.4.

I have seen very few specimens of typical *M. m. rufina* from Sitka, but have little hesitancy in referring the Queen Charlotte bird to this form. There seems to be no appreciable difference in color and very little, if any, in size. The measurements of 12 males from the Queen Charlotte Islands average as follows: Wing 73, tail 69.5, exposed culmen 14.7, tarsus 24.6. Average of 6 females: Wing 68.5, tail 64, exposed culmen 14, tarsus 23.7.

Passerella iliaca townsendi (Nutt.). Townsend Fox Sparrow.

Common, but, as usual, exceedingly shy. Occasionally a bird would be seen pouring out a wealth of song from the top of an alder or willow near the shore, but more frequently they skulked away through the brush before one could get a fair sight of them. Represented by 10 specimens, 6 adult and 4 immature. These are not identical with breeding birds from Sitka, and perhaps should be considered intermediate between *townsendi* and *fuliginosa*. The young particularly are more dusky than young from Sitka. In the adults the spotting on the lower parts is heavier and duskier and in general there is less of the deep rufescent shades than in typical *townsendi*.

Hirundo erythrogastra Bodd. Barn Swallow.

A few barn swallows were always found about the numerous deserted Indian villages and their nests were frequently noticed on the big cedar beams which are the framework of the Haida houses. Only one specimen was secured, and this is evidently not full grown, as the tail is not as long nor the color as rich as in the fully adult western birds I have examined.

Tachycineta bicolor (Vieill.). Tree Swallow.

One or two swallows supposed to be this species were seen among the barn swallows in Cumshewa Inlet. Mr. Keen reports it from Massett, and has noted the time of its spring arrival as follows: 1891, April 30; 1892, April 24; 1893, May 12; 1895, May 3; 1896, April 7; 1897, April 15; 1898, April 27.

Tachycineta thalassina (Swains.). Violet-green Swallow.

A bird thought to be this species was seen by Heller in Cumshewa Inlet June 30.

Helminthophila celata lutescens (Ridgw.). Lutescent Warbler.

Occasionally seen or heard. Two specimens were taken in Cumshewa Inlet June 15.

Dendroica æstiva rubiginosa (Pallas). Alaska Yellow Warbler.

Rather rare; seen twice in Cumshewa Inlet. Reported by Mr. Keen from Massett.

Dendroica townsendi (Towns.). Townsend Warbler.

One taken in Cumshewa Inlet June 15, and five at Skidegate July 14; very few others seen. Mr. Keen found it at Massett, and noted its spring arrival there as follows: 1891, May 30; 1893, April 28; 1894, May 15; 1896, April 20; 1898, April 17.

Wilsonia pusilla pileolata (Pallas). Pileolated Warbler.

Two specimens were taken and several seen in the mountains near the head of Cumshewa Inlet June 22–24. They were not seen elsewhere, but the species is noted in Keen's Massett list.

? Anthus pensilvanicus (Latham). American Pipit.

A bird thought to be this species was seen on a snow field in the mountains of Moresby Island June 23.

Cinclus mexicanus Swains. American Dipper.

A dipper was seen and heard several times along a stream emptying into West Arm of Cumshewa Inlet.

Anorthura hiemalis pacifica (Baird). Western Winter Wren.

Very common on all the islands. It is practically the only bird to be found in the deep forest away from the seashore. On the occasions when we attempted to penetrate the labyrinth of undergrowth toward the interior of the islands, we were always greeted, even in the darkest places, by the tiny wren's bright bubbling song or scolding chatter. It is always in motion and utterly regardless of the weather. During continuous rains while we were camped at the head of Cumshewa Inlet a wren would appear every few hours near the front of the tent and, after scolding us for awhile, move on through the wet brush cheerfully and oblivious of the drenching rain.

I flushed a bird from an empty nest in the upturned roots of a large fallen cedar June 15. I visited this nest frequently and flushed the bird from it each time, but up to June 28 it still contained no eggs.

Four specimens only were collected, two adults and one young from Cumshewa Inlet, and one young from Skidegate. These do not differ from specimens from the adjacent mainland of British Columbia and from Puget Sound near the type locality of *Anorthura h. pacifica.*

Certhia familiaris occidentalis Ridgw. Western Creeper.

One specimen was taken and several others were seen in Cumshewa Inlet June 20.

Sitta canadensis Linn. Red-breasted Nuthatch.

Two specimens were taken in Cumshewa Inlet June 18 and June 22, respectively. No others were seen during our stay.

Parus rufescens Towns. Chestnut-backed Chickadee.

Abundant. Seven specimens were taken.

Regulus satrapa olivaceus Baird. Western Golden-crowned Kinglet.

Common. An adult male was taken in Cumshewa Inlet June 20.

Regulus calendula grinnelli Wm. Palmer. Sitka Kinglet.

Reported by Mr. Keen. Not seen by us.

Hylocichla ustulata (Nutt.). Russet-backed Thrush.

Common. Eight specimens were taken in various parts of the islands. It was very abundant at Clew on the north side of Cumshewa Inlet, but was not seen at all at our camp at the head of the inlet, where we found *H. a. verecunda.* Mr. Keen found it common at Massett, and noted its annual arrival for seven years as follows: 1891, May 29; 1892, May 23; 1893, May 17; 1894, May 19; 1895, April 25, 1896, April 11; 1898, April 26.

Hylocichla aonalaschkæ verecunda Osgood.[1] Coast Hermit Thrush.

Rather rare. Two adult females were taken at the head of Cumshewa Inlet, and one male at Prevost Island. These Queen Charlotte specimens have the extreme development of the characters of this form, being rich brownish olivaceous, although in breeding plumage.

Merula migratoria propinqua Ridgw. Western Robin.

Common. No specimens preserved. Mr. Keen notes the spring arrival of the robin at Massett as follows: 1891, March 12; 1892, March 16; 1893, March 6; 1894, February 20; 1895, March 1; 1896, February 21; 1898, February 24.

Hesperocichla nævia (Gmelin). Varied Thrush.

Occasionally seen or heard. Seen at Massett (Keen).

---

[1] Auk, XVIII, 183, April, 1901.

NATURAL HISTORY OF THE COOK INLET REGION, ALASKA.

By Wilfred H. Osgood.

INTRODUCTION AND ITINERARY.

The region about Cook Inlet was, at the beginning of the field season of 1900, the only general district of consequence on the Pacific coast of Alaska that had not been recently visited by naturalists. The important bearing which collections from this region would have on problems connected with the general natural history of Alaska was strongly realized, and, accordingly, after the completion of work on the coast farther south, I was directed to proceed to Cook Inlet and make as thorough a biological reconnoissance of the region as time and circumstances would permit. On this trip, as earlier in the season, I had the efficient assistance of Mr. Edmund Heller. We entered the region August 21, making stops of a few hours each at Seldovia and Homer on the southwestern end of the Kenai Peninsula. From Homer we continued up the inlet and into Turnagain Arm, and landed at the mining camp of Hope City August 23. The lower coast country about Hope occupied us until August 31, when we moved on into the mountains at the head of Bear Creek, a medium-sized stream that empties into Turnagain Arm near Hope. A week later we left Turnagain Arm for the northwest side of the inlet at Tyonek, and there spent the remaining time from September 13 to September 28. From this it may be seen that most of the work was done in but two general localities, the vicinity of Hope and the vicinity of Tyonek. Short stops at Seldovia, Homer, Kenai, and Sunrise, however, were of considerable value, and information received from prospectors gave some general information about the Knik and Sushitna districts. If more time had been available it could doubtless have been spent profitably in these districts. The vicinity of Seldovia also seemed promising, but we were obliged to pass it by on our way into the inlet and could not return to it.

PHYSIOGRAPHY.

Cook Inlet is the first important indentation of the Alaskan coast east of the Alaska Peninsula. It is a long narrow inlet bifurcated at its upper end into two large arms, Knik Arm and Turnagain Arm. The first of these, Knik Arm, is about 15 miles long, and at its upper end receives the waters of a large stream, the Matanuska. The other, Turnagain Arm, is 30 miles or more in length, and extends inland until within about 5 miles of the waters of Prince William Sound. West of Knik Arm is the delta of the Sushitna River, the largest stream emptying into the inlet. South of Turnagain Arm, and connected with the mainland only by the 5 miles of glacier between the head of the arm and Prince William Sound, is the Kenai Peninsula Numerous relatively small streams enter both sides of Turnagain Arm and both sides of the main inlet as well, so that in addition to the great volume received from the Sushitna there is a large secondary supply of fresh water. This abundance of fresh water, much of which carries quantities of silt in suspension, makes the inlet unsuitable for an extensive marine fauna. The tides are very strong and the rise and fall very great, particularly in Knik and Turnagain Arms, where the flood is accompanied by a bore. Navigation by either large or small craft is difficult and often dangerous.

Except in Turnagain Arm, the country bordering Cook Inlet is low and comparatively level, though high mountains from 10 to 60 miles inland can be seen on all sides. The upper end of the main inlet, in the region of the Sushitna delta, is of course low and more or less swampy. The east shore along the Kenai Peninsula, from the mouth of Turnagain Arm nearly to Kachemak Bay, is also low and comparatively flat, but is for the most part heavily wooded. The northwest side from Mount Sushitna, near the mouth of the Sushitna River, westward for nearly 100 miles consists of a slightly rolling coastal plain, varying possibly from 20 to 60 miles in width. This country is broken here and there by rather sluggish streams, most of which head in the mountains farther back or in the small lakes which abound between the coast and the mountains. The small trading station and native village of Tyonek is situated on a low sandspit at the base of this plain about 20 miles west of the mouth of the Sushitna. About Turnagain Arm the mountains reach the coast, and except for a few small tide flats at the mouths of relatively narrow valleys, the shore is steep. Hope City, in the vicinity of which our work in Turnagain Arm was done, is situated at the mouth of Resurrection Creek, a stream of sufficient size to have made quite a wide cut through the mountains. On each side of Resurrection Creek rugged mountains rise to an altitude of 5,000 feet or more and from their canyons many small rushing streams pour into Resurrection Creek or Turnagain Arm near Hope. Bear Creek (see Pl. VI, fig. 2) and Palmer Creek, which come from the east side, are the most important of these.

Although the region north and east of Cook Inlet is exceedingly mountainous and quite imperfectly known, it is evident that there is a gap of some consequence between the Coast or Saint Elias Range, which practically culminates in the upper Matanuska region, and the

FIG. 1.—PEAT BOG AND MIXED WOODS NEAR TYONEK.

FIG. 2.—LOOKING TOWARD TURNAGAIN ARM FROM HEAD OF BEAN CREEK.

so-called Alaskan Range which lies north and northwest of the inlet and includes the lofty peak of Mount McKinley. The effectiveness of this gap in its relation to the geographic distribution of animals and plants is of great interest.

FLORA.

The flora of the Cook Inlet region is quite different in its general character from that of the coast farther south, although many species are common to both regions. The difference is largely in the reduction of the number of coniferous trees in the Cook Inlet region and the corresponding increase in deciduous trees; but other features somewhat transitional between the heavy saturated forest of the southern coast and the treeless tundra of the north are numerous. The flora of the mountainous district about Turnagain Arm is, of course, different from that of the coastal plains of other parts of the inlet. The low country near Hope consists of a grassy tide flat, about 50 acres in extent, and a few miles of forest and occasional small swamps along the lower part of Resurrection Creek. Balsam poplars, paper birches, alders, and willows abound near the streams, and spruces (*Picea canadensis* and *Picea sitchensis*) and hemlocks (*Tsuga mertensiana*) are common on the slopes and slightly elevated flats. A third species of spruce (*Picea mariana*) is found in the small peat bogs, where smaller Hudsonian plants, such as Labrador tea (*Ledum*), crowberry (*Empetrum*), and dwarf birch (*Betula glandulosa*) are in profusion. The hemlock is much the most abundant of the large trees, but it is exceeded in individual size by the spruces. The conifers ascend the mountain slopes to about 2,000 feet but above that point rapidly disappear. Beyond this elevation are alder thickets, small patches of dwarf willows and birches, and vast stretches of waving grass from 1 to 3 feet high. Still higher, the slopes and rounded backs of the ridges are cushioned with a mass of heather and heather-like shrubs, chiefly *Empetrum nigrum*. This extends up to an approximate altitude of 5,000 feet, above which there is very little or no plant growth. The whole country is characterized by the abundance of high grass; otherwise it is a typical Hudsonian-Alpine region.

The flora on the northwest side of the inlet in the vicinity of Tyonek is somewhat different in character. With the exception of considerable areas occupied by lakes and peat bogs, the whole country is covered with comparatively open forest (see Pl. VI, fig. 1). Deciduous trees greatly outnumber conifers, of which but two species occur, *Picea canadensis* and *Picea mariana*, and one of these, *P. mariana*, is quite rare and local. The paper birch (*Betula papyrifera*) is by far the most abundant tree, and next in rank are the poplars, of which there are two species, *Populus balsamifera* and *Populus tremuloides*. Alders and willows are found along the streams and sparingly through the forest. The underbrush is not heavy; it consists mainly of *Menziesia* and *Viburnum*, with an occasional clump of devil's club (*Echinopanax*) in wet places. Long grass grows luxuriantly in numerous pretty open glades in the birch woods. The September aspect of the forest is very attractive. From a little distance the birches on the low, rolling slopes appear as a mass of golden and rusty yellow, punctured here and there by the dark-green spruce tops. The foliage of many of the smaller plants, such as *Viburnum*, *Cornus*, *Ribes*, and *Epilobium*, is bright red, and adds greatly to the general effect. On the whole, it reminds one very much of the autumn woods of New England, and is quite unlike anything I have seen elsewhere in Alaska.

Unfortunately, we made no collection of plants in the Cook Inlet region, hence an authoritative list can not be given here. The following list, with brief annotations copied from my field notes, gives a general idea of the important trees and woody plants that occur. Specimens of a few species were preserved, and these have been identified by Mr. Frederick V. Coville; the remainder are field identifications only.

Tsuga mertensiana. Alpine Hemlock.

This is the most abundant tree from the seacoast to timberline all about Turnagain Arm. It was not found elsewhere in the Cook Inlet region.

Picea sitchensis. Sitka Spruce.

A few trees of this species were found at Hope. Specimens preserved.

Picea canadensis. White Spruce.

Found at all points visited, including Homer, Hope, and Tyonek. It is very common about Turnagain Arm, and is found on the mountains up to an altitude of about 2,000 feet. It is practically the only conifer to be found at Tyonek. Specimens preserved.

Picea mariana. Black Spruce.

Found in limited numbers in peat bogs at Hope, Sunrise, and Tyonek. Specimens preserved.

Empetrum nigrum. Black Crowberry.

This is by far the most common heather-like shrub. It abounds in all the peat bogs in the low country, and there are miles and miles on the mountains where one could not walk without treading on it.

Populus balsamifera. Balsam Poplar.

Very abundant. Large groves stand on the flat near the mouth of Resurrection Creek and trees of smaller stature are numerous in all the Turnagain Arm country; also abundant at Tyonek.

Populus tremuloides. Aspen.

A few trees supposed to be this species were found with the balsam poplars at Tyonek. Not seen elsewhere.

Salix sp. Willow.

Four or more species of willows occur in the Cook Inlet region, including several dwarf species only found above or near timberline.

Alnus sinuata. Alder.

Abundant all about the inlet. In the low country it is found along streams, and on the mountains it forms into dense thickets extending in altitude far above the coniferous trees.

Betula papyrifera. Paper Birch.

Abundant all about the inlet. Its vertical range is about coextensive with that of the conifers. At Tyonek it surpasses all other trees in point of numbers and grows to a slightly larger size than at Hope. Judging from the size of some of the Indian birch baskets trees a foot or more in diameter are to be found.

Betula glandulosa. Dwarf Birch.

Common. In the low country it is most common in peat bogs; high on the mountains it is found on open slopes in company with the dwarf willows.

Ledum grœnlandicum. Labrador Tea.

Rather common, but more or less confined to peat bogs and wet heather meadows. Specimen preserved.

Ledum palustre. Dwarf Labrador Tea.

Less common than the preceding species, with which it is found.

Menziesia ferruginea. Menziesia.

A large percentage of the underbrush is composed of this species. It was found at all points visited, but most commonly at Tyonek.

Phyllodoce glanduliflora. Heather.

Found in limited quantities above 2,000 feet altitude in the higher mountains near Hope. Specimen preserved.

Cassiope tetragona. Cassiope.

Rather rare. It was occasionally found in the beds of *Empetrum* in the high mountains near Hope. Specimen preserved.

Cassiope stelleriana. Cassiope.

Quite common above timberline in the mountains near Hope. Specimen preserved.

Arctostaphylos uva-ursi. Bearberry.

A few plants were found on a rocky point near Hope. It was not observed elsewhere about the inlet.

Vaccinium vitisidæa. Mountain Cranberry.

Very abundant from the coast to the upper limit of plant growth.

Vaccinium sp. Huckleberry.

Several species are abundant.

Sorbus sambucifolia. Mountain Ash.

Common, both in the mountains near Hope and in the low hills at Tyonek.

Viburnum pauciflorum. Highbush Cranberry.

Very common at all points visited.

Sambucus racemosus. Elderberry.

Abundant in the mountains near Hope; occasionally seen near Tyonek.

Cornus canadensis. Bunchberry.

Excessively abundant.

Echinopanax horridum. Devil's Club.

A few clumps of under-sized devil's club were occasionally found in damp shady places about Cook Inlet.

Ribes laxiflorum. Blue Currant.

Occurs sparingly in Turnagain Arm.

Ribes rubrum. Red Currant.

Quite common in Turnagain Arm.

Rosa acicularis. Wild Rose.

Abundant all about the inlet; especially so at Tyonek.

Amelanchier alnifolia. Serviceberry.

A single bush was found at Tyonek. The species was not seen elsewhere about the inlet.

Rubus strigosus. Raspberry.

Abundant at Hope and Tyonek.

Spiræa betulæfolia? Spiræa.

Abundant.

FAUNA.

The mammals of the Cook Inlet region are essentially the same as those of the interior of Alaska. Nearly all the species of the lower

Yukon Valley are found among them, and none show any marked peculiarities not possessed in their interior habitat. With the exception of widely distributed species, such as the black bear, no species are common to the Cook Inlet region and the Sitkan region. Thus, while the mammals of Cook Inlet are not peculiar to the region, the mammal fauna, as a whole, is peculiar, as contrasted with that of the coast farther south. Two new species, *Microtus miurus* and *Sorex eximius*, were found, but both are new, not only to Cook Inlet, but to Alaska as well, and will undoubtedly be found in other parts of the Territory. Considering the latitude, both large and small mammals are numerous in species as well as individuals. Moose, bear, and mountain sheep are the principal big game, and although they have already been hunted to a considerable extent, it is probable that they are more abundant than in any equally accessible place in North America. Fur-bearing animals are well represented, but, as elsewhere in the North, have been much reduced in numbers. The smaller, less conspicuous mammals are such as are generally found throughout northern Alaska, and are well represented on account of the varied conditions offered by the mountains of the Kenai Peninsula and the low country on the northwest side of the inlet. Our collection of mammals from Cook Inlet numbers 240 specimens, the majority of which, of course, are species of small size, such as shrews and mice, since we made no special effort to secure big game.

Birds were not found in great numbers. Owing to the lateness of the season at the time of our arrival in the inlet, those seen were permanent residents or fall stragglers, the summer residents being missed almost entirely. Land birds, with the exception of grouse, which were fairly common, were not numerous in species or individuals. Water birds, particularly littoral or semi-pelagic forms, are noticeably uncommon, probably on account of the brackish water of the inlet and the comparative absence of marine invertebrates. Ducks and geese, however, and birds which feed in fresh water are locally quite abundant. As in the case of the mammals, no birds are peculiar to the Cook Inlet region, but several interior species are found which do not occur on the Alaskan coast south of Cook Inlet.

The only other land vertebrate is a frog, collected by Heller at Tyonek. The species has very kindly been determined by Dr. L. Stejneger as *Rana cantabrigensis latiremis*. The land vertebrates may be summed up as follows:

MAMMALS.

1. Rangifer stonei.
2. Alces gigas.
3. Ovis dalli.
4. Oreamnos kennedyi.
5. Sciuropterus sp.
6. Sciurus hudsonicus.
7. Spermophilus empetra subsp.
8. Arctomys caligatus.
9. Castor canadensis.
10. Evotomys dawsoni.
11. Microtus operarius kadiacensis.
12. Microtus miurus.
13. Fiber spatulatus.
14. Synaptomys dalli.

15. Zapus hudsonius alascensis.
16. Erethizon epixanthus myops.
17. Ochotona collaris.
18. Lepus americanus dalli.
19. Lynx canadensis.
20. Canis occidentalis.
21. Vulpes kenaiensis.
22. Ursus americanus.
23. Ursus middendorffi.
24. Lutra canadensis.
25. Lutreola vison.
26. Putorius kadiacensis.
27. Putorius rixosus.
28. Mustela americana.
29. Gulo luscus.
30. Sorex personatus.
31. Sorex alascensis.
32. Sorex eximius.
33. Myotis lucifugus.

BIRDS.

1. Gavia imber.
2. Gavia lumme.
3. Uria troile californica.
4. Stercorarius parasiticus.
5. Rissa tridactyla pollicaris.
6. Larus sp.
7. Larus philadelphia.
8. Sterna paradisaea.
9. Diomedea albatrus.
10. Phalacrocorax pelagicus.
11. Anas boschas.
12. Dafila acuta.
13. Aythya marila nearctica.
14. Somateria v-nigra.
15. Oidemia perspicillata.
16. Branta canadensis subsp.
17. Olor columbianus.
18. Ardea herodias.
19. Grus canadensis.
20. Phalaropus lobatus.
21. Gallinago delicata.
22. Macrorhamphus griseus scolopaceus.
23. Tringa couesi.
24. Tringa bairdi.
25. Tringa alpina pacifica.
26. Ereunetes occidentalis.
27. Limosa haemastica.
28. Totanus melanoleucus.
29. Actitis macularia.
30. Numenius hudsonicus.
31. Squatarola squatarola.
32. Canachites canadensis osgoodi.
33. Lagopus rupestris.
34. Lagopus leucurus.
35. Circus hudsonicus.
36. Accipiter atricapillus striatulus.
37. Haliaeetus leucocephalus alascanus.
38. Falco columbarius.
39. Bubo virginianus saturatus.
40. Nyctea nyctea.
41. Ceryle alcyon.
42. Picoides americanus fasciatus.
43. Selasphorus rufus.
44. Contopus borealis.
45. Pica pica hudsonica.
46. Cyanocitta stelleri.
47. Perisoreus canadensis fumifrons.
48. Corvus corax principalis.
49. Scolecophagus carolinus.
50. Loxia curvirostra minor.
51. Loxia leucoptera.
52. Acanthis linaria.
53. Spinus pinus.
54. Calcarius lapponicus alascensis.
55. Ammodramus sandwichensis alaudinus.
56. Zonotrichia leucophrys gambeli.
57. Zonotrichia coronata.
58. Spizella monticola ochracea.
59. Junco hyemalis.
60. Melospiza melodia kenaiensis.
61. Melospiza lincolni.
62. Passerella iliaca annectens.
63. Lanius borealis.
64. Helminthophila celata lutescens.
65. Dendroica coronata.
66. Dendroica striata.
67. Anthus pensilvanicus.
68. Cinclus mexicanus.
69. Certhia familiaris montana.
70. Parus atricapillus septentrionalis.
71. Parus hudsonicus.
72. Regulus satrapa olivaceus.
73. Regulus calendula.
74. Hylocichla ustulatus almae.
75. Hylocichla aonalaschkae.
76. Merula migratoria.
77. Hesperocichla naevia.

BATRACHIAN.

Rana cantabrigensis latiremis.

Two zones are evident in the Cook Inlet region, the Hudsonian and the Arctic-Alpine. All the low country about the inlet and also the mountain sides up to timberline may be considered as Hudsonian, and the region above timberline on the mountains as Arctic-Alpine. The Hudsonian region has the same general features as the great interior transcontinental Hudsonian belt, and is doubtless imperfectly connected with it. This Hudsonian belt is not particularly marked by characteristic forms, since most of the species of plants and the genera of mammals and birds are also found in the Canadian zone; but its distinction consists in the absence of many of the forms which are characteristic of the Canadian zone or which range from the south up into that zone. A notable feature of the Hudsonian flora of Cook Inlet, which is to some extent an exception to the statement just made, is the abundance of *Tsuga mertensiana* at sea level on the shores of Turnagain Arm. This tree is exceedingly characteristic of the Hudsonian zone, and except at this point has been found only high on mountains in the vicinity of timberline, as its name, the alpine hemlock, implies. The other large trees of Cook Inlet, *Picea canadensis*, *Picea sitchensis*, *Picea mariana*, *Populus balsamifera*, *Populus tremuloides*, *Alnus sinuata*, and *Betula papyrifera*, are such as are generally found in a northern Hudsonian zone, but all are also found in the Canadian. Such trees as *Pinus*, *Abies*, *Thuja*, etc., which are represented in the Canadian zone of the Sitkan district, are entirely absent in Cook Inlet. The mammals and birds of the Hudsonian district of Cook Inlet are, like the trees, nearly all species which are found in the Hudsonian of the interior of Alaska, but which also range, at least to some extent, into the Canadian. All the genera of mammals belong to this category, as well as many species, such as *Sciurus hudsonicus*, *Evotomys dawsoni*, *Fiber spatulatus*, *Synaptomys dalli*, *Erethizon epixanthus myops*, *Putorius rixosus*, and *Sorex personatus*.

The Arctic-Alpine district includes the summits of nearly all the mountains on both sides of Turnagain Arm (see Pl. VII), and in the interior of the Kenai Peninsula. Spruce and hemlock timber ceases between 2,000 and 3,000 feet elevation, and the higher slopes are clothed only with matted masses of low shrubs or wide expanses of tall grass. In the gulches thickets of alders hold control and a few stunted individuals often straggle well up toward the snow line. The characteristic mammals of this Arctic-Alpine district are the Dall mountain sheep (*Ovis dalli*), the hoary marmot (*Arctomys caligatus*), and the Alaska mountain vole (*Microtus miurus*). The only characteristic birds found at the time of our work were the ptarmigans (*Lagopus rupestris* and *L. leucurus*), but pipits (*Anthus*), golden-crowned sparrows (*Zonotrichia*), and rosy finches (*Leucosticte*) doubtless occur in the breeding season.

Taken as a whole, the plant and animal life of Cook Inlet is very closely similar to that of the Yukon Valley, or, in more general terms, to that of the interior of Alaska. This condition is the more noteworthy, since the fauna and flora of the same coast south of Cook Inlet are in marked contrast to those of the interior in the same latitude. Since coast influences are usually conducive to life that is relatively more boreal than that of the interior, large faunal regions of the interior seldom extend to the actual coast, except with considerable modification. This is true in most cases even when no immense masses of mountains separate coast and interior as they do in the Pacific region from southern British Columbia northwestward. The contrast, however, would be less if no mountains intervened, or if low passes permitted communication; moreover, the climatic conditions of coast and interior would approximate each other more closely in a northern latitude than in a southern. It seems probable, therefore, that this contrast in plant and animal life is minimized in the Cook Inlet region both on account of the northern latitude and the existence of some degree of connection with the interior.

PREVIOUS WORK.

Very little natural history work has been done in the Cook Inlet region. In 1869, Ferdinand Bischoff made a small collection of birds and mammals at Fort Kenai which was sent to the U. S. National Museum; but though casual references to individual specimens have occasionally appeared, no account of the collection, as a whole, has been published. The entire collection is recorded in the catalogues of the Museum, but many of the specimens have been exchanged or distributed to educational institutions; enough still remain, however, to be of considerable value in making a faunal list. A few species of birds from Cook Inlet have been recorded by Dr. Tarleton H. Bean,[1] who made brief stops about the mouth of the inlet while connected with an expedition of the U. S. Coast and Geodetic Survey, and the specimens collected are deposited in the U. S. National Museum. A few specimens were also taken near the mouth of the inlet by Messrs. C. H. Townsend and B. W. Evermann during a brief stop of the U. S. Fish Commission steamer *Albatross*. Numerous sportsmen have, in recent years, been attracted by the large game in the vicinity of the inlet and in some cases have published accounts[2] of their trips containing many valuable notes on the natural history and general features of the region. The most prominent of these are Messrs. Dall De Weese and Andrew

[1] Proc. U. S. Nat. Museum, V, 144–173, 1882.

[2] Many narratives of hunting trips in Alaska, particularly about Cook Inlet, may be found in Recreation, Forest and Stream, and American Field.

FIG. 1.—PEAK AT HEAD OF BEAR CREEK.

Habitat of *Arctomys caligatus* and *Microtus miurus*.

FIG. 2.—CANYON OF UPPER BEAR CREEK.

J. Stone. Mr. De Weese collected and preserved an excellent series of moose[1] and Dall sheep for the U. S. National Museum, and Mr. Stone secured many fine specimens, including the type of *Rangifer stonei*,[2] for the American Museum of Natural History, New York.

The topography and geology of the region have been studied by parties from the U. S. Geological Survey and in their reports may be found a few notes regarding animals and plants as well as much other matter of general interest.[3]

The report of the War Department on the 'Sushitna Expedition,' under Capt. E. F. Glenn, also contains numerous general notes of value.[4]

MAMMALS OF THE COOK INLET REGION.

Alces gigas Miller. Alaska Moose.

According to report the moose has but recently appeared in the Cook Inlet region; the older Indians say no moose were there when they were boys; and even within the memory of white men it has moved westward, now being known as far out on the Alaska Peninsula as Katmai.

It is quite common in many places about Cook Inlet, but is hunted most successfully in the Knik district, and on the north shore of the Kenai Peninsula, from Kussilof and Fort Kenai to Point Possession at the mouth of Turnagain Arm. A few Indians hunt moose here practically all the year round, making a living by selling the meat in the mining camps of Hope and Sunrise. Several carcasses were brought in during our stay and the meat was quickly sold at 10 cents a pound. On the northwest side of the inlet moose are less common than on the Kenai Peninsula, but occur sparingly.

Rangifer stonei Allen. Stone's Caribou.

Rangifer stonei Allen, Bul. Am. Mus. Nat. Hist. New York, XIV, 143–148, May 28, 1901.

Caribou are rare on the Kenai Peninsula. I saw a pair of weather-beaten antlers said to have been picked up on the peninsula side near the mouth of Turnagain Arm and heard an unsatisfactory report of the killing of a large buck, but beyond this could obtain no evidence of the animal's occurrence in this region near the coast. Mr. Stone, who secured the type of the species, also received reliable reports of

[1] See Miller, Proc. Biol. Soc. Wash., XIII, 57–59, May 29, 1899.

[2] See Allen, Bul. Am. Mus. Nat. Hist., XIV, 143–148, 1901.

[3] See Dall, Report on Coal and Lignite. <17th Ann. Rept. U. S. Geol. Survey, Pt. I, pp. 771–908, 1896; Neocene of North Am., <Bul. No. 84, U. S. Geol. Survey, pp. 234–238, 1892; Eldridge, Reconnaissance Sushitna Basin, etc., <20 Ann. Rept. U. S. Geol. Survey, Pt. VIII, pp. 6–29, 1900; Mendenhall, Reconnaissance Resurrection Bay to Tanana River, Ibid, pp. 271–340.

[4] Reports of explorations in the Territory of Alaska, Adjt. Gen. Office, Doc. No. 102, pp. 5–289, 1899.

the occurrence of caribou in the southern and western part of the Kenai Peninsula, but stated that they are "already very scarce and will doubtless soon be exterminated." (See Allen, loc. cit.) Two specimens, male and female, shot by Mr. Harry E. Lee on the Kenai Peninsula, have been recorded by Mr. D. G. Elliot.[1] They are more or less common a short distance in the interior and are often killed near the Sushitna River, whence their skins are brought to the coast to be traded.

The characters which distinguish Stone's caribou from the mountain caribou (*Rangifer montanus*) seem to be slight, and the claim of *stonei* to full specific rank has been questioned. The statement in this connection that "it is very evident that our knowledge of western and northwestern caribou is very imperfect and unsatisfactory, our material having been altogether insufficient" is not only true, but should be very significant; for if more specimens of this rare animal are not obtained for our museums in the near future, the question of its specific distinctness may never be decided beyond question.

Oreamnos kennedyi Elliot. Alaska Mountain Goat.

Oreamnus kennedyi Elliot. Field Columbian Mus., Chicago. Pub. 46, Zool. Series III, 3–5, June, 1900.

I could obtain no evidence of the occurrence of goats on any part of the Kenai Peninsula, but I learned from T. W. Hanmore, who has been the Alaska Commercial Company's agent at Tyonek for the past eleven years, that a small band is known to inhabit a district between the headwaters of the Knik and Matanuska rivers. Mr. Hanmore knows the animals thoroughly and says he has seen skins and horns from this place and often heard of them from the Indians who hunt in that vicinity. As far as I can learn, this is the northernmost occurrence of the mountain goat.

Ovis dalli (Nelson). Dall Sheep.

Dall sheep were formerly common in the mountains on both sides of Turnagain Arm, but since active mining began there they have retreated to the interior of the peninsula, where they still occur in large numbers. They are also common in the mountains near the Knik River, from which place several heads were brought in by Indians while we were at Sunrise City. According to apparently reliable report, these sheep in the interior of the Kenai Peninsula gather into very large flocks in fall, as many as three hundred individuals having been seen together at one time. There are several routes into the sheep country, the easiest and the one most frequently used being that via the Kussilof River to Kussilof Lake and thence into the mountains.

---

[1] Field Columbian Mus. Zool. Ser., III, 59–62, Pls. XI–XIII, July, 1901.

Sciuropterus sp.

Flying squirrels are said to have been taken in the Knik district, but are unknown elsewhere about Cook Inlet. The numerous miners and woodsmen about Turnagain Arm were unable to give us any information as to the occurrence of flying squirrels, except in the Knik district.

Sciurus hudsonicus Erxleben. Hudson Bay Red Squirrel.

Common at all points visited. A few were seen at Homer, larger numbers at Tyonek, and in the low country about Hope they were excessively abundant. Sixteen specimens were taken, fourteen at Hope and two at Tyonek. These are indistinguishable from specimens in the same pelage taken near the west coast of Hudson Bay, as well as from those taken in the interior of Alaska, and show but very slight tendency toward *Sciurus h. petulans;* hence they are referred without hesitation to *Sciurus hudsonicus.*

Spermophilus empetra subsp.? Ground Squirrel.

Spermophiles do not occur near Turnagain Arm or at Tyonek. They are said to be abundant on the Barren Islands near the mouth of the inlet and are evidently so in the mountains lying some 60 miles back of Tyonek. While we were at Tyonek an Indian brought in a lot of one hundred spermophile skins from these mountains to be used in making parkas and other articles of native wearing apparel.

Arctomys caligatus Eschscholtz. Hoary Marmot.

Abundant in the mountains about Turnagain Arm, where they are known to the miners as 'whistling pigs.' In the mountains at the head of Bear Creek we found them living in grassy meadows above timberline and on open hillsides rather than in rocky places. Their burrows in these places differ from those of spermophiles only in size. Wide well-beaten paths through the grass connect different burrows and diverge from them here and there over the slopes in the same manner as those of spermophiles. The vicinity of the burrow is usually very filthy with excreta.

Although the hoary marmots of these mountains seldom see a human being even at a distance, they are exceedingly shy and disappear at the slightest alarm, on which account it was very difficult to get even within rifle range of them.

Castor canadensis Kuhl. American Beaver.

According to report which I received from a trapper at Hope, three beaver were secured by Indians near the mouth of Turnagain Arm in the fall of 1899. A limited number are secured every season along streams in the mountains about 60 miles inland from Tyonek. A trading station on the lower Sushitna River also obtains a small quota annually. Compared with former receipts, however, the number now obtained is lamentably small.

Mus norvegicus Erxleben. Norway Rat.

A few rats have established themselves about the wharf and stores at Sunrise. They have also occasionally escaped from vessels at Hope and Tyonek, but have not increased in numbers.

Evotomys dawsoni Merriam. Dawson Red-backed Mouse.

Abundant in mossy places and about decayed logs in the woods; only rarely taken in *Microtus* runways on the tide flats; quite numerous about the houses and stores in the villages and in the miners' cabins in the backwoods. Thirty-eight specimens were secured as follows: Hope, 21; mountains at the head of Bear Creek, near Hope, 4; Tyonek, 13. These seem to be intermediate in character between *Evotomys dawsoni* and *Evotomys dawsoni orca*,[1] including specimens which might be referred to each form. As the majority, however, are more similar to *dawsoni*, they are recorded under that name.

Microtus operarius kadiacensis (Merriam). Kadiak Vole.

Microtus kadiacensis Merriam, Proc. Biol. Soc. Wash., XI, 222, July 15, 1897; Bailey, North Am. Fauna, No. 17, 41–42, June 6, 1900.

This vole is rather rare at Hope, but very abundant at Tyonek. It is common at Homer also, as numerous signs of it were seen on the grass-grown sandspit near there. The runways are very large and were usually found in coarse grass (*Elymus mollis*) on low, sandy stretches near tide level. Great quantities of grass cuttings were always found in the runways. In one place near Tyonek these mice had invaded a potato patch and made well-beaten paths in and out under the vines, but the amount of damage they inflicted on the crop was apparently slight.

Thirty-one specimens were taken at Tyonek and five at Hope. These agree very closely with topotypes of *M. kadiacensis*, but in some respects tend toward *M. operarius*. The molar teeth and the audital bullæ are very slightly smaller than in *kadiacensis*, yet not quite so small as in *operarius*. In size the Cook Inlet specimens are about equal to those from Kadiak and somewhat larger than topotypes of *operarius*.

Microtus miurus[2] sp. nov. Alaska Mountain Vole.

Type from head of Bear Creek in mountains near Hope City, Turnagain Arm, Cook Inlet, Alaska. ♂ ad. No. 107175, U. S. Nat. Mus., Biological Survey collection. Collected Sept. 4, 1900, by W. H. Osgood and E. Heller. Original No. 1349.

Characters.—Size small, tail short, color buffy, underparts suffused with body color, molar enamel pattern as in *Microtus abbreviatus;* skull distinctive.

---

[1] *Evotomys orca* Merriam, Proc. Wash. Acad. Sci., II, 24–25, Mar. 14, 1900.

[2] Miurus=curtailed.

Color.—Fresh fall pelage (No. 107167): Upperparts uniform pale tawny, lightly mixed with black; face, sides, and flanks exactly like back; underparts strongly washed with buffy; tail dusky above, buffy below and on sides. *Worn summer pelage* (Type, No. 107175): Upperparts pale buffy gray with plumbeous under-fur showing through in places; underparts whitish gray faintly suffused with buffy. *Young:* Similar to adults but grayer, with buff somewhat intensified about the ears and base of tail.

Skull.—Size small; dorsal outline nearly straight, very slightly depressed in interorbital region; nasals of moderate length, somewhat expanded anteriorly; naso-frontal suture slightly emarginate; lachrymal shelf prominent, with decided dorsal depressions between fronto-maxillary sutures; zygomata rounded anteriorly with scarcely any prezygomatic notch; interparietal slightly produced anteriorly, about twice as wide as long; audital bullæ full and rounded, their inner surfaces nearly parallel; lateral pits of palate shallow, interpterygoid fossa rather wide. Teeth rather light; upper incisors slightly sulcate; molar enamel pattern essentially as in *Microtus abbreviatus;* $\underline{\text{m}}^3$ with but 2 closed triangles; $\overline{\text{m}}^1$ with 5 closed triangles and 5 inner and 4 outer salient angles.

Measurements.—Type: ♂ ad. Total length 153; tail vertebræ 31; hind foot 20. Average of three females: 133; 23; 19. Skull: Basal length 26; zygomatic breadth 15; mastoid breadth 11.4; nasals 7; alveolar length of upper molar series 6.2.

Remarks.—In a general way *Microtus miurus* is a miniature of *M. abbreviatus*, which is undoubtedly its nearest known relative, but detailed differences are very numerous. External characters other than size are its relatively longer tail and its more ochraceous color; cranial characters most appreciable are its full, rounded, and nearly parallel audital bullæ and the nearly straight dorsal outline of the skull. The peculiar enamel pattern like that of *M. abbreviatus* is sufficient to distinguish *miurus* from all other members of the subgenus *Microtus.* The discovery of a mainland relative of the insular species *abbreviatus* is very interesting and indicates in a slight way how much is still to be learned of the small mammals of Alaska. In the high mountains of the interior other similar forms undoubtedly remain to be discovered. Eleven specimens were secured, including five adults and six young. These were all taken in 'meadows' above timberline in the mountains near Hope on the south side of Turnagain Arm. In these places it was only by very careful and diligent trapping that they were secured, for though many runways were found but very few showed signs of recent use. The burrows instead of opening vertically in the side of a slight eminence, as usual, have entrances which are flush with the floor of the runway. The grass in which these runs are made is very short, as it grows in rocky soil near the upper limit of vegetation in small hollows and basins. Snow lies in these places all the time except a few months in summer.

Fiber spatulatus Osgood. Northwest Muskrat.

Fiber spatulatus Osgood, N. Am. Fauna No. 19, pp. 36–37, Oct. 6, 1900.

Muskrats are not known to occur about Turnagain Arm, but they are rather common about small ponds in the peat bogs near Tyonek. They also occur at Kenai, as is shown by two specimens from there collected by Bischoff in 1869 and now in the National Museum. These, as well as one that we secured at Tyonek, are typical *Fiber spatulatus*, having the small molars and expanded nasals exactly as in the type of the species.

Synaptomys dalli Merriam. Dall Lemming Mouse.

One adult male was taken in a small peat bog near Hope August 26, but persistent trapping in the same locality failed to secure more, and none were found elsewhere about the inlet. The single specimen secured is essentially the same as an adult male from Lake Lebarge, Yukon Territory. The Cook Inlet specimen is slightly lighter, particularly on the head and shoulders, where there is less admixture of black. The skull agrees perfectly with the Yukon specimen and both agree fairly well with the type of *dalli*, all having much larger audital bullæ than the coast form *wrangeli*. The flesh measurements of the Cook Inlet specimen are as follows: Total length 131; tail vertebræ 22; hind foot 20. Skull: Basal length 25; mastoid breadth 14.8; zygomatic breadth 15.6; nasals 7.5; alveolar length of upper molar series 7.9.

Zapus hudsonius alascensis Merriam. Alaska Jumping Mouse.

A jumping mouse in good condition was found floating in a sunken water barrel near Tyonek September 13. The entire vicinity was assiduously trapped, but no more could be obtained, from which it seems that the species is rare, though possibly it may have gone into early hibernation. It also occurs in Turnagain Arm, for a miner at Hope accurately described one to me that he had seen there several years ago.

Erethizon epixanthus myops Merriam. Alaska Porcupine.

Porcupines are abundant in the Turnagain Arm region, but are very rare at Tyonek. Mr. T. W. Hanmore, who has lived at Tyonek for eleven years, says that he has seen but one porcupine there in that time. The natives on the Kenai Peninsula use porcupine flesh as food and prize it very highly. They prepare the animals by first plucking out all the quills, then singeing off the hair, then roasting entire. I did not have an opportunity to taste the flesh cooked in this way, but found it very palatable when fried. Porcupines are eaten by various carnivorous animals, particularly wolves, though doubtless only in extreme hunger. Old trappers and hunters say that the majority of the wolves taken in this region have porcupine quills in their stomachs and under the skin about their heads.

Ochotona collaris (Nelson). Alaska Pika.

Pikas do not occur in the mountains on the peninsula side of Turnagain Arm, but I was told by a miner from Knik River that they had been seen in the mountains near there.

Lepus americanus dalli Merriam. Dall Varying Hare.

A few signs of rabbits were seen at Hope, but no specimens were secured; they are said to be very abundant in winter. Six specimens were obtained at Tyonek, but only by persistent and careful trapping. All were caught in steel traps set in runways in the thickets or in the peat bogs. Rabbits are very seldom seen here in the daytime, and dependence on gun alone would result in but a small bag. Although the summer pelage of *dalli* is unknown and there is some possibility that topotypes would be slightly different in color from the Cook Inlet specimens, I have little hesitancy in referring the latter to *dalli*, since the skulls are nearly typical, being but slightly smaller and shorter. Most of the specimens from Tyonek are slightly immature and the color of the upperparts is strongly mixed with black. In one of the oldest (No. 107611) the predominating color of the upperparts is cinnamon; the middle of the back is heavily mixed with black, which becomes less intense laterally until at the edge of the pure white under parts there is no trace of it, and a clear cinnamon lateral line remains. The throat is also cinnamon with very faint signs of black. The outer sides of the fore legs and fore feet are russet and dusky and the inner sides are creamy white. The hind feet are white well mixed with cinnamon and dusky. The ears are nearly white, except in front, where there is a strong cinnamon and dusky admixture. The flesh measurements of the six specimens average as follows: Total length 441; tail vertebræ 39; hind foot 139.

Lynx canadensis (Kerr). Canada Lynx.

Lynxes are evidently still fairly common. Mr. George Coon, a reliable woodsman of Hope, told me that in a season of two and one-half months' trapping in the winter of 1899 he secured fifteen lynxes near the mouth of Turnagain Arm.

Canis occidentalis Richardson. Wolf.

Wolves are considered rather common in the Cook Inlet region. Mr. Coon, of Hope, told me that during the winter of 1899 he secured fourteen with poison. Among these were six in the black phase.

Vulpes kenaiensis Merriam. Kenai Fox.

Vulpes kenaiensis Merriam, Proc. Wash. Acad. Sci., II, 670, Dec. 28, 1900.

Red, cross, and black foxes are taken annually in limited numbers. This species, which, as far as known, is the only one found in the region, is the largest fox known to North America. The skins secured here are usually of very good quality and blacks or 'silver grays' commanding high prices are not uncommon. Some attempt at 'farming' these large foxes has been made, but so far has generally proved unsuccessful on account of the vicious habit the males have of injuring or devouring the young. It seems possible, however, that while this might interfere with such extensive breeding farms as are feasible in the case of the blue fox, it might be controlled and the business made profitable if a few pairs were kept at each of a large number of localities.

Ursus americanus Pallas. Black Bear.

Black bears are moderately common in the Cook Inlet region. A few are killed about Turnagain Arm each year, but they are less common in the lower country on the northwest side of the inlet. While returning from our traps on upper Bear Creek on the evening of September 1, we discovered a bear crossing a grassy place between two alder thickets on a hillside near us. I hurried to camp and returned with my rifle and after a long stalk succeeded in securing it. It proved to be a young female. The fur was short, but even and glossy black. The animal had been feeding on berries entirely and its stomach was found filled to distension, chiefly with black crowberries (*Empetrum nigrum*). These were clean and absolutely free from twigs and leaves and so tightly packed that there hardly seemed room for another mouthful. The feeling of satisfaction enjoyed by the possessor of this well-filled paunch was very evident. Before shooting it I had an opportunity to watch it feeding and was amused at its exhibition of exuberant spirits. It would browse leisurely for a few minutes then would suddenly give a bound and roll over and over down a little heather-grown glade to the bottom and then jump up to gallop at full speed up and down and around in a circle, apparently impelled by nothing but sheer joy. This bear measured in the flesh as follows: Total length 1,310; tail vertebræ 166; hind foot 213.

Ursus middendorffi Merriam. Kadiak Bear.

Large bears are still very often seen both on the Alaska Peninsula side of Cook Inlet and on the mountainous Kenai Peninsula. According to report they were very abundant about ten years ago, but in the short time since have been so constantly pursued that their numbers have been greatly reduced. Nearly every old prospector has one or more stories to tell about personal experiences with big brown bears, and often is able to show the skins as evidence of his truthfulness.

Both whites and natives distinguish several varieties of large bears according to color. One of these, which is called the 'big white bear,' and of which I examined specimens, is creamy white about the neck, shoulders, and back, and pale brownish about the haunches and legs. Nearly every degree of gradation from these 'white' bears to the dark brown ones may be found, however, so that it does not seem probable that more than one species is represented.

Mr. T. W. Hanmore, of Tyonek, says the brown bear generally goes into hibernation early in October, but that a few years ago he saw the track of one that had plowed through 2 feet of snow down to the beach near Tyonek in the middle of November.

Latax lutris (Linnæus). Sea Otter.

Sea otters are said to have been seen in Cook Inlet, but owing to the very muddy water it is probable that they were never numerous there, even in times of their greatest abundance elsewhere.

Lutra canadensis (Schreber). Land Otter.

Apparently rather uncommon, though a few are said to be taken every winter.

Lutreola vison energumenos Bangs. Pacific Mink.

Moderately common. A few mink tracks were seen along some of the small streams. Several skins of poor quality were offered for sale by miners at Hope. One specimen, a male in good pelage, was taken on a small stream near Tyonek September 16. It is not fully adult, and its skull shows no characters of value, but its color is very dark. On this account it is referred to *energumenos*.

Putorius kadiacensis Merriam. Kadiak Weasel.

One specimen was secured at Hope August 30. It was shot while in the act of making away with some scraps of meat that had been thrown out near the door of our cabin. This specimen is not quite adult, but agrees with specimens of *kadiacensis* of the same age in size, color, and cranial characters. Its flesh measurements are as follows: Total length 326; tail vertebræ 91; hind foot 45.

Putorius rixosus Bangs. Bangs Weasel.

One adult female was taken in a swampy place near Tyonek September 19. It was caught in a small mouse trap in a *Microtus* runway and doubtless would have escaped had it not thrashed into a pool of water and drowned. This specimen differs but very slightly from the type of *rixosus*, and shows no definite approach to subspecies *eskimo*. The color of the upperparts is uniform vandyke brown, slightly darker than the type of *rixosus;* the tail is the same color, with a faint paleness on the underside and no trace of black anywhere; the under parts, including the under and inner sides of the forelegs and the fore-feet, are pure white: the toes and one-third of the hind feet are white. The skull of the Cook Inlet specimen is about the same size as that of the type of *rivosus;* the braincase is slightly flatter and more elongate: the teeth are identical. The flesh measurements are as follows: Total length 165; tail vertebræ 18; hind foot 21. Skull: Basal length 27.5; palatal length 10.8; zygomatic breadth 14; breadth across post-orbital processes 9; length of audital bullæ 10.

The natives regard the capture of one of these rare animals as a piece of great good fortune. One old Indian who frequently visited our cabin told us that his brother who had caught one when a small boy had in consequence become a 'big chief;' and he assured me that since I had caught one I must surely be destined to become a man of great wealth and power.

Mustela americana Turton. American Marten.

Martens are only moderately common. George Coon, a reliable trapper of Hope, told me that in a season of about two and one-half months in 1899, near the mouth of Turnagain Arm, he took but 15 martens. Two marten skulls in the Biological Survey collection, collected by Dall De Weese on the Kenai Peninsula, are not referable to either *M. a. caurina* or *M. a. actuosa*, but seem to be very nearly like typical *americana.* The skulls and teeth are about the same size as in *americana* from the Adirondack Mountains, New York, and the shape of the last upper molar also agrees with that of *americana*, being of almost equal width internally and externally.

Gulo luscus (Linnæus). Wolverine.

Apparently rather common, as a number of skins are said to be secured annually. All of these are shipped via St. Michael to trading posts on the Yukon River, where they are sold to the Yukon natives, who value them very highly for making trimmings for their fur clothing. The natives and older prospectors tell many stories of the wolverine's skill and cunning in discovering and securing caches of provisions.

Sorex personatus Geoffroy. Common Shrew.

Very common at both Hope and Tyonek. Twenty specimens were taken in the vicinity of Hope and 19 at Tyonek. These are smaller and lighter colored than topotypes of *streatori* from Yakutat, but can hardly be referred to *arcticus.* I have not been able to find any characters in which they differ from *personatus* of the eastern United States. The measurements of 20 Cook Inlet specimens average as follows: Total length 101; tail vertebræ 40; hind foot 12.

Sorex alascensis shumaginensis Merriam. Shumagin Shrew.

Sorex alascensis shumaginensis Merriam. Proc. Wash. Acad. Sci., II, 18, Mar. 14, 1900.

Seventy-six specimens of a shrew almost indistinguishable from *S. shumaginensis* were taken, 27 near Hope and 49 near Tyonek. These are slightly smaller than *S. alascensis* and decidedly paler colored, thus approaching *S. shumaginensis*. They do not show as much light peppery spotting as *shumaginensis*, but otherwise do not differ from it. The skulls are slightly smaller than those of *alascensis* and practically identical with those of *shumaginensis*.

Sorex (Microsorex) eximius,[1] sp. nov.

Type from Tyonek, Cook Inlet, Alaska. ♀ ad. No. 107126, U. S. National Museum, Biological Survey collection. Collected September 14, 1900, by W. H. Osgood and E. Heller. Orig. No. 1395.

Characters.—Similar to *Sorex hoyi*, but larger and paler colored; skull widely different.

Color.—Head, back, and sides uniform pearly sepia, slightly paler than in *S. hoyi;* underparts pale drab, not strongly contrasted with upperparts; tail bicolor.

Skull.—Rostrum and interorbital region narrow and elongate; braincase much higher than in *S. hoyi* and more compressed anteriorly, distinctly elevated above plane of rostrum; palate long, narrow, and excavated. Mandibles longer and relatively more slender than in *S. hoyi.* Dentition much heavier than in *S. hoyi;* relative size of fourth unicuspid, as compared with first and second, quite small; inferior cusp of first upper incisor long and slightly decurved.

Measurements.—Type: Total length 98; tail vertebræ 31; hind foot 11. Skull: Basal length (inferior lip of foramen magnum to front of middle incisors) 15; palatal length 6.5; mastoid breadth 7.1; antorbital breadth 4.3.

Remarks.—The specimen which is the basis of the foregoing description is the only one of its kind among nearly 150 shrews caught at Hope and Tyonek. It is of extreme interest, not only as representing a very distinct new species, but as the only specimen of the subgenus *Microsorex* recorded from Alaska. Its dentition is essentially as in the only other species of the subgenus, *Sorex hoyi*, but the form of its skull is entirely different and much more like the general type found in the subgenus *Sorex*.

Myotis lucifugus (Le Conte). Little Brown Bat.

A few bats were seen at Hope, but no specimens were secured.

---

[1] Eximius=excellent, extraordinary.

Gavia imber (Gunn.). Loon.

An old skin of a loon was seen at Hope; otherwise the species was not observed by us.

Gavia lumme (Gunn.). Red Throated Loon.

Five specimens of '*Colymbus septentrionalis*' are recorded in the catalogue of the National Museum among Bischoff's birds from Fort Kenai. I have been unable to find any of these in the National Museum.

Uria troile californica (Bryant). California Murre.

Bean records specimens taken at Chugachik Bay (=Kachemak Bay) June 30, 1880, and reports the species as abundant;[1] and a specimen taken by George Palmer at Knik Station is in the National Museum. The species was not seen by us in August and September.

Stercorarius parasiticus (Linn.). Parasitic Jaeger.

Several were seen at Homer August 22. Not seen elsewhere about the inlet.

Rissa tridactyla pollicaris Ridgw. Pacific Kittiwake.

A few were seen at Homer August 22. Not noticed elsewhere in the inlet.

Larus sp.

A few large gulls in immature plumage were occasionally seen, but they were noticeably uncommon. This scarcity I found was due to the fact that for the past two years gulls have been systematically slaughtered for millinery purposes. A trader offered from 10 to 20 cents each for them, and consequently the Indians and half-breeds have killed every one that has come within range of their guns.

Larus philadelphia (Ord). Bonaparte Gull.

Seen in considerable numbers at Homer August 22, but not elsewhere about the inlet.

Sterna paradisæa Brünn. Arctic Tern.

The National Museum catalogues show that Bischoff secured 4 specimens of this species in May and June, 1869. We did not meet with it.

Diomedea albatrus Pallas. Short-tailed Albatross.

In the summer of 1880 Dr. T. H. Bean found this species common about the mouth of Cook Inlet, and a specimen was secured near Fort

---

[1] Proc. U. S. Nat. Mus., V, 172, 1882.

Alexander. We did not see it when we were in this vicinity, in August and September, 1900.

Phalacrocorax pelagicus Pallas. Pelagic Cormorant.

A single, lonely-looking cormorant was several times seen at Tyonek flying up the inlet close to the shore. Others were seen at Homer.

Anas boschas Linn. Mallard.

Common at Tyonek, where 7 immature birds of the year were shot by E. Heller in September.

Dafila acuta (Linn.). Pintail.

A specimen was taken by Bischoff at Fort Kenai.

Aythya marila nearctica Stejn. American Scaup Duck.

A flock of 6 scaup ducks was seen on a pond near Tyonek September 17.

Somateria v-nigra Gray. Pacific Eider.

A young bird and four eggs were secured by Dr. T. H. Bean at Chugachik Bay (=Kachemak Bay) July, 1880.

Oidemia perspicillata (Linn.). Surf Scoter.

Several flocks of scoters supposed to be this species were seen at Homer August 22. A specimen was taken by Bischoff at Fort Kenai July, 1869.

Branta canadensis subsp? Canada Goose.

Small flocks were frequently seen at Tyonek, but no specimens were secured.

Olor columbianus (Ord). Whistling Swan.

The miners about the inlet say that swans are often seen there and that several have been killed.

Ardea herodias Linn. Great Blue Heron.

A great blue heron was seen at Hope by E. Heller.

Grus canadensis (Linn.). Little Brown Crane.

Immense flocks of migrating cranes are said to pass over Cook Inlet annually. Three specimens were taken at Fort Kenai by Bischoff.

Phalaropus lobatus (Linn.). Northern Phalarope.

Six specimens were taken by Bischoff at Fort Kenai May, 1869.

Gallinago delicata (Ord). Wilson Snipe.

One specimen was taken by Bischoff at Fort Kenai May 5, 1869.

Macrorhamphus griseus scolopaceus (Say). Long-billed Dowitcher.

Four specimens were taken at Fort Kenai May 4–9, 1869, and one July 20, 1869. Two of these are still in the National Museum.

Tringa couesi (Ridgw.). Aleutian Sandpiper.

An Aleutian sandpiper was seen on the beach near Homer August 22.

Tringa bairdi (Coues). Baird Sandpiper.

A sandpiper, thought to be this species, was seen at Homer August 22.

Tringa alpina pacifica (Coues). Red-backed Sandpiper.

One specimen taken at Fort Kenai by Bischoff May 16, 1869.

Ereunetes occidentalis Lawr. Western Sandpiper.

Three specimens were taken at Fort Kenai by Bischoff May 12–16, 1869. One of these (No. 58470) has been examined; it is perfectly typical of the large-billed form, *occidentalis*.

Limosa hæmastica (Linn.). Hudsonian Godwit.

Nine specimens were taken by Bischoff at Fort Kenai. At least two of these are still in the National Museum—one an adult in breeding plumage, the other in fall plumage.

Totanus melanoleucus (Gmel.). Greater Yellow-legs.

Taken at Fort Kenai by Bischoff May and June, 1869; specimen still in National Museum.

Actitis macularia (Linn.). Spotted Sandpiper.

Several were seen along Resurrection Creek near Hope, and one was taken.

Numenius hudsonicus Lath. Hudsonian Curlew.

A specimen was taken by Bischoff at Fort Kenai May 18, 1869, but can not now be found in the National Museum.

Squatarola squatarola (Linn.). Black-bellied Plover.

One taken by Bischoff at Fort Kenai May 6, 1869; specimen examined in National Museum.

Canachites canadensis osgoodi Bishop. Alaska Spruce Grouse.

'Black grouse' or 'fool hens,' as they are locally termed, are very common in all the Cook Inlet region. They are easily killed, and many thus find their way to the miner's frying pan. The Indians and half-breeds also hunt them to a considerable extent. When flushed from the ground, they rise quickly and fly swiftly, but only to light in the nearest spruce. When this is but a few yards away, they immediately flounder into the thickest part of it; but if a long stretch of birches, poplars, or small deciduous bushes intervenes, they continue winding in and out until they reach the requisite spruce. In September at Tyonek they were often found in small flocks of eight or ten individuals. When flushed each would betake itself to a separate tree and after a brief interval start a subdued clucking, so that all could be easily located. At Hope they were found ranging from sea level to timberline; on one occasion I flushed a flock of grouse and but a short distance farther on a flock of ptarmigan. Their crops were usually found to contain spruce needles and *Vaccinium* and *Viburnum* berries, and in one case heads of *Equisetum*. Cook Inlet specimens agree perfectly with typical *osgoodi* from the Yukon Valley.

Lagopus rupestris (Gmelin). Rock Ptarmigan.

A few small flocks were seen in the mountains on the north side of Bear Creek, and three females were taken. They have been hunted more or less by the miners in this vicinity, and we found them very wild and hard to secure.

Lagopus leucurus Swains. & Rich. Northern White-tailed Ptarmigan.

While setting small mammal traps in a few patches of grass at the extreme head of Bear Creek September 5, I suddenly became aware of a subdued clucking apparently from a rockslide not far away. Upon following up the sound I was soon able to distinguish some gray birds moving over the rocks, but so closely did they resemble the background that I could not see them unless they moved. They were quite tame and allowed me to approach within a few yards, so that I was easily able to make out an old female white-tailed ptarmigan and a brood of seven nearly full-grown young. The old bird was almost as solicitous for her charges as if they had been downy chicks, and led them away very adroitly, keeping up a continuous purring cluck and making herself as conspicuous as possible. Although evidently much alarmed at my presence, flight as a means of escape did not seem to enter their heads, and it was not until I fired on the old bird that the young took wing. I had only my rifle with me, and so was obliged to shoot the ptarmigan with a load intended for bear, but fortunately the bird was not irreparably mutilated and I was able to make a fairly good specimen of it. This specimen was found to be different from the white-tailed ptarmigan of the Colorado mountains, and examination of the original description of *Lagopus leucurus* showed that the northern bird was the one originally described by Swainson.[1]

Circus hudsonius (Linn.). Marsh Hawk.

One was seen flying back and forth near Homer over a meadow thickly populated with *Microtus*. Another was seen at Hope.

---

[1] Cf. Auk, XVII, 180, April, 1901.

Accipiter atricapillus striatulus Ridgw. Western Goshawk.

Goshawks were frequently seen near Tyonek, and two immature birds were taken September 18. Remains of *Sciurus* were found in their craws.

Haliæetus leucocephalus alascanus Towns. Alaska Bald Eagle.

Said to occur; not seen by us.

Falco columbarius Linn. Pigeon Hawk.

Rather common; several were seen at Hope and also at Tyonek. An immature bird was shot by E. Heller at Hope August 30. Its craw contained parts of crossbills.

Bubo virginianus saturatus Ridgw. Dusky Horned Owl.

Very common; they were heard nightly at Hope and occasionally at Tyonek. One was seen on a dark day in the deep birch woods back of Tyonek and an adult female was shot at Hope August 30. A specimen taken by George Palmer at Knik Station is in the National Museum. These two specimens are quite different from typical *saturatus*, but they are nearer to it than to any other described form. They are considerably lighter than *saturatus* and do not have barring extending down on the feet to the toes, as is usual in that form; also the bars on the sides are not blended, but separated by distinct light areas.

Nyctea nyctea (Linn.) Snowy Owl.

The miners and traders about Cook Inlet say that snowy owls have frequently been killed there in winter.

Ceryle alcyon (Linn.) Belted Kingfisher.

Common along streams. Its loud, clattering cry was heard frequently along Resurrection Creek, near Hope, when the thick growth of trees and shrubs prevented seeing the bird.

Picoides americanus fasciatus Baird. Alaska Three-toed Woodpecker.

Represented by eight specimens as follows: Hope, four; Tyonek, one; Fort Kenai, three. It was found to be quite common in the Turnagain Arm region, but at Tyonek, where coniferous trees are scarcer, only one bird was seen. Three specimens taken by Bischoff at Fort Kenai in 1869 are in the National Museum.

Selasphorus rufus (Gmelin). Rufous Hummingbird.

Mr. T. W. Hanmore, who has been stationed at Tyonek for eleven years, says that he has seen humming birds there several times. This is doubtless near the limit of the range of the species, as the bird has not been recorded farther north.

Contopus borealis (Swains.). Olive-sided Flycatcher.

A specimen from Fort Kenai is in the National Museum. It is an adult male taken by Bischoff May 26, 1869.

Pica pica hudsonica (Sab.). American Magpie.

The miners at Sunrise City told us that magpies had been seen in that vicinity frequently, but we did not observe them there or at any other point in the inlet. Specimens taken in Graham Harbor by C. H. Townsend in 1892 are in the National Museum.

Cyanocitta stelleri (Gmelin.). Steller Jay.

Several specimens taken at Graham Harbor in 1892 by C. H. Townsend and B. W. Evermann are in the National Museum. This is apparently the northern limit of the species, as we did not find it farther up the inlet in Turnagain Arm, nor on the north side at Tyonek.

Perisoreus canadensis fumifrons Ridgw. Alaska Jay.

Occasionally seen. One morning, after a light fall of snow, a small party of jays visited our camp in the mountains near Hope. A few were also seen at Tyonek. A large series was taken by Bischoff at Fort Kenai.

Corvus corax principalis Ridgw. Northern Raven.

Only moderately common. The trappers say they are very abundant in winter and a great nuisance to them, since they systematically spring their traps or take the bait from them.

Scolecophagus carolinus (Müller). Rusty Blackbird.

Two males were shot by Heller at Tyonek September 23. No others were seen during our stay, but the birds undoubtedly breed in the vicinity, for two specimens were taken by Bischoff May 28 and July 4, respectively. An examination of the material in the National Museum shows a slight difference in size between eastern and western birds of this species. The bill especially is constantly a trifle shorter and lighter in specimens from Alaska.

Loxia curvirostra minor (Brehm). American Crossbill.

A specimen taken at Graham Harbor in 1892 by C. H. Townsend and B. W. Evermann is in the National Museum.

Loxia leucoptera Gmelin. White-winged Crossbill.

Common. They were not seen in large flocks, however, but generally in pairs. Four specimens were taken at Hope August 25–28.

Acanthis linaria (Linn.). Redpoll.

Large flocks were seen frequently both at Hope and Tyonek, and one immature specimen was taken at Hope. Two summer adults are in the National Museum, one taken by Bean at Chugachik Bay (=Kachemak Bay), and one by Bischoff at Fort Kenai.

Spinus pinus (Wilson). Pine Siskin.

Three specimens were secured from a large flock at Tyonek September 22. They were not seen elsewhere about the inlet.

Calcarius lapponicus alascensis Ridgw. Alaska Longspur.

An adult male in breeding plumage was taken by Bischoff at Fort Kenai in May, 1869.

Ammodramus sandwichensis alaudinus (Bonap.). Western Savanna Sparrow.

Evidently an abundant breeder, as numerous specimens were taken in summer by Bischoff and Bean at Fort Kenai and Chugachik Bay. At the time of our work in August and September very few were seen. Four specimens were taken, three at Hope August 26, 28, and 29, respectively, and one at Tyonek September 18.

Zonotrichia leucophrys gambeli (Nutt.). Intermediate Sparrow.

Evidently a common breeder, as Bischoff took a number of specimens at Fort Kenai in May, 1869, at least one of which is still in the National Museum. The species was not observed by us, but it may have been overlooked among the immature birds seen at Hope and all supposed to be *Zonotrichia coronata.*

Zonotrichia coronata (Pallas). Golden-crowned Sparrow.

Common in the low second-growth brush about the village of Hope; also occasionally seen in the mountains near there. Four birds collected by Bischoff at Fort Kenai are recorded in the National Museum catalogue as *Zonotrichia querula.* None of these are now at hand, but the entries doubtless refer to *Z. coronata.*

Spizella monticola ochracea Brewst. Western Tree Sparrow.

A specimen is recorded taken by Bischoff at Fort Kenai May 19, 1869, but it can not now be found in the National Museum. As the occurrence of the species is altogether probable, however, there seems no reason to doubt the identification.

Junco hyemalis (Linn.). Slate-colored Junco.

Common. Three specimens were taken at Hope August 26–28.

Melospiza melodia kenaiensis Ridgw. Kenai Song Sparrow.

The type of this subspecies was taken by C. H. Townsend at Port Graham on the Kenai Peninsula April 9, 1892. Two specimens were also taken at this locality by Dr. Bean July 4, 1880. A specimen taken at Hope August 26 differs from the type of *kenaiensis* to such a degree that it hardly seems possible that it merely represents the difference between fall and summer plumage. It is characterized by very sooty coloration; the dark markings about the head, neck, and breast are very intense and the streaks on the back are very prominent. In size it is intermediate between *kenaiensis* and *caurina*, and as no fall specimens of either are at hand it seems best to refer it to *kenaiensis*, which is geographically near. It measures as follows: Wing 74; tail 70; exposed culmen 14; bill from nostril 10; tarsus 24.5.

Melospiza lincolni (Aud.). Lincoln Sparrow.

An adult male was taken at Hope August 28, and a few others were seen while we were there. The specimen taken shows none of the characters attributed to *Melospiza lincolni striata*.

Passerella iliaca annectens Ridgw. Yakutat Fox Sparrow.

Rather common, but very shy, as usual, and hard to secure. Two specimens were taken at Hope and one at Tyonek. These seem to be intermediate between *P. i. annectens* and *P. i. insularis*, as they have the smaller bill and more dusky underparts of *annectens* and the lighter upperparts of *insularis*.

Lanius borealis Vieill. Northern Shrike.

An immature bird in the brown plumage was shot by E. Heller at Hope September 6. Several were seen near Homer. An adult from Fort Kenai, collected by Bischoff, is in the National Museum. At present I am unable to find sufficient characters to warrant use of the name *invictus*[1] for these birds. The question is further complicated by *Lanius borealis sibiricus*, which, judging from three specimens in the National Museum, differs from Alaskan birds only in having the vermiculations on the breast nearly obsolete.

Helminthophila celata lutescens (Ridgw.). Lutescent Warbler.

Three specimens taken by Bischoff at Fort Kenai May 22–26, 1869, are still in the National Museum. Examination shows them to be typical *lutescens*.

Dendroica coronata (Linn.). Myrtle Warbler.

The National Museum catalogue records this bird taken at Fort Kenai by Bischoff.

Dendroica striata (Forster). Black-poll Warbler.

One specimen was taken at Fort Kenai by Bischoff, but is not now to be found.

---

[1]Grinnell, Birds of Kotzebue Sound, Pac. Coast Avifauna, I, 54–55, Nov., 1900.

Anthus pensilvanicus (Latham). American Pipit.

Comparatively few pipits were seen. One specimen was taken at Tyonek September 18.

Cinclus mexicanus Swains. American Dipper.

Several were seen in the mountains near Hope, and a specimen was taken there September 3.

Certhia familiaris montana Ridgw. Rocky Mountain Creeper.

An adult female taken at Hope August 31 is in fine fall plumage and typical of this subspecies. A few individuals were seen at Tyonek.

Parus atricapillus septentrionalis (Harris). Long-tailed Chickadee.

Very common both about Turnagain Arm and at Tyonek. An adult male was taken at Hope September 5. This specimen seems to be referable to *P. a. septentrionalis* rather than to *P. a. turneri*.

Parus hudsonicus Forster. Hudsonian Chickadee.

Very common at Tyonek, but rarely seen at Hope. Two specimens were taken at Fort Kenai by Bischoff. Cook Inlet specimens do not seem to differ from those of the Yukon and Kowak valleys. I am also unable to find any appreciable differences between them and three birds recently collected by E. A. Preble near the type locality of *hudsonicus*. Consequently I do not agree that the specimens at present available warrant the recognition of *Parus hudsonicus evura*.[1] From a rather hasty examination of the material in the National Museum there seems to be an average difference in the length of the tail between the Alaska birds and the birds from the extreme northeastern United States. The birds from the west side of Hudson Bay, however, are intermediate and apparently nearer to the Alaska birds. In other words, as far as present material goes, there are just as good grounds for the recognition of *Parus hudsonicus littoralis* Bryant, 1863, from Nova Scotia, as for *P. h. evura* Coues, 1884, from Alaska.

Regulus satrapa olivaceus Baird. Western Golden-crowned Kinglet.

Moderately common.

Regulus calendula (Linn.). Ruby-crowned Kinglet.

An adult male taken by Bischoff at Fort Kenai May 9, 1869, is in the National Museum. Examination of this specimen does not show any characters that approach those of *Regulus calendula grinnelli*, which is found on the coast only a short distance farther south.

[**Hylocichla aliciæ** (Baird). Gray-cheeked Thrush.

This is one of the entries in the Bischoff collection from Fort Kenai. The specimen has not been found in the National Museum collection,

---

[1] Cf. Rhoads, Auk, X, 331, 1893, et Bishop, Auk, XVII, 118–119, 1900.

and as the possibility of misidentification is considerable, the species is not unequivocally admitted to this list.]

Hylocichla ustulatus almæ Oberholser. Alma Thrush.

Two male birds in fresh fall plumage were taken at Hope August 26 and August 29, respectively. These are very olivaceous on the upperparts and agree with a bird taken at Circle City August 18, 1899.

Hylocichla aonalaschkæ (Gmelin). Dwarf Hermit Thrush.

Two specimens were taken at Hope and Tyonek September 7 and September 14, respectively. These are in fresh fall plumage, and are somewhat more olivaceous than fall birds from Kadiak.

Merula migratoria (Linn.). Robin.

The miners that we met at Hope and Sunrise reported that the 'regular eastern robin' had often been seen there. We did not observe it ourselves in the month of August, the time of our stay at those points.

Hesperocichla nævia (Gmelin). Varied Thrush.

Not abundant; occasional individuals were seen or heard.

INDEX.

O

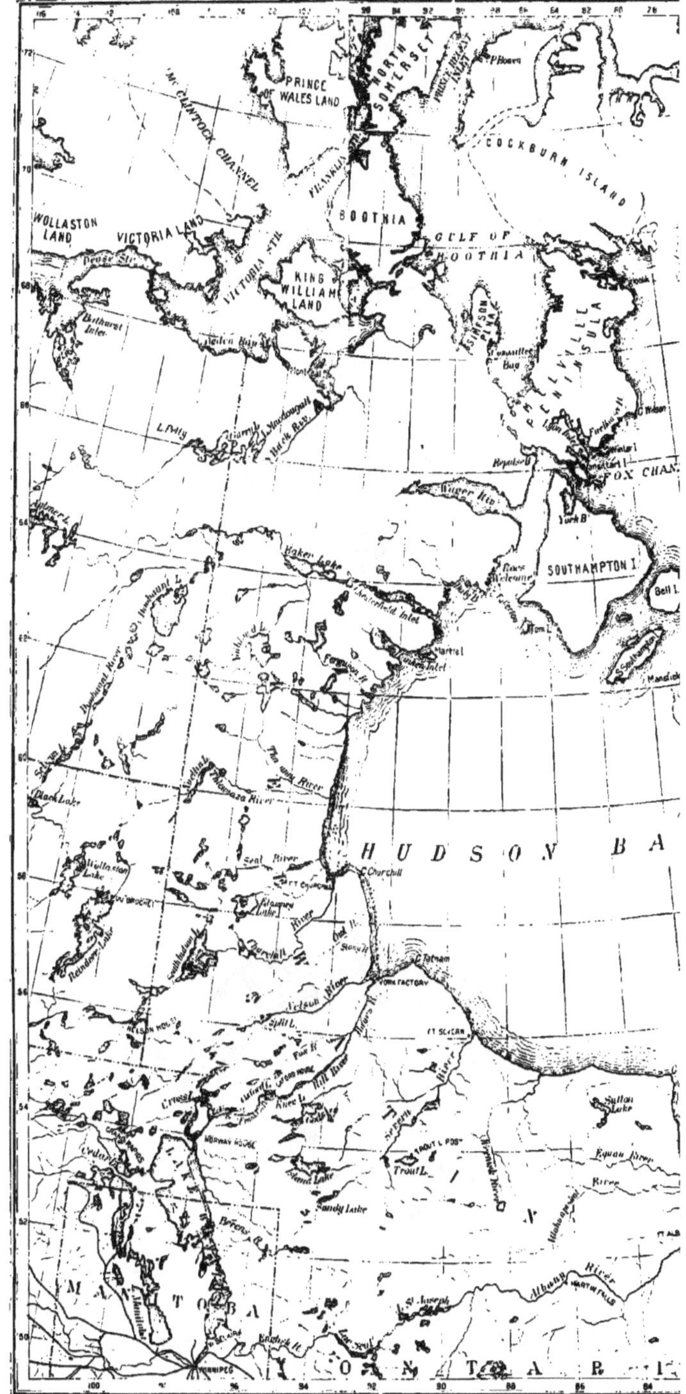
PRINCE OF WALES LAND
NORTH SOMERSET
PRINCE REGENT INLET
COCKBURN ISLAND
M'CLINTOCK CHANNEL
FRANKLIN
BOOTHIA
WOLLASTON LAND
VICTORIA LAND
GULF OF BOOTHIA
VICTORIA STR.
KING WILLIAM LAND
SIMPSON PENA.
MELVILLE PENINSULA
Bathurst Inlet
FOX CHAN.
SOUTHAMPTON I.
Bell I.
Baker Lake
Chesterfield Inlet
Black Lake
HUDSON BA
C. Churchill
Seal River
Nelson River
YORK FACTORY
NORWAY HOUSE
Sandy Lake
Albany River
MANITOBA
WINNIPEG
ONTARIO

KEEWATIN.

U. S. DEPARTMENT OF AGRICULTURE
DIVISION OF BIOLOGICAL SURVEY

NORTH AMERICAN FAUNA

No. 22

[Actual date of publication, October 31, 1902]

A BIOLOGICAL INVESTIGATION OF THE HUDSON BAY REGION

BY

EDWARD A. PREBLE
ASSISTANT BIOLOGIST, BIOLOGICAL SURVEY

Prepared under the direction of

Dr. C. HART MERRIAM
CHIEF OF DIVISION OF BIOLOGICAL SURVEY

WASHINGTON
GOVERNMENT PRINTING OFFICE
1902

STATISTICAL DEPARTMENT
NOV 12 1902
BOSTON PUBLIC LIBRARY

LETTER OF TRANSMITTAL.

United States Department of Agriculture,
Washington, D. C., August 18, 1902.

Sir: I have the honor to transmit for publication, as North American Fauna No. 22, a paper on the natural history of the Hudson Bay region, by my assistant, Edward A. Preble.

C. Hart Merriam,
Chief, Biological Survey.

Hon. James Wilson,
Secretary of Agriculture.

PREFATORY NOTE.

A century or more ago the employees of the Hudson's Bay Company sent collections of birds and mammals from Hudson Bay to London. Some of these specimens came from the shores of Hudson Bay, others from trading posts in the distant interior: but many were not labeled to show where they were obtained. They were examined by the naturalists of the time and a number of species were described and named as new. In most cases the original specimens have disappeared and modern naturalists have been greatly perplexed in attempting to ascertain just what the species really were. No modern museum possessed anything approaching a representative collection of the mammals and birds of Hudson Bay, and specimens for comparison with related forms from other parts of Boreal America were not to be had. The resulting embarrassment was most keenly felt when the Biological Survey secured large collections from Alaska. In many instances it was impossible to tell whether certain Alaska species were identical with or distant from related forms previously described from Hudson Bay. In order to obtain the long-needed material it was determined to send an expedition to Hudson Bay. Edward A. Preble was placed in charge of this expedition: his report shows how well and faithfully his duties were performed. His successful trip, in an open boat, in inclement fall weather, from Fort Churchill to the Barren Grounds near Cape Eskimo, in search of topotypes of the Hudson Bay ground squirrel (often known as Parry's marmot) deserves special commendation.

I take pleasure also in referring to the uniform courtesies and facilities extended by the officers and employees of the Hudson's Bay Company, particularly by Mr. C. C. Chipman, commissioner of the company, at Winnipeg.

C. H. M.

CONTENTS.

ILLUSTRATIONS

A BIOLOGICAL INVESTIGATION OF THE HUDSON BAY REGION.

By Edward A. Preble.

INTRODUCTION AND ITINERARY.

In 1610 Henry Hudson, while searching for a northwest passage, entered and partially explored the great inland sea that bears his name. In 1670 the Hudson's Bay Company was organized to trade for furs with the natives of the great unexploited territory adjacent to the Bay. This company first established several trading posts at the mouths of the rivers tributary to the Bay and then gradually extended its field of operations inland. By this means the southern and western shores of Hudson Bay and the principal rivers emptying into it on the west had become fairly well known at a time when immense areas in North America, apparently more favorably located and more accessible, were still unexplored. As a natural result the birds and mammals of this semiarctic region were early brought to the attention of naturalists, and many species whose ranges extend over a very large area were first described from specimens sent to Europe from Hudson Bay. This was mainly due to the labors of the employees of the Hudson's Bay Company, who, residing at trading posts and coming in contact for purposes of trade with practically all the natives of the region, were able to secure natural history specimens with comparative ease, especially the larger species. The many collections thus made were conveniently sent to England by the ships which paid annual visits to the posts. In this way a number of mammals and many birds, mainly littoral and pelagic species, first became known to science. As time went on, however, less attention was given to the fauna of this region, while most other parts of North America were ransacked for natural history material, so that the close of the nineteenth century found Hudson Bay one of the most neglected fields of modern zoological research. Some species, originally described from poor specimens, and in the loose and inaccurate style of a hundred years ago, were known by these descriptions alone, while others were represented in museums only by poorly stuffed and faded specimens, entirely inadequate to meet the requirements of modern scientific methods. This lack of material, in connection with the absence of definite knowledge as to the boundaries of the life zones, made it desirable that a collection, as thorough as possible, be made in the region.

Early in the summer of 1900, therefore, I was detailed to make a biological reconnaissance of as much of the region immediately to the west of Hudson Bay as it would be practicable to cover in a single season. My brother, Alfred E. Preble, of Tufts College, Massachusetts, accompanied me as assistant.

The Hudson's Bay Company still maintains trading posts throughout the region we were to visit, and the officials of the company compose almost its entire white population. These posts are situated on the usual lines of travel, and constitute the only bases of supplies available; hence it was considered advisable to arrange with the company for food and means of transportation.

This we did on our arrival at Winnipeg on June 13, and obtaining a canoe from the company, set out the next day down the Red River. The following morning we took the Northwest Navigation Company's steamer *Princess* at West Selkirk, and on June 17 arrived at Norway House, near the north end of Lake Winnipeg, where we were to begin operations. Here we collected until June 23, when our northern trip was resumed. We took two Indians for guides, boatmen, and camp hands, and a large Peterborough canoe, in which our collecting and camp outfit and provisions were carried.

We passed down the east channel of Nelson River, and ascending the Echimamish, followed the usual boat route to York Factory, stopping to collect at favorable points. At the head of the Echimamish proper, which terminates abruptly at a rock about 30 yards broad called the Painted Stone, we made a portage and launched our canoe in a small lake. A stream flows eastward from this lake and we thus had the advantage of the current for the remaining distance to Hudson Bay. Beyond Painted Stone Portage the route passes successively through the Robinson lakes, Franklin River, and Pine, Windy, Oxford, Knee, and Swampy lakes. These different lakes vary from a few miles to forty in length, and the channels connecting them contain numerous rapids. Hill River forms the outlet of Swampy Lake, the last of the chain, and unites with Fox River to form Steel River. This in turn unites with the Shamattawa, and the resulting stream, known as Hayes River, empties into Hudson Bay at York Factory. On reaching the Bay we exchanged our canoe for a sailboat and made our way up the west coast to Fort Churchill, at the mouth of the river of that name. Here, after a few days' stay, I left my brother to complete the collection, while I pushed northward well into the Barren Grounds. This trip consumed three weeks, and on my return to Fort Churchill we immediately started on the homeward journey in order

FIG. 1.—GENERAL VIEW OF NORWAY HOUSE.

FIG. 2.—SHORE OF CHANNEL NEAR NORWAY HOUSE.

Fig. 1.—Spruce Thicket near Norway House.

Fig. 2.—Aspen Thicket near Norway House.

to complete it before navigation closed. We passed down the coast to York Factory in a sailboat and retraced our way to Norway House in our canoe. The trip up the rapid streams with our heavily loaded boat was a very arduous one, but we reached Norway House without accident or delay on September 16, having completed a journey by canoe and sailboat of more than 1,200 miles, much of it through very difficult water. We took a steamer from Norway House on September 19 and arrived at Winnipeg on September 22.

During our trip to Hudson Bay we were placed under many obligations to a number of officers of the Hudson's Bay Company, to whom our cordial thanks are hereby extended. Through the courtesy of C. C. Chipman, commissioner of the Hudson's Bay Company at Winnipeg, we were able to make arrangements to secure supplies and transportation at the different trading posts of that company on our route; without this aid the trip could hardly have been accomplished. Among others who assisted us in various ways are Messrs. William Clark, W. C. King, and Roderick MacFarlane, of Winnipeg; J. K. MacDonald, of Norway House; William Campbell, of Oxford House; G. B. Boucher, of York Factory; Ashton Alston, of Fort Churchill; and especially Dr. Alexander Milne, of Winnipeg (formerly of York Factory), who has given me many notes on the distribution of the larger mammals. To the Rev. Mr. Chapman, a missionary at Fort Churchill, we were also indebted for information and various courtesies. The Rev. W. A. Burman, of Winnipeg, kindly furnished us with a list of the principal trees and shrubs occurring about Winnipeg. From Colonel Scobell, C. E., of Winnipeg, we obtained much detailed and valuable information concerning the boat route to York Factory.

In preparing this report I have received many courtesies from Mr. Robert Ridgway, curator, and Dr. C. W. Richmond, assistant curator of birds in the U. S. National Museum, who have not only permitted the unrestricted use of the collection under their charge but have helped me in many other ways. Mr. Gerrit S. Miller, jr., assistant curator of mammals in the National Museum, has extended similar favors in regard to the mammals; and Dr. Leonhard Stejneger, curator of reptiles in the National Museum, has aided in identifying the frogs collected. Thanks are also due to Frederick V. Coville, botanist of the Department of Agriculture, and his assistants, for identifying the plants collected. Finally, I am indebted to Dr. C. Hart Merriam and Dr. T. S. Palmer, chief and assistant chief of the Biological Survey, for various courtesies extended during the progress of the work.

GENERAL ACCOUNT OF THE REGION TRAVERSED.

Winnipeg is situated at the junction of the Red and Assiniboine rivers, on the site of old Fort Garry. To the westward stretch the plains, but the vicinity of the rivers is well wooded with elm (*Ulmus americana*), mossy-cup oak (*Quercus macrocarpa*), basswood (*Tilia americana*), ash-leaved maple (*Acer negundo*), and other species, with an undergrowth composed principally of viburnums, hazel (*Corylus americana* and *rostrata*), wolfberry (*Symphoricarpos occidentalis*), hawthorn (*Crataegus coccinea*), etc.

The Red River below (to the north of) Winnipeg is very winding and is inclosed between banks of clay and limestone which at first are rather high and steep and are fairly well wooded, though the woods seldom extend far back from the river. But just beyond West Selkirk (a village about 20 miles below Winnipeg, near the site of the historic Selkirk Settlement) the banks become lower and the woods gradually yield to willow thickets. Farther down, a few miles from the mouth of the river, these willow thickets in turn disappear, and Lake Winnipeg is approached through a marsh which extends as far as the eye can reach, and where numberless coots and other marsh-loving birds find a congenial home.

Soon after we entered the waters of Lake Winnipeg, about 42 miles from our starting point, our course carried us too far from shore to permit observations as to forest conditions, and such was the case throughout much of our voyage up the lake, though a few opportunities for notes were offered. At The Narrows we could see that the western shore was well wooded with birch and conifers, a character of forest which, we were told, continues south nearly to the mouth of Red River.

At Bull Head, off which we anchored early on the morning of June 16, the forest consisted mainly of spruce, tamarack, a species of pine (probably *Pinus divaricata*), birch (*Betula papyrifera*), and poplar (*Populus balsamifera*), the deciduous species predominating. The shores of the northern part of the lake are low and sandy with numerous outcrops of gneiss, and many low islands of the same rock occur. Great Playgreen Lake, the body of water next traversed, lies just east of the northern part of Lake Winnipeg, with which it is connected at its southern end by a rocky channel. This channel is entered at Mossy Point, the southern extremity of a slender strip of land separating the two lakes, on which Norway House was originally situated. Nelson River issues from Great Playgreen Lake by two main channels, known as East and West rivers, which, coming together at Cross Lake, inclose Ross Island, 50 miles in length. East River, on leaving Great Playgreen Lake, divides into several minor channels encircling small islands, then expanding forms Little Playgreen Lake, about 25 miles from the outlet of Lake Winnipeg, and 300 miles from the southern end. On one of these islands, at the southern margin of Little Playgreen Lake, stands Norway House (Pl. II, fig. 1). Two miles distant, on the eastern shore of the lake, is Rossville Mission.

Like most of the region between Lake Winnipeg and Hudson Bay, the country about Norway House consists largely of swamps, mainly

FIG. 1.—ROCKY SHORE NEAR NORWAY HOUSE, SHOWING STORE OF FUR TRADER.

FIG. 2.—SEA RIVER FALLS, LOOKING DOWN.

FIG. 1.—OXFORD HOUSE.

Photographed by William Campbell.

FIG. 2.—INDIAN CAMP, OXFORD HOUSE.

grown up to willows and tamaracks. Numerous elevated places occur, rocky 'islands' on which has accumulated a rather thin covering of soil, supporting a moderately heavy growth of black and white spruce (*Picea nigra* and *P. alba*, Pl. III, fig. 1), balsam fir (*Abies balsamea*), Banksian pine (*Pinus divaricata*), aspen poplar (*Populus tremuloides*, Pl. III, fig. 2), balsam poplar (*Populus balsamifera*), canoe birch (*Betula papyrifera*), and tamarack (*Larix laricina*). These species form the bulk of the forest between Lake Winnipeg and Hudson Bay.

From the vicinity of Norway House to the Sea River Falls (Pl. IV, fig. 2), about 20 miles below, the shores are rather low (see Pl. II, fig. 2; Pl. IV, fig. 1). Then for the few remaining miles before the mouth of the Echimamish is reached the route lies through a channel bordered by rather high banks and forested with birch and poplar. The water of the Echimamish, which flows into Nelson River from the eastward, is very dark and contrasts markedly with that of the Nelson, which carries the whitish waters of Lake Winnipeg. The course here leaves East River and ascends the Echimamish, a short distance from the mouth of which it passes through Hairy Lake, a broad, shallow sheet of water a few miles in length, in which grow extensive patches of bulrush (*Scirpus lacustris*). Above this lake for more than 20 miles the Echimamish is a winding, sluggish stream, with an east and west trend, and averages about 50 feet in width. Its banks are low and marshy, and on the lower part extensive swamps border it on either side. Occasional outcrops of gneiss occur, dry 'islands,' which form the only available camping places. The forest consists mainly of spruce, tamarack, and willow, the latter usually predominating. Mosquitos, which swarm over the entire region, are here almost unbearable, and as the shallowness of the water, which is barely deep enough to float a canoe, makes paddling very difficult, the ascent of this river was perhaps the least pleasant part of our journey. Three dams, at one of which —the second— we did some collecting, are kept up for the purpose of holding back a sufficient amount of water to permit the passage of boats. The stream flows through a flat country and in several places in the upper part of its course, broadens and forms small ponds. In its comparatively still waters the yellow pond lily (*Nymphæa*) grows abundantly.

At the Painted Stone, about 36 miles from Norway House, the stream comes to an abrupt termination, and boats are carried across a rock and launched in a small lake with high, rocky shores. From this lake issues a stream generally considered a part of the Echimamish, which in the Cree language signifies 'the river that flows each way.' It would appear that the small lake is fed from some underground source, and that some of its waters escape into the western part of the Echimamish. The vicinity of Painted Stone Portage proved a very good collecting ground. The eastern part of the Echimamish is deep and bordered by high, rocky banks, on which *Potentilla tridentata* grows abundantly. Seven miles from the Painted Stone the stream unites with White Water River, which discharges the waters of Little Lake Winnipeg, and from this point to Oxford Lake the stream is called Franklin River.

The Robinson lakes, the southern shores of which are rather marshy and the northern shores higher, are next passed, and then 12 miles from the junction of the two streams Robinson Portage is reached. Here a portage of about three-fourths of a mile is necessary to avoid Robinson Rapids, where the river plunges through a deep ravine in a series of falls and rapids, with a total descent of about 50 feet. Deep mossy woods border this gorge, the excessive moisture from the rapids causing a luxuriant growth. From the lower or northern end of Robinson Portage, which lies nearly north and south, extends a line of lakes with marshy shores and supporting an abundant growth of sweet flag (*Acorus calamus*). A short stop was made and some collecting done at the north end of the portage. Immediately below these lakes the river plunges with considerable rapidity through a rocky gorge called Hell Gate. A short portage is made at its entrance, and after being launched in a surging pool at the foot of the rapid, the canoe is borne swiftly through the gorge. In some places the rocky walls rise nearly perpendicularly without a break; in others the bank consists of a succession of steep mossy terraces, the homes of several eagles.[a] Throughout most of its course of 7 miles through the gorge, the river is confined within narrow limits, and the smooth but impetuous current bears the voyager rapidly onward, constantly bringing fresh vistas to his view. In a few places a portion of the rocky walls has fallen, partially damming the stream, and the canoe is run through short, rapid chutes, the perpendicular walls preventing a landing, however desirable it might appear. Farther down the rocky banks are not so high and the surrounding country is seen to consist of rugged rock masses scantily clothed with Banksian pines. Here the voyager may land to see the 'kettles'—deep, rounded potholes of various sizes, which have been worn in the rock during past ages.

A short distance beyond the lower end of Hell Gate Gorge, 23 miles from Robinson Portage, lies Pine Lake, a small, irregularly outlined body of water containing numerous islands and environed by rocky but fairly well wooded shores. Ten miles farther on, below a succession of small ponds and channels with marshy shores, lies Windy Lake. Here the banks are moderately high and formerly were well wooded; but within the past few years they have been partially denuded by fire. The head of Windy Lake is 12 miles distant from Oxford Lake, near

[a] Compare Franklin, who says, in speaking of this gorge, "The brown fishing-eagle had built its nest on one of the projecting cliffs." Narrative of a Journey to the Polar Sea, p. 39, 1823.

FIG. 1.—RAPID BELOW WINDY LAKE.

FIG. 2.—CANOE ENTERING RAPID, TROUT RIVER.

FIG. 1.—RAPID, TROUT RIVER.

FIG. 2.—SHORE OF KNEE LAKE NEAR SOUTH END.

the northern end of which, 30 miles farther, is situated Oxford House, the only post or habitation of any kind on this route between Norway House and Hudson Bay (Pl. V, figs. 1 and 2). In the short stream connecting the two lakes four rapids occur, at two of which portages are necessary (Pl. VI, fig. 1).

Oxford Lake extends southwest and northeast, and its 30 miles are marked by irregular shores and many islands. The shores are mainly of rock and are generally well forested. The locality about Oxford House we found favorable for collecting and a stop of a few days was made. The promontory on which the post is situated was probably well wooded originally, but its western half has been entirely cleared and is mainly covered with grass, with a few patches of willows and other shrubs. East of this cleared area the ground slopes gently to Back Lake, about a quarter of a mile distant, and is fairly well covered with spruce, fir, tamarack, poplar, and willow. The soil is a stiff clay, and potatoes and other garden vegetables of fine quality are raised.

The waters of Oxford Lake flow into Back Lake through a short, narrow channel, and those of Back Lake find their outlet in Trout River (Pl. VI, fig. 2; Pl. VII, fig. 1), which runs southeastward to Trout Falls, 9 miles from Oxford House. At Trout Falls the river makes a plunge of about 12 feet into a deep pool. A short distance below is Knee Lake, a rather narrow body of water 40 miles in length and extending in a general southwest and northeast direction, with two somewhat abrupt bends. The shores of the southern part are high and well wooded (Pl. VII, fig. 2), and many wooded islands of various sizes dot the surface of the lake. At about a third of its length from the upper end the lake contracts and its shores become low and swampy. In this narrow part is Magnetic (or Magnetite) Island, a low, bare, irregular rock which offers considerable attraction to the magnetic needle. The shores and islands of the northern or larger part of the lake are low by comparison with those of the southern part, but are, like those, well wooded.

The next lake in the series is Swampy Lake, which is connected with Knee Lake by Jack River, a stream about 10 miles in length, containing four rapids in its brief course. Swampy Lake is 13 miles long and has low shores, especially on the eastern side, where the Hudson's Bay Company formerly had a post, long since abandoned.

Various species of water milfoil (*Myriophyllum*) and pondweed (*Potamogeton*) grow in the shallow portions of all these lakes, sometimes so profusely as to seriously interfere with navigation; and the beautiful water arum (*Calla palustris*) is frequently seen near the margin of lake or river. The larger species of birds are noticeably scarce and wild, in consequence of the incessant warfare waged by the natives, who eat anything wearing fur or feathers, and never willingly allow a large bird of any kind to escape. For this reason the gulls, terns, and ducks, which were nesting as we passed, were excessively shy. On the lakes and rivers off the main route these conditions probably do not prevail.

Swampy Lake finds its outlet in Hill River, a rapid, winding stream, containing a great many willow-covered islands, and characterized during the first 30 miles of its course by numerous rapids. These necessitate frequent portages, half the entire number on the route, but fortunately they are all short. Each of these portages has its significant name—White Mud Portage (Pl. VIII, fig. 1), Mossy Portage, Seeing Portage, etc. The particular significance of the last named is that from the portage thus known Brassy Hill, a notable landmark, is seen for the first time on the way to the Bay (Pl. VIII, fig. 2). This hill, which is also responsible for the name of the river, is a remarkable gravelly elevation 390 feet high and three-quarters of a mile east of the river. As it is the highest point of land anywhere in the whole region between Lake Winnipeg and Hudson Bay, the natives naturally regard it as a veritable mountain.

About 15 miles below the 'Hill,' Rock Portage, the last on the route, is reached. Here a large flat rock divides the channel, and on each side is a fall of about 5 feet. Boats and baggage are carried over the rock. The Hudson's Bay Company formerly maintained a trading post near this point, but abandoned it many years ago. Between Brassy Hill and Rock Portage banks of clay gradually make their appearance. These, at first low, increase in size and in the vicinity of the Rock have attained considerable height (Pl. IX, fig. 1). From this point to Hudson Bay the character of the country and of the river remains much the same. The clayey banks continue on both sides nearly all the way and vary from a few feet to two hundred in height. They are marked by numerous gullies, cut by the many small streams that enter the main river, and, owing to frequent landslides, are continually giving way, precipitating uprooted trees into the river (Pl. IX, fig. 2). In many places they are covered with a rank growth of willows and grasses, amid which are various orchids, violets, polygonums, and other small plants.

Several species of scouring-rush (*Equisetum*) grow abundantly in the shallow water and often on the banks. Along upper Hill River sweet gale (*Myrica gale*) is common, and at the mouth of Fox River, 30 miles below Rock Portage, buckthorn (*Rhamnus alnifolia*), honeysuckle (*Lonicera glaucescens*), silverberry (*Elæagnus argentea*), small-flowered viburnum (*Viburnum pauciflorum*), and Canadian buffaloberry (*Lepargyræa canadensis*) were collected. None but the last two were noted farther north. Banksian pine (*Pinus divaricata*) and canoe birch (*Betula papyrifera*) also find their northern limit in this region near the confluence of Hill and Fox rivers; and the aspen poplar (*Populus tremuloides*) was not noted beyond this point, though it may possibly extend farther north on this route. In some places

FIG. 1.—WHITE MUD RAPID, HILL RIVER.

FIG. 2.—HILL RIVER NEAR SEEING PORTAGE; BRASSY HILL IN DISTANCE.

FIG. 1.—LEFT BANK OF HILL RIVER FROM ROCK PORTAGE.

FIG. 2.—CLAY BANKS, LOWER HILL RIVER.

the valley of Hill River is narrow and the view of the traveler is confined to the immediate banks; in others it is broad, and its gradual, well-wooded slopes afford more extended prospects.

The stream resulting from the junction of Hill and Fox rivers is called Steel River. After a course of 30 miles this in turn unites with the Shamattawa to form what is known as Hayes River, a broad, shallow stream on whose shores gravelly beaches, absent on the deep and narrow Steel River, alternate with high clay banks. In the next 50 miles the character of the country varies but little. Back from the river are mossy swamps, which support a growth of black crowberry (*Empetrum nigrum*), Labrador tea (*Ledum*), dwarf birch (*Betula glandulosa*), and associated species. Spruces and tamaracks are rather stunted. This characteristic Hudsonian country first makes its appearance in large areas on the east side of Swampy Lake, about 100 miles from Hudson Bay in a direct line, and is probably continuous from that point to the Bay, although for some distance the immediate banks of the river continue to show a more southern element.

In the lower part of Hayes River are a number of low, sandy islands nearly devoid of vegetation from being overflowed and ice swept during a large part of the year. Arctic terns and semipalmated plovers, which breed abundantly on some of the islands, were here met with for the first time.

Six miles from the point where Hayes River empties into the Bay is York Factory, a post of the Hudson's Bay Company. It is on a strip of land, here 3 or 4 miles wide, lying between the Hayes and the Nelson. In the old days it was an important and well-peopled post, and was formerly the base from which all the supplies for the great interior region, brought from England by fleets of sailing vessels, were distributed by boats.

The ground is low and swampy and is covered by the usual rather stunted growth of spruces, tamaracks, and thickets of willow. The soil is of a spongy character and remains frozen a few feet below the surface throughout the year.[a] The woods extend about 5 miles beyond the post and are succeeded by a mile of very wet marsh intersected by many sloughs and channels. This marsh, which is called Point of Marsh or Beacon Point, supports a rank growth of grass and water plants, among which bog-bean (*Menyanthes trifoliata*) and various species of pondweed (*Potamogeton*) are especially conspicuous.

[a] Concerning this subject Richardson says:

"At York Factory, on Hudson's Bay, in lat. 57°, in October, 1835, recent frosts had penetrated eight inches into the soil; the thaw due to the summer heat extended twenty-eight inches beyond this, beneath which a frozen bed seventeen and a half feet thick reposed on thawed mud which had a temperature of 33° F. The mean annual heat of this place is 25½ F., being equal to that of Fort Simpson, which lies five degrees further north." Arctic Searching Expedition, p. 217, 1851. (See also Richardson, Edin. New Phil. Journ., XXX, p. 117, Jan., 1841.)

Various ducks and marsh sparrows and the elusive yellow rail find here a congenial habitat, and here, during their semiannual migrations, the various geese, ducks, and shore birds which breed in myriads to the northward stop for rest and food. Mosquitos become more abundant as the Bay is neared and are extremely troublesome at this point.

During our stay at York Factory—July 11 to 17—collecting was difficult, owing to the almost incessant rain. More time was needed, but the short season and the distance still to be covered impelled us to proceed. Temporarily abandoning our canoe, therefore, we left in a sailboat for Fort Churchill, 150 miles up the coast.

Contrary winds and periods of calm conspired to delay us, and the trip occupied six days. On the afternoon of the second day, being unable to proceed, we pushed in as far as possible toward the shore at high tide, and during the ebb were able to go ashore by taking a 3-mile walk over the bouldery, weed-strewn beach, where, on every hand, flocks of shore birds of various species were hastily seeking a feeding place on the broad expanse left bare by the ebbing tide. On reaching the shore we found the Barren Grounds on a small scale lying before us. Gravelly ridges, the remains of old sea beaches, extended in various directions at a few feet above the general level, the intervening depressions occupied by small ponds or marshes. Occasional stunted spruces on the ridges and dwarf birches and straggling willows on the lower ground were the only fair-sized shrubs, though various small shrubby plants were abundant. Hundreds of curlews, godwits, phalaropes, plovers, and sandpipers of different species swam or waded about the shallow ponds in their never-ending search for food. A den on a gravelly hillock a foot or two higher than the general level was occupied by a litter of half-grown Arctic foxes, and not far away was seen a pair of willow ptarmigan with young just able to fly.

These patches of tundra are found all along the coast between York Factory and Fort Churchill. They seem to be roughly semicircular in shape, the woods that bound them extending much nearer the coast on the banks of the rivers than elsewhere. At the point where we landed, between Stony and Owl rivers, the forest was just visible from the shore of the Bay. Similar conditions are said to exist farther south toward the Severn, though in all probability fewer Barren Ground animals are found in that region.

No other stop was made until we reached the mouth of Churchill River. Here the physiographic conditions are different from those found at any other points visited on the shore of the Bay. A ridge of greenish-gray sandstone or quartzite (Pl. XI, fig. 1) extends to the coast on each side of Churchill River, and on the eastern side stretches eastward along the coast several miles toward Cape Churchill. These

FIG. 1.—FORT CHURCHILL.

FIG. 2.—MEADOWS, LOOKING SOUTHWEST FROM FORT CHURCHILL.
Habitat of *Calcarius pictus*.

FIG. 3.—SEAL (ERIGNATHUS BARBATUS), FORT CHURCHILL.

FIG. 1.—LEDGE OF QUARTZITE AT FORT CHURCHILL.

FIG. 2.—LOW TIDE AT FORT CHURCHILL.

Feeding ground of various shore birds.

ridges, particularly on the western side of the Churchill, consist of a succession of rounded hills, which attain a maximum altitude of about 100 feet, and support a shrubby, herbaceous growth with many mosses. Over this rocky area are scattered numerous shallow ponds with outlets flowing to the sea through narrow ravines that are scantily clothed with dwarfed spruces and willows. In these sparsely wooded ravines the Harris sparrow was common, the parent birds accompanying young just from the nest. A low, gravelly point extending seaward from the hills forms the western bank of the river immediately at its mouth. On this point lie the ruins of old Fort Prince of Wales, destroyed by the French in 1782. The bank of the river immediately opposite is composed of high rocks rising abruptly from the water.

Fort Churchill (Pl. X, fig. 1) is situated on the west side of the tidal lagoon which comprises the lower part of Churchill River. It is about 4 miles from the mouth of the river. To the south and west extends a broad, level meadow, only a foot or two above high-water mark, clothed with a low, shrubby growth in which appears an occasional dwarfed spruce or tamarack (Pl. X, fig. 2). This meadow is a favorite place for Smith longspurs and horned larks, and on its drier portions we found a few burrows of lemmings (*Dicrostonyx*). Numbers of seals (Pl. X, fig. 3) of several species frequent the mouth of Churchill River, attracted by the abundance of fish at that point.

As it was very desirable to do some collecting on the Barren Grounds, I left Fort Churchill on July 30 in a small sailboat, accompanied by three Indians, my brother remaining at Churchill to complete the collection. On account of the low coast, the tide in many places going out from 6 to 8 miles (see Pl. XI, fig. 2), traveling in a small boat is very difficult. We could not land except at high tide, and were obliged to embark at the same stage of water. Owing to the build of our boat, sailing was impossible unless the wind was fair or nearly so, and rowing was very difficult.

On the afternoon of July 31 a few hours were spent on the shore of Button Bay. Here the spruce woods nearly reach the shore at one point. North of the woods a broad grassy plain, intersected by many channels connecting small, shallow ponds, extends for several miles along the shore. Over this area a great many shore birds and ducks were feeding, some species accompanied by young, evidently reared in the vicinity, but by far the greater number associated in large, restless flocks, showing that the southward movement had commenced.

That evening we rowed several miles along the coast and encamped after dark on a small, sandy islet, just above high-water mark, where Arctic terns were breeding. The next evening our camping place was a sandy point near the mouth of Seal River, the position of which is indicated by a conspicuous rounded mound that stands near its banks.

Along the coast here the woods are visible from the Bay, and scattered dwarf spruces and tamaracks extend to the shore. Before Hubbart Point is reached, however, the tree limit curves inland so rapidly that the forest disappears from view altogether, although, according to Tyrrell, it can be seen with a glass from the summit of Hubbart Point.[a]

This point, which we passed on the afternoon of August 2, is a high, grassy headland used as a burial place by the Eskimos, and is the most conspicuous landmark on this part of the coast, the mound near Seal River, just mentioned, being next in importance. Egg Island, which is mentioned as a breeding place for many sea birds, is apparently not conspicuous, for we failed to identify it either time we passed, probably being too far off shore to see it. At dark on August 2 we anchored behind a small, rocky islet somewhat north of Egg Island, and at daylight next morning were again on our way.

By noon we had reached a sandy point near Thlewiaza River, which proved so favorable a spot for collecting that I remained there several days (Pl. XII, fig. 1). From the shore to a number of rocky and gravelly ridges a few feet in height, which were several miles inland, the country was nearly level, and was mostly wet and filled with small hummocks. Near the shore were many broad, shallow ponds and muddy flats. Occasional dry areas, apparently raised sea beaches, were covered with rounded boulders of various sizes, and were inhabited by lemmings of the genus *Lemmus*, the burrows of which also occurred in the drier portions of the adjacent grassy meadows. A large species of meadow mouse (*Microtus*) was also found here, but was more abundant in the patches of coarse beach grass (*Elymus mollis*) which grew on the sandy ridges near the shore. On the gravelly ridges back from the immediate shore, pied lemmings (*Dicrostonyx*) were fairly abundant, and a number were secured. Dwarf shrubs, none of them exceeding a few inches in height, abounded; the most conspicuous were black crowberry (*Empetrum nigrum*), dwarf birch (*Betula nana*), Labrador tea (*Ledum palustre*), and several species of dwarf willows, including *Salix anglorum* and *S. phylicifolia*. The scene was one of absorbing interest. On the beach and mud flats and about the shallow ponds thousands of shore birds of a dozen species circled and fed, the larger kinds, mainly Hudsonian curlews and godwits, keeping at a little distance, the smaller kinds almost oblivious of my presence. In the deeper ponds among the ridges back from shore red-throated and Pacific loons, which later made night hideous by their cries, were feeding their unfledged young. Pomarine and parasitic jaegers harried about the tundra or sat motionless on the knolls, apparently asleep. Willow ptarmigan led their broods about in search of food, and horned larks, Lapland longspurs, tree and savanna sparrows, and redpolls flitted from boulder to boulder.

---

[a] Ann. Rept. Can. Geol. Surv., 1896 (new ser.), IX, p. 90F. (1897).

FIG. 1.—CAMP ON BARREN GROUNDS, 50 MILES SOUTH OF CAPE ESKIMO.

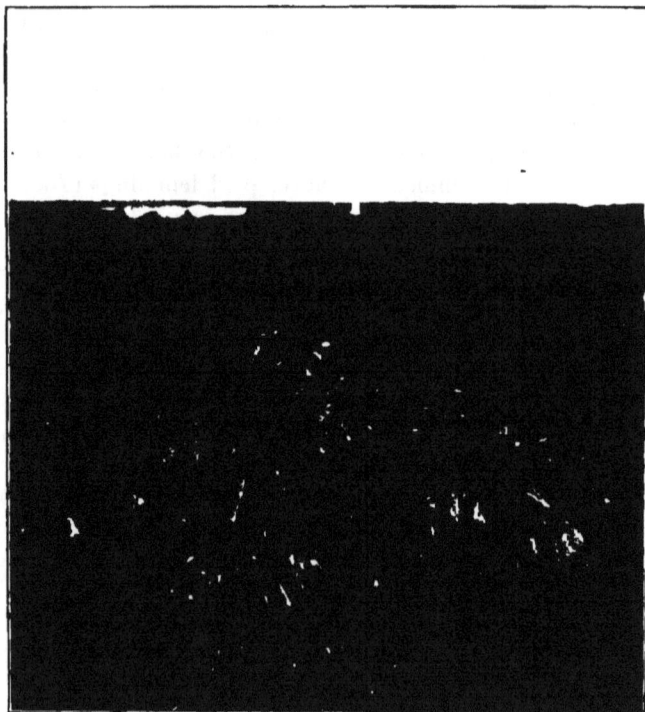

FIG. 2.—BURROW OF LEMMING (LEMMUS TRIMUCRONATUS), BARREN GROUNDS.

FIG. 1.—CAMP ON BARREN GROUNDS, 25 MILES SOUTH OF CAPE ESKIMO.

FIG. 2.—ESKIMO GUIDE AT NORTHERN CAMP.

Leaving on the afternoon of August 8, I pushed northward, accompanied by an Eskimo (Pl. XIII, fig. 2) who had promised to guide me to a place where ground squirrels could be secured. After a great deal of hard work and exposure we landed, on the evening of August 9, in a small, shallow bay at the mouth of a stream about 25 miles south of Cape Eskimo (Pl. XIII, fig. 1). Here I remained until August 13.

The country was similar to that farther south and supported a similar flora, but wet ground was less common and sandy ridges were more frequent. The Barren Ground caribou had commenced their southward movement and one was secured. Some of the shore birds had departed, and the daily lessening numbers of other species had an obvious significance. On the sandy ridges and hillocks were scattering burrows of ground squirrels, but the absence of colonies indicated that I had reached only the border of their range. Two polar bears were seen in the vicinity, but we were unable to secure either.

The distance to be covered and the difficulties of navigation to be overcome admonished me that my return journey must be begun, and on the morning of August 13 I started southward. The wind was against us the first day, and at the close of the second we had progressed no farther than my former camp, about 25 miles. Two more days of very laborious work took us a few miles below Hubbart Point. Here willows attain a respectable size, and on the boulder-covered ridges the buffalo-berry (*Lepargyræa canadensis*) is a conspicuous shrub. A few Arctic hares were found here, and on the morning of August 17 two were secured.

On August 17 and 18 slow progress was made, and we were only able to reach a point about 15 miles above Seal River, the high mound near its mouth being visible from the ridges close to camp. Tamaracks grew in the more sheltered places, and the edge of the forest was only a few miles back from the coast.

On the morning of August 19 a fair, strong wind that lasted until noon carried us within sight of the Beacon and the high rocks near Fort Churchill, which we reached that evening. During my absence my brother had made a good collection, and we left Fort Churchill on the afternoon of August 21, arriving at York Factory on the evening of August 26. In the course of this uneventful voyage we spent a few hours ashore on the afternoon of August 24 about 20 miles below Cape Churchill, where the conditions were somewhat similar to those between Stony and Owl rivers (see p. 18). Lemmings (*Dicrostonyx*) inhabited the sandy ridges near the shore.

We left York Factory on the afternoon of August 28 in our canoe, which was loaded with the outfit and the entire summer's collection, and, making further collections on the way, arrived at Norway House September 16, after a very laborious but pleasant trip.

Hayes, Steel, and Hill rivers as far as the Rock Portage were ascended by tracking—the men walking along the shore pulling the canoe by a line, while our efforts were directed toward steering and avoiding the rocks, though we sometimes relieved them on the line. The passage of the various rapids on Hill River was accomplished with much difficulty. Many we were able to ascend by poling and paddling; at others it was necessary to wade waist deep in the seething water, dragging the canoe by hand (Pl. XIV, fig. 1); and often neither of these methods was possible and we were obliged to unload the boat and carry both canoe and baggage around the rapids. To avoid the ascent of the river through Hell Gate Gorge (Pl. XIV, fig. 2), we made a detour, as is usual on the upstream journey. Leaving Pine Lake by a narrow passage, we followed the windings of a tortuous lake for several miles in a direction approximately parallel to the course of the river, and then made a portage over a low divide to the shore of an arm of one of the lakes a short distance above Hell Gate Rapids. About twenty-five portages in all were required on the return journey, during which we retraced, with the exception of the slight deviation just mentioned, the exact route followed on our northward journey.

The trip down Lake Winnipeg was made by steamer, and we arrived at Winnipeg September 22.

LIFE ZONES OF THE REGION.

Our route from Winnipeg to the northernmost point reached, a short distance south of Cape Eskimo, passed successively through the Canadian and Hudsonian zones and entered well into the Arctic.

In the fauna and flora of Winnipeg the Canadian element seems to predominate, though the presence of *Quercus macrocarpa*, *Acer negundo*, and *Ulmus americanus* among trees, *Blarina brevicauda* among mammals, and *Zenaidura macroura* and *Icterus galbula* among birds, indicate that there is a strong tinge of Transition. It is probable that this Transition element disappears a short distance to the northward, but no very definite data regarding the country bordering Lake Winnipeg is available. In the region about Norway House the fauna is pure Canadian, as evidenced by the presence of *Parus hudsonicus*, *Perisoreus canadensis*, *Hylocichla u. swainsoni*, *Canachites canadensis*, *Picoides arcticus*, *Regulus satrapa*, *Zonotrichia albicollis*, *Mustela pennanti*, *Mustela americana*, *Lynx canadensis*, and other characteristic Canadian species. In the cleared and burnt tracts *Chordeiles virginianus* is common.

Between Norway House and Swampy Lake there is little change in the fauna. On the eastern shore of Swampy Lake the true Hudsonian begins and is probably continuous to Hudson Bay. Here *Zonotrichia albicollis* is gradually replaced by *Z. leucophrys*, *Chordeiles* occurs only as a straggler, and *Peromyscus* is much reduced in numbers. This

FIG. 1.—DRAGGING CANOE UP RAPID, TROUT RIVER.

FIG. 2.—HELL GATE GORGE, LOOKING UP.

area appears to be a sort of coastal plain, and occupies a strip at this point about 100 miles wide on the borders of the Bay. It is comparatively level, and the beds of the streams are free from rocks. The spruce, tamarack, balsam poplar, aspen poplar, Banksian pine, and canoe birch, which have formed the bulk of the forest, are here reduced in size, and the last three practically disappear near this point. These conditions prevail until the Bay is reached at the mouth of Hayes River. How far this Hudsonian strip extends southward along the coast of Hudson Bay is not apparent, but the presence at Moose Factory of *Ampelis cedrorum*, *Dendroica maculosa*, *Wilsonia canadensis*, *Sialia sialis*, *Carpodacus purpureus*, and *Condylura cristata* would seem to indicate that at that point the Canadian element must predominate over the Hudsonian. To the northward from York Factory along the coast, patches of tundra, where the fauna is typically Arctic, are first found not far above Nelson River. Similar barren areas occur on the coast between York Factory and Severn River, but too little is known about the fauna of that region to determine whether the Arctic or the Hudsonian element predominates. These barren areas, from York Factory as far north as Fort Churchill, lie mainly between the mouths of the numerous rivers. Above Cape Churchill the Arctic Zone is continuous on the coast, except in the sparsely wooded area about the mouth of Churchill River and on part of the shores of Button Bay, where a considerable admixture of Hudsonian forms occurs. Among the birds and mammals which breed in the Arctic zone may be mentioned various species of *Tringa*, *Phalaropus lobatus*, *Lagopus lagopus*, *Numenius hudsonicus*, *Limosa hæmastica*, *Anthus pensilvanicus*, *Calcarius lapponicus*, and *Calcarius pictus*, together with the Arctic fox and Richardson lemming, and farther north the musk-ox and Barren Ground caribou.

PREVIOUS WORK.

The earliest important work referring to the natural history of the Hudson Bay region is Edwards's Natural History,[a] published in four volumes from 1743 to 1751, though a few notes on the more conspicuous birds and mammals of the region appeared about the same time in the narratives of the voyage of the ship *California*.

Edwards figured nearly forty species of birds and a few mammals from Hudson Bay, the largest part of which were thus first brought to the attention of the scientific world. His colored figures were accompanied by very good descriptions, but were designated only by English names. Linnæus bestowed binomial names on most of these species, in some cases referring exclusively to Edwards's figures, but in others citing other authors in addition. Concerning the source of

[a] For full references to publications, see Bibliography, p. 27.

the Hudson Bay material on which Edwards's figures were based, Richardson says:

> The first collections of Hudson's Bay birds of which I can find any record, are those formed by Mr. Alexander Light, who was sent out, ninety years ago, by the Hudson's Bay Company, on account of his knowledge of natural history; and by Mr. Isham, who, during a long residence, as Governor of various forts or trading posts, employed his leisure hours in preparing the skins of beasts, birds and fishes.[a]

While at Hudson Bay Mr. Light seems to have sent some birds to Edwards, which were figured in the first volume of Edwards's work, published in 1743; and on his return to England in 1745 he appears to have turned over to Edwards, through the Royal Society, other birds for illustration. Mr. Isham returning to England at nearly the same time, about 30 undescribed species passed into Edwards's hands from this source, and were figured, together with Mr. Light's collection, in the third volume, published in 1747.

In 1768 the Royal Society of London received from Mr. Graham, of Severn River, a collection of mammals and birds from the west coast of Hudson Bay. A report on this collection was published in 1772 by J. R. Forster, which included descriptions of a number of species new to science.

Mr. Hutchins, an officer of the Hudson's Bay Company, who spent many years on Hudson Bay, mainly at the trading post at the mouth of Severn River, wrote a short time previous to 1785 a manuscript account entitled 'Observations on Hudson's Bay,' which contains many notes on the habits of the birds.[b] Speaking of this manuscript, Richardson says:

> His observations, which, in fact, embrace almost all that has been recorded of the habits of the Hudson's Bay birds up to the present time, being communicated to Latham and Pennant, are incorporated in the 'General Synopsis of Birds' and in 'Arctic Zoology.'[c]

During the period from 1769 to 1772 Samuel Hearne made three journeys overland to the northwest of Fort Prince of Wales (Churchill) in search of copper mines and for purposes of exploration. On his first two attempts he was forced to return before proceeding very far.

---

[a] Fauna Boreali-Americana, II, Introduction, p. ix, 1831.

[b] Ernest E. Thompson, who has examined this manuscript, cites it as follows:

"1782. Hutchins, T. Observations on Hudsons Bay, 651 pp.: pp. 45–180 treats of birds.

"An interesting unpublished manuscript volume in the library of the Hudson's Bay Company at London, with marginal annotations by Pennant. The author was for twenty-five years in the employ of the Hudson's Bay Company. There is no date or title page to the volume, but the last date mentioned is July 10, 1782, and it seems safe to conclude that it was issued about this date, from the fact that Pennant, in his second volume of Arctic Zoology (p. 222), published in 1785, refers to Mr. Hutchins's manuscript as though it had just reached him in a complete state." Proc. U. S. Nat. Mus. XIII, 1890, p. 639 (May 29, 1891).

[c] Fauna Boreali-Americana, II, Introduction, p. xi, 1831.

His third, however, was more successful, and he reached a point near the mouth of the Coppermine River. The narrative of these trips, which he published in 1795, contains much information concerning the animals of the region visited, but as this lies for the most part beyond the boundaries of the region now under consideration (see p. 38), few of the observations made on his journey are of use in the present connection. In his closing chapter, however, he gives a detailed account of most of the animals with which he had become familiar during a long residence in the country, and some of these observations have been utilized.

Sir John Franklin, in 1819, on his first journey to the Polar Sea, passed through Keewatin from York Factory to Norway House, following the route now usually traveled between these points. And in 1822, on his return to the Bay to sail for England, he again traversed this route. A considerable collection of birds was made at York Factory. Some specimens were apparently taken on the first arrival at the post in the early autumn of 1819, but a larger number between the middle of July and 1st of September, 1822, while the expedition was waiting to return to England. Apparently referring to the first York Factory collection, Joseph Sabine says:

> The specimens collected on the first arrival of the travellers at York Factory [a] were sent as soon as received in England, to the British Museum, where they became mixed with other collections from Hudson's Bay and Baffin's Bay * * *. It being impossible to separate them, these have been entirely excluded from the account, which consequently contains a much less number of sea birds than would otherwise have appeared in it. [b]

The valuable collection made at York Factory in the late summer of 1822 was almost equally unfortunate, as Richardson says:

> This was the only autumn collection made on either Expedition, and we regret that we have not been able to avail ourselves of it, so much as we could have wished, in drawing up the present work. Exclusive of the specimens above alluded to as having been entirely lost, many were destroyed by moths in London; and the only portion of the collection which I can now trace are forty specimens, which were presented to the Museum of the University of Edinburgh, and are still in good condition. [c]

The voyages of Capts. Edward Parry and John Ross, between 1819 and 1833, added much to the knowledge of the fauna of the northern part of the Hudson Bay region, and reports on the natural history were published in the appendices to their narratives.

---

[a] This seems to be the only evidence that a collection was made at York Factory on the arrival of the expedition at that post in the early autumn of 1819. No mention of this collection occurs in the narrative of the journey given by Franklin (nor, indeed, is the second collection alluded to), and Richardson intimates that no such collection was made.

[b] Franklin's Narrative of a Journey to the Polar Sea, Appendix, p. 670, 1823.

[c] Fauna Boreali-Americana, II, Introduction, p. xv, 1831.

Between 1829 and 1837 appeared Fauna Boreali-Americana, by Swainson, Richardson, and others, the first volume treating of mammals, the second of birds, and the others of fishes and insects. The material accumulated on the voyages of Parry, the first voyage of Ross, and the journeys of Franklin, together with much additional information on the natural history of the northern parts of North America, is elaborated in this valuable publication.

The expedition to the mouth of the Great Fish River under Captain Back (1833 to 1835) was productive of much new information regarding the natural history of the interior of British America, owing chiefly to the labors of Richard King, surgeon and naturalist to the expedition. A few notes appear in Back's narrative of the expedition, which also contains a list by Richardson, unfortunately often without annotations, of the specimens collected. King's narrative of the journey is replete with information on the birds and mammals collected and observed, and the localities and dates of many specimens merely listed in Back's narrative are thus put on record. A few observations of the fauna of the lower part of the Great Fish River and of the region between Lake Winnipeg and York Factory, the route followed by King on his return journey, were made within the boundaries of the present Province of Keewatin.

In 1846 and 1847 John Rae made a journey of exploration from York Factory to the Arctic coast, wintering at Repulse Bay. A great many notes on the natural history of the region appear in his narrative, and the 'Appendix' contains an annotated list, by J. E. Gray, of the mammals secured, and nominal lists, by G. R. Gray, of the birds, fishes, plants, and geological specimens collected. Many of the birds, obtained mainly at Repulse Bay, are still in the British Museum. A few small woodland species, recorded in the British Museum Catalogue of Birds as taken at Repulse Bay, were probably collected farther south during the early part of the expedition.

Thomas Blakiston made some observations in the region between York Factory and Fort Carlton, the results of which appeared in several papers published from 1857 to 1863.

A collection of birds and mammals gathered by officers of the Hudson's Bay Company, mainly from the west coast of Hudson Bay, was reported by Andrew Murray in 1859.

Between 1864 and 1869 C. F. Hall lived among the Eskimos in northern Keewatin, with headquarters at Repulse Bay, whence he made several overland journeys, during which he visited Igloolik, King William Land, and other points. His principal object was to search for traces of Sir John Franklin. The narratives of his expedition contain a great deal of information regarding the game animals of the region

The catalogues of the United States National Museum contain the records of many specimens—mainly birds—collected at Moose Factory, James Bay, by James MacKenzie, C. Drexler, William MacTavish, and others between 1860 and 1870, and a few from other localities. Some of these have been published; others are published for the first time in the present report. Some of the specimens have apparently been lost or exchanged.

In 1878 and 1879 Dr. Robert Bell made some observations and collections in the region between Lake Winnipeg and Hudson Bay, and in 1884 made another collection on the west coast of the Bay. The results of his work were published in the reports of the Canadian Geological Survey.

In 1879 and 1880 Lieut. Frederick Schwatka made an overland journey from Camp Daly, near the mouth of Chesterfield Inlet, to King William Land to search for traces of Sir John Franklin. The narrative of this expedition contains much information concerning the game mammals of the country, on which the party depended chiefly for subsistence.

In the summer of 1881 a small collection of birds and mammals—afterwards acquired by the United States National Museum—was made at Moose Factory by Walton Haydon.

In the Auk for 1890 W. Eagle Clarke records a collection of birds made at Fort Churchill previous to 1845 by Dr. Gillespie, jr.

John Macoun, in the first part of a catalogue of Canadian birds, published in 1900, records for the first time a few birds from the region and contributes much original information regarding their distribution, derived mainly from observations and collections made by himself and other members of the Canadian Geological Survey during various surveying and exploring trips.

BIBLIOGRAPHY.

The following titles include the principal works and articles bearing more or less directly on the natural history of Keewatin, which have been consulted in the preparation of the present report:

1743–51. EDWARDS, GEORGE. A Natural History of Uncommon Birds and of some other rare and undescribed animals. 4 vols. Vol. I, 1743; Vol. II, 1747; Vol. III, 1750; Vol. IV, 1751. 4to. London.

A number of undescribed birds and mammals from Hudson Bay figured in colors, and accompanied by short descriptions and biographical remarks.

1744. DOBBS, ARTHUR. An Account of the Countries adjoining to Hudson's Bay, in the North-West Part of America. 1 vol. 8vo. London.

Mainly historical, but a few notes on natural history *passim*.

1748. [DRAGE, or SWAINE, CHARLES].[a] An Account of a Voyage for the Discovery of a North-West Passage by Hudson's Streights to the Western and Southern Ocean of America. Performed in the year 1746 and 1747 in the Ship California, Capt. Francis Smith, Commander. By the Clerk of the California. 2 vols. 8vo. Vol. I, pp. i–vii, 1–237; Vol. II, pp. 1–326. London.

Vol. I contains much information on the mammals and birds, based mainly on observations made at the mouth of Hayes River, where the expedition passed the winter of 1746–47. The natural history notes are mainly on pp. 174–178, but may also be found here and there throughout the volume. Vol. II, which has not been seen, completes the account and is said to contain natural history notes.

1752. ROBSON, JOSEPH. An Account of six years residence in Hudson's Bay, from 1733 to 1736, and 1744 to 1747. 12mo. London.

A general account, chiefly concerning Fort Churchill and York Factory, with a few observations regarding the larger mammals and birds.

1758 and **1766.** LINNAEUS, CAROLUS. Systema Naturæ per Regna Tria Naturæ * * *

Many species described from Hudson Bay.

1771. PENNANT, THOMAS. Synopsis of Quadrupeds.

Contains much information on mammals of the Hudson Bay region.

1772. BARRINGTON, DAINES. Investigation of the Specific Characters which distinguish the Rabbit from the Hare. <Phil. Trans. London, LXII, pp. 4–14.

Refers especially to the 'Hudson Bay Quadruped,' and concludes that it is a hare.

1772. FORSTER, JOHN REINHOLD. Account of several quadrupeds sent from Hudson's Bay. <Phil. Trans. London, LXII, pp. 370–381.

A report on a collection of mammals sent by Mr. Graham to the Royal Society of London.

1772. FORSTER, J. R. An account of the Birds sent from Hudson's Bay; with observations relative to their Natural History, and Latin descriptions of some of the most uncommon. <Phil. Trans. London, LXII, pp. 382–433.

Report on a collection sent by Mr. Graham to the Royal Society of London. Fifty-eight species treated, the following seven described as new: *Falco spadiceus*, *Strix nebulosa*, *Emberiza leucophrys*, *Fringilla hudsonias*, *Muscicapa striata*, *Parus hudsonicus*, *Scolopax borealis*.

1778. PALLAS, PETER SIMON. Novae Species Quadrupedum e Glirium Ordine.

Original description of *Mus empetra* (p. 75) evidently based on a specimen of the Canadian woodchuck.

1780. ZIMMERMANN, EBERHARD AUGUST WILHELM. Geographische Geschichte des Menschen, und der vierfusigen Thiere. Vol. II.

Original description of *Bos moschatus* (p. 86) and of *Dipus hudsonius* (p. 358).

1781. LATHAM, JOHN. A general synopsis of Birds.

Contains much matter from Hutchins's manuscript article on Hudson Bay.

1784, 1785. PENNANT, THOMAS. Arctic Zoology. Vol. I: Introduction; Class I, Quadrupeds, 1784. Vol. II: Class II, Birds, 1785.

Contains much original matter concerning Hudson Bay, communicated by Hutchins, Hearne, and Graham; also much material derived from the writings of Ellis, Forster, Latham, etc.

---

[a] The name of the author is not given; it is stated by some writers to be 'Drage,' by others 'Charles Swaine.'

1790. LATHAM, JOHN. Index Ornithologicus, sive Systema Ornithologiæ * * *
Contains original description of *Numenius hudsonicus*, based mainly on a specimen from Hudson Bay received from Hutchins.

1795. HEARNE, SAMUEL. A Journey from Prince of Wales' Fort in Hudson's Bay to the Northern Ocean. 1 vol., 4to, pp. i–xliv, 1–458. London, 1795. There is also a Dublin edition, 8vo, 1796, and another London edition, 1807.

Contains notes throughout the narrative on the mammals and birds of the region, Chapter X, pp. 358–458, treating exclusively of the natural history of those parts of British America known to the author.

1814. LEACH, WILLIAM ELFORD. [Description of] *Arvicola xanthognatha*. <Zoological Miscellany; being descriptions of new, or interesting animals. Vol. 1, p. 60. Plate XXVI.

"Habitat ad Hudson's Bay."

1819. LEACH, W. E. Notice of some animals from the Arctic Regions. <Thomson's Annals Philos., XIII, pp. 60–61.

A list, slightly annotated, "of the mammalia and birds that have been received from the Northern Expeditions and which have since been sent to the British Museum by the Admiralty." Includes specimens received from Ross and Parry.

1819. ROSS, JOHN. A Voyage of Discovery, made under the orders of the Admiralty, in His Majesty's ships Isabella and Alexander, for the purpose of exploring Baffin's Bay and inquiring into the probability of a Northwest Passage. 1 vol, 4to, pp. i–xl, 1–252, i–cxxxvi. London.

Appendix II, pp. xlviii–lx (author not stated) contains an article on the birds and mammals observed about Baffin Bay.

1819. SABINE, JOSEPH. An Account of a new Species of Gull lately discovered on the West Coast of Greenland. <Trans. Linn. Soc. London, Vol. XII, Art. XXXII, pp. 520–523.

Original description of *Larus sabinii*. Mention made of a black-headed gull from Hudson Bay, "certainly an undescribed species, bearing a strong resemblance to the *Larus Sabini* except that it has an even tail and is without the dark collar round the neck." (Probably *L. philadelphia*).

1821. PARRY, WILLIAM EDWARD. Journal of a voyage for the discovery of a N. W. Passage from the Atlantic to the Pacific, 1819–20, in her Majesty's Ships Hecla and Griper. 4to. London.

A few observations relative to the natural history of Melville Island. (See also Supplement to the Appendix, 1824).

1823. FRANKLIN, JOHN; and SABINE, JOSEPH. Narrative of a Journey to the Shores of the Polar Sea, in the years 1819, 20, 21 and 22. By John Franklin. >Zoological appendix by Joseph Sabine. 1 vol., 4to, pp. i–xv, 1–783. London.

A few species from York Factory mentioned in the appendix. The narrative contains a few references to the natural history of Keewatin.

1824. LYON, G. F. The Private Journal of Captain G. F. Lyon, of H. M. S. Hecla, during the recent Voyage of Discovery under Captain Parry. [*Parry's Second Voyage.*] 12mo. London.

Contains numerous notes on natural history, relating especially to the vicinity of Winter Island and Igloolik, Melville Peninsula.

1824?. MACGILLIVRAY, WILLIAM. Description, Characters, and Synonyms of the Different Species of the Genus *Larus*, with critical and explanatory Remarks. <Memoirs Wernerian Nat. Hist. Soc., V, Part I.

The currently accepted original description of the Ross Gull occurs on page 249.

1824. PARRY, W. E. Journal of a Second Voyage for the discovery of a North West passage from the Atlantic to the Pacific in his Majesty's Ships Fury and Hecla. 1821–23. 4to. London.

Contains a few notes on the natural history of Melville Peninsula. (See also Appendix, 1825.)

1824. SABINE, EDWARD. A supplement to the Appendix of Captain Parry's [first] Voyage for the Discovery of a North-west passage, in the years 1819–20. >Vertebrata, by Edward Sabine.

Collections and observations made mainly about Melville Island.

1825. LYON, G. F. A Brief Narrative of an unsuccessful attempt to reach Repulse Bay through Sir Thomas Roe's "Welcome," in His Majesty's Ship Griper, in the year MDCCCXXIV. 8vo, pp. 198. London.

Contains a few notes on the natural history of Southampton Island and of the mainland in the vicinity of Cape Fullerton.

1825 (1827). RICHARDSON, JOHN. Appendix to Captain Parry's Journal of a Second Voyage for the discovery of a north-west passage from the Atlantic to the Pacific. 4to, pp. 1–432. London. >Zoological Appendix. Account of the Quadrupeds and Birds, by John Richardson.

Refers mainly to natural history of Melville Peninsula, particularly Winter Island and Igloolik.

1826. PARRY, WILLIAM EDWARD, and ROSS, JAMES CLARK. Journal of a Third Voyage for the discovery of a North-west passage from the Atlantic to the Pacific, performed in the years 1824–25, by William Edward Parry. 1 vol., 4to, pp. i–xxviii, 1–186, 1–151. London. Appendix, Zoology. By Lieut. James Clark Ross.

The zoological appendix refers mainly to the natural history of Port Bowen, where the expedition wintered, and other points about Prince Regent Inlet. A few notes on natural history occur in the narrative.

1828. RICHARDSON, JOHN. Short Characters of a few quadrupeds procured on Captain Franklin's late expedition. <Zool. Journ. III, No. 12, pp. 516–520, Jan. to Apr., 1828.

Original description of *Sorex palustris*, "Hab. Marshy places, from Hudson's Bay to the Rocky Mts." *Cricetus talpoides* described from Hudson Bay (specimen probably from Manitoba or Saskatchewan).

1829. DOUGLAS, DAVID. Observations on some Species of the Genera *Tetrao* and *Ortyx*, natives of North America; with Descriptions of Four new Species of of the former and Two of the latter Genus. <Trans. Linn. Soc. London, XVI, pp. 133–149.

Mentions abundance of *Pediœcetes phasianellus* about Oxford Lake, and of *Canachites canadensis* on the streams which flow into Hudson Bay.

1829. RICHARDSON, JOHN. Fauna Boreali-Americana, Part First. Quadrupeds. 4to, pp. i–xlii, 1–300. London.

Contains much original matter on the mammals of the Hudson Bay region.

1831. SWAINSON, WILLIAM, and RICHARDSON, JOHN. Fauna Boreali-Americana. Part Second. The Birds. 4to, pp. i–lxvi, 1–524. London.

Contains much original matter on the birds of the Hudson Bay region.

1835. ROSS, SIR JOHN. Narrative of a Second Voyage in search of a North-west Passage, and of a residence in the Arctic Regions during the years 1829, 1830, 1831, 1832, 1833. 1 vol., 4to, pp. i–xxxiv, 1–740. London.

Contains many notes on the zoology of the country to the northward of Hudson Bay, particularly about Felix Harbor, Boothia.

1835. Ross, James Clark. Appendix to the Narrative of a Second Voyage in search of a North-west Passage, etc. By Sir John Ross. 1 vol., 4to, pp. i–xii, 1–120, i–ciii. London. >Report on mammals and birds by J. C. Ross, pp. vii–xlv.

A systematic account of the collections and observations made on Ross's second voyage.

1836. Back, George; and Richardson, John. Narrative of the Arctic Land Expedition to the mouth of the great Fish River, and along the shores of the Arctic Ocean, in the years 1833, 1834, and 1835; by Captain Back, R. N., Commander of the Expedition. 1 vol., 8vo, pp. i–x, 1–663. London. Appendix No. 1. Zoological Remarks by John Richardson, pp. 477–518.

In the narrative mention is made of the observation of a number of species within the present boundaries of Keewatin. The Zoological Appendix contains a nominal list of the mammals collected, and a list of the birds in which the localities of a few species are indicated, followed by a commentary "respecting those which are objects of chase to the Indian hunter, either for food or for the sake of their fur," with "a few brief remarks on the specimens of the other species when they serve for the elucidation of doubtful points of their history."

1836. King, Richard. Narrative of a Journey to the shores of the Arctic Ocean in 1833, 1834, and 1835; under the command of Capt. Back, R. N. 2 vols. 12mo. Vol. I, pp. i–ix, 1–312. Vol. II, pp. i–viii, 1–321. London.

Contains many notes on the natural history of the Great Fish River and on the route from the headwaters of that river to York Factory via the Slave, Athabasca, Clearwater, Churchill, Saskatchewan, Nelson, Echimamish, Hill, Steel, and Hayes rivers.

1837. Richardson, John. Report on North American Zoology. <Report Sixth Meeting Brit. Assn. Adv. Sci. for 1836, V, pp. 121–224.

Mentions occurrence of many species of birds and mammals in the Hudson Bay region.

1841. Barnston, George. Observations on the Progress of the Seasons as affecting Animals and Vegetables at Martin's Falls, Albany River, Hudson's Bay. <Edin. New Phil. Journ., XXX, pp. 252–256.

Records the times of arrival and departure of many birds, and times of breeding of different birds and mammals, spawning of fish, and many other natural phenomena.

1850. Rae, John; Gray, George Robert; Gray, John Edward. Narrative of an Expedition to the Shores of the Arctic Sea in 1846 and 1847; by John Rae. 1 vol, pp. 248. London. List of birds by G. R. Gray. Mammals by J. E. Gray.

Besides the systematic lists of the specimens collected, the narrative contains much information on the natural history of the northwest coast of Hudson Bay, particularly about Repulse Bay.

1857. Blakiston, Thomas. Notes of a Naturalist on his Passage towards the Far West. <Zoologist, XV, pp. 5840–5843.

Natural history notes made on a voyage from England to York Factory, Hudson Bay, in which a few species of birds noted in Hudson Straits and on Hudson Bay are mentioned.

1858. Baird, Spencer Fullerton; Cassin, John; and Lawrence, George Newbold. Vol. IX of the Pacific R. R. reports. 4to, pp. l–lvi, 1–1005. Washington.

This volume, usually quoted as 'Birds of North America,' contains records of a few birds from Nelson River and Hudson Bay.

1858, 1859. MURRAY, ANDREW. Contributions to the Natural History of the Hudson's Bay Company's Territories. <Edin. New Phil. Journ. (new series), VII, pp. 189–210, Part I, Reindeer, April, 1858. IX, pp. 210–220. Part II, Mammalia, concluded; pp. 221–231, Part III, Aves; April, 1859.

Annotated lists of a number of mammals and of 82 species of birds from the west coast of Hudson Bay and adjoining region.

1859. BLAKISTON, THOMAS. Scraps from the Far West. <Zoologist, XVII, pp. 6318–6325, 6373–6376.

Field notes on birds observed on a journey from York Factory, Hudson Bay, to Fort Carlton, on the Saskatchewan.

1860. BARNSTON, GEORGE. Recollections of the Swans and Geese of Hudson's Bay. <Ibis, II, pp. 253–259, July, 1860.

Remarks on migration and other habits of the common species as observed by the writer at Moose Factory and other posts on Hudson Bay, during a long residence in the country.

1860. SCLATER, PHILIP LUTLEY. [Exhibition of some bird skins procured by Captain Herd at Port Churchill, Hudson's Bay]. <Proc. Zool. Soc. London, XVIII, p. 418.

Mentions a crane, probably *Grus canadensis*, and a goose referred to *B. hutchinsii.*

1861. BARNSTON, GEORGE. Recollections of the Swans and Geese of Hudson's Bay. <Canadian Naturalist and Geologist, VI, Article XXIV, pp. 337–344.

The same as his article under the same title, published in the Ibis for 1860, as above cited, but containing new matter.

1861, 1862. BLAKISTON, THOMAS. On Birds collected and observed in the Interior of British North America. <Ibis, III, pp. 314–320, 1861; IV, pp. 3–10, 1862.

Contains a few references to birds of Hudson Bay, and the region between York Factory and Lake Winnipeg.

1863. BLAKISTON, T. On the Birds of the Interior of British North America. <Ibis, V, pp. 39–87, 121–155.

A nearly complete list of the birds of the interior of Canada, east of the Rocky Mountains, compiled from Fauna Boreali-Americana, and the writings of Murray, Ross, etc., together with the author's own observations.

1864–1866. BAIRD, S. F. Review of American Birds in the Museum of the Smithsonian Institution. Part I, pp. 1–450. [Issued in installments from June, 1864, to June, 1866.]

Contains many records of birds from Hudson Bay. These papers are brought together in Smithsonian Miscellaneous Collections, Vol. XII, 1874.

1869. BAIRD, S. F. Occurrence of the Barnacle Goose in North America. <Am. Nat., II, p. 49.

Records a specimen taken by B. R. Ross near Rupert House, James Bay.

1871. HARTING, J. E. Catalogue of an Arctic Collection of Birds presented by Mr. John Barrow, F. R. S., to the University Museum at Oxford. <Proc. Zool. Soc., London, XXXIX, pp. 110–123.

Catalogues a collection made by officers of various arctic expeditions between 1848 and 1855, which includes specimens from region about Melville Peninsula.

1874. BAIRD, SPENCER FULLERTON; BREWER, THOMAS M.; and RIDGWAY, ROBERT. A History of North American Birds. Land Birds. Vols. I, II, III. 4to. Boston.

Contains much original matter on birds from the Hudson Bay region.

1874–1898. Sharpe, R. Bowdler; Saunders, Howard; Salvin, Osbert; Ogilvie-Grant, W. R.; Sclater, P. L.; and others. Catalogue of Birds in the British Museum. Vols. I to XXVII.

Contains many records of specimens from the Hudson Bay region.

1876. Ridgway, Robert. Studies of the American Falconidæ. <Bull. U. S. Geol. and Geog. Surv. Terr., Vol. II, Bull. 2, pp. 91–182.

Records specimens of *Nisus fuscus* (=*Accipiter velox*) from Moose Factory.

1877. Coues, Elliott; and Allen, Joel Asaph. Monographs of North American Rodentia. =Vol. XI, Geological Survey of the Territories.

Contains some original matter based on material from Hudson Bay.

1877. Coues, Elliott. Fur-bearing Animals: A Monograph of North American Mustelidæ. =Misc. Pub. No. 8, U. S. Geological Survey of the Territories, 8vo, pp. i–xiv, 1–348. Washington.

Many specimens listed from Fort Churchill and other localities on Hudson Bay.

1878. Starbuck, Alexander. History of the American Whale Fishery from the earliest inception to the year 1876. <Rept. of Commissioner of Fish and Fisheries for 1875–1876. Appendix A, pp. 1–779.

Besides a few incidental allusions to whaling on Hudson Bay, a "Table showing returns of whaling-vessels sailing from American ports" gives (on p. 581 et seq.) the records of 47 voyages made to Hudson Bay, mainly from New England.

1879. Bell, Robert. Report on the country between Lake Winnipeg and Hudson's Bay. <Rept. Prog. Can. Geol. Survey, 1877–78, pp. 1cc–31cc, 1879.

On the physiography and geology of the region, with observations on timber, climate, etc.

1879. Nourse, J. E. Narrative of the Second Arctic Expedition made by Charles F. Hall: His voyage to Repulse Bay, sledge journeys to the Straits of Fury and Hecla and to King William's Land. 1 vol., 4to, pp. 644. Washington.

Contains many notes on the game animals of the region.

1880. Bell, Robert. Report on Explorations on the Churchill and Nelson rivers, and around God's and Island Lakes, 1879. <Rept. Prog. Can. Geol. Survey, 1878–1879, pp. 1c to 72c, including Appendices I to VII. (1880.)

A geological and general description of the region. App. II. List of plants collected. App. VI. Birds or eggs obtained. Other appendices on Fossils, Molluscs, Insects, etc.

1881. Gilder, William H. Schwatka's Search. 8vo, pp. 1–316. New York.

An account of the overland journey made in 1879–80, by Lieut. Frederick Schwatka, from Daly Bay, near Chesterfield Inlet, Hudson Bay, to King William Land and return. Contains many notes on the game animals of the region.

1883. Bell, R. Notes on the Birds of Hudson's Bay. <Proc. Royal Soc. Canada, 1882, I, Sec. IV, pp. 49–54.

A running commentary on birds observed or taken, mainly by the writer, on or near Hudson Bay, with remarks on the migration of various northern birds.

1883. Schwatka, Frederick. A Musk-ox Hunt. <Century Magazine, September, 1883, pp. 671–679.

A popular account of musk-ox hunting during his overland journey between Daly Bay and King William Land.

1884. Baird, S. F.; Brewer, T. M.; and Ridgway, R. The Water Birds of North America. 2 vols., 4to. Boston.

Contains much original matter on birds from the Hudson Bay region.

1884. Langille, J. Hibbert. Our Birds in their Haunts: A popular treatise on the birds of eastern North America. 8vo., pp. 618. Boston.

Contains some original information furnished by a correspondent of the author regarding the ornithology of Hudson Bay.

1884. Nourse, J. E. American Explorations in the Ice Zones. 1 vol., pp. 578. Boston. >Hall's Second Arctic Expedition, Chapter VII, pp. 199–268.

A concise account, prepared chiefly from official sources, of Hall's expedition in northern Keewatin and adjacent lands in 1864–1869, in search of records, relics, and other traces of Sir John Franklin's last expedition. The account contains many notes on the game animals of the region.

1884. Ridgway, Robert. Note on the *Anas hyperboreus*, Pall., and *Anser albatus*, Cass. Proc. Biol. Soc., Wash., II, pp. 107–108, April 28, 1884.

Anas nivalis of Forster revived for the Hudson Bay snow goose, which is called "*Chen* (or *Anser*) *hyperboreus nivalis* (Forst.)."

1884. Stejneger, Leonhard. Remarks on the Species of the Genus *Cepphus*. Proc. U. S. Nat. Mus., VII, pp. 210–229.

Cepphus mandtii recorded from St. (=Ft.) George and Moose Factory, Hudson Bay.

1885. Bell, Robert. Observations on the Geology, Mineralogy, Zoology and Botany of the Labrador Coast, Hudson's Strait and Bay. <Rept. Prog. Can. Geol. Survey, 1882-3-4, pp. 5DD–62DD, including five appendices.

An account of an expedition sent to Hudson Bay by way of the Labrador Coast and Hudson Straits. Landings were made and more or less collecting was done in Hudson Bay at Nottingham, Mansfield, Digges, South Southampton and Marble islands, Fort Churchill, and York Factory. Appendix II: List and notes of Mammals of the vicinity of Hudson Bay and Labrador, by Dr. R. Bell. Appendix III: List and notes of Birds of the vicinity of Hudson Bay and Labrador, by Dr. R. Bell. Other appendices on plants, crustaceans, insects, etc.

1885. Seton, Ernest E. T. The Swallow-tailed Flycatcher in Manitoba and at York Factory. <Auk, II, p. 218.

Quotes Bell's note of the occurrence of the species at York Factory.

1885. Turner, Lucien M. List of the Birds of Labrador, including Ungava, East-main, Moose, and Gulf districts of the Hudson's Bay Company, together with the island of Anticosti. <Proc. U. S. Nat. Mus., VIII, pp. 233–254.

Several species recorded for the first time from Moose Factory, Ontario.

1886. Bell, Robert. Observations on the Geology, Zoology and Botany of Hudson's Strait and Bay made in 1885. <Annual Report Geological and Natural History Survey of Canada, I (new series), 1885, part DD, pp. 5DD to 27DD, including two appendices. (1886.)

On the physiography and geology of the west coast of Hudson Bay.

1886. Seton, Ernest E. T. The Birds of Western Manitoba. <Auk, III, pp. 145–156, 320–329. Addenda, p. 453.

A number of species recorded from Norway House and many from Lake Winnipeg.

1887. Chamberlain, Montague. A Catalogue of Canadian Birds, with Notes on the Distribution of the Species. 1 vol., 8vo, pp. i–v, 1–143. St. John, N. B.

Quotes many notes, mainly from Bell, on birds of Hudson Bay region.

1887. Fielden, Henry W. On the Zoology of Captain Markham's Voyage to Hudson's Bay in the summer of 1886. <Transactions of the Norfolk and Norwich Naturalists' Society, IV, pp. 344–353.

A report on observations made and specimens collected by Capt. Albert Markham during a voyage from Halifax to York Factory in 1886. Notes the occurrence of many species at Fort Churchill and York Factory.

1888. Rae, John. Notes on some of the Birds and Mammals of the Hudson's Bay Company's Territories and the Arctic Coast. <Canadian Record of Science, III, No. 3, pp. 125–136.

Observations on the distribution and habits of a few species of mammals and birds of Hudson Bay and of the coast of Melville Peninsula and adjacent lands.

1889. Cope, Edward Drinker. The Batrachia of North America. <Bulletin of the U. S. National Museum, No. 34.

Records *Rana palustris* and *R. cantabrigensis* from James Bay, and several species of salamanders from the same region.

1890. Clarke, William Eagle. On a collection of birds from Fort Churchill, Hudson Bay. <Auk, VII, pp. 319–322.

A list, with some annotations, of a collection of birds taken by Dr. Gillespie, jr., at Fort Churchill, Hudson Bay, some time previous to 1845.

1890? Rae, John. Notes on some of the Birds and Mammals of the Hudson's Bay Company's Territory, and of the Arctic Coast of America (communicated by G. J. Romanes). <Journ. Linn. Soc. London, Zool. XX, pp. 136–145, 1886 to 1890. (Read Feb. 16, 1888).

The same, with a few omissions and minor changes, as the article published in the Canadian Record of Science for 1888, above cited.

1890. Dwight, Jonathan, jr. The Horned Larks of North America. <Auk, VII, pp. 138–158 (with map), 1890.

Otocoris alpestris recorded from Moose Fort [Moose Factory], Hudson Bay region.

1890. Thompson, Ernest E. The Birds of Manitoba. <Proc. U. S. Nat. Mus., XIII, pp. 457–643.

Contains some references to birds from the Hudson Bay region.

1892. Bendire, Charles. Life Histories of North American Birds, with special reference to their breeding habits and eggs, with twelve lithographic plates. Smithsonian Contributions to Knowledge, Vol. XXVIII, or Special Bulletin No. 1, U. S. National Museum. 4to, pp. 446. Washington.

This portion of the work treats of the gallinaceous birds, pigeons, and birds of prey, and contains some original information concerning the ornithology of the Hudson Bay region.

1892. Mearns, Edgar A. A Study of the Sparrow Hawks (subgenus *Tinnunculus*) of America, with especial reference to the continental species (*Falco sparverius* Linn.) <Auk, IX, pp. 252–270.

Falco sparverius recorded from Moose Factory.

1893. Nutting, C. C. Report on Zoological Explorations on the Lower Saskatchewan River. <Bulletin from the Laboratories of Natural History of the State University of Iowa. Vol. II, No. 3, Article IV, pp. 235–293. January, 1893.

A report on collections and observations made at Grand Rapids and Chemawawin, on the Lower Saskatchewan, during the summer of 1891.

1893. Rhoads, Samuel N. The Hudsonian Chickadee and its Allies, with remarks on the Geographic Distribution of Bird Races in Boreal America. <Auk, X, pp. 321–333. October, 1893.

Parus hudsonicus recorded from Moose Factory, James Bay.

1895. Merriam, C. Hart. Synopsis of the American Shrews of the Genus *Sorex.* <N. Am. Fauna No. 10, pp. 57–98. Dec. 31, 1895.

Sorex sphagnicola recorded from Shamattawa River.

1895. Southwell, Thomas. Notes on the Seal and Whale Fishery, 1894. <Zoologist, Third Series, XIX, pp. 91–95.

Note on white whales taken by whaling ship *Balaena* in Elwin Bay, Prince Regent Inlet, during August, 1894.

1896. Bangs, Outram. A Review of the Weasels of Eastern North America. <Proc. Biol. Soc., Wash., IX, pp. 1–24, February 25, 1896.

Putorius rixosus recorded from Moose Factory and Fort Albany.

1896. Bangs, Outram. A Review of the Squirrels of Eastern North America. <Proc. Biol. Soc., Wash., X, pp. 145–167, December 28, 1896.

Sciuropterus sabrinus recorded from Moose Factory.

1896. Bendire, Charles. Life Histories of North American Birds from the Parrots to the Grackles, with special reference to their breeding habits and eggs, with seven lithographic plates. Smithsonian Contributions to Knowledge, Vol. XXXII, or Special Bulletin No. 3, U. S. National Museum. 4to, pp. 518. Washington.

This portion of the work, which, though dated 1895, did not appear until 1896, treats of various families of North American birds, and contains some information concerning the ornithology of the Hudson Bay region not before published. The following families are included: Psittacidæ, Cuculidæ, Trogonidæ, Alcedinidæ, Picidæ, Caprimulgidæ, Micropodidæ, Trochilidæ, Cotingidæ, Tyrannidæ, Alaudidæ, Corvidæ, Sturnidæ, and Icteridæ.

1896. Bishop, Louis B. Descriptions of a new Horned Lark and a new Song Sparrow, with remarks on Sennett's Nighthawk. <Auk, XIII, pp. 129–135, April, 1896.

Mentions two specimens of *Otocoris a. hoyti* from Depot Island, Hudson Strait (=Hudson Bay).

1896. True, Frederick W. A revision of the American Moles. <Proc. U. S. Nat. Mus., XIX, pp. 1–112.

Condylura cristata recorded from Moose Factory, Ontario.

1897. Coues, Elliott. A North American snipe new to the A. O. U. List. <Auk, XIV, p. 209, April, 1897.

Attention called to a specimen of *Gallinago major* from Hudson Bay in the British Museum.

1897. Coues, Elliott. Status of *Helodromas ochropus* in the A. O. U. List. <Auk, XIV, pp. 210–211, April, 1897.

Attention called to a specimen from Hudson Bay in the British Museum.

1897. Coues, Elliott. Status of the Redshank as a North American Bird. <Auk, XIV, pp. 211–212, April, 1897.

Attention called to a specimen of *Totanus calidris* said to be from Hudson Bay, described in Fauna Boreali-Americana.

1897. Miller, Gerrit Smith, jr. Revision of the North American Bats of the Family Vespertilionidæ. =N. Am. Fauna No. 13, October 16, 1897.

Myotis lucifugus recorded from James Bay.

1897. Tyrrell, J. Burr. Report on the Doobaunt, Kazan and Ferguson Rivers and the North-West Coast of Hudson Bay and on two overland routes from Hudson Bay to Lake Winnipeg. Ann. Rept. Geol. Surv. Canada, IX (new series), 1896, Part F. (1897.)

Chiefly geographical and geological. Notes on fauna of the region, pp. 164F to 167F, and *passim* throughout the report.

1898. Allen, J. A. Revision of the Chickarees, or North American Red Squirrels (Subgenus *Tamiasciurus*). <Bull. Am. Mus. Nat. Hist., XIV, pp. 249–298. [Author's edition issued July 22, 1898.]

Sciurus hudsonicus recorded from Hudson Bay.

1898. Russell, Frank. Explorations in the Far North. Being a Report on an Expedition under the Auspices of the University of Iowa during the years 1892, '93 and '94. Published by the University. 8vo., pp. 290. Iowa City.

Contains observations on natural history of Grand Rapids, at the mouth of the Saskatchewan River.

1898. Tyrrell, J. W. Across the Sub-Arctics of Canada. 8vo, pp. 280. New York.

A narrative of a journey of exploration by way of Athabasca Lake, Telzoa River, and Chesterfield Inlet to Hudson Bay, returning by way of Fort Churchill, York Factory, Oxford House, and Norway House. Contains many notes on the natural history of the region.

1899. Howe, Reginald Heber, jr. North American Wood Frogs. <Proc. Boston Soc. Nat. Hist., vol. 28, No. 14, pp. 369–374. February, 1899.

Contains references to *Rana cantabrigensis* and *R. septentrionalis* from James Bay.

1899. Preble, Edward A. Revision of the Jumping Mice of the Genus Zapus. =N. Am. Fauna No. 15, August 8, 1899.

Zapus hudsonius recorded from James Bay and Fort Churchill.

1900. Bailey, Vernon. Revision of American Voles of the Genus Microtus. =N. Am. Fauna No. 17, June 6, 1900.

Microtus xanthognathus recorded from Nelson River; *M. drummondi* from Fort Churchill.

1900. Macoun, John. Catalogue of Canadian Birds. Part I, Water Birds, Gallinaceous Birds and Pigeons. =Publication No. 692, Geological Survey of Canada. [Introduction dated March 26, 1900.]

Contains much original matter relating to the birds of the Hudson Bay region.

1900. Merriam, C. Hart. Descriptions of Twenty-six new Mammals from Alaska and British North America. <Proc. Wash. Acad. Sci., II, pp. 13–30, March 14, 1900.

Dicrostonyx richardsoni described from Fort Churchill, Hudson Bay.

1900. Southwell, Thomas. Notes on the Seal and Whale Fishery, 1899. <Zoologist, Fourth Series, IV, pp. 65–73.

Notes on walruses and musk-oxen obtained by the whale ship *Active*, from Dundee, in the northern part of Hudson Bay, in 1899. Contains also allusions to whales in northern Hudson Bay.

1901. Allen, J. A. The Musk-Oxen of Arctic America and Greenland. <Bull. Am. Mus. Nat. Hist., XIV, pp. 69–86. [Author's edition issued March 30, 1901.]

Main facts in regard to distribution of *Ovibos moschatus* given.

1901. Howe, Reginald Heber, jr. An additional note on the Genus *Macrorhamphus*. <Auk, XVIII, p. 272. July, 1901.

Specimens from Fort Churchill referred to *Macrorhamphus g. scolopaceus*.

1901. Howell, Arthur H. Revision of the Skunks of the Genus Chincha. =N. Am. Fauna, No. 20, August 31, 1901.

Chincha mephitis [=*Mephitis mephitis*] recorded from Moose Factory, Oxford House, and Pine Lake.

1902. Brewster, William. An Undescribed Form of the Black Duck (*Anas obscura*). <Auk, XIX, pp. 183–188.

Refers the black duck of the Hudson Bay region to subspecies *rubripes*, and records specimens from Moose Factory, Fort Churchill and Severn River.

1902. Oberholser, Harry C. A Review of the Larks of the Genus Otocoris. <Proc. U. S. National Museum, XXIV, pp. 801–884. June, 1902.

Otocoris alpestris recorded from Moose Fort [Moose Factory], Ontario; *Otocoris alpestris hoyti* recorded from three points in Keewatin—Depot Island, Fort Churchill, and Cape Eskimo, 'Northwest Territory.'

NOTE ON BOUNDARIES OF THE REGION TREATED.

The present report relates to the birds, mammals, and batrachians of the Province of Keewatin, in Canada. But though it has seemed best to thus limit by geographic boundaries the region treated, some notes from localities outside these boundaries have been included, where such a course has seemed desirable. Thus many references to the birds and mammals observed and collected on the voyages of Ross and Parry have been utilized. It is believed, however, that all the species thus included will be found to occur within the boundaries of Keewatin. A few old references to the occurrence of species on Hudson Bay can also be safely referred to Keewatin, since most of the early collections and observations were made on the west coast of the Bay. The lists also contain a few species which have been recorded only from the southern extremity of Hudson Bay, mainly at Moose Factory, a few birds seen by us only on the lower Red River, and a number of birds recorded from Grand Rapids, at the mouth of the Saskatchewan River. A few Old-World species, straggling individuals of which have been recorded from Hudson Bay, have also been included. Since each record speaks for itself, it has not been deemed necessary to indicate 'extralimital' species by placing them in a separate category or otherwise differentiating them from those seen or taken strictly within the geographical boundaries of Keewatin.

Additional work, especially in the southeastern part of the Province, will doubtless add a considerable number of birds and a few mammals to the list of species known to inhabit Keewatin. Thus a number of northern plains species will undoubtedly be found to occur regularly near the Manitoba border, and many birds which are known to occur at a much higher latitude to the westward will probably be found to breed regularly in southern Keewatin.

Among the numerous published notes relative to many species only those which best illustrate their geographic distribution or supplement our own observations have been selected.

NEW SPECIES.

The following six new species and subspecies of mammals are described in the present report:

Microtus aphorodemus.
Fiber zibethicus hudsonius.
Lepus arcticus canus.
Lutreola vison lacustris.
Mustela americana abieticola.
Sorex (Microsorex) alnorum.

In addition to these, three others which properly come within the scope of this report have been described since the completion of the trip:

Phenacomys mackenzii.
Synaptomys (*Mictomys*) bullatus.
Vulpes lagopus innuitus

Balæna mysticetus Linn. Greenland whale.

Formerly found as far south as Churchill River, according to Hearne, who says that three were killed there in the course of twenty years. They were more plentiful to the northward, and the Hudson's Bay Company carried on a whale fishery in the vicinity of Marble Island for several years, in the latter part of the eighteenth century, which, however, proved to be unprofitable and was abandoned.[a] During the latter half of the nineteenth century American whalemen frequently visited Hudson Bay and vicinity. Starbuck gives the records of forty-seven voyages to Hudson Bay by whaling vessels sailing mainly from New Bedford, Mass., and New London, Conn., between 1861 and 1874. These vessels brought home 21,810 barrels of whale oil and 353,740 pounds of whalebone.[b] The bark *Pioneer*, which sailed from New London June 4, 1864, returned from Hudson Bay September 18, 1865, with a cargo of whale oil and whalebone worth $150,000. This, it is claimed, is the best voyage on record. Detailed information in regard to whaling vessels sailing from other countries and in regard to American vessels sailing in recent years is not at hand, but Marble Island is still used to a considerable extent as a wintering post for whaling vessels.

J. C. Ross recorded Greenland whales from the western shore of Prince Regent Inlet, where they were found in considerable numbers. A few were also seen about Boothia.[c] Captain Lyon saw many whales, probably of this species, near Duke of York Bay, Southampton Island.[d] Southwell records that the whale ship *Active*, from Dundee, while in the northern part of Hudson Bay, in the summer of 1899, spoke an American vessel which had on board the produce of sixteen whales, presumably killed in the Bay.[e] It is possible that other species occur in the region, but I find no specific reference to them.

Monodon monoceros Linn. Narwhal.

Said to be occasionally killed about the northern part of the Bay.[f] During Parry's second voyage many were seen near Duke of York

---

[a] Hearne, Journey * * * to the Northern Ocean, p. 392, 1795.

[b] Report of the Commissioner of Fish and Fisheries, 1875–76, Part IV, p. 581 et seq., 1878.

[c] Appendix to Ross's Second Voyage, p. xxiv, 1835.

[d] Lyon's Private Journal, p. 48, 1824.

[e] Zoologist, Fourth Series, IV, p. 71, 1900.

[f] Rept. Prog. Can. Geol. Surv. 1877–78, p. 29c (1879).

Bay, on the north side of Southampton Island,[a] and at other points in the upper part of Hudson Bay.

Delphinapterus catodon (Linn.). White Whale. Beluga.

Common all along the coast and seen daily whenever we were on the water. On July 13, while returning from Beacon Point, I saw a school of about a dozen white whales in Hayes River halfway between the mouth of the river and York Factory. They were passing out with the ebbing tide, having ascended the river a short distance during the flow, as is their custom, and were emitting a variety of sounds from a shrill scream to a hoarse snort or grunt.

The mouth of Churchill River is a favorite place for these whales, and at the time of our visit the Hudson's Bay Company kept several men constantly at work capturing them. The method generally employed is as follows: A large net is sunk to the bed of the river—usually at the mouth of some natural basin—and after a school has passed over it into the inclosed area it is raised, imprisoning some of the animals, so that when the tide falls they are at the mercy of their captors. The oil is extracted for export and the meat is used for food for the dogs. In former years white whales were shot from 'whale stands' maintained at the mouth of the Hayes and other rivers, but this method seems to be less in vogue at present.

A number of the animals, the largest about 14 feet in length, were rather hurriedly examined at the 'whale fishery' near the mouth of the Churchill July 30. Their color ranged from a deep blue-black to silvery white. It is usually stated that the dark ones are the young, but this is not invariably the case, since some, at least half grown, were nearly black, and others, 3 or 4 feet in length, were as white as the largest. An embryo about 7 inches long and nearly pure white was obtained at Churchill.

This species seems to abound in all parts of Hudson Bay, and has also been recorded from several localities to the northward.

Southwell reports that the whaling ship *Balaena* took 820 white whales in Elwin Bay, Prince Regent Inlet, during the month of August, 1894.[b]

Rangifer caribou (Gmel.). Woodland Caribou.

The woodland caribou is found throughout the region traversed between Norway House and Hudson Bay. It seems to be more common toward the Bay, but is occasionally killed (usually in the winter) near Norway House. Mr. William Campbell, of Oxford House, reported the species much less common than formerly; it is said to have become scarcer in some localities, as the moose extends its range, apparently being driven out by that animal. On our way to the Bay

---

[a] Lyon's Private Journal, p. 44, 1824.
[b] The Zoologist (Third Series), XIX, p. 94, 1895.

we saw tracks of woodland caribou several times on Steel River, once coupled with the tracks of a wolf that had evidently been trailing the caribou, and on our return trip a caribou was killed on Steel River by a party which ascended a few days in advance of us.

Between York Factory and Fort Churchill a few small bands are found throughout the year on the 'Barrens.' Tyrrell saw them here early in the winter of 1893.[a] We saw none, but noted a great many tracks on the 'Barrens' between Stony and Owl rivers July 19, and were told by some Indians we met that they had killed several within a week.

Dr. Milne informs me that he has seen them between Fort Churchill and Cape Churchill, and that the latter point is considered a good place for hunting them at any time of the year. He thinks these small bands form the "northern fringe of the bands which migrate to the coast in spring, the great majority of which in their journey cross to the south of Nelson River," an opinion which has weight from his fourteen years' residence at York Factory. A favorite crossing point on the Hayes River is about 40 miles above York Factory, though they sometimes cross much closer to that post. Their return movement occurs from about the middle of October to the last of November. During these semiannual movements the animals are much pursued, especially in the fall, when the weather is usually cold enough to preserve the meat for winter use.

Rangifer arcticus (Richardson). Barren Ground Caribou.

The presence of Barren Ground caribou was first noted August 3 at a point about 50 miles south of Cape Eskimo, where we saw their tracks on landing. During the next few days the Indians made several trips in quest of the animals, but found none, although comparatively fresh tracks were observed in every direction. While we were encamped about 25 miles south of Cape Eskimo August 10 to 13 we frequently heard wolves howling in pursuit of caribou, and occasionally saw a few of the latter, of which we killed two, a rather young doe and a buck about three years old.

The animals were evidently just commencing their usual fall journey to the southward. They showed a tendency to seek the vicinity of the shore on account of the protection gained from mosquitos and other insects, which were less numerous there, owing to the wind. Even under favorable conditions they were attended by swarms of insects, and when feeding were almost constantly moving.

Soon after leaving our camp on the morning of August 13 we saw several near the shore. A young buck on a point of land was approached as closely as the depth of water would permit—about 200 yards. He showed little fear, trotting along the shore abreast of our

[a] Across the Sub-Arctics of Canada, p. 226, 1898.

boat for about a quarter of a mile. He would frequently stop and wade a short distance toward the boat, at short intervals spreading and contracting the white patch on his throat laterally into an oval disk, so abruptly as to give the appearance of flashes of light. He finally grew tired of following us and dropped behind. This was the last one seen.

James Clark Ross recorded reindeer from Cape Warrender, north shore of Barrow Strait, and from the coast of North Somerset,[a] and observed them in great numbers on the Isthmus of Boothia.[b] Dr. Rae observed them migrating northward about the 1st of March, near Repulse Bay,[c] and found them on the west coast of Melville Peninsula as far as Fraser Bay.[d] Lyon recorded them from Duke of York Bay, Southampton Island.[e] Schwatka's party killed large numbers between Camp Daly and King William Land in 1879 and 1880.[f] During the summer of 1893 the Tyrrell brothers, of the Canadian Geological Survey, saw on the shores of Carey Lake, about 450 miles northwest of Fort Churchill, a herd which they estimated to contain from one to two hundred thousand individuals.[g] On their exploring trip northward through the interior of Keewatin, in 1894, they first met with Barren Ground caribou, near Ennadai Lake, on August 14. The animals were then moving southward in large numbers.[h]

The southern range of Barren Ground caribou, on the west coast of Hudson Bay, may be said to be limited by Churchill River. Even in former years these caribou were seldom known to cross that river,[i] and they are still killed within a few miles of Fort Churchill. Farther inland they reach the south end of Reindeer Lake.[j]

Description.—Adult male in summer pelage, killed on the Barren Grounds about 25 miles south of Cape Eskimo August 10: General color of upperparts and head dull brown; face dull reddish brown; legs dusky brown with an indistinct ashy stripe on inner side of fore leg, and of hind leg below the heel; a lateral stripe of dusky brown where the hairs of the belly and sides meet, separated from the color of upperparts by an indistinct ashy stripe; chest dusky; belly and ventral surface of tail white; a small white patch on rump, divided by a narrow stripe of brown extending from dorsal area to tip of tail. A white disk on throat 15 inches long and (when spread) about 10 inches

---

[a] Parry's Third Voyage, Appendix, p. 94, 1826.

[b] Appendix to Ross's Second Voyage, p. xvii, 1835.

[c] Narrative of an Expedition to the Shores of the Arctic Sea, p. 93, 1850.

[d] Ibid., p. 149, 1850.

[e] Lyon's Private Journal, p. 46, 1824.

[f] Gilder, Schwatka's Search, Introduction, p. viii, and elsewhere, 1881.

[g] Ann. Rept. Can. Geol. Surv., 1896, IX (new ser.), p. 165F (1898).

[h] Ibid., p. 19F (1898).

[i] Hearne, Journey * * * to the Northern Ocean, p. 225, 1795.

[j] Tyrrell, Forest and Stream, XLIII, No. 4, p. 70, July 28, 1894.

wide, consisting of hairs 5 inches in length; an indistinct stripe of ashy extending from this white area forward to chin; feet around edge of hoof white; nose and edges of lips whitish; ears mainly ashy.

The winter skins seen were more or less suffused with white or hoary, which in some cases was the predominating color.

Alces americanus Jardine. Eastern Moose.

This species occurs in suitable places throughout the region traversed from Lake Winnipeg nearly to Hudson Bay. While ascending the Echimamish, which is a noted locality for moose, we frequently saw places where the animals had crossed the stream, and the bordering swamps were intersected by a network of their tracks. Many of the tracks in the bed of the stream were so recent that the mud had not yet settled; but no animals were seen, though we were constantly on the lookout for them. A young one was killed by a party of Indians at the outlet of the Echimamish a few days before we ascended the river, and the cranium of a female, probably killed during the preceding winter, was found at the middle dam. The extensive swamps below Robinson Portage are also much frequented by the animals, one of which was trailed for some distance by my Indian guides.

At Oxford House, 60 or 70 miles beyond Robinson Portage, the moose was formerly almost unknown, according to information received independently from several officers of the Hudson's Bay Company, but is extending its range toward Hudson Bay and is now frequently killed near that post. On Steel River, as we were ascending it September 1, we saw a fresh track at the mouth of a small stream a few miles below Fox River, and at York Factory I was shown a skin which had been brought from Shamattawa River, where the York Factory Indians now go regularly to hunt moose.

Farther inland the moose ranges to the northwest. I saw a head at Norway House from the vicinity of Split Lake, and J. B. Tyrrell met with moose on Stone River about 400 miles west of Fort Churchill.[a]

Ovibos moschatus (Zimm.). Musk-Ox.

The musk-ox was first described from the region between Seal and Churchill rivers and formerly ranged in winter more or less regularly to about that latitude. Hearne saw its tracks within a few miles of Fort Churchill, and on his first journey toward the northwest met with it within a hundred miles of that place.[b] Edward Sabine recorded musk-oxen from Melville Island, where they arrived in May from the southward. They crossed on the frozen sea, and recrossed it on leaving in September.[c] Ross recorded them from Felix Harbor,

---

[a] Ann. Rept. Can. Geol. Surv., 1896, IX (new ser.), 165F (1897).

[b] Journey * * * to the Northern Ocean, p. 135, 1795.

[c] Suppl. to Appendix to Parry's First Voyage, p. clxxxix, 1825.

Boothia.[a] Dr. Rae found them in the vicinity of Repulse Bay,[b] east of which they seem never to have been recorded. Frederick Schwatka found a herd between Wager and Back rivers in the winter of 1879.[c] Tyrrell saw fresh skins in the possession of the Eskimos near the head of Chesterfield Inlet in the summer of 1893.[d]

A skull that had been brought by Eskimos from the vicinity of the head of Chesterfield Inlet was obtained by us at Fort Churchill. The tips of the horns of this specimen, an adult male, curve forward in nearly parallel planes, instead of diverging, as is usually the case.

Through the kindness of Dr. Alexander Milne, of York Factory, I obtained what is probably the most southern authentic record of the occurrence of this animal. Several times during the summer of 1897 parties of Indians reported seeing a pair of musk-oxen on the 'Barrens' about halfway between York Factory and Fort Churchill. The male was finally killed in August and the head was brought by the Indians to York Factory. This pair had probably wandered beyond the limits of their normal range during the preceding winter, and for some reason failed to return northward.

The Eskimos who trade at Fort Churchill hunt the musk-ox in the Barren Grounds several days' journey northwestward from Cape Eskimo.

Sciuropterus sabrinus (Shaw). Hudson Bay Flying Squirrel.

Said to be found throughout the region traversed between Lake Winnipeg and Hudson Bay. It has been recorded by Bell from Nelson River House, on Churchill River,[e] and by Bangs from Moose Factory.[f] We obtained several hunters' skins at Norway House and one at Oxford House, and learned that the species is rather common on Shamattawa River. It sometimes becomes a pest to trappers on account of the frequency with which it is caught in traps set for martens and other fur-bearing animals.

Several winter skins from the vicinity of Norway House differ as follows from skins of *Sciuropterus s. macrotis* in corresponding pelage from New England: Upperparts slightly darker (less yellowish); tail with duskier color toward tip; face and cheeks darker; feet similar in color, but much more heavily furred; color beneath not noticeably different; hind foot (measured dry), *sabrinus*, 38–40 mm.; *macrotis*, about 36 mm. Breadth of tail with hairs spread naturally, *sabrinus*, 60–65 mm.; *macrotis*, about 45 mm. These specimens from Norway House and Oxford House can probably be considered typical *sabrinus*, which was described from Severn River. An imperfect skin in the

[a] Ross's Second Voyage, p. 337, 1835.

[b] Narrative of an Expedition to the Arctic Sea, p. 49, 1850.

[c] Gilder, Schwatka's Search, p. 67, 1881.

[d] Ann. Rept. Can. Geol. Surv., 1896, IX (new ser.), p. 165F (1897).

[e] Rept. Prog. Can. Geol. Surv. 1882–3–4, App. II, p. 48DD (1885).

[f] Proc. Biol. Soc. Wash., X, p. 163, 1896.

National Museum from Hudson Bay closely resembles those from Norway House, but is slightly more tinged with yellowish brown beneath. An albinistic specimen from Norway House is nearly white above, slightly tinged with light brown, with a brownish area on the middle of the back: the tail is very light brownish above, and lighter, about normal, beneath. No skulls from the Hudson Bay region are available for comparison.

Sciurus hudsonicus (Erxleben). Hudson Bay Red Squirrel.

We found the red squirrel abundant and generally distributed throughout the region to within a few miles of York Factory, where the trees dwindle to such an insignificant size as not to afford the animals a congenial home. At Fort Churchill I saw a specimen which had been secured a few miles up the river, where good-sized spruce trees occur and the animals are said to be fairly abundant. An adult male taken at Norway House June 18 is in nearly full winter pelage, the summer coat just beginning to appear in patches on the face and back, the lower parts still showing the heavy grayish vermiculations, and the red median dorsal stripe of the winter pelage being present. The tail is concolor with the back, and is edged with yellowish gray. All the other specimens taken at Norway House and at various points between that post and Oxford House late in June and early in July, and on the return trip between Steel River and Norway House during the first half of September, are in summer pelage and appear to represent two phases of color. By far the greater number are very dark reddish above, the red suffusion tingeing the lower parts quite appreciably in many cases; the tails are bordered with grayish or reddish indiscriminately.

Tamias striatus lysteri (Richardson). Northeastern Chipmunk.

A specimen collected by C. Drexler at James Bay is recorded in the U. S. National Museum catalogue. It can not now be found, but notes regarding it, made years ago by Dr. C. Hart Merriam, indicate that it is referable to the present form.

Eutamius neglectus (Allen). Lake Superior Chipmunk.

We obtained about a dozen specimens which were collected by Mr. William Campbell at Oxford House during July, August, and the early part of September. Chipmunks were reported to be rather common about Oxford Lake, but we failed to see any at that point, or at any other on our entire trip except the shore of Pine Lake, where we observed one September 13.

The specimens secured were preserved in formalin and skinned on our return in September. The average measurements of 6 are as follows: Total length 217; tail vertebræ 101; hind foot 31.[a]

This series agrees essentially with a series taken by Gerrit S.

[a] All measurements are in millimeters unless otherwise stated.

Miller, jr., in September and the early part of October at Nepigon and Peninsula Harbor, Ontario, which are in the type region of *neglectus*. The Oxford House series shows some seasonal variation, those taken in July having grayer rumps than the September specimens.

Dr. Bell reported '*Tamias quadrivittatus*' to be common along the Nelson and Churchill rivers,[a] but it is probable that it is not found on the lower portions of these rivers.

Spermophilus parryi (Richardson). Hudson Bay Spermophile.

Arctomys Parryii Rich., Appendix to Parry's Second Voyage, p. 316, 1825 (1827).

The name *empetra*, usually since 1877 applied to this species, was apparently based on a specimen of the Canadian form of *Arctomys monax* (see p. 47), and will have to be replaced by *parryi* of Richardson, based on specimens collected at Five Hawser Bay, Lyon Inlet, Melville Peninsula, on Parry's second voyage, which seems to be the next available name.

Ground squirrels have been stated to inhabit the coast of Hudson Bay south to Fort Churchill, but I ascertained that the animal does not occur on the coast south of a point about 25 miles below Cape Eskimo—about 150 miles to the northward of Fort Churchill. I was unable to find any in the vicinity of my camp near Thlewiaza River. On starting northward in search of them I was fortunate in securing as guide one of the most intelligent of the Eskimos of the region, who was perfectly familiar with the entire coast and who promised to take me to the nearest point at which the animals were to be found. We left camp near the Thlewiaza on the afternoon of August 8 and on the evening of the next day succeeded in reaching our objective point—the mouth of a stream which enters the sea at the head of a shallow bay thought to be about 25 miles south of Cape Eskimo. One of the animals was seen a few minutes after we landed, but they proved to be rather rare, though the Eskimo gave me to understand that farther north they were more common. They inhabited the scattered sandy ridges and hillocks, but with the exception mentioned none were seen abroad, and careful trapping during the next three days at all the burrows found yielded only four specimens. The apparent inactivity of the animals and the fact that most of the burrows were closed from the inside led me to conclude (somewhat hastily) that the animals had commenced to hibernate, but from the fact that they have been taken at points farther north much later in the season it is probable that such was not the case. The burrows usually had several entrances, and, judging from the amount of earth removed, were quite extensive.

The four specimens taken agree closely in color and may be thus described: Fur of back, sides, breast, and belly dusky plumbeous at

---

[a] Rept. Prog. Can. Geol. Surv. 1882-3-4, App. II, p. 48DD (1885).

base, on breast, belly, and sides tipped with rather bright rufous; legs, feet, and throat rufous to base of hairs, this color deepest on upper side of legs; sides and thighs more or less flecked with black hairs; fur on back from nape to base of tail with a broad zone of yellowish-brown, which forms the ground color, succeeded by a subterminal zone of yellowish gray and tipped with black. The black tips and the wholly black hairs which are interspersed through the fur predominate in places, causing the grayish zone of color to be broken up into more or less distinct spots; top and sides of head varying from light rufous to dark chestnut rufous, deepest on cheeks and top of head and interspersed with many black hairs; ears light rufous; hairs of upper side of tail dark gray at base, this color succeeded by a narrow zone of black, a narrow zone of dull yellowish, and a broad zone of black, and narrowly tipped with rufous; pencil of tail, comprising about a third of its length, nearly all black; most of under side of tail bright rufous.

The three males average: Total length 386.6; tail vertebræ 117; hind foot 64.3. The female measured: 392; 114; 57.

Arctomys monax empetra (Pallas). Canadian Woodchuck.

Mus empetra Pallas, Novæ Species Quad. e Glirium Ordine, p. 75, 1778.

The name *Mus empetra* of Pallas, usually of late years applied in a general way to the spermophiles inhabiting the Barren Grounds and the northern Rocky Mountains, was evidently based on a specimen of the Canadian woodchuck, as has been recognized by Sabine,[a] Richardson,[b] and others. Pallas cites (as *Marmota quebekana*) the 'Quebec Marmot' of Pennant,[c] qualifying the reference to Pennant's figure with the word *bona*. Pennant's 'Quebec Marmot,' apparently based on a live specimen in the possession of Mr. Brooks, is unquestionably a woodchuck.[d] Forster's 'Quebec Marmot,'[e] which Pallas also cites, is as certainly a spermophile. Pallas's description of *Mus empetra* follows his citations. Dr. C. Hart Merriam and I have gone carefully over this description with specimens of both *Arctomys* and *Spermophilus* from Hudson Bay before us, and he agrees with me that the description applies to the woodchuck. The description of the color of those parts which are most unlike in the two animals (the head and feet) applies unquestionably to the woodchuck, and the entire description fits that animal far better than it does the spermophile.

The 'wenusk,' as the animal is called in the Hudson Bay country,

---

[a] Trans. Linn. Soc. London, XIII, p. 584, 1822.

[b] Fauna Boreali-Americana, I, p. 147, 1829.

[c] Synopsis of Quad., p. 270, pl. 24, fig. 2.

[d] Pennant, Hist. Quad., 3d ed., 1793, p. 129, says: "It has lately been described by Pallas under the name of *Mus empetra*."

[e] Phil. Trans., LXII, p. 378, 1772.

is apparently found throughout the region as far north as York Factory, but is said to be more common to the southward. A hunters' skin, taken near Oxford House in the spring of 1900, was obtained from Mr. William Campbell, who reported the animal as not common in the vicinity. At York Factory we obtained an immature specimen and a hunters' skin, taken on Hayes River. Several skins from Trout Lake, where the animals are said to be numerous, were also obtained.

The skins secured, representing immature individuals and adults of both sexes, vary but little in coloration. The prevailing color of the upperparts is rufous, the subapical zone of black being restricted and contributing little to the general color; the lower parts are rufous, this color varying somewhat in intensity; the tail is very dark brown or black, the hairs tipped with rufous; the feet are nearly black.

The only skull available for comparison was taken at Oxford House June 10, 1901. It is that of a female, and though the teeth are not much worn, the development of the occipital crest and the general appearance indicate maturity. It is much smaller than skulls of woodchucks of about the same age from Washington, D. C., and Maryland (the type locality of *monax*), and shows other differences as follows: Rostrum less deflected, the nasals flatter and more narrowed posteriorly; basal portion of zygomatic process of squamosal broader and flatter, its posterior edge, when viewed from above, being practically continuous with the general curve of the zygoma and lacking the conspicuous concavity usually seen in typical *monax*. The teeth are proportionally weaker, the incisors especially being more slender. This skull measures as follows: Occipito-nasal length 80; zygomatic breadth 55; breadth of rostrum immediately in front of zygoma 15; length of nasals 33; breadth of posterior end of nasals 7.

Allen recorded specimens of *Arctomys monax* from James Bay and Nelson River.[a]

Castor canadensis Kuhl. Canadian Beaver.

Owing to persistent trapping, the beaver is becoming scarce throughout the region, but skins are annually traded at all the posts visited. We saw the remains of a beaver house between Pine and Windy lakes and a comparatively recent dam on a small stream which empties into Hayes River about 15 miles above York Factory.

A number of skins were seen at Fort Churchill. These had been taken on the Lower Churchill River. Several black pelts were among the furs at Norway House.

A fine large ligamentary skeleton was obtained at York Factory from Dr. Milne, who said that the animal it belonged to had been taken in the vicinity. Its measurements (approximate) are as follows:

[a] Mon. N. Am. Rod., p. 919, 1877.

Total length 1,000; tail vertebræ 380; hind foot 190. Skull: Occipito-nasal length 137; zygomatic breadth 99; interorbital constriction 25; length of nasals 51; breadth of nasals 24; length of upper molar series (crowns) 30.

Nine adult skulls of both sexes, collected in the vicinity of Oxford House, average as follows: Occipito-nasal length 131.3; zygomatic breadth 93.5; interorbital constriction 25; length of nasals 47.3; breadth of nasals 23.6; length of molar series (crowns) 28.2.

Dr. Bell reports that a family of beavers was found by Indians on North River, a stream that flows into the Bay about 15 miles above Fort Churchill.[a] Hearne relates that the Indians who accompanied him on his first exploring trip killed beavers on Seal River, the mouth of which is about 40 miles north of Fort Churchill.[b] The point at which Hearne seems to have crossed Seal River, and presumably where the beavers were taken, is not far from the Bay.

[**Mus musculus** Linn. House Mouse.

Referring to this species, Richardson says: "I have seen a dead mouse in the storehouse at York Factory."[c] J. E. Gray records a specimen of the house mouse collected by Dr. Rae at York Factory, with the remark that it was probably introduced from Europe.[d] The species does not seem to have ever become established, though individuals should apparently be landed occasionally with goods from England.

We trapped large numbers of mice about the various buildings, but took only the native species.]

Peromyscus canadensis umbrinus Miller.[e] Clouded White-footed Mouse.

We found this form common and apparently quite generally distributed between Norway House and York Factory, and took specimens at the following localities: Norway House, Sea Falls, Echimamish River (Middle Dam and Painted Stone Portage), Robinson Portage, Pine Lake, Oxford Lake (south end), Oxford House, Trout Falls, Hill River (mouth of Fox River and near Rock Portage), Steel River, Hayes River, and York Factory.

I have compared this series with the type series of *P. canadensis umbrinus* from Peninsula Harbor, Ontario. The specimens of the type series are mainly in fall pelage, while most of our Keewatin specimens were taken during the summer; but enough are comparable

[a] Rept. Prog. Can. Geol. Surv. 1882–3–4, App. II, p. 49DD (1885.)

[b] Journey from Prince of Wales Fort * * * to the Northern Ocean, p. 9, 1795.

[c] Fauna Boreali-Americana, I, p. 141, 1829.

[d] Rae's Narrative of an Expedition to the Shores of the Arctic Sea, Appendix, p. 199, 1850.

[e] Proc. Bost. Soc. Nat. Hist., 28, No. 1, p. 23, April, 1897.

to make it reasonably certain that they are all referable to the same form. In detailed measurements the two series agree very well.

Seven adults from Painted Stone Portage average: Total length 186; tail vertebræ 92; hind foot 20. Average of seven adults from Oxford House: 186, 92, 20. Average of seven adults from York Factory: 183, 91, 20.7. This mouse constitutes the house mouse of the region, and infests all the houses and stores within its range.

At York Factory almost all our specimens were taken in or about the post buildings; the animal seemed to be rare in the surrounding swamps.

[(?) **Neotoma drummondi** (Richardson). Drummond Wood-Rat.

A wood-rat received about 1860 from W. MacTavish, and supposed to have been taken at Fort Churchill,[a] is in the U. S. National Museum. It was mounted for many years, but is now made into a study skin. The color has become so changed by exposure that no dependence can be placed on it, and the skull lacks the greater part of the braincase and is otherwise defective. Comparison of this imperfect specimen with specimens taken at Jasper House in the type region of *drummondi* shows no essential differences. As Jasper House seems to be the nearest point to Fort Churchill that the genus *Neotoma* has been recorded, and as the animal is unknown to the inhabitants of Fort Churchill, it seems most likely that the specimen came from some point in the interior, and was erroneously included in a lot of specimens from Hudson Bay.]

Phenacomys mackenzii Preble. Mackenzie Phenacomys.

Phenacomys mackenzii Preble, Proc. Biol. Soc. Wash., XV, p. 182, August 6, 1902. (Fort Smith, Mackenzie.)

Found only at Fort Churchill, where an adult female and two immature individuals were secured July 26 to 28. Two of these were taken on a dry hummock in the meadow west of the post, and the third in a grassy place among the rocks. The rostral portion of a skull, with a part of the skin attached, was found in a fissure among the rocks, where the animal had evidently been eaten by a weasel.

Evotomys gapperi (Vigors). Common Red-backed Mouse.

We found this species to be rather common throughout the region between Norway House and Hudson Bay, and took a large series embracing specimens from the following localities: Norway House, Sea Falls, Echimamish River, Robinson Portage, Pine Lake, near head of Oxford Lake, Oxford House, Knee Lake (near outlet and on island

[a] In a letter to Sir John Richardson, extracts of which were published, Professor Baird says: "From * * * our other northern collectors and correspondents we have already received many interesting specimens, though the best are still on the way. We have already had *Neotoma cinerea* from Fort Churchill." (Edin. New Phil. Journ. (new ser.), XIII, p. 164, 1861.)

near 'Narrows'), Swampy Lake, Hill River (at various points), Steel River, Hayes River, and York Factory. One was also taken by my brother on Churchill River, 15 miles above Fort Churchill. Mossy spruce woods seemed to be their favorite habitat, but we also frequently trapped them in deciduous or mixed woods, and occasionally in willow thickets in swamps. They inhabited the larger wooded islands in Knee Lake. On lower Hill River as well as on Steel and Hayes rivers we took only an occasional one, and at York Factory a week's trapping yielded but three, all of which were caught about the post buildings. None were found at Fort Churchill, or anywhere on the tundra, and several nights' careful trapping by my brother on Churchill River above the post yielded but one.

Of this series of nearly seventy specimens all but six are in the normal red-backed pelage. The exceptions may be briefly referred to. One, from Robinson Portage, has a few black hairs scattered over the back, perceptibly darkening the dorsal stripe. One, from Oxford Lake, taken September 12, has the dorsal stripe dusky, sides only slightly darker than normal, and belly yellowish white. One, taken on upper Hill River September 5, has the dorsal stripe dusky brown, the belly and sides about normal, and the back nearly concolor with sides; the dorsal stripe is indicated only by a slight admixture of red. Of the three specimens taken at York Factory, one is normal; another, probably in left-over winter pelage, differs from the average only in being slightly brighter. The third is yellowish white beneath, with nearly normal sides, and dorsal stripe indicated only by a slight darkening of the dorsal area. One, taken on Churchill River, 15 miles above Fort Churchill, has the dorsal area dull red and only faintly indicated, with dull-brown sides and ashy-brown lower parts. It is immature and measures: 133; 30; 20. A skin from Fort Churchill in the U. S. National Museum (No. 4206) has a seal-brown dorsal stripe, which extends forward nearly to the eyes, and dull, yellowish-brown lower parts and sides.

Three adults from Norway House average: Total length 134; tail vertebræ 39; hind foot 18. Four from Robinson Portage average: 135.5; 38; 18. Five from Knee Lake average: 138; 42; 19. Three from York Factory average: 141; 40; 19.

Microtus drummondi (Aud. and Bach.). Drummond Vole.

We found this vole rather common in suitable locations throughout the region between Norway House and Hudson Bay. It seemed most abundant in the partially cleared areas near the posts, and as it readily takes up a residence in the houses and stores, it becomes a decided pest. A large series was collected, including specimens from the following localities: Norway House, Echimamish River, Robinson Portage, Oxford House, Steel River, York Factory, Cape Churchill, and Fort Churchill.

This series averages a little larger than typical *drummondi*, but on the whole is fairly typical. Seven adults from Norway House average: Total length 152; tail vertebræ 42; hind foot 19. Seven from Oxford House average: 153; 47; 19. Seven from Fort Churchill (selected as being nearly typical of this form): 156; 40; 19.

Microtus aphorodemus sp. nov. Barren Ground Vole.

Type from Barren Grounds, about 50 miles south of Cape Eskimo, Keewatin (near mouth of Thlewiaza River). ♀ ad. (skin and skull), No. 106422, U. S. Nat. Mus., Biological Survey collection. Collected August 5, 1900, by Edward A. Preble. Original number, 3208.

General characters.—Similar to *Microtus drummondi*, but larger, with much larger and heavier skull.

Color.—Upperparts dark yellowish bister, as in *drummondi*, but usually with an admixture of yellowish-tipped hairs, imparting a coarse appearance to the pelage; adults varying beneath from nearly white to light plumbeous, sometimes tinged with light brownish. *Young:* Varying but little from adults, lighter in color than young of *drummondi*, especially beneath.

Skull.—Much larger than that of *drummondi;* rostrum proportionally heavier; molar series longer. Compared with that of *pennsylvanicus* the skull is much larger, less arched, and with more widely spreading zygomata; teeth about as in *pennsylvanicus*.

Measurements.—Type: Total length 190; tail vertebræ 50; hind foot 20. Average of six adults from type locality: 182; 49; 20.3. Average of four adults from near Hubbart Point: 181; 49; 20.2. Skull of type: Basal length 28; nasals 7.6; zygomatic breadth 16.5; mastoid breadth 12.6; alveolar length of upper molar series 7. Skull, No. 106274, ♀ ad. (largest in series): Basal length 29; nasals 8.5; zygomatic breadth 17; mastoid breadth 12.5; alveolar length of upper molar series 7.

Remarks.—I found these voles common on the Barren Grounds near the mouth of Thlewiaza River early in August, and collected a large series. They frequented patches of wild rye (*Elymus mollis*) on the sandy raised sea beaches near the shore, and burrowed extensively in the sand and to some extent in the semidecayed masses of seaweed which had accumulated at high-water mark. They were most active in the morning and evening, but were taken at all hours without difficulty in traps set in their well-worn trails. Their principal food seemed to be the culms of the wild rye. I found many cavities beneath boulders which they had utilized for storehouses and had filled with the stalks of this coarse grass, cut into short sections.

Along the coast north of Fort Churchill at several points where the wild rye grew luxuriantly I found abundant traces of these voles. A series of twenty-two was collected on a point about 10 miles north of Hubbart Point on the morning of August 16, while I was waiting for the tide to come in that we might resume our journey southward. This series includes young and adults of all ages and both sexes, the females slightly predominating. Several females contained embryos which varied in number from seven to ten.

Ten specimens taken on the 'Barrens,' near Cape Churchill, August 25, are referable to this form, and the large series of *drummondi* taken at Fort Churchill includes some specimens that seem to incline toward *aphorodemus*.

Microtus xanthognathus (Leach). Chestnut-cheeked Vole.

This species was originally described from Hudson Bay. During our trip, though constantly on the lookout for the species, we saw no traces of voles which we could attribute to *xanthognathus* and secured no specimens.

Coues recorded *Arvicola xanthognathus* from Fort Churchill.[a] The fact that the table in which this specimen is listed contains only skins with chestnut cheeks, as shown by the context, makes it evident that the identification is correct, a fact further supported by the length of the hind foot (1 inch). This specimen can not now be found. Bailey records *Microtus xanthognathus* from Nelson River.[b]

Fiber zibethicus hudsonius subsp. nov. Hudson Bay Muskrat.

Type from Fort Churchill, Keewatin, Canada. ♂ ad. (skin and skull), No. 106881, U. S. Nat. Mus., Biological Survey collection. Collected August 9, 1900, by Alfred E. Preble. Original number, 3081.

General characters.—Rather small, about the size of *Fiber spatulatus*, but differing in color and cranial characters; smaller than and otherwise different from *Fiber zibethicus*.

Color.—Above, dull yellowish brown, darker on middle of back and head; fur of lower parts tipped with yellowish brown, throat yellowish white; middle of chin dusky; cheeks yellowish fawn. Compared with *spatulatus*, *hudsonius* is more yellowish brown, the back lacking the dusky tinge of *spatulatus*, and the sides are uniformly lighter.

Skull.—Smaller and more angular than in *Fiber zibethicus*, with highly developed interorbital crest and small molars, as in *spatulatus*, but differing from skull of *spatulatus* as follows: Bullæ more inflated; parieto-squamosal suture irregular; lateral face of zygoma, especially jugal, broader; nasals narrower anteriorly, but otherwise similar.

Measurements.—Type: Total length 539; tail vertebræ 225; hind foot 74. Average of four adults from Echimamish River: Total length 542.5; tail vertebræ 236.2; hind foot 75.7. Average of two adults from York Factory: Total length 556; tail vertebræ 246.5; hind foot 74.5. Skull of type: Occipito-nasal length 56; zygomatic breadth 38; interorbital breadth 6; greatest length of nasals 20; greatest breadth of nasals 9.5; length of upper molar series (crowns) 13.5.

---

[a] Mon. N. Am. Rod., p. 201, 1877. (No. 8356 in table.)

[b] N. A. Fauna, No. 17, p. 58, 1900.

Remarks.—This species is common and quite generally distributed throughout the region between Lake Winnipeg and Hudson Bay. Though its skin is worth but a few cents in trade, the abundance of the animal and the ease with which it is trapped cause it to be much sought for, and many thousands are annually taken.

While ascending the Echimamish we found muskrats abundant and tame, and we also saw many in the marshy lakes in the neighborhood of Robinson Portage. On account of the abundance of 'rats' these two localities are favorite trapping grounds of the Indian hunters. The sweet flag (*Acorus calamus*), a favorite food, abounds in the lakes below Robinson Portage, and the great number of muskrat houses seen there showed the locality to be a favorite resort. We saw many muskrats while we were descending Hill and Steel rivers. Here they live entirely in burrows in the banks, as the deep water and swift current render house building impracticable. Piles of mussel shells, showing where the animals had been feasting, were frequently noticed on the banks. The point of marsh below York Factory seemed to be the home of a considerable number of muskrats, but as it is frequently overflowed, they are said to seldom survive the winter. In the vicinity of Fort Churchill they were found in but one place- a stream entering the Churchill River a few miles above the post, and there they were rather uncommon.

A winter following a dry season is said to be very destructive to muskrats. On account of the low water the animals construct their burrows and houses correspondingly low and are often forced out by floods at a season when they are unable to procure food and have no protection from their enemies.

We collected specimens on the Echimamish, and at Robinson Portage, York Factory, and Fort Churchill. Young of the year in fresh pelage are darker than adults, but the entire series we collected shows much less rich brown than *zibethicus* from New England.

Synaptomys (Mictomys) bullatus Preble. Northern Lemming Vole.

Synaptomys (*Mictomys*) *bullatus* Preble, Proc. Biol. Soc. Wash., XV, p. 181, August 6, 1902. (Trout Rock, near Fort Rae, Mackenzie.)

We trapped only two specimens of this lemming vole—one at Norway House June 21 and one, a female with six embryos, in the swamps bordering Echimamish River June 25.

Lemmus trimucronatus (Richardson). Back Lemming.

We found this fine species at but one locality, near the mouth of Thlewiaza River, where it was common and where a series of about seventy, comprising adults and young of both sexes, was secured August 4 to 8. A succession of low, flat, boulder-covered areas, which lay between the shore and some shallow lagoons a few hundred yards inland, was occupied by the animals. The ground was dry and well covered with short, thick grass, through which their runways extended in every direction. They burrowed extensively, sometimes beneath boulders, but as often in the sides of tiny terraces or from a flat surface. Their holes seemed to be connected in an endless labyrinth. We captured several by suddenly overturning some of the boulders, but most were taken in traps set in their well-trodden roads. They paid no attention to bait, but were readily caught in runway traps. When taken alive they showed considerable ferocity for animals of their size, snarling and biting vigorously. The breeding season seemed to be nearly over, but a few females contained from four to six embryos. The teats are eight in number, four inguinal and four pectoral.

No specimens of *Lemmus* from the vicinity of Point Lake, the type locality of *trimucronatus*, are available for comparison, but it is not likely that the Hudson Bay animal differs appreciably. A few specimens, in poor condition, from the Anderson River region, show no apparent differences.

Color of adults.—Fur everywhere plumbeous at base, on head and shoulders tipped with black and grayish in about equal proportion; lower parts and sides rusty ochraceous, this color extending forward on cheeks and lips and sometimes tingeing slightly the head and shoulders; lower part of back bright reddish brown, the fur very thick and long; tail usually distinctly bicolor, yellowish below, and dusky brown above; pencil about 9 mm. in length; feet dusky brown above, lighter below. In some of the half-grown specimens in fresh pelage the head and shoulders are considerably flecked with the color of the sides and there is an indistinct dusky stripe extending from between the ears to the middle of the back. Younger specimens are nearly unicolor throughout, the plumbeous fur tipped with yellowish brown.

Measurements.—One of largest specimens: Total length 160; tail vertebræ 22; hind foot 20. Average of ten adults: 151; 21.8; 20. Skull: Average of four adults: Basal length 30.6; zygomatic breadth 21.5; interorbital breadth 3.1; mastoid breadth 15.5.

Remarks.—*Arvicola trimucronatus* was recorded from Igloolik, Melville Peninsula, by Richardson,[a] and was said by J. C. Ross to be common on the shores of Boothia Felix.[b]

Dicrostonyx richardsoni Merriam. Richardson Lemming.

Dicrostonyx richardsoni Merriam, Proc. Wash. Acad. Sci., II, p. 26, March 14, 1900.

About 120 specimens of this interesting lemming were collected. These ranged in age from young a few days old to fully adult individuals. About 80 were taken, mainly by my brother, at Fort Churchill,

[a] Appendix to Parry's Second Voyage, p. 311, 1825 (1827).

[b] Appendix to Ross's Second Voyage, p. xiv, 1835.

the type locality of the species, and the remainder at two localities on the Barren Grounds, and on the 'Barrens' below Cape Churchill. Many burrows were untenanted and there was usually nothing in their appearance to indicate the fact. On this account trapping was found to be less satisfactory in securing specimens than digging. A few minutes' digging usually disclosed whether or not a burrow was occupied.

Gravelly ridges, the remains of old raised sea beaches, occur throughout the country bordering the Bay, and are found inland many miles from the present coast line. Richardson lemmings frequent mainly these ridges. Their burrows differ widely from those of *Lemmus* and *Microtus*. Each seemed to be distinct and occupied by only a single individual, except in the case of females accompanied by young. The hole sometimes has its entrance beneath a piece of driftwood or at the base of a dwarfed spruce. It usually proceeds at an angle of about 45° for a foot or so, and then extends nearly horizontally for 2 or sometimes 3 feet to the nest of grass and moss, which occupies a circular chamber 4 inches in diameter. A side gallery a foot or two in length usually branches off from the main burrow not far from the nest. This is without a terminal chamber and is evidently used as a place of refuge. In this retreat we usually found the owner of the burrow if he was at home.

The sand and gravel dug from the burrow is usually pushed out into a long pile extending sometimes 2 feet from the entrance. No runways are made, even when the burrows are near soft ground; the animals evidently range indiscriminately over the ground. The only food observed in the burrows was a few leaves of bear-berry (*Arctostaphylos uva-ursi*).

Three young at a birth seems to be the usual number. Every litter we found consisted of three, and in each pregnant female we secured were three embryos. The breeding season seemed to be nearly over, however, so that pregnant females were not common. One, captured on the Barren Grounds August 12, besides containing the usual three small embryos, was suckling three young.

Several young were kept for a few days by my brother. They were very readily tamed and took food (rolled oats and crumbs of bread) within a few hours of their capture, allowing him to hold them on his hand while they ate. They sat on their haunches and held the food in their fore feet like squirrels. These young lemmings were very gentle and interesting in their ways, but the old ones fought viciously when captured, and their sharp incisors and strong jaws made them somewhat formidable.

In the immediate vicinity of the post at Fort Churchill we found only a few lemmings, but the sandy ridges on the south side of the river and on the point near the ruins of Fort Prince of Wales proved fruitful collecting grounds. At my two camps on the Barren Grounds south of Cape Eskimo they were abundant in their favorite locations; and on a long sandy ridge below Cape Churchill, where we landed on the afternoon of August 24, we found many burrows and captured about 20 lemmings. Only one was actually seen away from its burrow—a female which, when surprised by us some distance from her home, ran into a shallow deserted burrow for refuge. The animals seem mainly nocturnal in their habits, though a few were taken at the mouths of holes during the daytime. They are known by the natives of Churchill as 'Husky' or 'Huckey' (i. e., Eskimo) mice, because of their northern habitat.

Compared with skulls of *Dicrostonyx hudsonius* from the coast of Labrador, those of *D. richardsoni* exhibit the following conspicuous differences: Braincase broader; interparietal broader and squarish in outline (in *hudsonius* small and triangular); parietals broader; bullæ much more inflated.

In color *D. richardsoni* is much darker than *hudsonius*, owing to a much greater proportion of chestnut- and brown-tipped hairs in the pelage. The specimens taken by us vary remarkably, but the grayest show more brown than any specimens of *hudsonius* examined.

Topotypes of *D. richardsoni* in summer pelage may be described as follows: Fur light plumbeous at base; that of back and sides tipped with chestnut or rich yellowish brown, occasionally with these colors and gray, sides usually lighter than back. A dusky stripe, rarely continuous but usually interrupted on face, extending from nose to tail; this stripe nearly obsolete in some very old specimens, but very distinct in young, and in some about half grown more than 5 mm. in width. Orifice of ears surrounded by black hairs tipped with chestnut. Lower parts varying from yellowish white to rich yellowish brown, variation according with the richness of color of upperparts; throat and chest always darker than rest of lower parts. Soles and forelegs lighter than rest of lower parts, usually nearly white; tail usually unicolor, yellowish white, but sometimes dusky above; pencil yellowish white.

The young vary but little in color. Those perhaps a week old are grayish brown above, with the very distinct dusky dorsal stripe; beneath practically naked, but with a trace of white down. Others a little older are grayish brown above, varying a little even in the same litter, in the same way as the adults, but in a less degree; beneath yellowish white, brownish between forelegs. Young half grown and larger resemble adults, but are generally grayer, and never show the extreme richness of coloring exhibited by adults.

Measurements.—Average of 10 adults: Total length 139.5; tail vertebræ 17.8; hind foot 18.4. One of the largest specimens: Total length 150; tail vertebræ 20; hind foot 20. Average of 8 adult skulls: Basal length (incisor to occipital condyle) 28.3; zygomatic breadth 19.2; interorbital breadth 4; mastoid breadth 14.4. A large skull measures: Basal length 29; zygomatic breadth 20; interorbital breadth 4; mastoid breadth 15.

This lemming has been several times recorded from this region, usually under the name *Arvicola hudsonius*. J. C. Ross reported it from Port Bowen, Prince Regent Inlet,[a] and obtained specimens from Boothia Felix, where it was active throughout the winter;[b] Richardson states that it inhabits Melville Peninsula;[c] specimens were taken by Parry's party on Melville Island on June 13, on which date the pelage was turning brown;[d] and Lyon observed it on Duke of York Bay, Southampton Island.[e] Rae says, probably referring to this species:

> Occasionally large numbers of lemmings are found drowned along the shores of James's Bay, but as they are generally seen after a very high tide, it is uncertain whether they are then migrating, or merely caught by the high tide on their native grounds.[f]

Zapus hudsonius (Zimm.). Hudson Bay Jumping Mouse.

We took *Zapus* at Norway House, on Echimamish River, at Robinson Portage, at Oxford House, on Steel River (near the mouth of the Shamattawa), and at York Factory. It was especially common in the grassy thickets about Oxford House, where the greater part of the series was taken. At York Factory our traps yielded but two specimens, and at Fort Churchill none, though at the latter place we obtained an imperfect skin from the natives. The last one taken was trapped on Steel River near the mouth of the Shamattawa August 31. At Oxford House, where these jumping mice had been very plentiful in July, we set out traps on September 10 and 11 in the same places as before, without success; but as heavy frosts had occurred for some time it is probable that the animals were hibernating. An adult female taken at Oxford House June 30 is suffused above with buffy clay color and is lighter and less bright than the others secured; but the series taken as a whole agrees essentially in color with a series from northern Minnesota, and no differences are noticeable in the skulls. Six adults from Oxford House average: Total length 209.3; tail vertebræ 126; hind foot 30.3. Two from York Factory average: 212; 129.5; 30. Embryos from five to eight in number were noted in several instances.

Zapus hudsonius was originally described from a specimen sent from Hudson Bay, probably from Severn River, by Mr. Graham. It has been recorded from James Bay and Fort Churchill.[g]

---

[a] Parry's Third Voyage, Appendix, p. 93, 1824.
[b] Appendix to Ross's Second Voyage, p. xiv, 1835.
[c] Fauna Boreali-Americana, p. 132, 1829.
[d] Journal of Parry's (First) Voyage, p. 202, 1821.
[e] Lyon's Private Journal, p. 47, 1824.
[f] Journ. Linn. Soc. London, Zool., XX, p. 141, 1888.
[g] Preble, N. A. Fauna, No. 15, p. 17, 1899.

Erethizon dorsatum (Linn.). Canada Porcupine.

The porcupine occurs throughout the region between Lake Winnipeg and Hudson Bay, but is nowhere abundant. In a country where the life of the native is a constant struggle for food, the ease with which this animal is taken is a sufficient reason for its scarcity.

According to Hearne, the porcupine was formerly found north of Churchill River, but was scarce.[a] Bell says:

> Mr. Isbister, of the Nelson River House on the Churchill, informs me that it was once abundant there. It is rare between Lake Winnipeg and Hudson's Bay, but an individual is occasionally found as far north as York Factory.[b]

Forster recorded it from Severn River.[c] Dr. Milne informed me that he had known it to occur but once at York Factory, but that it is reported by the Indians to be common in certain places about the headwaters of the Shamattawa.

Lepus americanus Erxleben. Hudson Bay Varying Hare.

This species is quite generally distributed throughout the region between Lake Winnipeg and Hudson Bay. Unmistakable evidences of its presence were observed all along our route. We obtained a pair of adults and two immature specimens at Oxford House early in July. At York Factory the animal was stated to be rather uncommon. Bell reports it to be common some years in the neighborhood of Fort Churchill.[d]

An adult female taken at Oxford House July 1 is yellowish gray dorsally, much flecked with black, which predominates along center of back; color of sides extending down on upper part of thighs; head and face yellowish brown; outer side of legs yellowish fawn; inner side of legs dull white; upper side of feet dull white, slightly tinged with yellowish. A male taken at Oxford House July 3 has much less dusky on the back, and the back and sides are much suffused with grayish fawn. The measurements of these specimens are as follows: Male: Total length 430; hind foot 117; female: Total length 450; tail vertebræ 43; hind foot 133. Skull of female: Occipito-nasal length 74; zygomatic breadth 36; breadth across postorbital processes 25; length of nasals 30.5; breadth of nasals 26.5.

Lepus arcticus canus subsp. nov. Keewatin Arctic Hare.

Type from Barren Grounds near Hubbart Point, about 75 miles north of Fort Churchill, Keewatin. ♂ yg. ad. (skin and skull), No. 106860, U. S. Nat. Mus., Biological Survey collection. Collected August 17, 1900, by Edward A. Preble. Original number, 3347.

General Characters.—Differing from *Lepus arcticus* in assuming a

---

[a] Journey * * * to the Northern Ocean, p. 381, 1795.

[b] Rept. Prog. Can. Geol. Surv. 1882-3-4, App. II, p. 49DD (1885).

[c] Phil. Trans., LXII, p. 374, 1772.

[d] Rept. Prog. Can. Geol. Surv., 1882-3-4, App. II, p. 49DD (1885).

gray summer pelage; similar to *Lepus labradorius*, but differing slightly in color and in cranial characters.

Color.—Fur of upperparts light plumbeous at base, succeeded by a broad zone of yellowish fawn and a narrow zone of dusky and tipped with white; throat and rump light plumbeous without the fawn-colored zone and white tips; legs and feet white, but the light plumbeous of rump extending down a short distance on thighs; soles clay-color—probably stained; chin grayish white, lighter than throat; head grizzled fawn, brightest on nose and around eyes; nape grayish; occiput dusky; ears dusky grayish anteriorly, edged, except at tip and anteriorly toward the base, with white, dusky tips about 15 mm. long, mesial surface white; fur on inside of ears dusky, tipped with white. An imperfect winter skin obtained near York Factory has the entire fur pure white to base of hairs, except at extreme tips of ears, where the fur is black, becoming yellowish brown toward base.

Lepus labradorius presents a more bluish appearance than *Lepus arcticus canus*, as the plumbeous element of its pelage is more conspicuous. In *labradorius* the plumbeous extends down on outside of leg from thigh nearly to heel.

Skull.—No skulls of adult *arcticus* are available for comparison. Compared with skulls of *labradorius*, those of *canus* differ as follows: Bullæ flatter, rising but slightly above level of basioccipital; jugal broader. A skull of a fully adult animal obtained at Fort Churchill shows the cranial characters of the species better than the Hubbart Point specimens, which are younger. The Fort Churchill skin resembles very closely those taken north of that post.

Measurements.—Type: Total length 570; tail vertebræ 80; hind foot 154. A topotype: Total length 550; tail vertebræ 83; hind foot 155. Skull of adult from Fort Churchill: Greatest length 98; zygomatic breadth 50; breadth across postorbital processes 39; alveolar length of upper molar series 19.

Remarks.—These fine hares occur sparingly in summer throughout the Barren Grounds from Fort Churchill northward. A few breed near Fort Churchill, and one was obtained there August 12. In winter they migrate to a slight extent, reaching the neighborhood of York Factory and perhaps farther.

Though a few signs of this species were noted on a rocky area near my camp 50 miles south of Cape Eskimo early in August, I saw none of the animals themselves until the morning of August 17, when an opportunity was afforded for hunting over a number of low morainic ridges just below Hubbart Point. Tracks made during the previous night were found along the sandy beach, from which they led toward the ridges where I was sure the animals would be found. Carefully scanning the ground, which was covered with gray rounded boulders with occasional bunches of dwarfed willows, I hunted back and forth over the ridges. Hundreds of Lapland longspurs flitted from boulder to boulder, but otherwise few signs of life were evident on the semibarren tracts, and I had walked several miles before my attention was attracted by what at first appeared to be a boulder on which a small restless bird was perched. A second glance showed that the object was an Arctic hare whose ears, twitching slightly, completed the resemblance that had deceived me. Another was afterwards started from beneath a dwarfed willow near by. Both were secured and proved to be males, evidently young of the year, but full grown.

Lynx canadensis Kerr. Canada Lynx.

Found throughout the region between Lake Winnipeg and Hudson Bay, but not common in the vicinity of York Factory. The abundance of lynxes from season to season is said to depend on the abundance of rabbits. We obtained a summer skin from the vicinity of Island Lake and a number of skulls from Cross Lake and Oxford House. The skin is apparently that of an adult. It differs considerably from the winter pelage, and may be described as follows: General color on back and sides yellowish brown, the hairs tipped with light grayish brown; a median dorsal stripe reaching from between ears nearly to tail dusky brown, the hairs tipped with light brown; a few obscure spots on sides; beneath dirty yellowish white, with a few spots of dusky on chest and belly; head and neck colored like sides, but tips of hairs more whitish; 'chin beard' white, with a conspicuous black blotch; ears grayish, edged and tipped with black with a few white hairs intermixed; legs concolor with sides; tail yellowish brown above, indistinctly banded with whitish, lighter below and tipped with black.

The average measurements of six adult skulls from Oxford House are as follows: Occipito-nasal length (measured to anterior point of nasals) 123; zygomatic breadth 93.3; breadth of braincase 58.6.

Dr. Bell says:

> This animal in its apparently erratic migrations does not reach the verge of the forest. A few skins are obtained at Fort George on the East-main coast and at York Factory. It has been occasionally rather numerous about Oxford House.[a]

Hearne saw its tracks near Fort Churchill.[b]

Canis albus (Sabine). Barren Ground Wolf.

Wolves were fairly common in the vicinity of our camp on the Barren Grounds 25 miles south of Cape Eskimo, and several were seen. They were dirty yellowish white in color, and were conspicuous on the barren ridges. Their howling frequently reached our ears, especially at night, when their wild cries seemed peculiarly in keeping with that

---

[a] Rept. Prog. Can. Geol. Surv., 1882-3-4, App. II, p. 49DD (1885).

[b] Journey * * * to the Northern Ocean, p. 366, 1795.

lonely and desolate waste. Sometimes they were heard pursuing caribou. Owing to the limited time, I was unable to obtain specimens, though I saw many skins, mostly nearly pure white, at Fort Churchill, where they are traded by the Eskimos and Chippewyans.

The Barren Ground wolf has been recorded by the different Arctic expeditions from various points in northern Keewatin.

Canis occidentalis Richardson. Gray Wolf.

Gray wolves are found more or less commonly throughout the region between Lake Winnipeg and Hudson Bay, and numbers are traded at all the posts. We saw tracks in several places on Steel River, and a wolf was seen a few miles above Fort Churchill during our stay there. A skull of a female from the vicinity of Norway House was obtained from Mr. MacDonald, who informed me that the animal was one of a pair which for some time during the winter of 1900–1901 lived on small fish which they caught at a certain place in the river where ice did not form. Mr. Campbell, of Oxford House, informed me that during the winter of 1899–1900 several were killed within a few rods of the buildings.

Vulpes fulvus (Desmarest). Common Red Fox.

Foxes occur rather plentifully throughout the region between Lake Winnipeg and Hudson Bay. Many skins were seen in the storehouses of the company at Norway House, Oxford House, and York Factory. A few are collected at Fort Churchill. The 'cross' phase of pelage seems to predominate, and the normal or red phase to be next in abundance; but many black, or 'silver,' foxes are taken. Certain districts are said by the traders to produce black foxes almost exclusively; in others these are rarely taken. Large numbers of skins seen at the different posts exhibited every possible degree of variation from the normal red phase to almost pure black. A series of skulls, including specimens from Oxford House, Cross Lake, and Split Lake, was obtained from Mr. Campbell and Mr. MacDonald.

Vulpes lagopus innuitus Merriam. Continental Arctic Fox.

Vulpes lagopus innuitus Merriam, Proc. Biol. Soc. Wash., XV, p. 170, August 6, 1902.

We first met with this species July 19 on the 'Barrens' between Stony and Owl rivers, about 75 miles north of York Factory. Here, on a slightly elevated part of the tundra, we found a burrow occupied by a family of Arctic foxes. This burrow was typical—an underground labyrinth with several entrances. In the vicinity were scattered the bones and feathers of various birds, principally ptarmigans, and well-worn trails leading in various directions gave evidence of the activity of the mother fox in providing for her family. A young one was enticed from the depths of the burrow and secured. Its color may be described as follows: Head, back, base of tail dorsally, and stripe extending down on outer side of legs, seal brown; face and legs seal brown, flecked with white; shoulders and thighs seal brown, flecked with fawn; sides and lower parts, including ventral surface of tail, light fawn, deepest on sides; proximal two-thirds of tail above, hair brown, strongly overlaid with fawn. The color of the adults is stated to be essentially the same as that of the young. On the Barren Grounds north of Fort Churchill tracks of Arctic foxes were frequently seen. Their dens were found at both of my camps south of Cape Eskimo, and an immature individual, which closely resembled the specimen above described, was trapped at the southernmost of these camps.

The Continental Arctic fox occurs throughout northern Keewatin and the adjacent islands of the Arctic Sea. In summer it seems to be found chiefly on the seacoast, and breeds on the west coast of Hudson Bay as far south at least as the vicinity of York Factory.[a] It was recorded by Edward Sabine from the North Georgia Islands, where it remained throughout the year,[b] and by J. C. Ross from Port Bowen.[c] Richardson described a specimen in the blackish-brown phase, which was killed December 16, on Winter Island, Melville Peninsula.[d] Lyon recorded the species from Duke of York Bay, Southampton Island.[e]

In winter large numbers of these foxes pass down the coast of Hudson Bay. Many are still taken at Fort Churchill, though fewer than in former years. Dr. Milne informs me that at York Factory a few were taken every winter during his residence there, and that the species reaches Severn River, a fact formerly recorded by Hearne.[f] I saw a skin which had been taken during the winter of 1899–1900 near a large lake about 75 miles north of Oxford House, and Mr. William Campbell informed me that he trapped one some years ago at the outlet of Oxford Lake.

The winter of 1900–1901 was remarkable for the great number of these animals which came southward, and for their wide dispersion in the interior. Mr. J. K. MacDonald, of Norway House, wrote me that on account of the light fall of snow the foxes followed up the rivers from the Bay to the vicinity of that post.

The normal phase of color seems to predominate throughout the region. Among the many winter skins seen at York Factory and Fort Churchill was only one 'blue' one—probably the winter pelage of the sooty phase—and the color is said to be of very rare occurrence in the region.

---

[a] Joseph Sabine, Narrative of a Journey to the Shores of the Polar Sea, Appendix, p. 658, 1823.

[b] Suppl. to Appendix Parry's First Voyage, p. clxxxvii, 1824.

[c] Parry's Third Voyage, Appendix, p. 92, 1826.

[d] Fauna Boreali-Americana, I, p. 89, 1829.

[e] Lyon's Private Journal, p. 46, 1824.

[f] Journey * * * to the Northern Ocean, p. 364, 1795.

Ursus americanus Pallas. Black Bear.

Black bears are rather numerous throughout the region between Lake Winnipeg and Hudson Bay, and many skins are annually traded at all the posts. Toward the northward they become less common. One was seen near Robinson Portage by Mr. W. C. King, who passed this point a day or two ahead of us on his way toward York Factory. This bear was feeding on the piles of Mayflies (Ephemeridæ), which perish in myriads and are washed up on the shores in long 'windrows.' These are said to constitute a favorite food of the bear

A number of skulls were obtained from Oxford Lake, where the black bear is rather common. On our return journey we obtained an immature skull at York Factory, and saw a bear's track on the banks of Steel River a few miles below the mouth of Fox River.

Dr. Bell killed a black bear on Churchill River, about 100 miles from its mouth.[a]

Ursus richardsoni Swainson. Barren Ground Bear.

While at Fort Churchill I made inquiries in regard to the Barren Ground bear, but the official in charge, Mr. Alston, knew nothing of such a species. Dr. Bell, speaking of the Barren Ground bear, says:

> In the barren grounds to the northwest of Hudson's Bay, I have been told that a large bear is found, which the Eskimo consider a variety of the polar bear, which has adopted a terrestrial life, and to which they have given the name of "blue" or "grey" bear. * * * This bear is found in the barren grounds south of Hudson's Strait.[a]

While on his journey of exploration Hearne saw the skin of an enormous grizzled bear at the tents of the Indians on the Copper River.[b] This was probably the skin of this species. If the animal extends its range to the vicinity of Hudson Bay it must be very rare.

Thalarctos maritimus (Phipps). Polar Bear.

We obtained several skulls at York Factory through the kindness of Dr. Milne and Mr. G. B. Boucher. While we were traveling between York Factory and Fort Churchill a party of Indians reported seeing a white bear, which swam out to sea on perceiving them. On August 9, below Cape Eskimo, while we were awaiting the rise of the tide so that we could land, we saw a polar bear on the tundra. He was rapidly quartering the rolling ground in search of food. A few days later another was seen several miles north of our camp. While traveling down the coast we saw several places where the animals had lain in the rank beach grass. Mr. Boucher killed a female and her two cubs on the coast between York Factory and Cape Tatnam about the middle of August.

---

[a] Rept. Prog. Can. Geol. Surv., 1882-3-4, App. II, p. 51DD (1885).

[b] Journey * * * to the Northern Ocean, p. 372, 1795.

Polar bears occur regularly all along the coast of Keewatin as far south at least as Severn River, and probably to the head of James Bay. Bell reports that a few have been seen at Moose Factory.[a] The female 'dens up' in a snowdrift in the winter, brings forth her young about March, and soon afterwards leads them to the sea. The male is said to pass the winter at sea. The animals are frequently seen during late summer swimming in the Bay, and Bell records that one was killed by the captain of one of the Hudson's Bay Company ships in open water about the middle of the Bay.[b] Polar bears have been met with by various expeditions to the north of Hudson Bay.

Lutra canadensis (Schreber). Canada Otter.

Otters seem to be found throughout the region, as we saw skins at all of the posts visited. They are not found in the immediate vicinity of Fort Churchill, but are said to be taken not far up the Churchill River. Hearne states, probably referring to the interior, that they used to frequent the rivers to the north of Churchill as far as latitude 62°.[c] We frequently saw their tracks while we were ascending the upper part of Hill River, and the Indians often spoke of the abundance of otters in the vicinity. At York Factory we obtained the skin of a young one about the size of a small mink and dull dark brown in color.

Skulls composing a series recently obtained from Oxford House, Norway House, and Cross Lake are apparently not separable from skulls from Godbout, Quebec, assumed to be typical *canadensis*.

Mephitis mephitis (Schreber).[d] Canada Skunk.

A large number of skins were seen in the storehouse at Norway House, and the animal is occasionally taken about Oxford House, where we obtained a hunters' skin.

While paddling up the channel between Windy and Pine lakes on September 12 we saw a skunk swimming across the stream a hundred yards in front of our canoe. On seeing us he redoubled his exertions, but we overtook and shot him just as he reached the shore. This was the only one seen on our trip. This specimen, as well as the Oxford House skin, has been recorded by Howell.[e]

Dr. Bell gives the skunk as occurring on both sides of James Bay.[a] Forster records one sent from Severn River by Mr. Graham.[f] Howell records a specimen from Moose Factory.[e] Among the large number of skins at Norway House were several which seemed to be referable to the Northern Plains skunk (*M. hudsonica*).

---

[a] Rept. Prog. Can. Geol. Surv., 1882-3-4, App. II, p. 50DD (1885).

[b] Ibid., p. 50DD (1885).

[c] Journey * * * to the Northern Ocean, p. 374, 1795.

[d] See Science (new ser.), XVI, No. 394, p. 114, July 18, 1902.

[e] N. A. Fauna, No. 20, p. 23, 1901.

[f] Phil. Trans., LXII, p. 374, 1772.

Lutreola vison lacustris subsp. nov. Keewatin Mink.

Type from Echimamish River (near Painted Stone) Keewatin, Canada. ♂ yg. ad. (skin and skull), No. 106872, U. S. Nat. Mus., Biological Survey collection. Collected September 14, 1900, by Alfred E. and Edward A. Preble. Original number, 3518.

General characters.—Similar to *Lutreola vison* but larger; skull more angular.

Skull.—Compared with skulls of *vison* from the Adirondack Mountains and New England (assumed to be typical) a large series from Oxford House and the surrounding region differ as follows: More angular and much larger, those of males of *vison* hardly equaling those of females of *lacustris;* sagittal ridge highly developed; zygomata strongly bowed outward; dentition heavier than in *vison;* bullæ large and rather flat, inner anterior corner produced toward pterygoids and ending in a rather acute angle.

Color. Type: Upperparts very dark chocolate brown; underparts lighter; chin and a narrow, irregular, and partially broken stripe extending to lower part of breast, white; a little white between hind legs.

Measurements.—Type: Total length 560; tail vertebræ 190; hind foot 67. Adult female from Swampy Lake: 540; 180; 60. Average of five immature but well-grown individuals of both sexes: 549; 153; 64.8. Skull: Average of ten adults (probably males) from Oxford House: Occipito-nasal length 62.5; zygomatic breadth 41.25; interorbital constriction 11.9; breadth across postorbital processes 18.25; mastoid breadth 36.1.

Remarks.—Minks seem to be common and quite generally distributed throughout the region between Lake Winnipeg and Hudson Bay. They are annually traded at all the posts, and at all except Fort Churchill in large numbers. A few skins were seen in the storehouse at Fort Churchill, and Mr. Chapman, the resident missionary, saw a mink a few miles up Churchill River late in July. We trapped an adult female at Oxford House and another on the shore of Swampy Lake, and while traveling through the lakes on our return journey early in September shot several that were swimming across rivers or between the islands of the lakes. These were mainly immature. The Indians say that when traveling through the lakes at this season they always see minks, and that these are young of the year that are leaving the place of their birth and starting out for themselves. We were usually able to approach quite near before they took alarm and dived, and as they came to the surface within 15 or 20 yards they fell an easy prey.

The specimens composing the series show more white than is usual in *vison*, the average excess appearing in the type.

Putorius cicognani (Bonaparte). Bonaparte Weasel.

Weasels, or 'ermines,' as they are usually styled in the north, were reported as common throughout the region between Lake Winnipeg and the Bay, and many are caught in traps set for more desirable species. We failed to trap any while traveling through the district, but saw many skins at the trading posts and obtained a number of skulls from the natives at Oxford House. The skulls prove to be typical *cicognani*, and as two skins in the U. S. National Museum collection from Fort Albany, Hudson Bay, are referable to this form, the Bonaparte weasel is probably the common weasel of the region.

Two winter skins from York Factory seem referable to this species, but in the absence of skulls I find it impossible to decide whether they are nearest to *cicognani* or to *richardsoni*.

Putorius cicognani richardsoni (Bonaparte). Richardson Weasel.

This form probably replaces *cicognani* throughout the northern parts of the region as far north as the limit of trees and perhaps farther. An immature male taken on Churchill River a few miles above Fort Churchill is dark brown above, as in *cicognani*, and white beneath, very slightly tinged with yellow. The black on the tail occupies its terminal third, inclusive of the hairs. Another male, also immature, shot on the edge of the Barren Grounds below Hubbart Point August 17, resembles the Churchill River specmen, but is strongly tinged with sulphur yellow beneath.

Putorius arcticus Merriam. Tundra Weasel.

Weasels collected by Ludwig Kumlien at Cumberland Gulf in 1878 prove to be referable to this species, the type of which came from Point Barrow. This indicates that the range of the animal may extend across the northern part of the continent, and makes it probable that the following notes referring to *erminea* properly relate to *arcticus*. J. C. Ross mentions the occurrence of the animals on the shores of Boothia Felix. He states that they assumed their winter dress in September and turned brown again toward the end of May. Their tracks were seen at intervals throughout the winter, and the accompanying tracks of lemmings showed that the weasels were in pursuit of what was found to be their chief prey.[a] They are also recorded from the west side of Baffin Bay,[b] and were found on the east side of Melville Peninsula on Parry's second voyage.[c]

Putorius rixosus Bangs. Bangs Weasel.

A specimen of this diminutive weasel from Moose Factory is in the U. S. National Museum collection, and has been several times recorded.

---

[a] Appendix to Ross's Second Voyage, p. x, 1835.

[b] Appendix to Ross's First Voyage, p. xliii, 1819.

[c] Lyon's Private Journal, p. 54 (and elsewhere), 1824.

I made inquiries of the fur traders in regard to the species, but none knew it. It is probable, however, that it is found throughout the southern part of the region, and that their ignorance of it results from its fur being considered valueless and consequently not offered in trade.

Forster recorded a small weasel sent from Hudson Bay by Mr. Graham, which was undoubtedly this species. Its total length is given as 7 inches, and that of its tail as about 1 inch.[a] Bangs recorded a specimen from Fort Albany.[b]

Mustela americana abieticola subsp. nov. Hudson Bay Marten.

Type from Cumberland House, Saskatchewan. ♂ ad. (skin and skeleton), No. $\frac{12356}{31962}$, U. S. Nat. Mus. Collected February, 1890, by Roderick MacFarlane.

General characters.—Much larger than *Mustela americana;* dentition heavier.

Skull.—Much larger than in *M. americana* from the Adirondack Mountains and New Brunswick; more angular, the sagittal crest being more highly developed; zygomata more bowed outward posteriorly; dentition much heavier except last upper molar, which is usually about the same size.

Color. Type in winter pelage: General color of body rich dark yellowish brown, darkest on middle of back; legs and tail darker, the latter almost black at tip; an irregular blotch and a small spot on chest, ochraceous; face and cheeks grayish brown; ears edged with whitish.

Measurements.—Type (estimated from cleaned skeleton): Total length 640; tail vertebræ 210; hind foot 95. Skull (average of six adults from Oxford House): Occipito-nasal length 78.9; zygomatic breadth 47.8; interorbital constriction 16.7; breadth across post-orbital processes 22.5; mastoid breadth 37.8.

Remarks.—The marten is fairly common throughout the region north to the tree limit, but is most abundant in the heavy spruce forests of the southern part. Many skins were seen at Norway House, Oxford House, and York Factory, and a few at Fort Churchill that were said to have been taken on lower Churchill River. A large series of skulls from Oxford House, Norway House, and Cross Lake, collected in the winter of 1900–1901, and a few from York Factory and Fort Churchill collected by our party, are in the Biological Survey collection. These differ from skulls of typical *americanus* to such a degree as to require subspecific separation. A fine skin and skeleton from Cumberland House in the U. S. National Museum has been selected as the type. This form approaches in some of its characters *Mustela a. actuosa* from Alaska, but though smaller has heavier dentition than that species and consequently more crowded molars.

[a] Phil. Trans., LXII, p. 373, 1772.

[b] Proc. Biol. Soc. Wash., X, p. 22, 1896.

Mustela pennanti Erxleben. Fisher.

Found sparingly throughout the southern part of the region. We saw many skins at Norway House and Oxford House. Dr. Milne, of York Factory, informed me that he had known of one fisher being taken about 60 miles east of York Factory, toward Cape Tatnam, and had on one occasion seen a track about 50 miles southeast of York Factory. Farther south more are taken; about thirty or forty are annually traded at Trout Lake and a few at Severn River.

Skulls from Oxford House and Cross Lake, collected during the winter of 1900–1901, resemble closely skulls from Godbout, Quebec, assumed to be typical *pennanti*.

Gulo luscus (Linn.). Hudson Bay Wolverine.

Edwards figured and gave a short account of the 'Quick Hatch or Wolverene' from an individual which was brought alive from Hudson Bay, and which lived for several years.[a] This specimen formed the basis of Linnaeus's description of the species, which he named *luscus*, in allusion to the beast having lost an eye. The wolverine is still found throughout the region between Lake Winnipeg and Hudson Bay, but is rather rare in the southern districts, though a few skins were seen at all the posts visited. To the northward it is more abundant, its range extending over the Barren Grounds and some of the islands of the Polar Sea.

In Hearne's day the natives to the north of Fort Churchill killed many wolverines.[b] Edward Sabine recorded a skull found on Melville Island on Parry's first voyage,[c] and on his second voyage the remains of wolverines were found on Melville Peninsula.[d] Wolverines were killed in midwinter near Felix Harbor and Victoria Harbor during Ross's second voyage.[e] Skins are still obtained in the vicinity of Marble Island.

Odobenus rosmarus (Linn.). Atlantic Walrus.

Occasionally found about the northern part of the Bay, to which section it is probably confined. The body of one had been brought to Fort Churchill by the Eskimos and skinned there a few days before our arrival. A small quantity of walrus ivory, which forms one of the articles of trade, was seen at that post.

Dr. Bell says:

> The walrus is killed by the Eskimo, principally about the entrance to Hudson's Straits and around the Belcher Islands. In former years this animal is reported to have been seen occasionally as far south as Little Whale River. On the opposite side of Hudson's Bay walruses are said to have been seen near Cape Henrietta Maria.[f]

---

[a] Nat. Hist., II, pl. 103, 1747.

[b] Journey * * * to the Northern Ocean, p. 372, 1795.

[c] Suppl. to Appendix to Parry's First Voyage, p. clxxxiv, 1824.

[d] Appendix to Parry's Second Voyage, p. 293, 1825 (1827).

[e] Appendix to Ross's Second Voyage, p. viii, 1835.

[f] Rept. Prog. Can. Geol. Surv., 1877–78, p. 29c (1879).

The animal was formerly abundant at Walrus Island and other points to the north of Fort Churchill. Hearne relates that in 1767 while passing Sea Horse 'Island' he and his party "saw such numbers of those animals lying on the shore, that when some swivel guns loaded with ball were fired among them, the whole beach seemed to be in motion."[a]

J. C. Ross recorded walruses from the northern part of Prince Regent Inlet.[b] According to the same author they were unknown about Boothia, but abounded at Repulse Bay.[c] Dr. Rae mentions seeing many on a small island near Whale Island, below Wager Bay.[d]

Phoca hispida Schreber. Rough Seal.

A number of skins were seen at Fort Churchill and the animal was said to be rather common in the vicinity.

J. C. Ross reported the species from the seas on both sides of the Isthmus of Boothia,[e] and from Port Bowen.[f]

Phoca grœnlandica Erxleben. Harp Seal.

Dr. Bell gives this species as the commonest seal in all parts of Hudson Bay at all seasons.[g] I saw no skins, but the species was reported at Fort Churchill. J. C. Ross reported it from the west side of the Isthmus of Boothia.[b]

Cystophora cristata (Erxleben). Hooded Seal.

Dr. Bell reported this seal from Hudson Straits, where a few skins were seen in the possession of the Eskimos.[g] Its range may extend to other parts of the Bay.

Erignathus barbatus (Erxleben). Bearded Seal.

An adult female was killed July 20 about 75 miles north of York Factory. Its total length was 2,300 mm. (about 7½ feet), and it was so heavy that we had great difficulty in getting it on board. Its general color is grayish; lighter, almost silvery on sides and head, darker on nape and middle of back.

We saw numbers of these seals both to the north and south of Fort Churchill, and the species probably reaches all parts of the Bay. The so-called 'clapmatch' lines, used in place of heavy rope for various uses, are made largely from the skins of this species in the region we

[a] Journey * * * to the Northern Ocean, p. 388, 1795.
[b] Appendix to Ross's Second Voyage, p. xxi, 1835.
[c] Ibid, p. xxii, 1835.
[d] Expedition to the Shores of the Arctic Sea, p. 180, 1850.
[e] Appendix to Ross's Second Voyage, p. xix, 1835.
[f] Appendix to Parry's Second Voyage, p. 94, 1825 (1827).
[g] Rept. Prog. Can. Geol. Surv., 1882-3-4, App. II, p. 52DD (1885).

visited. A long line of surprising strength is made by cutting the skin of the animal in a spiral manner.

Dr. Bell saw skins of the bearded seal in the possession of the Eskimos in Hudson Strait in 1884, and a few of the animals were observed by his party. He also saw bearded seals on the Eastmain coast in 1877, and killed a large one at the mouth of Moose River.[a] J. C. Ross reported the species to be found in summer on the shores of Boothia.[b]

Phoca vitulina Linn. Harbor Seal.

Doubtless found in all parts of the Bay. We saw it in numbers all along the coast visited, especially at the mouth of Churchill River. We saw one August 28, several miles above York Factory, as we were ascending Hayes River. A specimen of a rather small female that measured 1,500 mm. in length was secured 50 miles south of Cape Eskimo August 14. Its general color is light yellowish, irregularly and obscurely spotted on the back with brownish.

Neosorex palustris (Richardson). Marsh Shrew.

This species is represented in our collection by a series of seven collected between Echimamish River and the upper portion of Hill River. We trapped them in the grassy margins of marshes or in wooded swamps. As this series shows considerable variation in color some of the specimens may be noted in detail. Four taken at Robinson Portage June 27 agree very closely in color—back dusky, very finely flecked with silvery white; beneath grayish white with a tinge of brown, grading insensibly on sides into color of upperparts; throat and chin noticeably lighter than remainder of lower parts, this light color sharply divided from the dusky of the face; inner surface of forelegs and fringes of feet silvery white; tail rather sharply bicolored nearly to tip; hind feet dark on outer side, lighter on inner side. A specimen taken on Hill River, near Swampy Lake, September 5, agrees closely with the June specimens. One taken at Robinson Portage September 14 is evidently in winter pelage, the fur being very soft and full, and glossy black dorsally, much darker than the June specimens. One taken on Echimamish River September 15 agrees with the June specimens dorsally, but the ventral surface is strongly clouded with brownish. The skulls, compared with skulls from Minnesota and South Edmonton, Alberta, assumed to be typical *palustris*, show no differences of value.

The series shows little variation in size. The seven specimens average: Total length 157.3; tail vertebræ 72; hind foot 19.4. A specimen received from Mr. J. K. MacDonald, of Norway House, who obtained it from the Indians, is in full winter pelage. It resembles

---

[a] Rept. Prog. Can. Geol. Surv., 1882–3–4, App. II, p. 52DD (1885).

[b] Appendix to Ross's Second Voyage, p. xxi, 1835.

the specimen from Robinson Portage taken in September, except that the fur of the back is tipped with whitish. The species, it is stated, is locally called the 'beaver mouse,' from being found in the houses of the beaver.

Sorex personatus I. Geoffroy. Common Eastern Shrew

About twenty-five specimens were collected from various localities between Norway House and York Factory. These average lighter in color beneath and slightly larger than specimens of *personatus* in corresponding pelage from the Northeastern States. The males average a little darker dorsally than the females. Sometimes there is a distinct dark area on the back separated from the ashy of the lower parts by a lighter lateral stripe. Among my specimens this dark area seems to increase in intensity with age, and a very old male with excessively worn teeth from York Factory has a conspicuous tricolor pattern. This specimen has an abnormally short tail and measures: Total length 90; tail vertebræ 32; hind foot 12. Eight of the largest specimens average: 104.1; 40.6; 12.

King records a specimen of *Sorex forsteri* which was found near the mouth of Great Fish River.[a]

Sorex (Microsorex) alnorum sp. nov. Keewatin Shrew.

Type from Robinson Portage, Keewatin, Canada. ♀ ad. (skin and skull), No. 107014, U. S. Nat. Mus., Biological Survey collection. Collected June 27, 1900, by Alfred E. and Edward A. Preble. Original number, 2662.

General characters.—Larger than *Sorex hoyi* with larger skull; lower parts not tinged with buffy.

Color.—Type: Upperparts sepia brown as in *S. hoyi;* lower parts ashy gray, not tinged with buffy; hind foot dusky on outer and whitish on inner half; tail dusky above, whitish below, becoming dusky toward tip. Compared with *Sorex* (*Microsorex*) *eximius* from Cook Inlet, Alaska, the type is slightly darker above and slightly grayer beneath.

Skull.—Compared with skulls of *S. hoyi* from Elk River, Minnesota, and Red River Settlement, Manitoba, the skull of the type is much larger; rostrum about the same; braincase much broader and higher, both actually and proportionally; mandible longer and slenderer; teeth more heavily pigmented. Compared with *Sorex* (*Microsorex*) *eximius* from Cook Inlet, *alnorum* has rostrum slightly broader, longer and higher; unicuspids more crowded.

Measurements.—Type: Total length 98; tail vertebræ 35; hind foot 12. Skull: Greatest length 16.5; width of braincase 7; length of mandible 10.

Remarks.—We found this species only at Robinson Portage, where the type was trapped, and on the Echimamish. At the latter point a

[a] Narrative of a Journey to the Shores of the Arctic Ocean, II, p. 17, 1836.

Sorex was so eaten in the trap as to be unfit for a skin, but the skull was saved and proves referable to *Microsorex*. It was badly injured and is valueless for comparison, but it is assumed to belong to this species.

Sorex richardsoni Bachman. Richardson Shrew.

This shrew is represented in our collection by a series of eleven specimens. Excepting one taken in June at Robinson Portage and one trapped on Swampy Lake, near its outlet, September 6, all are from Norway House, where the species was abundant. Four of these Norway House specimens, taken June 18 to 22, are still in the dark-backed winter pelage; all the others have the dorsal area less distinctly indicated. One taken at Norway House June 22 contained six embryos. The Swampy Lake specimen, which was trapped in spruce woods, has an abnormally long tail and measures: Total length 120; tail vertebræ 48; hind foot 14. The rest of the series average: 116.6; 41.5; 14.

Under the name *Sorex araneus*, Forster recorded a specimen sent from Hudson Bay, probably referable to the present species.[a]

Sorex sphagnicola Coues. Coues Shrew.

A shrew collected by Dr. Bell on Shamattawa River was referred to this species by Dr. C. Hart Merriam.[b] *Sorex sphagnicola* is very imperfectly known, but is undoubtedly most nearly related to *richardsoni*.

Condylura cristata (Linn.). Star-nosed Mole.

Dr. Bell speaks of this species as being common at Moose Factory.[c] A specimen in the U. S. National Museum, collected at that post in 1881 by Dr. Walton Haydon, and one taken at Rupert House, James Bay, have been recorded by True.[d] I have recently examined the Moose Factory specimen. It was kept in alcohol, but has spoiled in the fluid so that only the skull and part of the skeleton are intact. The skull resembles closely specimens from Pennsylvania and the Northern States.

Myotis lucifugus (Le Conte). Little Brown Bat.

Two specimens from James Bay are recorded by Miller.[e] We saw no bats of any kind during our trip.

---

[a] Phil. Trans., LXII, p. 380, 1772.
[b] N. A. Fauna, No. 10, p. 65, 1895.
[c] Rept. Prog. Can. Geol. Surv. 1882–3–4, App. II, p. 48DD (1885).
[d] Proc. U. S. Nat. Mus., XIX, p. 84, 1896.
[e] N. A. Fauna, No. 13, p. 63, 1897.

Colymbus holbœlli (Reinh.). Holbœll Grebe.

The Holbœll Grebe was first described by Reinhardt from Greenland. Joseph Sabine mentions having received specimens from Hudson Bay.[a] MacFarlane took its eggs in the vicinity of Fort Anderson;[b] and as the species is recorded from other northern points and breeds in Manitoba it probably occurs throughout the Hudson Bay region.

Colymbus auritus Linn. Horned Grebe.

Several grebes thought to belong to this species were seen near the the mouth of Red River June 15, and a specimen was collected at Sea Falls, 20 miles north of Norway House, September 16.

Andrew Murray recorded the species from Trout Lake,[c] and Dr. Bell has collected it at Fort Severn and York Factory, and found it breeding at Fort Churchill,[d] from which point Clarke records a specimen in summer plumage.[e] A specimen collected by Dr. Rae at Repulse Bay is recorded in the British Museum Catalogue.

Podilymbus podiceps (Linn.). Pied-billed Grebe.

Bell records a specimen from York Factory collected and presented to him by Dr. Matthews.[f] This seems to be the only published record of the capture of the pied-billed grebe in the Hudson Bay region. Nutting records specimens taken at Chemawawin, Saskatchewan, an Indian village about 60 miles west of the mouth of the Saskatchewan, where the species was breeding abundantly.[g]

Gavia imber (Gunn.). Loon.

We found this species rather common on the lakes and deeper parts of the rivers on our route between Norway House and York Factory, and its wild notes were heard nearly every night. Several were seen near the mouth of Churchill River July 30, and several more near North River July 31. On our return trip we saw one or more on nearly every lake between Oxford House and Norway House.

---

[a] Franklin's Narrative of a Journey to the Polar Sea, Appendix, p. 692 (*Podiceps rubricollis*), 1823.

[b] Proc. U. S. Nat. Mus., XIV, p. 415, 1891.

[c] Edin. New Phil. Journ., (new ser.), IX, p. 231, 1859.

[d] Proc. Royal Soc. Canada, 1882, I, Sec. IV, p. 49 (1883).

[e] Auk, VII, p. 320, 1890.

[f] Rept. Prog. Can. Geol. Surv., 1882–3–4, App. III, p. 56DD (1885).

[g] Nat. Hist. Bull. Univ. Iowa, II, p. 249, 1893.

King recorded the 'great northern diver' from the mouth of Back River,[a] and Murray received a specimen from Severn House.[b] The catalogue of birds in the U. S. National Museum collection contains the record of a specimen collected at Moose Factory in 1881 by Walton Haydon.

Gavia adamsi (Gray). Yellow-billed Loon.

Under the name *Colymbus glacialis*, James Clark Ross records three loons, which from his description were undoubtedly of this species, obtained about Boothia during (John) Ross's second voyage.[c] In speaking of a loon given him by James Clark Ross, which had been procured in a very high latitude, and which had a yellow bill, Audubon probably refers to one of these specimens.[d]

This record extends the previously recognized range of the species considerably to the eastward.

Gavia arctica (Linn.). Black-throated Loon.

Swainson and Richardson speak of this species as common on Hudson Bay,[e] and Murray recorded it from Severn House.[f] Clarke recorded an adult in summer plumage from Fort Churchill.[g] Some of these records may refer to *pacifica*.

Gavia pacifica (Lawr.). Pacific Loon.

First seen on Hudson Bay about 25 miles north of York Factory July 17, and rather common northward. A fine adult male was secured at Fort Churchill July 25. On the Barren Grounds below Cape Eskimo, August 4 to 13, the species was abundant on the shallow ponds on the tundra, where the young are raised. The old birds were often seen flying to and from the Bay, where most of their food seemed to be secured. The howl of a wolf, or any unusual sound, was generally followed by a chorus of their wild, weird calls, lasting for several minutes. The species was last seen by us below Cape Churchill August 24.

Edwards's plate and description of the 'Speckled Diver, or Loon,'

---

[a] Narrative of a Journey to the Shores of the Arctic Ocean, II, p. 21, 1836.

[b] Edin. New Phil. Journ. (new ser.), IX, p. 231, 1859. Severn House does not appear on most maps. Thompson, who has had access to the maps and records of the Hudson's Bay Company in London, locates it "on Severn Lake, at 54° 5′ north latitude and 92° 30′ west longitude." (Proc. U. S. Nat. Mus., XIII, p. 463, 1890). In Stieler's Hand Atlas it is located at the same place. If this information is correct (as it probably is) Severn House should not be confounded with Fort Severn, the post at the mouth of Severn River, though it is barely possible that Murray's records refer to Fort Severn.

[c] Appendix to Ross's Second Voyage, p. xlii, 1835.

[d] Birds of America, VII, p. 291.

[e] Fauna Boreali-Americana, II, p. 475, 1831.

[f] Edin. New Phil. Journ. (new ser.), IX, p. 231, 1859.

[g] Auk, VII, p. 320, 1890.

plainly represent this species. He gives the length of the bill to the angle as 3 inches, and the color of the hinder part of head and neck as light ash.[a] Specimens from Winter Island, Melville Peninsula, and from Repulse Bay are recorded in the British Museum Catalogue.

Gavia lumme (Gunn.). Red-throated Loon.

First seen July 21 at Fort Churchill, where it was fairly common. It was abundant August 4 to 13 on the shallow ponds on the Barren Grounds below Cape Eskimo, where the old birds were feeding young that were still unable to fly. An adult male and a young one in the dusky downy plumage were collected 50 miles below Cape Eskimo August 4. At our camp 25 miles south of Cape Eskimo the species was more abundant than *G. pacifica*, and fully as noisy. While returning we saw two and shot one on Knee Lake September 9.

Cepphus mandti (Licht.). Mandt Guillemot.

Two were seen on Hudson Bay about 50 miles south of Cape Eskimo August 3. One of these sat motionless on the water while the boat passed by within a few feet. On August 19 I saw one on Button Bay near Fort Churchill.

Two specimens from Melville Peninsula are recorded in the British Museum Catalogue.

Andrew Murray records the black guillemot (*C. grylle*) from Severn House,[b] and Richardson states that it was very numerous off Melville Peninsula.[c] In his narrative Dr. Rae speaks of finding it nesting on an island off Knaps River, Hudson Bay.[d] These records of *grylle* probably refer to the present form.

Uria troile (Linn.). Murre.

Richardson describes a specimen collected at York Factory, Hudson Bay.[e]

Uria lomvia (Linn.). Brünnich Murre.

Swainson and Richardson say that this species frequents Hudson Bay.[f] A specimen from Hudson Bay, received from the Hudson's Bay Company, is recorded in the British Museum Catalogue. Though I have been unable to find any other record, there seems to be no reason that this murre should not occur there regularly. The species is recorded by Ross from Port Bowen, Prince Regent Inlet, where the birds arrived early in June.[g]

---

[a] Nat. Hist., III, pl. 146, 1750.

[b] Edin. New Phil. Journ. (new ser.), IX, p. 231, 1859.

[c] Appendix to Parry's Second Voyage, p. 377, 1825 (1827).

[d] Narrative of an Expedition to the Shores of the Arctic Sea, p. 22, 1850.

[e] Fauna Boreali-Americana, II, p. 477, 1831.

[f] Ibid., p. 477, 1831.

[g] Parry's Third Voyage, Appendix, p. 107, 1826.

Alle alle (Linn.). Dovekie.

This species is recorded as abundant in Baffin Bay, Davis Strait, and other parts of the Arctic seas, and it winters southward on the coast, hence it probably occurs, at least in migrations, about the northern part of Hudson Bay.

Megalestris skua (Brünn.). Skua.

Larus keeask of Latham,[a] stated to inhabit Hudson Bay, and based on Hutchins's 'Esquimaux Keeask,'[b] probably refers to the present species. Though it has apparently not since been recorded from Hudson Bay, the facts of its presence in Hudson Straits and its breeding at Lady Franklin Island, north of the straits, render its occurrence on the waters of the Bay probable.

Stercorarius pomarinus (Temm.). Pomarine Jaeger.

Several seen near the mouth of Churchill River July 21. They were pursuing the terns and small gulls with remarkable agility, evidently to rob them of their prey. A male was collected below Cape Eskimo, where the birds were common August 3 to 8. They were generally flying about over the tundra somewhat after the manner of a marsh hawk, frequently hunting in small, noisy companies; but sometimes several would be seen standing on a small knoll, apparently asleep. They were easily decoyed by the imitation of the cry of a bird in distress. Despite the frequency of melanism in this genus, I saw no black jaegers of any species, though my Indian guide reported seeing one at our camp 25 miles below Cape Eskimo.

Richardson records the species from Igloolik, Melville Peninsula.[c]

Stercorarius parasiticus (Linn.). Parasitic Jaeger.

First met with about 50 miles north of York Factory, where several were seen, and a female was shot July 19. The species was common on the Barren Grounds below Cape Eskimo, August 4 to 13. In habits it resembles *S. pomarinus*.

Clarke records a melanistic specimen from Fort Churchill.[d]

Stercorarius longicaudus Vieill. Long-tailed Jaeger.

Edwards's plate of the 'Arctick Bird' represents this species. In his description he says the tail feathers are 13 inches long.[e] Dr. Bell records a specimen which was shot near York Factory by Dr. Matthews, who presented it to him.[f] The catalogue of birds in the U. S. National Museum collection records a specimen of *S. richardsoni* taken at Fort

---

[a] Index Ornithologicus, II, p. 818, 1790.

[b] Latham, Synopsis, III, Part 2, p. 389, 1791.

[c] Appendix to Parry's Second Voyage, p. 361, 1825 (1827).

[d] Auk, VII, p. 320, 1890.

[e] Nat. Hist., III, pl. 148, 1750.

[f] Rept. Prog. Can. Geol. Surv., 1882–3–4, App. III, p. 56DD (1885).

Churchill by W. W. Kirkby. A specimen from Duke of York Bay is recorded in the British Museum Catalogue.

Pagophila alba (Gunn.). Ivory Gull.

Richardson describes a specimen of the ivory gull killed at Hudson Bay.[a] This gull has also been recorded from Port Bowen and from other localities to the north of Keewatin.

Rissa tridactyla (Linn.). Kittiwake.

Sabine says this species abounds in Hudson Bay,[b] and Richardson gives a description of one killed on Melville Peninsula in July.[c] A specimen from the Savage Islands, Hudson Bay, is recorded in the British Museum Catalogue.

Larus glaucus Brünn. Glaucous Gull.

Doubtless found in all parts of Hudson Bay. It has been recorded from Melville Island, Felix Harbor, and other places in the Arctic regions, and has been found breeding in James Bay and at various points on the east coast of Hudson Bay.[d]

Larus leucopterus Faber. Iceland Gull.

Undoubtedly occurs on Hudson Bay during migrations, since many winter on the Great Lakes, and Arctic expeditions have obtained specimens in Davis Strait and Baffin Bay and at Melville Island.

Larus marinus Linn. Great Black-backed Gull.

Common on the coasts of Greenland and Labrador and frequent in winter on the Great Lakes. It is, therefore, like the Iceland gull, very probably to be found, at least during migrations, inhabiting Hudson Bay.

Larus argentatus Brünn. Herring Gull.

Common on Lake Winnipeg, Hudson Bay, and all the intermediate lakes and larger rivers, and breeding throughout the region. It is usually very shy, however, as it is shot for food by the natives whenever opportunity offers. At Fort Churchill, where we collected a specimen July 28, the eggs are gathered in large numbers in May and packed in salt, to be eaten in the late autumn and early winter. On our return trip during the latter part of August and fore part of September, many birds in the dark, immature plumage, sometimes almost sooty, were seen.

Larus delawarensis Ord. Ring-billed Gull.

Gulls referred to this species were rather common in June and July on the inland waters from Lake Winnipeg to Hudson Bay and north-

[a] Fauna Boreali-Americana, II, p. 419, 1831.

[b] Franklin's Narrative of a Journey to the Polar Sea, Appendix, p. 695, 1823.

[c] Fauna Boreali-Americana, II, p. 423, 1831.

[d] Catalogue Canadian Birds, Part I, p. 34, 1900.

ward to the Barren Grounds. An immature female was collected on the Barren Grounds 50 miles below Cape Eskimo August 8.

Larus franklini Sw. and Rich. Franklin Gull.

A specimen of this gull from Hayes River, Hudson Bay, is recorded in the British Museum Catalogue. It breeds about Lake Winnipeg, and probably at other points in southwestern Keewatin.

Larus philadelphia (Ord). Bonaparte Gull.

Common on Lake Winnipeg June 15 to 17. Several were seen near the outlet of Swampy Lake July 7, and one was collected on Hayes River July 9. They were also common on Hudson Bay, especially in the vicinity of Fort Churchill, where immature birds were numerous July 25 to 30. On our southward trip we saw a few of these gulls on Knee Lake September 6.

The species is recorded from Severn House by Murray,[a] and from several places on Hudson Bay by various other authors.

Rhodostethia rosea (Macgil.). Ross Gull.

The first known specimen of this beautiful species was killed at Alagnak, Melville Peninsula, by James Clark Ross, in June, 1823, during Parry's second voyage. Another was shot a few days later by another officer attached to the same expedition. These two birds served as the basis of Richardson's description of *Larus rossii*,[b] but this name is antedated by *Larus roseus*, inadvertently bestowed by Macgillivray previously. J. C. Ross later recorded it from Boothia Felix.[c] The species does not seem to have been since taken in the Hudson Bay region.

Xema sabinii (Sab.). Sabine Gull.

In the Zoological Appendix to Parry's Second Voyage, Richardson speaks of many specimens being procured at Winter Island, Melville Peninsula.[b] J. E. Harting mentions a pair in breeding plumage obtained off Melville Peninsula and presented to the University Museum at Oxford by John Barrow.[d] Fielden reports a specimen in full breeding plumage which was obtained by Captain Markham near York Factory in August, 1886.[e] The catalogue of birds in the U. S. National Museum collection records a specimen (No. 13715) which was collected at Norway House by Robert Kennicott. The original description was based on a specimen from the west coast of Greenland.

[a] Edin. New Phil. Journ. (new ser.), IX, p. 321, 1859.
[b] Appendix to Parry's Second Voyage, p. 360, 1825 (1827).
[c] Appendix to Ross's Second Voyage, p. xxxvi, 1835.
[d] Proc. Zool. Soc. of London, p. 111, 1871.
[e] Transactions Norwich Soc., IV, p. 351, 1887.

Sterna caspia Pallas. Caspian Tern.

A specimen procured at Moose Factory by J. McKenzie is recorded by Turner,[a] and Nutting records one taken near Grand Rapids, Saskatchewan, in the summer of 1891.[b] As the species occurs in the interior as far as Great Slave Lake, it is probable that it regularly reaches southern Keewatin.

Sterna hirundo Linn. Common Tern.

Common on Lake Winnipeg June 15 to 17, and seen on nearly all the lakes on our route to Hudson Bay. One was collected near Robinson Portage June 28. It may occur on Hudson Bay north to the Barren Grounds, in association with the Arctic tern, but was not identified with certainty this far north. It was recorded from Hudson Bay by Andrew Murray.[c]

Sterna paradisæa Brünn. Arctic Tern.

Common on Hudson Bay. A specimen was collected 50 miles north of York Factory July 19, and the species was seen daily at Fort Churchill. Young just able to fly were observed on the meadows bordering Button Bay July 31, and still smaller young on a sandy islet a few miles farther north August 1. When I started south from near Cape Eskimo August 13, the species had apparently commenced to migrate, and old and young were common on the Bay until we reached York Factory August 26. The catalogue of birds in the U. S. National Museum collection contains the record of an Arctic tern collected at Moose Factory in 1881.

Hydrochelidon nigra surinamensis (Gmel.). Black Tern.

Abundant on the marshes at the mouth of Red River June 15. Many were seen June 24 on a marsh near Hairy Lake and many more June 27 at the north end of Robinson Portage, where two specimens were collected.

Murray recorded the species from Severn House and Moose Factory.[c]

Fulmarus glacialis (Linn.). Fulmar.

Stated by James Clark Ross to be peculiarly numerous in Hudson Bay, Davis Strait, and Baffin Bay.[d]

Phalacrocorax dilophus (Swain.). Double-crested Cormorant.

A single immature bird was taken near Pine Lake September 12. It was swimming about in some rapids and was easily approached.

---

[a] Proc. U. S. Nat. Mus., VIII, p. 245, 1885.

[b] Nat. Hist. Bull. Univ. Iowa, II, p. 252, 1893.

[c] Edin. New Phil. Journ. (new ser.), IX, p. 231, 1859.

[d] Appendix to Ross's Second Voyage, p. xxxviii, 1835.

We saw cormorants, probably referable to this species, on Lake Winnipeg in June, and our Indian guide said he had often seen them at Norway House.

Pelecanus erythrorhynchos Gmel. White Pelican.

We saw several June 16 near Bull Head Point, Lake Winnipeg.

Forster recorded the species (as a variety of *P. onocrotalus*) from York Fort, Hudson Bay.[a] Andrew Murray also recorded a specimen from Hudson Bay,[b] though Blakiston says later that Mr. Murray was not sure where the specimen was killed.[c]

Merganser americanus (Cass.). Merganser.

A female, apparently with a brood of young in the vicinity, was seen near Sea Falls, about 20 miles north of Norway House, June 23. Several individuals were observed on Oxford Lake June 30. A female with moulting wing quills was taken on Hayes River a few miles above York Factory July 10. Though unable to fly, this bird dived so adroitly that we had some difficulty in securing it.

Merganser serrator (Linn.). Red-breasted Merganser.

A number were killed for food by the Indians at our camp on the Barren Grounds 50 miles south of Cape Eskimo August 3 to 8. While ascending Hill River September 2 we observed a female accompanied by young unable to fly, and near Pine Lake September 13 we met with a flock of the birds and obtained from it several specimens.

This species is recorded by Murray from Trout Lake and Severn House:[b] and a specimen collected at Repulse Bay by Dr. Rae is recorded in the British Museum Catalogue.

Lophodytes cucullatus (Linn.). Hooded Merganser.

A female was collected by my brother, Alfred E. Preble, on Churchill River about 15 miles above Fort Churchill August 6.

The species was recorded by Murray from Trout Lake and Hudson Bay;[b] and Bell saw the young going south on Nelson River in September.[d]

Anas boschas Linn. Mallard.

First seen near Pine Lake June 28, where a female followed by a brood of young was noted, and one of the brood collected. An adult female was taken near Oxford House July 3, and between this point and York Factory the species was several times observed. One was

[a] Phil. Trans., LXII, p. 419, 1772.

[b] Edin. New. Phil. Journ. (new ser.), IX, p. 231, 1859.

[c] Ibis, V, p. 151, 1863.

[d] Rept. Prog. Can. Geol. Surv., 1878–79, App. VI, p. 69c (1880).

seen on the marsh at Beacon Point, near York Factory, July 13, one on a small pond on the Barren Grounds north of Seal River August 18, and a female with a brood of unfledged young on Steel River August 31. During the first half of September the species was seen almost daily. A large flock was noted on Trout River September 9, and many were observed on the Echimamish September 14 and 15. During the first part of our journey, when we were going north, we saw several females with broods, and it was noticeable that the young, when startled, invariably took to the woods, where they easily concealed themselves. On our return we started up several large flocks, but more commonly found just three in a flock. The name given to this duck throughout the whole region visited is 'stock duck.' The catalogue of birds in the U. S. National Museum collection shows that specimens were collected at Moose Factory in 1881, and the bird undoubtedly occurs throughout the wooded portion of the region.

Anas obscura rubripes Brewst.[a] Red-legged Black Duck.

An adult male was collected at Fort Churchill July 28 and another had been shot by an Indian a day or two previously.

Blakiston records a specimen received from York Factory.[b] A specimen taken at Moose Factory and one from Cape Hope, Severn River, have been recorded by Brewster.[c] The species is called throughout the region the '*black* stock duck' to distinguish it from the mallard.

Chaulelasmus streperus (Linn.). Gadwall.

One was taken near Cape Churchill August 24, but was not preserved. Bell collected the species on Hudson Bay.[d]

Mareca americana (Gmel.). Baldpate.

Recorded by Murray from Hudson Bay.[e] A specimen from Nelson River is recorded by Baird as being in the Smithsonian collection.[f] Clarke records an adult male from Fort Churchill collected many years ago.[g]

Nettion carolinensis (Gmel.). Green-winged Teal.

An immature male was preserved from a lot killed for food by Indians at Fort Churchill July 25. Several green-winged teals were seen in small ponds about 15 miles north of Seal River August 18, and large flocks were started up from ponds between Fort Churchill and

---

[a] Auk, XIX, p. 184, 1902.
[b] Ibis, V, p. 146, 1863.
[c] Auk, XIX, p. 187, 1902.
[d] Proc. Royal Soc. Canada, 1882, I, Sec. IV, p. 50 (1883).
[e] Edin. New Phil. Journ. (new ser.), IX, p. 229, 1859.
[f] Birds of North America, p. 784, 1858.
[g] Auk, VII, p. 320, 1890.

Cape Churchill August 22. A small flock was seen between Robinson Portage and Echimamish River September 14.

The species is recorded from Severn River by Forster,[a] and from York Factory by Dr. Bell.[b] The British Museum Catalogue records a specimen from Repulse Bay collected by Dr. Rae.

Querquedula discors (Linn.). Blue-winged Teal.

A specimen from Repulse Bay, collected by Dr. Rae, is recorded in the British Museum Catalogue. Nutting found the species common and collected specimens at Chemawawin, Saskatchewan, in the summer of 1891.[c]

Spatula clypeata (Linn.). Shoveller.

Andrew Murray received this species from Moose Factory and Trout Lake,[d] and a specimen collected by James McKenzie at Moose Factory is recorded in the catalogue of birds in the U. S. National Museum. Seton, on the authority of Bell, records it from Lake Winnipeg.[e] Nutting reported it common at Chemawawin, Saskatchewan, where he took specimens in 1891.[c]

Dafila acuta (Linn.). Pintail.

First seen on the 'Barrens,' 50 miles north of York Factory, July 19, when an adult female was taken. Common on the meadows bordering Button Bay July 31. Hundreds were seen on the shallow ponds of the Barren Grounds, 50 miles below Cape Eskimo, August 4 to 8, and on our way back to Fort Churchill, August 13 to 19, numbers were seen whenever we landed. At Duck Point, Playgreen Lake, September 20, I saw one which had been shot by a hunter. Throughout the region the species is called the 'long-necked duck.'

Murray received the species from Trout Lake and Severn House,[d] and Bell records a specimen from York Factory,[f] and says the species breeds near Norway House.[b] A specimen taken at Moose Factory in 1881 is recorded in the catalogue of birds in the U. S. National Museum collection.

Aix sponsa (Linn.). Wood Duck.

The wood duck is recorded from 'Hudson Bay,' Moose Factory, and Trout Lake by Andrew Murray.[g] Two specimens collected at Moose

[a] Phil. Trans., LXII, p. 419, 1772.
[b] Rept. Prog. Can. Geol. Surv., 1878–79, App. VI, p. 69c (1880).
[c] Nat. Hist. Bull. Univ. Iowa, II, p. 257, 1893.
[d] Edin. New Phil. Journ. (new ser.), IX, p. 229, 1859.
[e] Auk, III, p. 149, 1886.
[f] Rept. Prog. Can. Geol. Surv. 1882–3–4, App. III, p. 55DD (1885).
[g] Edin. New Phil. Journ. (new ser.), IX, p. 230, 1859.

Factory by James McKenzie are recorded in the catalogue of birds in the U. S. National Museum.

Aythya americana (Eyt.). Redhead.

Nutting recorded this species and obtained specimens near Chemawawin, Saskatchewan, in the neighborhood of which "countless numbers of the red-head breed."[a]

Aythya vallisneria (Wils.). Canvas-back.

Nutting obtained a specimen near the mouth of the Saskatchewan in the summer of 1891.[b]

Aythya marila (Linn.). Scaup Duck.

A specimen was taken August 22 from a large flock on a pond near the shore of the Bay about midway between Fort Churchill and Cape Churchill. These birds were moulting their wing quills, for which purpose they had evidently congregated on this pond, where they were safe from the attacks of predatory mammals.

Aythya affinis (Eyt.). Lesser Scaup Duck.

The lesser scaup duck has been reported from a number of localities in the Hudson Bay region, though on account of its close resemblance to the greater scaup the accuracy of some of the records is doubtful. Baird recorded a specimen from Nelson River and considered Forster's record of *marila* from Severn River[c] to refer to the present species.[d] Murray records it from Severn House,[e] and Bell reports it from Fort Churchill and York Factory.[f]

Clangula clangula americana (Bonap.). Golden-eye.

Rather common on the larger lakes between Norway House and Oxford House. At the latter place we took an adult male July 3, but did not shoot any more on our northward trip, although we saw a flock containing about 200 near the outlet of Knee Lake July 6, and daily noted a few on our way to York Factory. After this we did not meet with the species again until we reached Hill River on our return, September 5. Here we found many, and on Knee Lake September 6 noted a large flock, probably the same we had observed there two months before. We saw a few more on the Echimamish September 15, one of which we collected, and found a large flock on Hairy Lake.

Macoun records specimens taken at Fort Churchill,[g] and Forster notes the occurrence of the species on Severn River.[h]

---

[a] Nat. Hist. Bull. Univ. Iowa, II, p. 257, 1893.

[b] Ibid., p. 258, 1893.

[c] Phil. Trans., LXII, p. 413, 1772.

[d] Birds of North America, p. 791, 1858.

[e] Edin. New Phil. Journ. (new ser.), IX, p. 230, 1859.

[f] Rept. Prog. Can. Geol. Surv. 1882–3–4, App. III, p. 55DD (1885).

[g] Catalogue Canadian Birds, Part I, p. 95, 1900.

[h] Phil. Trans., LXII, p. 417, 1772.

Clangula islandica (Gmel.). Barrow Golden-eye.

Bell intimates that this species, as well as the common golden-eye, is found on Hudson Bay.[a] Seton, on the authority of R. H. Hunter, records it from Lake Manitoba, Shoal Lake, and the mouth of Red River, Manitoba.[b]

Charitonetta albeola (Linn.). Buffle-head.

We saw an adult male on Red River, between Winnipeg and West Selkirk, June 14, and an immature bird on Hill River, near Swampy Lake, September 5.

The species is recorded by Forster from Severn River,[c] by Murray from Severn House, Moose Factory, and Trout Lake,[d] and by Clarke from Fort Churchill.[e] It probably ranges throughout southern Keewatin.

Harelda hyemalis (Linn.). Old-squaw.

Though not observed by us, this species has been recorded from a number of localities on the west shore of Hudson Bay. Edwards figured a 'Long-tailed Duck from Hudson's Bay;'[f] Forster recorded the species from Churchill River;[g] and Richardson described specimens killed at Winter Island, Melville Peninsula, on Parry's second voyage.[h] Rae speaks of finding this species breeding on an island off Knaps River,[i] and collected a specimen at Repulse Bay which is recorded in the British Museum Catalogue. Murray recorded the species from Severn House;[d] Clarke, an adult pair in summer plumage from Fort Churchill;[j] and Bell, a specimen from York Factory.[k]

Histrionicus histrionicus (Linn.). Harlequin Duck.

Forster recorded a specimen from the Hudson Bay region;[l] Blakiston says he examined one at York Factory;[m] and the catalogue of birds in the U. S. National Museum collection records a specimen taken in James Bay August 3, 1860.

Camptolaimus labradorius (Gmel.). Labrador Duck.

Joseph Sabine includes *Anas labradoria* among the species which at that time were found on Hudson Bay and its vicinity, but which were

[a] Proc. Royal Soc. Canada, 1882, I, Sec. IV, p. 50 (1883).
[b] Auk, III, p. 328, 1886.
[c] Phil. Trans., LXII, p. 417, 1772.
[d] Edin. New Phil. Journ. (new ser.), IX, p. 230, 1859.
[e] Auk, VII, p. 320, 1890.
[f] Nat. Hist. III, pl. 156, 1750.
[g] Phil. Trans., LXII, p. 418, 1772.
[h] Appendix to Parry's Second Voyage, p. 373, 1825 (1827).
[i] Narrative of an Expedition to the Shores of the Arctic Sea, p. 22, 1850.
[j] Auk, VII, p. 320, 1890.
[k] Rept. Prog. Can. Geol. Surv., 1882-3-4, App. III, p. 55DD (1885).
[l] Phil. Trans., LXII, p. 419, 1772.
[m] Ibis, V, p. 149, 1863.

not obtained on Franklin's Expedition.[a] An adult male, perhaps from Hudson Bay, was presented to the British Museum by the Hudson's Bay Company about the year 1835;[b] and it is likely that a specimen to which no definite locality is assigned that is recorded in the British Museum Catalogue is this same bird.

Somateria mollissima borealis (C. L. Brehm). Northern Eider.

This species was obtained on Parry's second voyage at Winter Island, being recorded as *mollissima*;[c] and Blakiston, writing of *S. mollissima*, speaks of having received specimens from Hudson Bay.[d] A specimen collected by Dr. Rae at Repulse Bay is recorded in the British Museum Catalogue.

Somateria dresseri Sharpe. American Eider.

First seen in small numbers about 50 miles north of York Factory July 19, and rather common from there northward. Flocks of a hundred or more were often seen north of Fort Churchill, and a female with young two or three weeks old was observed August 3, 50 miles south of Cape Eskimo. The species is said to breed in large numbers on certain rocky islands north of Fort Churchill. Quantities of the eggs are gathered by Indians and Eskimos in late April and early May, and brought to Fort Churchill, where they are packed in salt for later consumption. The bird is called by the Indians 'Husky' (i. e., Eskimo) duck.

Two specimens, collected by Dr. Bell at Fort Churchill, are recorded by Macoun.[e]

Somateria v-nigra Gray. Pacific Eider.

Murray recorded this species from Severn House.[f] As the species occurs on Great Slave Lake, it should be occasionally found in Keewatin.

Somateria spectabilis (Linn.). King Eider.

Edwards figures this species from Hudson Bay, calling it the 'Gray-Headed Duck.'[g] Linnæus cites Edwards's figure in his description of *spectabilis*, but does not base his description exclusively on it. It was recorded by Blakiston, who speaks of having seen specimens from Hudson Bay,[h] and by Richardson, who states that it was observed in numbers about Melville Peninsula on Parry's second voyage.[i] A

[a] Franklin's Narrative of a Journey to the Polar Sea, Appendix, p. 698, 1823.
[b] See Dutcher, Auk, VIII, p. 203, 1891.
[c] Appendix to Parry's Second Voyage, p. 370, 1825 (1827).
[d] Ibis, V, p. 150, 1863.
[e] Catalogue Canadian Birds, Part I, p. 105, 1900.
[f] Edin. New Phil. Journ. (new ser.), IX, p. 229, 1859.
[g] Nat. Hist., III, pl. 154, 1750.
[h] Ibis, V, p. 150, 1863.
[i] Appendix to Parry's Second Voyage, p. 373, 1825 (1827).

specimen from York Factory, collected by Dr. Bell, is recorded by Macoun.[a]

Oidemia americana Swains. American Scoter.

Swainson based his description of this species on Hudson Bay specimens;[b] a specimen collected by Dr. Rae at Repulse Bay is recorded in the British Museum Catalogue; Andrew Murray recorded the species from Trout Lake:[c] Blakiston says that he "received specimens from the west coast of Hudson's Bay;"[d] and an adult male is recorded from Fort Churchill by Clarke.[e]

Oidemia deglandi Bonap. White-winged Scoter.

We observed this species twice on Knee Lake—a small flock on July 7 and a few on September 8 as we were returning. We also noted a few about fifty miles north of York Factory July 19. Many breed about the borders of small ponds throughout the interior, and large numbers are killed for food before they are able to fly.

Murray recorded the species from Trout Lake, Moose Factory, and Severn House.[c]

Oidemia perspicillata (Linn.). Surf Scoter.

Rather common in Hudson Bay south of Cape Eskimo August 4 to 13. My brother took a specimen on Churchill River near Fort Churchill August 11, and we obtained another near Pine Lake September 13, as we were returning. On September 14 we saw several near Robinson Portage.

Edwards figured this species, which he called the 'Great Black Duck from Hudson's-Bay,'[f] and on this figure Linnaeus based his description of *Anas perspicillata*. The species has since been several times recorded from Hudson Bay by different writers, and a specimen from Repulse Bay, collected by Dr. Rae, is mentioned in the British Museum Catalogue.

Erismatura jamaicensis (Gmel.). Ruddy Duck.

Blakiston speaks of having examined a specimen at York Factory, on Hudson Bay,[d] and Bell records a specimen from the same place.[g] The species seems to be of regular occurrence about Lake Winnipeg and other points in Manitoba, and probably in the adjacent parts of Keewatin.

---

[a] Catalogue Canadian Birds, Part I, p. 108, 1900.
[b] Fauna Boreali-Americana, II, p. 450, 1831.
[c] Edin. New Phil. Journ. (new ser.), IX, p. 231, 1859.
[d] Ibis, V, p. 150, 1863.
[e] Auk, VII, p. 320, 1890.
[f] Nat. Hist., III, pl. 155, 1750.
[g] Rept. Prog. Can. Geol. Surv., 1878–79, App. VI, p. 69c (1880).

Chen hyperborea nivalis (Forst.). Greater Snow Goose.[a]

This bird was first described by Forster from Severn River, Hudson Bay.[b] Swainson and Richardson speak of its occurrence at Albany Fort and York Factory;[c] Murray records it from Moose Factory and Severn House;[d] and Bell characterizes it as abundant at Fort Churchill and York Factory during migrations,[e] and says that it began to arrive at the former place September 5.[f] Barnston says that at Martin Falls, Albany River, the species was generally passing in large flocks about May 10.[g] A specimen collected by Dr. Rae at Repulse Bay is recorded in the British Museum Catalogue; and one from Black Island, Lake Winnipeg, is recorded by Macoun.[h]

Chen cærulescens (Linn.). Blue Goose.

Edwards figured the 'Blue-Winged Goose' from Hudson Bay,[i] and on this figure Linnæus based his description of the species. A specimen from Repulse Bay is recorded in the British Museum Catalogue, and one taken at Moose Factory in August, 1860, by J. McKenzie, and one from Fort Churchill, appear in the catalogue of birds in the U. S. National Museum. It is said to be found principally about the southern part of the Bay, and according to Indian information breeds in northern Labrador.

Chen rossi (Cassin). Ross Snow Goose.

A specimen procured at Fort Churchill, Hudson Bay, is recorded by Macoun.[j] The 'Horned Wavey' of Hearne is doubtless this species.

Anser albifrons gambeli (Hartl.). White-fronted Goose.

Edwards figured a specimen procured from Hudson Bay, which he called the 'Laughing-Goose.'[k] Barnston says that this species is seldom seen in the southern part of Hudson Bay, but is less rare at York Factory, and is frequent at Fort Churchill.[l] Blakiston received a specimen from Hudson Bay,[m] and one from Repulse Bay, collected by Dr. Rae, is recorded in the British Museum Catalogue.

---

[a] In some of the cases cited the species has been recorded as *C. hyperborea*, but I have assumed that the eastern form is referred to.

[b] Phil. Trans., LXII, pp. 413, 433, 1772.

[c] Fauna Boreali-Americana, II, p. 467, 1831.

[d] Edin. New Phil. Journ. (new ser.), IX, p. 225, 1859.

[e] Rept. Prog. Can. Geol. Surv., 1878–79, App. VI, p. 69c (1880).

[f] Ibid., 1882–3–4, App., III, p. 55DD (1885).

[g] Edin. New Phil. Journ., XXX, p. 254, 1841.

[h] Catalogue Canadian Birds, Part I, p. 114, 1900.

[i] Nat. Hist., III, pl. 152, 1750.

[j] Catalogue Canadian Birds, Part I, p. 115, 1900.

[k] Nat. Hist., III, pl. 153, 1750.

[l] Ibis, II, p. 257, 1860.

[m] Ibid., V, p. 141, 1863.

Anser fabalis (Lath.). Bean Goose.

Under the name *segetum*, Richardson mentions this species as one of the geese which are known to visit the Hudson Bay region, but are rarely seen, being accidental visitors.[a]

Branta canadensis (Linn.). Canada Goose.

An island in the northern part of Lake Winnipeg, on which this bird is said to nest in considerable numbers, was pointed out to us. While descending Steel River July 9 we took a half-grown bird from a flock of five. These had probably come down Fox River, where the species is said to breed. Young geese unable to fly, probably of this species, were seen by Alfred E. Preble on Churchill River August 11. While ascending Steel and Hill rivers, August 31 to September 4, we saw one or two flocks daily. We shot one bird on Hill River September 4, but found its preservation impracticable.

Murray recorded the species from Moose Factory and Severn House,[b] and Bell says it breeds on Churchill River.[c]

In former times, when the posts on Hudson Bay supported a much larger population than at present, geese constituted a staple article of food, and this species and *B. c. hutchinsi* especially were shot in great numbers, both for immediate consumption and to be salted for winter. Barnston, from the number recorded at the different posts, estimated that at least 57,500 were annually killed on James Bay and the west coast of Hudson Bay.[d] At present the demand for the birds is less and their numbers are diminished; hence fewer are killed.

Branta canadensis hutchinsi (Rich.). Hutchins Goose.

Several flocks of geese referred to this species were seen on the Barren Grounds near Hubbart Point August 16.

Dr. Rae saw a female with a brood of young near Neville Bay.[e] Macoun records two specimens taken at Fort Churchill by Dr. Bell, and an egg obtained at Repulse Bay.[f] The species has been recorded from other points in northern Keewatin.

Branta bernicla (Linn.). Brant.

Said by Swainson and Richardson to breed in great numbers on the coast and islands of Hudson Bay.[g] A specimen from Repulse Bay, collected by Dr. Rae, is recorded in the British Museum Catalogue; and Dr. Bell records a specimen killed at York Factory.[h]

---

[a] Appendix to Parry's Second Voyage, p. 364, 1825 (1827).

[b] Edin. New Phil. Journ. (new ser.), IX, p. 225, 1859.

[c] Rept. Prog. Can. Geol. Surv., 1878–79, App., VI, p. 69c (1880).

[d] Ibis, II, p. 258, 1860.

[e] Narrative of an Expedition to the Shores of the Arctic Sea, p. 24, 1850.

[f] Catalogue Canadian Birds, Part 1, p. 120, 1900.

[g] Fauna Boreali-Americana, II, p. 469, 1831.

[h] Proc. Royal Soc. Canada, 1882, I, Sec. IV, p. 51 (1883).

Branta leucopsis (Bechst.). Barnacle Goose.

Said by Richardson to be rare and accidental in Hudson Bay.[a] A specimen in the U. S. National Museum collection was obtained near Rupert House, James Bay, by Bernard R. Ross.[b]

Olor columbianus (Ord). Whistling Swan.

Whistling swans visit the western shores of Hudson Bay in great numbers in the spring and fall, and their assembled thousands are said to present a beautiful and imposing spectacle during their semiannual visits. The broad expanse of Churchill River near its mouth is a favorite place of resort. A specimen collected at Fort Churchill and presented by J. R. Spencer is recorded by Dr. Robert Bell,[c] who also states that the species breeds on Nottingham Island, Hudson Bay.[d]

Olor buccinator (Rich.). Trumpeter Swan.

Richardson describes a specimen killed on Hudson Bay.[e] Barnston speaks of the breeding of swans near Eastmain Fort, on James Bay, and of eggs being brought to him from a nest on the banks of a lake near Norway House.[f] His remarks probably refer to this species, as it has a more southern breeding range, especially in the interior, than *O. columbianus*.

Botaurus lentiginosus (Montag.). Bittern.

We saw several flying over the marsh at the mouth of Red River June 15, heard the notes of one near Norway House June 19, and saw one near Sea Falls June 23. At Beacon Point, near York Factory, I started one from the marsh July 13. On our return trip we saw several near Oxford House September 10, and on the Echimamish September 15.

The species is recorded from Severn River by Forster,[g] and from Fort Churchill by Clarke.[h]

Ardetta exilis (Gmel.). Least Bittern.

Under the name *Botaurus minor* Dr. Robert Bell records this species from York Factory,[i] and later says, "Of the least Bittern (*Ardetta exilis* Gray) I have specimens from Manitoba and York Factory."[j] It has been taken near Winnipeg, but can scarcely reach Keewatin except as a straggler.

---

[a] Appendix to Parry's Second Voyage, p. 364, 1825 (1827).
[b] Baird, Brewer, and Ridgway, Water Birds of North America, I, p. 475, 1884.
[c] Rept. Prog. Can. Geol. Surv., 1878–79, App. VI, p. 70c (1880).
[d] Rept. Prog. Can. Geol. Surv., 1882–3–4, p. 30DD (1885).
[e] Fauna Boreali-Americana, II, p. 464, 1831.
[f] Ibis, II, p. 253, 1860.
[g] Phil. Trans., LXII, p. 401, 1772.
[h] Auk, VII, p. 320, 1890.
[i] Rept. Prog. Can. Geol. Surv., 1878–79, App. VI, p. 68c (1880).
[j] Proc. Royal Soc. Canada, 1882, I, Sec. IV, p. 51 (1883).

Ardea herodias Linn. Great Blue Heron.

One noted on Red River, near Lake Winnipeg, June 15.

Edwards figured a specimen from Hudson Bay, calling it the 'Ash-colour'd Heron from North America.'[a] His figure formed the principal basis for Linnæus's description of *Ardea herodias*. Turner records a specimen collected at Moose Factory, James Bay, August 29, 1860, by James McKenzie.[b] This specimen is recorded in the catalogue of birds in the U. S. National Museum, but can not now be found.

Grus americana (Linn.). Whooping Crane.

Edwards figured the 'Hooping-Crane from Hudson's-Bay.'[c] His figure formed a partial basis for Linnæus's description of *Ardea americana*. Hearne says: "This bird visits Hudson's Bay in the spring, though not in great numbers. They are generally seen only in pairs, and that not very often."[d]

Grus canadensis (Linn.). Little Brown Crane.

Noted several times on the Barren Grounds 25 miles south of Cape Eskimo, usually in companies of four or five. The distorting effect of the atmosphere sometimes gave them a strange appearance as they stood or walked about on some distant hillock.

Edwards's 'Brown and Ash-colour'd Crane,' figured from a Hudson Bay specimen,[e] formed the basis of Linnæus's description of *Ardea canadensis*. Forster recorded the species from Severn River,[f] Murray from Trout Lake,[g] and Richardson from Igloolik, Melville Peninsula.[h] It was seen at Montreal Island, near the mouth of Great Fish River, and recorded by King.[i]

Rallus virginianus Linn. Virginia Rail.

Dr. Bell records from York Factory a specimen presented by Dr. Matthews.[j] Nutting found it common about Chemawawin, Saskatchewan, and secured specimens.[k]

Porzana carolina (Linn.). Sora.

Edwards figured a sora rail from Hudson Bay,[l] calling it the 'Little American Water Hen,' and the species is recorded from Severn House

---

[a] Nat. Hist., III, pl. 135, 1750.
[b] Proc. U. S. Nat. Mus., VIII, p. 245, 1885.
[c] Nat. Hist., III, pl. 132, 1750.
[d] Journey * * * to the Northern Ocean, p. 422, 1795.
[e] Nat. Hist., III, pl. 133, 1750.
[f] Phil. Trans., LXII, p. 409, 1772.
[g] Edin. New Phil. Journ. (new ser.), IX, p. 225, 1859.
[h] Appendix to Parry's Second Voyage, p. 353, 1825 (1827).
[i] Narrative of a Journey to the Shores of the Arctic Ocean, II, p. 21, 1836.
[j] Rept. Prog. Can. Geol. Surv., 1882-3-4, App. III, p. 55DD (1885).
[k] Nat. Hist. Bull. Univ. Iowa, II, p. 262, 1893.
[l] Nat. Hist. III, pl. 144, 1750.

by Andrew Murray,[a] from Fort Churchill by Clarke,[b] and from York Factory by Bell.[c] Macoun records it from Moose Factory, James Bay, where Spreadborough found it breeding.[d] It undoubtedly breeds throughout the region.

Porzana noveboracensis (Gmel.). Yellow Rail.

During the afternoon of July 13, while collecting on the marsh at Beacon Point, near York Factory, I flushed five yellow rails, securing three, two males and a female. The males were discovered by following up the source of their notes, which, though rather low, have a penetrating quality that makes them distinctly audible at a distance of several hundred yards. A metallic cluck, five or six times repeated, constituted their call. It was given thus: First two notes uttered with scarcely an interval between them, then a pause of about a second, then three, or occasionally four, notes exactly like the first two. The birds flushed were in open, grassy places where I was usually able to approach near and, guided by the note, to start them almost from beneath my feet. I heard about a dozen, but was unable to flush more than five, for the others, which were in bushy portions of the marsh, seemed to hear me and take alarm before I could get near, and would cease calling.

Hutchins wrote a short account of the habits of the yellow rail as observed by him at the mouth of Severn River,[e] and Bell has recorded the species from Fort George, on the east coast of the Bay.[c]

Fulica americana Gmel. Coot.

Abundant in the marsh at the mouth of Red River June 15, and not again noted until we reached the same marsh on our return September 21, when many were seen. We were told it occurs about some of the marshy lakes north of Lake Winnipeg.

Crymophilus fulicarius (Linn.). Red Phalarope.

The red phalarope was fairly common on the shallow ponds below Cape Eskimo August 3 to 8, at which time they were feeding downy young. An adult male and a young bird in the down were collected August 6.

Edwards figured a specimen brought from Hudson Bay by Mr. Isham.[f] His figure formed a partial basis for Linnæus's description of *Tringa fulicaria*. Clarke records an adult pair in summer plumage

---

[a] Edin. New Phil. Jour., (new ser.), IX, p. 225, 1859.
[b] Auk, VII, p. 321, 1890.
[c] Proc. Royal Soc. Canada, 1882, I, Sec. IV, p. 51 (1883).
[d] Catalogue Canadian Birds, Part I, p. 140, 1900.
[e] MS. quoted by Richardson, Swainson, Thompson, Macoun, and others.
[f] Nat. Hist., III, pl. 142, 1750.

from Fort Churchill:[a] and a specimen, collected at Repulse Bay by Dr. Rae, is noted in the British Museum Catalogue. King observed the species near the mouth of Great Fish River.[b]

Phalaropus lobatus (Linn.). Northern Phalarope.

First seen on the marshes about 50 miles north of York Factory July 19, where a male was taken whose abdominal bareness showed that it had recently been sitting on eggs. The birds were present in numbers, and when I left the marshes, about an hour after sunset, were still active and noisy. The species was common at Fort Churchill July 21 to August 21, and abundant on the Barren Grounds below Cape Eskimo August 3 to 8, where an immature bird was collected. While returning down the coast of the Bay August 14 to 26 we saw the species daily, but after leaving York Factory did not again meet with it.

Macoun records it from an island in James Bay, where about a dozen, evidently breeding, were observed by Spreadborough June 16, 1896.[c]

Steganopus tricolor Vieill. Wilson Phalarope.

Murray records *Phalaropus lobatus* (Ord) from Severn House.[d] *Phalaropus lobatus* of Ord is supposed to be referable to *S. tricolor*, and as the species regularly reaches Lake Winnipeg, its occurrence at Severn House is not improbable.

Philohela minor (Gmel.). Woodcock.

Recorded by Dr. Bell, who says "I saw one specimen of the woodcock at York Factory in the end of August."[e] The woodcock may regularly reach southern Keewatin, but its presence as far north as York Factory must be merely accidental.

Gallinago delicata (Ord). Wilson Snipe.

Several seen on a marsh between Oxford and Back lakes, near Oxford House, September 10, and three or four started beside the Echimamish September 14.

Two specimens in summer plumage are recorded by Clarke from Fort Churchill,[a] and a specimen said to have been collected by Dr. Rae at Repulse Bay is listed in the British Museum Catalogue. The catalogue of birds in the U. S. National Museum collection shows that the species was taken at Moose Factory in 1881.

---

[a] Auk, VII, p. 321, 1890.
[b] Narrative of a Journey to the Shores of the Arctic Ocean, II, p. 21, 1836.
[c] Catalogue Canadian Birds, Part I, p. 146, 1900.
[d] Edin. New Phil. Jour. (new ser.), IX, p. 225, 1859.
[e] Rept. Prog. Can. Geol. Surv., 1878–79, App. VI, p. 70c (1880).

Gallinago major (Gmel.). Greater Snipe.

A snipe sent to the British Museum many years ago from 'Hudson's Bay,' and which served as the type of Swainson's *Scolopax leucurus*,[a] is identified by Dr. Sharpe as *Gallinago major*, and so recorded in the British Museum Catalogue.[b]

Macrorhampus scolopaceus (Say). Long-billed Dowitcher.

Abundant on the meadows bordering Button Bay near Fort Churchill, July 31. The species was then moving southward, and, with the exception of a few individuals seen about 50 miles below Cape Eskimo August 3, was not again noted. Two specimens were collected. These have been examined by Reginald Heber Howe, jr., in connection with his study of the genus, and he considers them practically typical *scolopaceus*.[c]

Micropalama himantopus (Bonap.). Stilt Sandpiper.

A male bird was shot on the mud beside a small pool on the tundra, about 50 miles north of York Factory, July 19. It was in company with a pectoral sandpiper which was killed by the same shot. It is probably a bird of the year, and has dusky back and crown, the feathers of which are edged with rusty brown, those of wing-coverts being edged with pale buffy; lower parts buffy, very faintly spotted on sides and flanks; superciliary stripe, cheeks and throat buffy whitish, very faintly spotted with brown. Several more individuals were noted on August 12 about 25 miles south of Cape Eskimo.

Richardson describes one from Hudson Bay,[d] and a specimen collected by Dr. Rae at Repulse Bay is recorded in the British Museum Catalogue.

Tringa canutus Linn. Knot.

A specimen from Melville Peninsula is described by Richardson.[e] One from Repulse Bay, collected by Dr. Rae, and one from 'Hudson Bay' are recorded in the British Museum Catalogue.

Tringa maritima Brünn. Purple Sandpiper.

Recorded by Richardson from Winter Island, Melville Peninsula, where at was obtained in June,[f] and by James C. Ross, who says it arrived at Port Bowen, Prince Regent Inlet, early in June.[g] Richardson gives a description of a specimen from Hudson Bay;[h] and one

---

[a] Fauna Boreali-Americana, II, p. 501, 1831.

[b] See Coues, Auk, XIV, p. 209, 1897.

[c] See Howe, Auk, XVIII, p. 272, 1901.

[d] Fauna Boreali-Americana, II, p. 381, 1831.

[e] Ibid., p. 387, 1831.

[f] Appendix to Parry's Second Voyage, p. 354, 1825 (1827).

[g] Narrative of Parry's Third Voyage, Appendix, p. 101, 1826.

[h] Fauna Boreali-Americana, II, p. 382, 1831. From the date of collection, July 29, 1822, it is almost certain that it was collected in the vicinity of York Factory.

collected by Dr. Rae at Repulse Bay is recorded in the British Museum Catalogue.

Tringa maculata Vieill. Pectoral Sandpiper.

First seen on the marshes about 50 miles north of York Factory July 19, where the species was rather common, and where a male was shot. Common on the meadows bordering Button Bay July 31, and abundant on the Barren Grounds south of Cape Eskimo August 3 to 13. Several seen near Oxford House September 10.

A specimen collected by Dr. Rae at Repulse Bay is recorded in the British Museum Catalogue, and the catalogue of birds in the U. S. National Museum collection records a specimen taken at Moose Factory in 1881 by Walton Haydon.

Tringa fuscicollis Vieill. White-rumped Sandpiper.

Rather common on the meadows bordering Button Bay July 31, where a specimen was collected. A number were noted on the Barren Grounds below Cape Eskimo August 3 to 8.

Tringa bairdi (Coues). Baird Sandpiper.

Macoun says: "Spreadborough saw three on a small island in James Bay on June 16, 1896, and believes they were breeding."[a]

The species probably occurs regularly in western Keewatin

Tringa minutilla Vieill. Least Sandpiper.

First met with at Fort Churchill, where adult birds accompanied by young were seen on the meadows July 26 and an immature bird was taken. Large numbers were observed on the shores of Button Bay July 31, and immense flocks were seen on the Barren Grounds south of Cape Eskimo August 3 to 13, though at the latter date their numbers had greatly diminished. The species was noted at nearly every place at which we landed on the way back to Fort Churchill, August 13 to 19; and it was doubtless present in flocks of small sandpipers seen between Fort Churchill and York Factory, though it was not again positively identified.

Tringa alpina Linn. Dunlin.

Blakiston speaks of having seen a specimen from Hudson Bay,[b] and the American Ornithologist Union Check List records it as accidental on the west side of Hudson Bay, though I have been unable to discover on exactly what information the record is based.

Tringa alpina pacifica (Coues). Red-backed Sandpiper.

First seen on the beach about 50 miles north of York Factory July 19, where numbers were observed and several specimens were taken. The birds seemed to be just commencing their southward migration.

---

[a] Catalogue Canadian Birds, Part 1, p. 162, 1900. [b] Ibis, V, p. 132, 1863.

They were seen in large numbers wherever we landed on the coast north of Fort Churchill, were present by thousands on the Barren Grounds south of Cape Eskimo August 3 to 13, and were still seen, though in diminishing numbers, wherever we landed on the way back to York Factory.

Ereunetes pusillus (Linn.). Semipalmated Sandpiper.

Rather common at Fort Churchill, where adults and young were taken on the meadows July 25. Common along the coast north of Fort Churchill, and seen in immense flocks on the Barren Grounds south of Cape Eskimo. As with the red-backed sandpipers, we found the species common on the beach wherever we landed on our return down the coast to York Factory August 13 to 26.

One taken at Moose Factory in 1881 is registered in the catalogue of birds in the U. S. National Museum collection.

Calidris arenaria (Linn.). Sanderling.

A number were seen at my camp 25 miles south of Cape Eskimo on the morning of August 13. They were flying southward with other species.

A specimen from Repulse Bay, collected by Dr. Rae, is recorded in the British Museum Catalogue, and Clarke records an adult in summer plumage from Fort Churchill.[a] According to Swainson and Richardson, Hutchins reported that the species bred on Hudson Bay as low as the fifty-fifth parallel.[b]

Limosa fedoa (Linn.). Marbled Godwit.

Edwards figured a specimen brought from Hudson Bay by Mr. Isham.[c] His figure formed the basis of Linnæus's description of *Scolopax fedoa*. Murray recorded the species from Hudson Bay,[d] and a specimen from Hayes River is recorded in the British Museum Catalogue. The catalogue of birds in the U. S. National Museum collection records a specimen from Moose Factory.

Limosa hæmastica (Linn.). Hudsonian Godwit.

We first met with this species on the beach about 50 miles north of York Factory July 19, where a number were seen and two were taken. It was common on the Barren Grounds south of Cape Eskimo August 4 to 8, and several were seen between Fort Churchill and Cape Churchill August 22, and below Cape Churchill August 24.

Edwards figured a specimen brought from Hudson Bay by Mr. Isham, calling it the 'Red-breasted Godwit,'[e] and on his figure Lin-

---

[a] Auk, VII, p. 321, 1890.
[b] Fauna Boreali-Americana, II, p. 336, 1831.
[c] Nat. Hist., III, pl. 137, 1750.
[d] Edin. New Phil. Journ. (new ser.), IX, p. 225, 1859.
[e] Nat. Hist., III, pl. 138, 1750.

næus based his description of *Scolopax hæmastica*. Forster recorded the species from Churchill River,[a] and Murray from Severn House.[b] A specimen collected by Dr. Rae at Repulse Bay is recorded in the British Museum Catalogue.

Totanus melanoleucus (Gmel.). Greater Yellow-legs.

One was seen at Oxford House July 3 and one on upper Hill River July 7. I saw several on the beach near York Factory July 13, and at Fort Churchill July 24. The species was very common on the shores of Button Bay July 31, and about 50 miles below Cape Eskimo August 4 to 8, and was noted wherever we landed on the way back to Churchill. While ascending Hayes River August 29 and 30 we observed a great many, usually in flocks of about a dozen, but after passing the mouth of the Shamattawa, which is in reality the main river, we saw none for several days. Steel and Hill rivers, which compose the other branch of Hayes River, are not so well adapted to the habits of the species and it seems probable that most of the yellow-legs, and probably other sandpipers, migrate up the Shamattawa, where low, sandy shores more often occur. We saw two or three on Knee Lake September 7, one at Oxford House September 10, and several on Playgreen Lake September 17 to 20.

The catalogue of birds in the U. S. National Museum collection contains the record of a specimen taken at Moose Factory in 1881.

Totanus flavipes (Gmel.). Yellow-legs.

First met with June 27 at Robinson Portage, where a pair were seen on the marsh and the male was collected. They probably had a nest in the vicinity, as they were very uneasy and frequently alighted on neighboring dead trees. The species was common at Fort Churchill July 22 to 30, and abundant at Button Bay July 31, and 50 miles below Cape Eskimo August 3 to 8. A few were noted on Hayes River August 30.

The species is recorded by Andrew Murray from Severn House,[b] and the catalogue of birds in the U. S. National Museum collection records a specimen taken at Moose Factory in 1881.

Totanus totanus (Linn.). Common Redshank.

The only evidence that this is a North American species is a description by Swainson and Richardson of a specimen from Hudson Bay of '*Totanus calidris*, the Redshank or Gambet,' which they said existed in the British Museum.[c] The species occupies a place in the 'Hypothetical List' of the American Ornithologists' Union Check List on the strength of this record.

---

[a] Phil. Trans., LXII, p. 411 (*Scolopax lapponica*), 1772.

[b] Edin. New Phil. Journ. (new ser.), IX, p. 225, 1859.

[c] Fauna Boreali-Americana, II, p. 391, 1831. (See also Coues, Auk, XIV, p. 211, 1897.

Helodromas solitarius (Wils.). Solitary Sandpiper.

We met with this species only while descending the Echimamish, where several were seen September 14 and 15 and one was taken.

A specimen collected at Repulse Bay by Dr. Rae is recorded in the British Museum Catalogue, and A. McKenzie took one at Moose Factory. The species probably occurs throughout the region.

Helodromas ochropus (Linn.). Green Sandpiper.

Swainson and Richardson described a specimen of *Totanus ochropus*, the 'White-tailed Tattler,' from Hudson Bay.[a] This specimen is recorded in the British Museum Catalogue.

Heteractitis incanus (Gmel.). Wandering Tattler.

Bell says: "I obtained a specimen of the Wandering Tattler (*Heteroceelus brevipes*, Vieill.), supposed to be a western species, on the Eastmain Coast.[b]

Bartramia longicauda (Bechst.). Bartramian Sandpiper.

Several were seen and one was shot on the Barren Grounds 50 miles below Cape Eskimo August 8, and several more were noted 25 miles farther north August 10 to 13.

Tryngites subruficollis (Vieill.). Buff-breasted Sandpiper.

I saw a number on the higher parts of the tundra 25 miles south of Cape Eskimo August 10 to 13, and noted others (collecting two) August 24 on some sandy ridges, the remains of old shore lines, below Cape Churchill. The birds had a soft, plaintive call, and were rather tame and unsuspicious.

A specimen taken by Dr. Rae at Repulse Bay is recorded in the British Museum Catalogue.

Actitis macularia (Linn.). Spotted Sandpiper.

This widely distributed species is found throughout the region as far north at least as Fort Churchill. It was common on Red River June 14, and seen daily between Norway House and York Factory June 23 to July 10. A deserted nest was found on the rocky island that constitutes Rock Portage, at the lowest of the Hill River rapids. While we were descending Steel and Hayes rivers, July 9 and 10, one or more pairs of this species were almost constantly in sight, and young ones unable to fly were often seen running along the shelving banks. The species was not seen on the shores of Hudson Bay, but we collected a specimen on Churchill River, near Fort Churchill, July 25, and my brother found it rather common on the same stream, about 20 miles from its mouth, early in August. While ascending

[a] Fauna Boreali-Americana, II, p. 392, 1831.

[b] Proc. Royal Soc. Canada, 1882, I, Sec. IV, p. 51 (1883).

Hayes, Steel, and Hill rivers on our return, August 28 to September 3, we found that many were still lingering there, and I saw several at Duck Point, Playgreen Lake, on September 20. The catalogue of birds in the U. S. National Museum collection contains the record of a specimen taken at Moose Factory in 1881 by Walton Haydon.

Numenius hudsonicus Lath. Hudsonian Curlew.

First observed about 50 miles north of York Factory, where a number were seen and one was shot July 19. Another was shot at Fort Churchill July 28. Quite a number were seen on the shores of Button Bay July 31, and to the north of that place on the following day. We saw a few daily while we were encamped on the Barren Grounds south of Cape Eskimo, August 10 to 13; and noted a few more below Hubbart Point August 18, between Fort Churchill and Cape Churchill August 22, and below the cape August 24. The birds were usually seen at low tide, when they flew about in small companies over the broad boulder-strewn flats in search of feeding grounds, uttering a loud, clear whistle. During high tide they resorted to the tundra and were less observable.

Latham based his description of this species mainly on a specimen received from Mr. Hutchins, probably from Severn River.[a] Murray recorded the species from Severn House.[b]

Numenius borealis (Forst.). Eskimo Curlew.

This species was first described by Forster from a specimen taken at Albany Fort, Hudson Bay.[c] It is recorded by Dr. Bell as abundant at Fort Churchill in August, 1879.[d]

Squatarola squatarola (Linn.). Black-bellied Plover.

A specimen in winter plumage, evidently a bird of the year, was killed on a small rocky island in Swampy Lake September 5.

Forster recorded the species from Severn River;[e] Richardson from Melville Peninsula, where he says it breeds;[f] and Bell from York Factory.[g] Clarke recorded two adults in summer plumage from Fort Churchill.[h]

In former years the various plovers, particularly this species and the golden plover, were much hunted at the various posts on the Bay, especially during their southward movement in August.

---

[a] Index Ornithologicus, II, p. 712, 1790. (See also Latham, Syn. Suppl., I, p. 243, 1787.)

[b] Edin. New Phil. Journ. (new ser.), IX, p. 225, 1859.

[c] Phil. Trans., LXII, pp. 411, 431, 1772.

[d] Rept. Prog. Can. Geol. Surv., 1878–79, App. VI, p. 69c (1880).

[e] Phil. Trans., LXII, p. 412, 1772.

[f] Appendix to Parry's Second Voyage, p. 352, 1825 (1827).

[g] Rept. Prog. Can. Geol. Surv., 1882–3–4, App. III, p. 55DD (1885).

[h] Auk, VII, p. 321, 1890.

Charadrius dominicus Müll. Golden Plover.

I found this species moving southward in small flocks at a point 50 miles south of Cape Eskimo August 4 to 8, and also, though in diminished numbers, 25 miles to the northward, August 10 to 13. A specimen was taken at the first point. It has the black of the lower parts varied by a number of whitish feathers, which predominate on the throat and the sides of the chest.

This species was recorded by J. C. Ross from Port Bowen, Prince Regent Inlet, where it arrived the middle of May;[a] by Murray from Trout Lake,[b] and by Bell from York Factory.[c] Clarke recorded an adult in summer plumage from Fort Churchill.[d] The catalogue of birds in the U. S. National Museum collection contains the record of a specimen taken at Moose Factory in 1881 by Walton Haydon.

Ægialitis vocifera (Linn.). Killdeer.

Several were seen on the grassy meadows about midway between Fort Churchill and the mouth of Churchill River on July 24. Though shy, they seemed very solicitous and probably had young in the vicinity.

Ægialitis semipalmata Bonap. Semipalmated Plover.

First noted as we were descending Hayes River July 10. Here, on a sandy island about 25 miles above York Factory, the birds were breeding and downy young were running about. We saw a number of old birds at Beacon Point, York Factory, July 13, and six days later, on landing at a spot 50 miles farther north, found the species common. At Fort Churchill it was abundant, and we collected, on July 24, both adults and downy young. As I passed up the coast, July 30 to August 8, it continued rather common; and it was still represented, though sparingly, at the northernmost point of the trip, about 25 miles south of Cape Eskimo, during my four days' stay there (August 10 to 13). On our return trip it had evidently migrated, as we saw nothing of it.

Andrew Murray recorded the species from Trout Lake and Severn House;[e] and a specimen from Repulse Bay, collected by Dr. Rae, is recorded in the British Museum Catalogue.

Ægialitis meloda circumcincta Ridgw. Belted Piping Plover.

Under the name *Charadrius melodus*, King records a bird, probably referable to the present form, which he shot on Lake Winnipeg, near the northern end, while he was traveling between the mouth of the Sas-

---

[a] Parry's Third Voyage, Appendix, p. 102, 1826.
[b] Edin. New Phil. Journ. (new ser.), IX, p. 229, 1859.
[c] Rept. Prog. Can. Geol. Surv., 1882–3–4, App. III, p. 55DD (1885).
[d] Auk, VII, p. 321, 1890.
[e] Edin. New Phil. Journ. (new ser.), IX, p. 225, 1859.

katchewan and the outlet of the lake.[a] This specimen is described by Richardson in the appendix to Captain Back's narrative.[b] Specimens referable to *circumcincta* have since been taken on Lake Winnipeg on several occasions.

Arenaria morinella (Linn.). Ruddy Turnstone.

The first turnstones we met with were feeding on the beach at the 'whale fishery,' Fort Churchill, July 30. On August 1 we again noted the species a few miles north of Fort Churchill, and August 10 to 13 observed many small flocks about 25 miles south of Cape Eskimo. On August 14 I took a specimen just after starting down the coast on my return, and from Fort Churchill to York Factory, August 21 to 26, we saw many flocks daily. Whenever the birds perceived our boat they would approach and circle about it very slowly, sometimes coming within a few yards of us. Whether they mistook the boat for a rock on which they designed to alight, or were impelled merely by curiosity, I could not tell.

Under the specific name of *interpres* this bird has been recorded by Forster from Severn River,[c] by Bell from York Factory,[d] and by Clarke from Fort Churchill.[e] Swainson and Richardson say it breeds on Hudson Bay,[f] as it probably does to the northward. The British Museum contains a specimen from Repulse Bay, collected by Dr. Rae.

Canachites canadensis (Linn.). Hudsonian Spruce Grouse.

A pair in worn breeding plumage were collected at Oxford House July 3. A flock was started as we were ascending Hayes River, August 30, and others were seen on Hill River September 3 and 4. A female that was shot September 4 was preserved, and also an adult male of several that were killed on the Echimamish, September 14 and 15.

Linnaeus based his description of *Tetrao canadensis* on Edwards's figures of a male and female from Hudson Bay, probably from the west coast:[g] Forster recorded the species from Severn River;[h] Murray received it from Trout Lake;[i] Bell reported it from York Factory;[j] and a specimen from Fort Churchill is recorded by Clarke.[e] Speci-

[a] Narrative of a Journey to the Shores of the Arctic Ocean, II, p. 229, 1836.

[b] Narrative of the Arctic Land Expedition to the Mouth of the Great Fish River, App., p. 509, 1836.

[c] Phil. Trans., LXII, p. 412, 1772.

[d] Rept. Prog. Can. Geol. Surv., 1878-79, App. VI, p. 68c (1880).

[e] Auk, VII, p. 321, 1890.

[f] Fauna Boreali-Americana, II, p. 371, 1831.

[g] Nat. Hist., II, pl. 71, female, 1747; III, pl. 118, male, 1750.

[h] Phil. Trans., LXII, p. 389, 1772.

[i] Edin. New Phil. Journ. (new ser.), IX, p. 224, 1859.

[j] Rept. Prog. Can. Geol. Surv., 1882-3-4, App. III, p. 55DD (1885).

mens collected at Moose Factory are recorded in the catalogue of birds in the U. S. National Museum collection, and the species doubtless ranges throughout the wooded portions of Keewatin.

Bonasa umbellus togata (Linn.). Canadian Ruffed Grouse.

We heard grouse drumming on the Echimamish during the night of June 24, and near the south end of Oxford Lake June 30. We did not note the species again until September 8, on our return trip, when a female was taken on Knee Lake. While making a portage on Trout River, September 9, I heard a grouse drumming, and entering the thicket soon located the bird on its drumming stand, a low rock, and shot it. Within a few minutes, as I carried the dead body on my hand toward our embarking place, I noticed that its wings had become raised over the back until they were nearly in contact and were firmly fixed in that position, apparently owing to the contraction of the muscles of the wings, which had been exercised so violently. As the body cooled the wings dropped to their normal position. Near the south end of Oxford Lake, September 11, a few were seen, and near Hairy Lake, September 15, several were shot, a male and a female of which were preserved.

Forster recorded the species from Albany Fort and Severn River,[a] and Bell reported it rare as far north as York Factory.[b]

Lagopus lagopus (Linn.). Willow Ptarmigan.

First seen and a fine pair shot on the tundra about 50 miles north of York Factory. Specimens were also taken July 24 in the vicinity of Fort Churchill, where the species was rather common. While encamped on Churchill River, about 15 miles above Fort Churchill, early in August, Alfred E. Preble observed them in considerable numbers. They were rather common on the Barren Grounds south of Cape Eskimo August 3 to 13, where two immature specimens were secured.

Forster recorded this species from Severn River,[c] and Swainson and Richardson state on the authority of Hutchins that 10,000 were captured in a single season at that place.[d] James C. Ross recorded it from Port Bowen, Prince Regent Inlet, where it remained as late as November 16, and returned about the middle of March.[e] Barnston states that it arrived at Martin Falls, on Albany River, about October 20,[f] and that it began to assume its summer plumage about March 20, at which time also it departed northward.[g] In the interior of Keewatin this species regularly goes south in winter as far at least as Norway

---

[a] Phil. Trans., LXII, p. 393, 1772.

[b] Rept. Prog. Can. Geol. Surv., 1878-9, App. VI, p. 68c (1880).

[c] Phil. Trans., LXII, p. 390, 1772.

[d] Fauna Boreali-Americana, II, p. 351, 1831.

[e] Parry's Third Voyage, Appendix, p. 99, 1826.

[f] Edin. New Phil. Journal, XXX, p. 256, 1841.

[g] Ibid., p. 253, 1841.

House, and forms a welcome addition to the winter bill of fare at the Hudson Bay posts. Definite data are lacking regarding the southern limit of its breeding range on the west coast of Hudson Bay, but on the east coast it breeds as far south as James Bay, where Spreadborough found it nesting at a point a short distance north of Fort George.[a]

Lagopus rupestris (Gmel.). Rock Ptarmigan.

First described by Gmelin, who based the name on Pennant's 'Rock Grouse' from Hudson Bay.[b] We did not meet with the species, as its summer home lies to the north of the region visited. The people of Fort Churchill say it occurs at that post regularly in winter, and Clarke has recorded from there a pair in winter plumage.[c] Swainson and Richardson, quoting Hutchins's manuscript, say that the species reaches York Factory and Fort Severn in very severe winters.[d] J. C. Ross says that the species left Port Bowen, Prince Regent Inlet, in October and returned in March.[e] Dr. Rae saw a rock ptarmigan with a brood of young near Cape Fullerton,[f] and specimens collected by him at Repulse Bay are recorded in the British Museum Catalogue.

Pediœcetes phasianellus (Linn.). Sharp-tailed Grouse.

Two males in worn breeding plumage were taken at Norway House June 18, and downy young at Oxford House early in July. We did not again note the species until September 14, when we saw a pair on the upper Echimamish. It was common at Norway House September 16 to 19, and a specimen in fall plumage was taken at that point. The tracks of a large flock were seen on the sandy shore at Duck Point, Playgreen Lake, September 19.

Edwards figured and described a specimen from Hudson Bay, calling it the 'Long-Tailed Grous from Hudson's Bay,'[g] and this figure and description formed the basis of Linnæus's description of *Tetrao phasianellus*. Forster recorded it from Albany Fort and Severn River,[h] and Murray from Trout Lake.[i] J. B. Tyrrell saw it near York Factory.[j] The catalogue of birds in the U. S. National Museum collection contains the record of one taken at Moose Factory in 1881.

---

[a] Macoun, Catalogue Canadian Birds, Part I, p. 206, 1900.

[b] Arct. Zool., II, p. 312, 1785.

[c] Auk, VII, p. 321, 1890.

[d] Fauna Boreali-Americana, II, p. 354, 1831.

[e] Parry's Third Voyage, Appendix, p. 99, 1826.

[f] Narrative of an Expedition to the Shores of the Arctic Sea, p. 29, 1850.

[g] Nat. Hist., III, pl. 117, 1750.

[h] Phil. Trans., LXII, pp. 394, 425, 1772.

[i] Edin. New. Phil. Journ. (new ser.), IX, p. 224, 1859

[j] Ann. Rept. Can. Geol. Surv., 1896 (new ser.), IX, p. 165F (1897).

Ectopistes migratorius (Linn.). Passenger Pigeon.

Forster received a specimen from Severn River.[a] Swainson and Richardson, quoting Hutchins's manuscript, say that a flock visited York Factory in 1775 and stayed two days.[b] James Clark Ross relates that while he was crossing Baffin Bay July 31, 1829, on Ross's second voyage, a passenger pigeon flew on board.[c] Dr. Bell saw small flocks on the upper part of Nelson River early in September, 1878.[d] Macoun records a small breeding colony found on Waterhen River, northern Manitoba, on June 23, 1881.[e] Turner records specimens collected at Moose Factory, August 16, 1860, by C. Drexler.[f] Macoun records a set of eggs taken at the same place in June, 1888, by Mr. Miles Spence,[g] and Clarke records an adult pair taken at Fort Churchill many years ago.[h] Barnston, writing in 1840, states that the migratory pigeon was frequently seen during August at Martin Falls, on Albany River, but disappeared about September 10.[i]

Zenaidura macroura (Linn.). Mourning Dove.

While descending Red River from Winnipeg to West Selkirk, June 14, we saw several birds of this species. It is probable that it regularly reaches southern Keewatin.

Circus hudsonius (Linn.). Marsh Hawk.

We saw five marsh hawks during our trip—the first, June 13, as we were descending Red River; the next, July 13, at Beacon Point, York Factory; another, August 19, at Fort Churchill; another, August 24, below Cape Churchill, and the last, August 29, on lower Hayes River. Those observed August 19 and 24 were immature birds and were secured.

Edwards figured and described this bird, which he called the 'Ringtail'd Hawk,' from a Hudson Bay specimen.[j] His figure and description form the basis of Linnæus's description of *Falco hudsonius*. Murray recorded the species from Moose Factory and Severn River.[k]

Accipiter velox (Wils.). Sharp-shinned Hawk.

One was seen at Norway House June 19, one at Oxford House early in July, and one as we were ascending Hill River September 3.

Richardson described a specimen from Moose Factory under the

[a] Phil. Trans., LXII, p. 398, 1772.
[b] Fauna Boreali-Americana, II, p. 363, 1831.
[c] Appendix to Ross's Second Voyage, p. xxix, 1835.
[d] Rept. Prog. Can. Geol. Surv., 1878-79, App. VI, p. 70c (1880).
[e] Catalogue Canadian Birds, Part I, p. 216, 1900.
[f] Proc. U. S. Nat. Mus., VIII, p. 245, 1885.
[g] Catalogue Canadian Birds, Part I, p. 217, 1900.
[h] Auk, VII, p. 322, 1890.
[i] Edin. New Phil. Journ., XXX, p. 255, 1841.
[j] Nat. Hist., III, pl. 107, 1750.
[k] Edin. New Phil. Journ. (new ser.), IX, p. 221, 1859.

name of *Accipiter pennsylvanicus*.[a] Fielden records one taken by Captain Markham near York Factory in August, 1886.[b]

Accipiter cooperi (Bonap.). Cooper Hawk.

A Cooper hawk darted into a flock of Canada grouse which we flushed while ascending Hayes River August 30, and though it did not succeed in capturing any it terrorized them so completely that we were unable to approach them.

Accipiter atricapillus (Wils.). Goshawk.

Richardson gives a description of a goshawk killed at York Factory, accompanied by a figure presumably drawn from the same bird.[c] Baird recorded one collected on Nelson River,[d] and Clarke an adult female collected at Fort Churchill many years ago.[e]

Buteo borealis (Gmel.). Red-tailed Hawk.

A number were seen July 8 and 9 as we descended Hill and Steel rivers, where they were undoubtedly nesting. They flew from tree to tree in advance of the canoe, with squeals of protest at our intrusion. When we were returning we saw several on Hill River September 3.

Dr. Bell recorded the species from Fort Churchill.[f]

Buteo lineatus (Gmel.). Red-shouldered Hawk.

Bell recorded a specimen from York Factory, collected and presented by Dr. Matthews.[g]

Buteo swainsoni Bonap. Swainson Hawk.

An adult specimen (sex not noted) in the dark plumage, collected at Moose Factory in 1881 by W. Haydon, is in the U. S. National Museum collection.

Buteo platypterus (Vieill.). Broad-winged Hawk.

A specimen taken at Moose Factory in 1862 by J. McKenzie is recorded by Turner.[h]

Archibuteo lagopus sancti-johannis (Gmel.). Rough-legged Hawk.

Two were seen at Swampy Lake September 5, swooping about over the wooded shores, evidently at play. A specimen collected early in August at Norway House was presented to us by J. K. MacDonald of that post.

---

[a] Fauna Boreali-Americana, II, p. 46, 1831.
[b] Transactions Norwich Society, IV, p. 349, 1887.
[c] Fauna Boreali-Americana, II, p. 43, 1831.
[d] Birds of North America, p. 16, 1858.
[e] Auk, VII, p. 322, 1890.
[f] Rept. Prog. Can. Geol. Surv., 1878–79, App. VI, p. 67c (1880).
[g] Ibid., 1882-3-4, App. III, p. 54DD (1885).
[h] Proc. U. S. Nat. Mus., VIII, p. 244, 1885.

This species is recorded by Murray from Severn House and Trout Lake,[a] and an apparently immature specimen, collected at Fort Churchill many years ago, is recorded by Clarke.[b]

Aquila chrysaëtos (Linn.). Golden Eagle.

We saw several of these birds as we were passing through Hell Gate Gorge June 28, and noticed at least two of their nests on its rocky walls.

Edwards figured a specimen from Hudson Bay,[c] and Linnaeus based his description of *Falco canadensis* on Edwards's figure. Sir John Franklin, who passed through Hell Gate Gorge in October, 1819, speaks of a nest of the brown fishing-eagle on one of the projecting cliffs.[d]

Haliæetus leucocephalus (Linn.). Bald Eagle.

We were informed that white-headed eagles were occasionally seen near Fort Churchill and nested in the vicinity, and we obtained the upper mandible of one that had been killed there. Mr. Campbell, of Oxford House, reported having seen the species near Oxford Lake.

Falco islandus Brünn. White Gyrfalcon.

The residents of Fort Churchill spoke of a white hawk, probably this species, which is sometimes seen there.

Richardson described a mature bird from Hudson Bay;[e] Murray recorded *F. candicans* from York Factory;[a] Ridgway described a specimen from Moose Factory;[f] and a specimen collected by Dr. Rae at Repulse Bay, is recorded in the British Museum Catalogue.

Falco rusticolus gyrfalco (Linn.). Gyrfalcon.

Forster described *Falco sacer*, probably identical with *gyrfalco*, from a specimen taken at Severn River;[g] Bell records a specimen (as *Falco sacer*) from York Factory;[h] and Clarke records two specimens from Fort Churchill.[b]

Falco peregrinus anatum (Bonap.). Duck Hawk.

One was seen on the Barren Grounds below Cape Eskimo about August 5.

The species is recorded by Murray from Trout Lake and Severn House,[i] and by Bell from York Factory,[h] and also from Marble

[a] Edin. New Phil. Journ. (new ser.), IX, p. 221, 1859.
[b] Auk, VII, p. 322, 1890.
[c] Nat. Hist., I, pl. 1, 1743.
[d] Franklin's Narrative of a Journey to the Polar Sea, p. 39, 1823.
[e] Fauna Boreali-Americana, II, p. 28, 1831.
[f] Land Birds of North America, III, p. 112, 1874.
[g] Phil. Trans., LXII, pp. 383, 423, 1772.
[h] Rept. Prog. Can. Geol. Surv., 1878–79, App. VI, p. 67c (1880).
[i] Edin. New Phil. Journ. (new ser.), IX, p. 271, 1859.

Island, where an adult and two young birds were killed September 1.[a] An adult male collected at Fort Churchill many years ago is recorded by Clarke.[b] The catalogue of birds in the U. S. National Museum collection contains the record of a specimen taken at Moose Factory in 1881.

Falco columbarius Linn. Pigeon Hawk.

A pigeon hawk which was moulting from the brownish plumage to that of the more mature bird was shot at a portage on Hill River July 7. From its actions and those of its mate, which was seen with food in its talons, it was probably nesting in the vicinity, but a careful search failed to locate the nest. Another bird was taken at Fort Churchill July 25, and on our return trip we saw one on Hill River September 1, and one at Oxford House September 10.

Richardson described a specimen shot at York Factory September 4, 1822;[c] Forster recorded the species from Severn River;[d] Baird from Nelson River;[e] and Bell from between Norway House and Fort Churchill.[f]

Falco richardsoni Ridgw. Richardson Merlin.

Nutting records this species from Grand Rapids, Saskatchewan, where a male was secured in the summer of 1891.[g]

Falco sparverius Linn. Sparrow Hawk.

We saw one while descending Red River June 14, one at Sea Falls, 20 miles north of Norway House, June 23, and several while ascending Hayes and Steel rivers August 30 to September 1. On September 15 we again noted the species at Sea Falls.

Baird recorded a specimen from Nelson River;[e] Bell reported the species at York Factory;[f] and Mearns lists a specimen from Moose Factory.[h]

Pandion haliaëtus carolinensis (Gmel.). Osprey.

At Robinson Portage June 26 we saw a pair, and on the shore of Windy Lake, June 29, another that had a nest containing good-sized young. While returning we saw one bird at York Factory August 27, one on Hill River September 4, one on Trout River September 9, and a number about Windy Lake September 12.

Bell reported several nesting along the Churchill and Grass rivers.[f]

---

[a] Rept. Prog. Can. Geol. Surv., 1882–3–4, App. III, p. 54DD (1885).
[b] Auk, VII, p. 322, 1890.
[c] Fauna Boreali-Americana, II, p. 361, 1831.
[d] Phil. Trans., LXII, p. 382, 1772.
[e] Birds of North America, p. 10, 1858.
[f] Rept. Prog. Can. Geol. Surv., 1878–79, App. VI, p. 67c (1880).
[g] Nat. Hist. Bull. Univ. Iowa, II, p. 269, 1893.
[h] Auk, IX, p. 262, 1892.

Asio wilsonianus (Less.). Long-eared Owl.

Thompson, quoting Hutchins's manuscript, says that this species was found at Severn Settlement, presumably Fort Severn, where Hutchins resided.[a]

Asio accipitrinus (Pall.). Short-eared Owl.

Rather common and quite generally distributed throughout the region wherever favorable ground occurs. One was seen at Beacon Point, near York Factory, July 13, and two were taken at Fort Churchill. One was seen on the Barren Grounds south of Cape Eskimo August 4, and one at Oxford House September 10.

Murray recorded the species from Trout Lake Station,[b] and Bell from York Factory and Fort Churchill.[c] The catalogue of birds in the U. S. National Museum collection contains the record of one taken at Moose Factory in 1881 by Walton Haydon.

Syrnium varium (Barton).[d] Barred Owl.

> *Strix varius* Barton, Frag. Nat. Hist. Penna., p. 11, 1799.
> *Syrnium nebulosum* authors (not *Strix nebulosa* Forster, Phil. Trans., LXII, pp. 386, 424, 1772, which is based on the great gray owl.

Several specimens of the barred owl taken at Moose Factory are recorded in the catalogue of birds in the U. S. National Museum. Nutting records the species from Chemawawin, Saskatchewan.[e]

Scotiaptex nebulosum (Forst.). Great Gray Owl.

> *Strix nebulosa* Forst., Trans. Phil. Soc. London, LXII, pp. 386, 424, 1772. (Severn River.)
> *Strix cinerea* Gmel., Systema Naturæ, I, p. 291, 1788; and of authors.

Forster based the name *Strix nebulosa* on a specimen of the great gray owl sent by Mr. Graham from Severn River, Hudson Bay.[f] His description, in part, is as follows:

> Strix capite lævi, corpore fusco, albido undulatim striato, remige sexto longiore apice, nigricante.
>
> *Description.*—Rostrum fusco flavum, mandibula superius magis flava.
>
> Oculi magna iridibus flavis. * * *
>
> *Pectus* albidum maculis longitudinalibus transversisque fuscis.
>
> *Abdomen* album superius uti pectus maculis longitudinalibus sed inferius striis transversis notatum.
>
> Latitudo pedum quattuor.

---

[a] Proc. U. S. Nat. Mus., XIII, p. 540, 1890.

[b] Edin. New Phil. Journ. (new ser.), IX, p. 222, 1859.

[c] Rept. Prog. Can. Geol. Surv., 1878–79, App. VI, p. 67c (1880).

[d] *Strix nebulosa* Forster plainly having been based on the great gray owl, the name *Strix varius* Barton seems to be the next name available for the barred owl The barred owls will, therefore, stand as follows:

Syrnium varium (Barton), Frag. Nat. Hist. Penna., p. 11, 1799.
Syrnium varium alleni (Ridgw.), Proc. U. S. Nat. Mus., III, p. 8, March, 1880.
Syrnium varium helveolum (Bangs), Auk, XVIII, p. 299, 1901.

[e] Nat. Hist. Bull. Univ. Iowa, II, p. 270, 1893.

[f] The fact that Forster's description of *Strix nebulosa* refers to the great gray owl was brought to my attention by Mr. R. Ridgway.

Blakiston received two specimens from York Factory,[a] and Turner has recorded the species from Moose Factory, where a specimen was collected by J. McKenzie.[b] Seton records it as resident in the woods about Lake Winnipeg.[c]

Nyctala tengmalmi richardsoni (Bonap.). Richardson Owl.

A specimen collected at Repulse Bay by Dr. Rae is recorded in the British Museum Catalogue. Fielden records the species from near York Factory, where it was obtained by Captain Markham in 1886.[d] It probably occurs throughout the Hudson Bay region.

Nyctala acadica (Gmel.). Saw-whet Owl.

Strix passerina, recorded by Forster from Hudson Bay, probably from Severn River,[e] is in all likelihood referable to this species. Turner recorded a specimen, which is still in the National Museum, collected at Moose Factory by J. McKenzie.[b]

Megascops asio (Linn.). Screech Owl.

George Barnston, writing in 1840, mentions "The small owl (Scops)" as being heard in April at Martin Falls, Albany River.[f]

Bubo virginianus arcticus (Swains.). Arctic Horned Owl.

The characteristic notes of the great horned owl were heard during the night of September 11, while we were encamped near the south end of Oxford Lake, September 13 at Robinson Portage, and the next day on the Echimamish.

The catalogue of birds in the U.S. National Museum collection records two specimens of the great horned owl collected at Moose Factory by J. McKenzie. One of these has been examined and is referable to this form. Clarke recorded a specimen probably referable to this form collected many years ago at Fort Churchill.[g] Fielden records a fine specimen obtained near York Factory in 1886 by Captain Markham.[d]

Nyctea nyctea (Linn.). Snowy Owl.

We did not meet with this species but saw wings at several posts, and learned from a young man at Fort Churchill that he had seen a snowy owl late in July.

Forster early recorded it from Churchill River,[e] and the reports of

---

[a] Rept. Prog. Can. Geol. Surv., V, p. 50, 1863.
[b] Proc. U. S. Nat. Mus., VIII, p. 243, 1885.
[c] Auk, III, p. 155, 1886.
[d] Transactions Norwich Society, IV, p. 349, 1887.
[e] Phil. Trans., LXII, p. 385, 1772.
[f] Edin. New. Phil. Journ., XXX, p. 253, 1841.
[g] Auk, VII, p. 322, 1890.

various Arctic expeditions note its occurrence at several points to the north and northwest of Hudson Bay. Its presence throughout the region in winter is attested by various observers.

Surnia ulula caparoch (Müll.). Hawk Owl.

"The Little Hawk Owl"[a] of Edwards, from Hudson Bay, formed the basis of Müller's *Strix caparoch*. Forster recorded the hawk owl, under the name *Strix funerea*, from Severn and Churchill rivers;[b] Murray received it from Trout Lake and Severn House;[c] Swainson and Richardson mention a specimen from York Factory;[d] and one collected by Dr. Rae at Repulse Bay is recorded in the British Museum Catalogue. The catalogue of birds in the U. S. National Museum collection contains the record of a specimen taken by Walton Haydon at Moose Factory in 1881.

Ceryle alcyon (Linn.). Belted Kingfisher.

We found the belted kingfisher common throughout the region between Norway House and Oxford House. A large clay bank on the shore of Oxford Lake near Oxford House, which we passed June 30, was perforated by the nesting holes of a numerous colony of bank swallows. In this bank were also several larger holes, which were probably those of kingfishers, judging from the anxiety manifested by at least four pairs of kingfishers that were flying about. During our return trip we saw several kingfishers on Steel and Hill rivers early in September, on Trout River September 9, and on the Echimamish September 14.

Clarke recorded an adult male from Fort Churchill,[e] and several have reported the species from York Factory. The catalogue of birds in the U. S. National Museum collection contains the record of a specimen taken at Moose Factory in 1881 by Walton Haydon.

Dryobates villosus leucomelas (Bodd.). Northern Hairy Woodpecker.

Forster recorded *villosus* from Severn River,[f] and Bell reported it from York Factory,[g] both probably referring to the present form. An adult male of this species from Fort Churchill is recorded by Clarke.[e] It is likely the bird ranges throughout the wooded portion of the region.

Dryobates pubescens medianus (Swains.). Downy Woodpecker.

A specimen collected at Moose Factory by Walton Haydon is in the U. S. National Museum. The downy woodpecker undoubtedly ranges throughout southern Keewatin.

---

[a] Nat. Hist., II, pl. 62, 1747. (See Stejneger, Auk, I, p. 362, 1884.)
[b] Phil. Trans., LXII, p. 385, 1772.
[c] Edin. New Phil. Journ. (new ser.), IX, p. 221, 1859.
[d] Fauna Boreali-Americana, II, p. 94, 1831.
[e] Auk, VII, p. 322, 1890.
[f] Phil. Trans., LXII, p. 388, 1772.
[g] Rept. Prog. Can. Geol. Surv., 1882–3–4, App. III, p. 54DD (1885).

Picoides arcticus (Swains.). Arctic Three-toed Woodpecker.

We collected two males in a tract of burnt spruce woods at Norway House June 19, and while ascending Hayes River August 30 saw another in spruce woods.

Baird recorded a pair from Hudson Bay obtained from John Gould.[a] Fielden recorded a specimen obtained near York Factory in August, 1886.[b] The catalogue of birds in the United States National Museum collection records a specimen taken at Moose Factory in 1881 by Walton Haydon.

Picoides americanus Brehm. Striped-backed Three-toed Woodpecker.

Forster recorded this species from Severn River;[c] Murray received one from Severn House;[d] Baird recorded a male from Hudson Bay,[e] supposed to be the one figured by Audubon, and Clarke recorded several adults of both sexes collected at Fort Churchill many years ago.[f]

Sphyrapicus varius (Linn.). Yellow-bellied Sapsucker.

An adult female collected at Fort Churchill more than fifty years ago is recorded by Clarke.[g] The U. S. National Museum collection contains a specimen taken at Moose Factory in 1881 by Walton Haydon.

Ceophlœus pileatus abieticola Bangs. Northern Pileated Woodpecker.

Thompson, quoting Hutchins's manuscript, reported this species from Albany River,[h] Baird recorded a specimen from Nelson River,[i] and the catalogue of birds in the U. S. National Museum collection records four specimens collected at Moose Factory in 1862.

Colaptes auratus luteus Bangs. Northern Flicker.

We found flickers rather common throughout the region between Lake Winnipeg and Hudson Bay, and saw several at Fort Churchill July 25.

Forster recorded this species from Albany Fort;[j] Murray from Trout Lake and Hudson Bay;[k] and others have recorded it from different points in the region covered by our observations. The catalogue of birds in the U. S. National Museum collection contains the record of a specimen taken at Moose Factory in 1881 by Walton Haydon.

---

[a] Birds of North America, p. 98, 1858.

[b] Transactions Norwich Society, IV, p. 348, 1887.

[c] Phil. Trans., LXII, p. 388 (*P. tridactylus*), 1772.

[d] Edin. New Phil. Journ. (new ser.), IX, p. 223 (*A. tridactylus*), 1859.

[e] Birds of North America, p. 100, 1858.

[f] Auk, VII, p. 322, 1890.

[g] Ibid., p. 322, 1890.

[h] Proc. U. S. Nat. Mus., XIII, p. 551, 1890.

[i] Birds of North America, p. 107, 1858.

[j] Phil. Trans., LXII, p. 387, 1772.

[k] Edin. New Phil. Journ. (new ser.), IX, p. 223, 1859.

Antrostomus vociferus (Wils.). Whip-poor-will.

Bell says: "The Whippoorwill was not seen nor heard north of Norway House,"[a] implying its occurrence at that point. Bendire, probably referring to the same record, says: "As far as I have been able to ascertain, this species reaches the extreme northern limits of its range on the north shore of Lake Winnipeg, near Norway House.[b]

The species probably occurs regularly in extreme southern Keewatin.

Chordeiles virginianus (Gmel.). Night-hawk.

Rather common at Norway House June 17 to 23, and several seen at Robinson Portage June 27, Oxford Lake June 30, and about the shores of Knee Lake July 5. They seemed to frequent entirely the districts which had been swept by fire. As we saw none on our return early in September, they had undoubtedly migrated by that time.

Murray recorded the species from Trout Lake.[c] Bell reported it from York Factory,[a] and a specimen collected years ago at Fort Churchill is recorded by Clarke.[d] Edward Sabine recorded one that was picked up dead on Melville Island.[e] The catalogue of birds in the U. S. National Museum collection contains the record of a specimen taken at Moose Factory in 1881 by Walton Haydon.

Muscivora forficata (Gmel.). Scissor-tailed Flycatcher.

Bell recorded a "specimen of *Milvulus forficatus* in the Government Museum shot at York Factory in the summer of 1880."[f] The species has also been reported from Manitoba.[g] It evidently occurs only as a rare straggler.

Tyrannus tyrannus (Linn.). Kingbird.

The catalogue of birds in the U. S. National Museum collection records a specimen collected at Moose Factory July 11, 1881, by Walton Haydon. Nutting found the species abundant at Grand Rapids, at the mouth of the Saskatchewan.[h] It should occur regularly in southern Keewatin.

Sayornis phœbe (Lath.). Phœbe.

A pair had a nest beneath the wharf at Norway House, and several more nests were observed on June 28 as we were passing through Hell Gate Gorge. They were placed on the face of cliffs overhanging the water, and contained young nearly ready to fly. Though the bird

[a] Rept. Prog. Can. Geol. Surv., 1878–79, App. VI, p. 68c (1880).

[b] Life Hist. N. A. Birds (U. S. Nat. Mus. Special Bull. 3), p. 146, 1895 (1896).

[c] Edin. New Phil. Journ. (new ser.), IX, p. 222, 1859.

[d] Auk, VII, p. 322, 1890.

[e] Suppl. to Appendix, Parry's First Voyage, p. cxciv, 1824.

[f] Proc. Royal Soc. Canada, 1882, I, Sec. IV, p. 52, (1883).

[g] Seton, Auk, II, p. 218, 1885.

[h] Nat. Hist. Bull. Univ. Iowa, II, p. 271, 1893.

should occur throughout southern Keewatin, I find no published records of its occurrence in the Province.

Contopus borealis (Swains.). Olive-sided Flycatcher.

Observed but once, on July 4, in a swamp bordering Trout River, between Oxford House and Knee Lake. The bird was perched on a tall dead tree, uttering at intervals its characteristic note.

Murray recorded the species from Hudson Bay.[a]

Empidonax trailli alnorum Brewst. Alder Flycatcher.

Flycatchers referred to *alnorum* were several times observed by us at Norway House, and while we were ascending the Echimamish, but various causes, including their extremely wary habits, conspired to prevent their collection.

Empidonax minimus Baird. Least Flycatcher.

A nest containing well-incubated eggs was collected near the south end of Oxford Lake on the morning of June 30. The female parent was secured at the same time. I have compared this bird with most of the specimens in the large series in the U. S. National Museum collection, and it proves to have a smaller bill than any of them.

Turner recorded a specimen collected by C. Drexler at Moose Factory.[b]

Otocoris alpestris (Linn.). Horned Lark.

An adult male horned lark, collected at Moose Factory June 18, 1863, and now in the U. S. National Museum collection, has been several times recorded. This specimen proves, on examination, to be referable to the typical form.

Otocoris alpestris hoyti Bishop. Hoyt Horned Lark.

We first met with this form on the 'Barrens' about 50 miles north of York Factory July 19, and found it common from there north as far as we went, especially at Fort Churchill, where adult birds and a young one not long from the nest were taken July 24 to 26. A specimen was collected 50 miles south of Cape Eskimo August 4. During our return we saw many horned larks on the clearing at Norway House, September 16 to 19, feeding in company with Lapland longspurs.

Bishop has recorded specimens of this form from Depot Island, 'Hudson Strait' [Hudson Bay].[c] J. C. Ross took a specimen near Felix Harbor, Boothia.[d] Richardson recorded a specimen taken July 10,

[a] Edin. New Phil. Journ. (new ser.), IX, p. 223, 1859.
[b] Proc. U. S. Nat. Mus., VIII, p. 242, 1885.
[c] Auk, XIII, p. 132, 1896.
[d] Appendix to Ross's Second Voyage, p. xxvi, 1835.

1822, near Cape Wilson, Melville Peninsula;[a] and Murray received specimens from York Factory and Severn House.[b] Forster recorded horned larks from Albany Fort,[c] but in the absence of specimens, it is impossible to decide whether *alpestris* or *hoyti* is referred to.

Pica pica hudsonia (Sab.). Magpie.

Forster recorded the magpie from Albany Fort under the name *Corvus pica.*[d] Thompson, quoting Hutchins's manuscript, says that one was caught in a marten trap at York Factory;[e] and Joseph Sabine, who described *hudsonia* from Cumberland House, speaks of having had a specimen from Hudson Bay in his possession some time before.[f] Fielden records a specimen procured at York Factory.[g]

Cyanocitta cristata (Linn.). Blue Jay.

Several were seen in the groves of *Quercus macrocarpa* at West Selkirk on the morning of September 22. According to Chamberlain, the species has been taken at Moose Factory.[h] Russell records it from Grand Rapids.[i]

Perisoreus canadensis (Linn.). Canada Jay.

This species was reported about Norway House, and was seen nearly every day on our journey between there and York Factory. Specimens were collected on the Echimamish, at Oxford House, and at York Factory; and Alfred E. Preble saw several near Fort Churchill.

Murray recorded the species from Severn House,[j] and Clarke an adult from Fort Churchill.[k] The catalogue of birds in the U. S. National Museum collection records a specimen taken at Moose Factory in 1881 by Walton Haydon.

Corvus corax principalis Ridgw. Northern Raven.

We saw several between Robinson Portage and Pine Lake June 28, and while descending Hill River July 8 noticed a pair flying about the face of a high clay bank. Except for one seen at Fort Churchill July 30, we did not again note the species.

James Clark Ross speaks of a pair which wintered at Port Bowen, Prince Regent Inlet.[l] Bell reported this species as breeding through-

---

[a] Appendix to Parry's Second Voyage, p. 343, 1825 (1827).

[b] Edin. New Phil. Journ. (new ser.), IX, p. 398, 1859.

[c] Phil. Trans., LXII, p. 398, 1772.

[d] Ibid., p. 387, 1772.

[e] Proc. U. S. Nat. Mus., XIII, p. 565, 1890.

[f] Franklin's Narrative of a Journey to the Polar Sea, Appendix, p. 671, 1823.

[g] Transactions Norwich Society, IV, p. 348, 1887.

[h] Catalogue Canadian Birds, p. 75, 1887.

[i] Explorations in the Far North, p. 264, 1898.

[j] Edin. New Phil. Journ. (new ser.), IX, p. 222, 1859.

[k] Auk, VII, p. 322, 1890.

[l] Parry's Third Voyage, Appendix, p. 97, 1826.

out the region between Norway House and Forts Churchill and York.[a]

Corvus americanus Aud. Crow.

A few seen about Lake Winnipeg and Norway House, and small numbers noted nearly every day between Norway House and York Factory. They were rather common at York Factory July 10 to 17. A few were seen about the mouth of Churchill River July 30, and another was noted 50 miles south of Cape Eskimo on the morning of August 14. On our return trip they were several times met with.

Murray recorded the crow from Trout Lake and Hudson Bay.[b]

Xanthocephalus xanthocephalus (Bonap.). Yellow-headed Blackbird.

Recorded by Murray from Hudson Bay.[c] Nutting found it breeding abundantly at Chemawawin, Saskatchewan.[d]

Agelaius phœniceus (Linn.). Red-winged Blackbird.

Common in the Red River Valley, and abundant about the marshes below Robinson Portage, where two specimens were collected June 27. A number were seen near Oxford House July 4, in the marsh between Oxford and Back Lakes.

Recorded by Murray from Hudson Bay.[b]

Sturnella magna neglecta (Aud.). Western Meadowlark.

A number seen along Red River between Winnipeg and West Selkirk June 14. Specimens procured at Winnipeg have been examined and prove referable to this form.

Icterus galbula (Linn.). Baltimore Oriole.

One seen flying across Red River, about midway between Winnipeg and West Selkirk, June 14.

Bell records a specimen from York Factory, collected and presented by Dr. Matthews.[e]

Scolecophagus carolinus (Müll.). Rusty Blackbird.

Several seen and a female taken June 25, near the head of the Echimamish, where they were undoubtedly breeding. One was taken from a large flock at Fort Churchill July 26. While encamped on Churchill River, about 15 miles above Fort Churchill, Alfred E. Preble found the species abundant, and took several specimens August 8. We found the species common as we ascended Hill River September 3 to 5, and saw several between Oxford and Windy lakes September 12. It was very common along the Echimamish September 14 and 15.

---

[a] Rept. Prog. Can. Geol. Surv., 1878–79, App. VI, p. 67c (1880).

[b] Edin. New Phil. Journ. (new ser.), IX, p. 222, 1859.

[c] Ibid., p. 222, 1859.

[d] Nat. Hist. Bull. Univ. Iowa, II, p. 274, 1893.

[e] Rept. Prog. Can. Geol. Surv., 1882–3–4, App. III, p. 54DD (1885).

Forster recorded it from Severn River,[a] his record being probably the earliest notice of the bird, which was then undescribed. Murray received specimens from Severn House and Trout Lake,[b] and Bell reported it from York Factory.[c] The catalogue of birds in the U. S. National Museum collection contains the record of specimens taken at Moose Factory in 1881 by Walton Haydon.

Scolecophagus cyanocephalus (Wagl.). Brewer Blackbird.

Common in the Red River Valley, between Winnipeg and West Selkirk, June 14. Nutting records this species and *S. carolinus* from the lower Saskatchewan, where both apparently breed.[d]

Quiscalus quiscula æneus (Ridgw.). Bronzed Grackle.

Several seen near Sea Falls and on the lower Echimamish June 24; rather common June 26 and 27 at Robinson Portage, where two specimens were collected; common at Oxford House June 30 to July 4; and a single bird seen on upper Hill River July 7. On the return trip several were seen on Trout River, near Oxford House, September 9.

Bell reported *Q. purpureus* from York Factory,[c] and Clarke recorded an adult male from Fort Churchill, collected many years ago.[e] The catalogue of birds in the U. S. National Museum collection contains the record of a specimen taken at Moose Factory in 1881 by Walton Haydon.

Coccothraustes vespertinus (Coop.). Evening Grosbeak.

Thompson, on the authority of R. H. Hunter, reports that this species was seen in autumn, winter, and early spring at Big Island, Lake Winnipeg.[f]

Pinicola enucleator leucura (Müll.). Pine Grosbeak.

A male was seen perched on a tree overhanging Hill River, near the mouth of Fox River, July 8. The species was reported to us by the residents of Fort Churchill.

Edwards figured a pair of these birds brought from Hudson Bay by Mr. Isham, who informed him that they wintered there.[g] Forster recorded it from Severn River;[h] Murray recorded it from Severn House;[b] Bell reported that it was frequently seen on the Churchill in the latter part of July;[i] Clarke recorded adults of both sexes, collected

---

[a] Phil. Trans., LXII, p. 400, 1772.
[b] Edin. New Phil. Journ. (new ser.), IX, p. 223, 1859.
[c] Rept. Prog. Can. Geol. Surv., 1878–79, App. VI, p. 68c (1880).
[d] Nat. Hist. Bull. Univ. Iowa, II, p. 274, 1893.
[e] Auk, VII, p. 322, 1890.
[f] Proc. U. S. Nat. Mus., XIII, p. 584, 1890.
[g] Nat. Hist., III, pls. 123 and 124, 1750.
[h] Phil. Trans., LXII, p. 402, 1772.
[i] Rept. Prog. Can. Geol. Surv., 1878–79, App. VI, p. 70c (1880).

at Fort Churchill many years ago;[a] and Fielden recorded a specimen obtained near York Factory in 1886.[b] The catalogue of birds in the U. S. National Museum collection contains the record of a specimen taken at Moose Factory in 1881 by Walton Haydon.

Carpodacus purpureus (Gmel.). Purple Finch.

The song of the purple finch was several times heard early on the morning of June 16 at Bull Head Point, Lake Winnipeg, where the steamer stopped for wood.

Turner recorded the species from Moose Factory,[c] and Walton Haydon collected specimens there in 1881.

Loxia curvirostra minor (Brehm). Red Crossbill.

A small flock was seen at our camp on the Echimamish June 25.

Forster recorded two specimens of '*Loxia curvirostra*' from Severn River.[d]

Loxia leucoptera Gmel. White-winged Crossbill.

Murray received this species from Hudson Bay, Severn House, and Trout Lake;[e] Baird recorded a specimen in the U. S. National Museum from Nelson River;[f] and adults of both sexes from Fort Churchill were recorded by Clarke.[a]

Acanthis hornemanni (Holb.). Greenland Redpoll.

Clarke recorded two adults collected many years ago at Fort Churchill.[a] Murray's record of *Linota borealis* from Severn House may be referable to the present form or to *exilipes*.[e]

Acanthis hornemanni exilipes (Coues). Hoary Redpoll.

Three specimens from York Factory and one from Fort Churchill, collected in July, are referable to this form.

Acanthis linaria (Linn.). Redpoll.

Eight specimens, including one in juvenal plumage, were collected July 12 to 16 at York Factory, where the birds were abundant, and a very bright male was taken at Fort Churchill July 23.

Forster recorded this species from Severn River.[g]

Acanthis linaria holbœlli (Brehm). Holbœll Redpoll.

A specimen (No. 89311) taken at Moose Factory in 1881 by Walton Haydon, and now in the U. S. National Museum collection, seems to be a typical example of this form.

---

[a] Auk, VII, p. 322, 1890.
[b] Transactions Norwich Society, IV, p. 348, 1887.
[c] Proc. U. S. Nat. Mus., VIII, p. 239, 1885.
[d] Phil. Trans., LXII, p. 402, 1772.
[e] Edin. New Phil. Journ. (new ser.), IX, p. 223, 1859.
[f] Birds of North America, p. 428, 1858.
[g] Phil. Trans., LXII, p. 405, 1772.

Spinus pinus (Wils.). Pine Siskin.

Nutting records the pine siskin from Grand Rapids, Saskatchewan, where two specimens were taken.[a] It probably occurs throughout southern Keewatin, but I find no published records.

Passerina nivalis (Linn.). Snowflake.

Edwards figured the 'Snow-Bird from Hudson's-Bay,'[b] and on this figure Linnæus partially based his description of *Emberiza nivalis*. Forster recorded migrants from Severn River;[c] Richardson says the species breeds on Melville Peninsula,[d] and records it as usually arriving at Fort Churchill from March 26 to April 6, and being very rarely seen in midwinter; Dr. Rae saw young near Neville Bay;[e] Swainson and Richardson speak of its breeding about Chesterfield Inlet;[f] Murray received specimens from Hudson Bay, Severn House, and Trout Lake;[g] Bell reported it from York Factory;[h] Clarke from Fort Churchill;[i] the British Museum Catalogue has a record of one collected by Dr. Rae at Repulse Bay; and the catalogue of birds in the U. S. National Museum collection records a specimen taken at Moose Factory in 1881 by Walton Haydon.

Calcarius lapponicus (Linn.). Lapland Longspur.

Rather common 10 miles north of Fort Churchill, on the shores of Button Bay, where an immature bird was taken July 31. Abundant on the Barren Grounds south of Cape Eskimo August 4 to 13, at which time the old birds were moulting and were almost invariably destitute of tail feathers. They were seen in great numbers near Hubbart Point August 17, and were common at Norway House September 19, where one was secured. At the latter point they were feeding in company with horned larks.

Forster recorded the species from Severn River;[j] Murray received specimens from Trout Lake and Severn House;[g] Clarke recorded adults and young collected at Fort Churchill;[i] and two specimens collected at Repulse Bay by Dr. Rae are mentioned in the British Museum Catalogue. The catalogue of birds in the U. S. National Museum collection contains the record of a specimen taken at Moose Factory in 1881 by Walton Haydon.

---

[a] Nat. Hist. Bull. Univ. Iowa, II, p. 275, 1893.
[b] Nat. Hist., III, pl. 126, 1750.
[c] Phil. Trans., LXII, p. 403, 1772.
[d] Appendix to Parry's Second Voyage, p. 344, 1825 (1827).
[e] Narrative of an Expedition to the Shores of the Arctic Sea, p. 24, 1850.
[f] Fauna Boreali-Americana, II, p. 246, 1831.
[g] Edin. New Phil. Journ. (new ser.), IX, p. 222, 1859.
[h] Rept. Prog. Can. Geol. Surv., 1878–79, App. VI, p. 68c (1880).
[i] Auk, VII, p. 322, 1890.
[j] Phil. Trans., LXII, p. 404, 1772

Calcarius pictus (Swains.). Smith Longspur.

Rather common on the meadows at Fort Churchill July 23 to 30. They were quite tame, but were hard to see on the mossy hillocks, their coloring rendering them very inconspicuous. Often the first intimation I had of their proximity was their note, consisting of several sharp chips uttered in rapid succession. Three adult males and a young male just from the nest were taken. The young male, which was secured July 24, may be described as follows: Back dusky, the feathers edged with deep buff and whitish; feathers of head and neck dusky, mostly edged with buff, collar of adult being indicated by white edgings; wing quills strongly edged with brown, coverts tipped with white; lower parts buffy, about as in adults, chest conspicuously streaked with black, and sides marked with obscure spots of dusky; white markings of head plainly indicated.

Murray recorded specimens from Severn House;[a] and a specimen collected by Dr. Rae at Repulse Bay is recorded in the British Museum Catalogue.

Poœcetes gramineus (Gmel.). Vesper Sparrow.

Nutting reports the vesper sparrow common at Grand Rapids, Saskatchewan, where specimens were taken in the summer of 1891.[b]

Ammodramus sandwichensis savanna (Wils.). Savanna Sparrow.

Common throughout the region wherever suitable ground occurred, especially in the vicinity of the posts. At Norway House June 22 we found young just beginning to fly, and took several specimens of these and the old birds. We collected other specimens at Oxford House June 30 to July 4, and at York Factory, where they were especially common on the marsh at Beacon Point. We collected two more at Fort Churchill and two in the juvenal plumage at my camp 50 miles south of Cape Eskimo August 4 to 8. The last were taken in traps set for voles. The catalogue of birds in the U. S. National Museum collection contains the record of a specimen taken at Moose Factory in 1881 by Walton Haydon.

Ammodramus bairdi (Aud.). Baird Sparrow.

Nutting records a specimen taken at Grand Rapids, Saskatchewan, in the summer of 1891.[c]

Zonotrichia querula (Nutt.). Harris Sparrow.

A number of specimens, including adults of both sexes and young just from the nest, were collected, July 23 to 30, at Fort Churchill, where the birds were rather common. They frequent the scattered patches of

[a] Edin. New Phil. Journ. (new ser.), IX, p. 223, 1859.
[b] Nat. Hist. Bull. Univ. Iowa, II, p. 275, 1893
[c] Ibid., p. 275, 1893.

dwarfed spruce that grow in the small valleys and ravines intersecting the extensive expanse of precipitous ledges along Churchill River in the vicinity of the post. They undoubtedly nest among these spruces, but no nests attributable to the species were found. We heard no song, but they had a loud metallic chip which was audible and easily recognized at a distance of several rods. Young just from the nest, taken July 24 and 25, may be thus described: Upperparts dusky black, the feathers edged with deep buffy and brown, the black predominating on crown, the brown on hind neck, and the black and brown about equally divided on back; outer wing quills edged with deep buffy, inner with brown; tail feathers edged and tipped with whitish; sides of head and lower parts buffy; chest and side streaked with black, which is most conspicuous on sides of chest and forms a prominent malar stripe; upper throat grayish white, with fine dusky markings.

Several were seen in a thicket bordering upper Hayes River August 30, and the species was rather common in a fire-swept tract between Robinson Portage and the Echimamish September 14.

Zonotrichia leucophrys (Forst.). White-crowned Sparrow.

First seen at York Factory, where it was abundant and where a small series, including adults of both sexes and young birds not long from the nest, was taken July 12 to 14. About the post at Fort Churchill it was extremely abundant July 23 to 30, but after passing that point we saw no more of it until we had repassed the post on our return and were ascending Hayes River August 30, when we observed a number in a thicket bordering the river.

This species was first described by Forster from Severn River, Hudson Bay. He also received it from Albany Fort, where it was said to breed;[a] and Murray received it from Severn House.[b] It probably breeds throughout the northern wooded portions of Keewatin, being confined mainly to the Hudsonian zone.

Zonotrichia albicollis (Gmel.). White-throated Sparrow.

Abundant throughout the region between Norway House and York Factory. It was especially numerous in the extensive tracts which had been devastated by fire, where its simple but beautiful song lent a charm to the gloomy surroundings. A few were noted, one of which was collected, at York Factory in July; and on our return trip a few were seen at Oxford Lake September 11.

Murray reported the species from Hudson Bay,[b] and Clarke recorded an adult pair from Fort Churchill.[c] It was taken at Moose Factory

[a] Phil. Trans., LXII, p. 403, 426, 1772.

[b] Edin. New Phil. Journ. (new ser.), IX, p. 223, 1859

[c] Auk, VII, p. 322, 1890.

in 1881 by Walton Haydon. Its breeding range in Keewatin is probably nearly co-extensive with the forest, though it is rare in that part that lies in the Hudsonian zone.

Spizella monticola (Gmel.). Tree Sparrow.

First seen at York Factory, where the species was rather common and a young bird not long from the nest was taken July 12. It was abundant at Fort Churchill July 24 to 30, and we took a series at that point. Many were noted on the Barren Grounds, 50 miles south of Cape Eskimo, August 4 to 8. On our return trip several were seen at Duck Point, Playgreen Lake, September 19.

Murray received specimens from Severn House;[a] and a specimen collected by Dr. Rae at Repulse Bay is recorded in the British Museum Catalogue. The catalogue of birds in the U. S. National Museum collection contains the record of a specimen taken at Moose Factory in 1881 by Walton Haydon.

Spizella socialis (Wils.). Chipping Sparrow.

A few chipping sparrows were seen about the post at Norway House, one of which was collected. We also met with them about the post buildings at Oxford House, and saw one or two on an island in Knee Lake July 5. It was peculiarly pleasant to meet with this friendly little bird in these wild northern forests; and it was interesting to note that it showed no absence of that social trait to which its name is due, but was usually found nesting near dwellings

Spizella pallida (Swains.). Clay-colored Sparrow.

Nutting records three specimens, evidently taken on the lower Saskatchewan.[b] The species probably barely reaches southwestern Keewatin.

Junco hyemalis (Linn.). Slate-colored Junco.

A specimen was taken June 20 at Norway House, where the species was common; another was secured July 3, one of a number seen in the spruce and tamarack woods about Oxford House, and the species was again observed while we were ascending Steel River, September 1.

Specimens were received from Severn River by Forster, who, supposing the species to be undescribed, renamed it *Fringilla hudsonia*.[c] Two specimens are recorded from Fort Churchill by Clarke.[d] Specimens were taken at Moose Factory in 1881 by Walton Haydon.

Melospiza melodia (Wils.). Song Sparrow.

Common at Norway House and in the shrubbery about the clearing at Oxford House, and a number observed about Knee Lake July 5

---

[a] Edin. New Phil. Journ. (new ser.), IX, p. 223, 1859.
[b] Nat. Hist. Bull. Univ., Iowa, II, p. 275, 1893.
[c] Phil. Trans., LXII, pp. 406, 428, 1772.
[d] Auk, VII, p. 322, 1890.

and 6. Not noted between Knee Lake and York Factory on our inward trip, though we found the species rather common on Hill River when we ascended it early in September.

Melospiza lincolni (Aud.). Lincoln Sparrow.

Rather common July 13 to 16 at York Factory, where three specimens were collected.

Melospiza georgiana (Lath.). Swamp Sparrow.

A few were seen in the shrubby woods back of the post at Oxford House, and an adult was taken July 3. At York Factory, where the species was rather common, two young, not long from the nest, were taken July 13 and 16.

Passerella iliaca (Merr.). Fox Sparrow.

First noticed on the afternoon of July 10, when its beautiful song was heard in the willow thickets bordering Hayes River a few miles above York Factory. While at York Factory we found fox sparrows fairly common in willow thickets, and took a pair July 16.

Zamelodia ludoviciana (Linn.). Rose-breasted Grosbeak.

We heard the song of this bird while descending Red River, a few miles below Winnipeg, June 14. King took one near the north end of Lake Winnipeg in the summer of 1835.[a]

Progne subis (Linn.). Purple Martin.

Edwards figured a bird brought from Hudson Bay by Mr. Isham, calling it the 'Great American Martin'.[b] Linnæus based his description of *Hirundo subis* on Edwards's figure.

Petrochelidon lunifrons (Say). Cliff Swallow.

Forster recorded a specimen sent from Severn River as 'Hirundo No. 35.'[c] This is probably the earliest notice of the species, which was not formally described until many years afterward. Baird records a specimen taken at Moose Factory May 27, 1860;[d] and Barnston mentions the species as arriving at Martin Falls, Albany River, by May 15.[e]

Hirundo erythrogastra Bodd. Barn Swallow.

On the morning of August 13, while I was encamped at the mouth of a river on the Barren Grounds, about 25 miles south of Cape Eskimo, a barn swallow that had evidently been following the course of the stream flew past the camp. When it reached the Bay it turned southward and soon disappeared from sight down the coast.

[a] Narrative of a Journey to the Shores of the Arctic Ocean, II, p. 225, 1836.

[b] Nat. Hist., III, pl. 120, 1750.

[c] Phil. Trans., LXII, p. 408, 1772.

[d] Review of American Birds, p. 290, May, 1865.

[e] Edin. New Phil. Journ., XXX, p. 254, 1841.

Tachycineta bicolor (Vieill.). Tree Swallow.

A few seen on lower Red River June 15, and two at Norway House June 19. Common between Norway House and Oxford House, and many nests observed in deserted holes of woodpeckers in trees at the water's edge, June 23 to 30. Common also at Oxford House, and noted on Steel River July 9. On our return through their haunts early in September we saw none. The catalogue of birds in the U. S. National Museum collection records specimens taken at Moose Factory, and the species probably ranges throughout southern Keewatin.

Riparia riparia (Linn.). Bank Swallow.

A small colony was found on Red River, a few miles below Winnipeg, June 14, and a large one in a high clay bank on the shore of Oxford Lake, near Oxford House, June 30. Several colonies were seen July 10 on Hayes River, a few miles above York Factory. The catalogue of birds in the U. S. National Museum collection records a specimen taken at Moose Factory in 1881 by Walton Haydon.

Ampelis garrulus Linn. Bohemian Waxwing.

Three were observed by Alfred E. Preble in the stunted spruce woods near Fort Churchill July 25.

Tyrrell speaks of seeing a flock "in a grove of birch trees near the shore of Theitaga Lake, on their breeding grounds."[a] This lake is situated about 300 miles slightly north of west of Fort Churchill.

Ampelis cedrorum (Vieill.). Cedar Waxwing.

Recorded by Baird from Moose Factory, where it was collected by Drexler August 26, 1860.[b] Walton Haydon took specimens at the same place in 1881. Nutting found it breeding abundantly at Grand Rapids and Chemawawin, Saskatchewan.[c]

Lanius borealis Vieill. Northern Shrike.

Two specimens were taken at Fort Churchill, where the birds were rather common July 23 to 30, and one was taken and another noted near Painted Stone Portage September 14.

Forster recorded *Lanius excubitor*, referring to the present species, from Severn River;[d] Murray received specimens from Trout Lake and Severn House;[e] and Bell reported it from York Factory.[f] The species was collected at Moose Factory in 1881 by Walton Haydon.

Vireo olivaceus (Linn.). Red-eyed Vireo.

Abundant about Lake Winnipeg, at Norway House, and between Norway House and Oxford House. In the vicinity of Oxford House

---

[a] Ann. Rept. Can. Geol. Surv., 1896 (new ser.), IX, p. 165F (1897).

[b] Review of American Birds, p. 408, May, 1866.

[c] Nat. Hist. Bull. Univ. Iowa, II, p. 277, 1893.

[d] Phil. Trans., LXII, p. 386, 1772.

[e] Edin. New Phil. Journ. (new ser.), IX, p. 223, 1859.

[f] Rept. Prog. Can. Geol. Surv., 1878–79, App. VI, p. 68c. (1880).

its song was heard almost continually. After leaving that point we saw nothing more of the bird.

Vireo philadelphicus (Cass.). Philadelphia Vireo.

A peculiar vireo song heard on Hill River July 8 was probably the song of this species, but I was unable to secure the bird.

Baird recorded the species from Moose Factory, where specimens were taken June 2, 1860, by C. Drexler.[a] Walton Haydon took specimens at the same place in 1881.

Vireo solitarius (Wils.). Blue-headed Vireo.

One (a male) taken at Oxford House July 3, and one heard singing in a swamp bordering Knee Lake July 5.

Mniotilta varia (Linn.). Black and White Warbler.

Recorded by Turner from Moose Factory, where Drexler took specimens May 13 and 31, 1860.[b] Nutting reported one specimen from Grand Rapids, Saskatchewan.[c]

Helminthophila celata (Say). Orange-crowned Warbler.

One was taken in a willow thicket at York Factory July 16, and the species was again noted near Pine Lake September 13, and at Duck Point, Playgreen Lake, September 19.

Helminthophila peregrina (Wils.). Tennessee Warbler.

Two females taken at Oxford House July 3. The species was fairly common at York Factory, where specimens were taken July 13, 14, and 16. Baird recorded specimens from Fort George and Moose Factory, collected by C. Drexler in 1860,[d] and the species was taken at Moose Factory in 1881 by Walton Haydon.

Dendroica tigrina (Gmel.). Cape May Warbler.

Recorded by Turner from Moose Factory, where one was taken May 28, 1860, by Drexler.[e] The collection of the U. S. National Museum contains other specimens from the same place.

Dendroica æstiva (Gmel.). Yellow Warbler.

Rather common at Norway House, Oxford House, and York Factory. Specimens taken at these three points have a slightly darker crown than is usual in eastern examples, but are referable to the typical form.

Specimens taken at Fort Churchill many years ago are recorded by Clarke,[f] and the catalogue of birds in the U. S. National Museum col-

---

[a] Review of American Birds, p. 341, May, 1866.
[b] Proc. U. S. Nat. Mus., VIII, p. 236, 1885.
[c] Nat. Hist. Bull. Univ. Iowa, II, p. 277, 1893.
[d] Review of American Birds, p. 179, April, 1865.
[e] Proc. U. S. Nat. Mus., VIII, p. 237, 1885.
[f] Auk, VII, p. 322, 1890.

lection records a specimen taken at Moose Factory in 1881 by Walton Haydon. The species probably breeds throughout the wooded portions of Keewatin.

Dendroica coronata (Linn.). Myrtle Warbler.

One was seen, in company with kinglets and chickadees, in the spruce woods bordering Hill River, September 2.

According to Turner, Drexler took the species at Moose Factory.[a] Clark records specimens from Fort Churchill.[b]

Dendroica maculosa (Gmel.). Magnolia Warbler.

One or two seen at Norway House, and one taken at Oxford House July 3.

Turner recorded the species from Moose Factory,[a] where specimens were also taken in 1881 by Walton Haydon, and the British Museum Catalogue records one collected by G. Barnston on Albany River.

Dendroica castanea (Wils.). Bay-breasted Warbler.

One was taken at Oxford House July 3.

Turner recorded the species from Moose Factory, where it was taken June 2, 1860, by Drexler.[a]

Dendroica striata (Forst.). Black-poll Warbler.

At Oxford House, June 30 to July 4, they were rather common and a female was collected. On July 10 we noticed a pair in a thicket that bordered Hayes River, a few miles above York Factory, and on arriving at that post we again found the birds rather common. At Fort Churchill, where they were also common, we took another specimen July 24. It proved to be a young one not long from the nest and still in the speckled plumage.

This species was first described by Forster from Severn River,[c] and was recorded by Murray from Trout Lake.[d]

Dendroica blackburniæ (Gmel.). Blackburnian Warbler.

This species is recorded by Murray from Severn House and Trout Lake under the name *Sylvicola parus*.[d]

Dendroica palmarum (Gmel.). Palm Warbler.

One was seen on a small willow-covered island in Hill River September 4.

Clarke recorded an adult specimen from Fort Churchill.[b]

---

[a] Proc. U. S. Nat. Mus., VIII, p. 237, 1885.

[b] Auk, VII, p. 322, 1890.

[c] Phil. Trans., LXII, pp. 406, 428, 1772.

[d] Edin. New. Phil. Journ. (new ser.), IX, p. 222, 1859.

Dendroica palmarum hypochrysea Ridgw. Yellow Palm Warbler.

Turner recorded a specimen taken at Moose Factory in July, 1860, by Drexler,[a] and Baird records one (under the name *palmarum*) taken at Fort George, on the east coast of James Bay, in 1861.[b]

Seiurus aurocapillus (Linn.). Oven-bird.

A specimen taken at Moose Factory by Walton Haydon in the summer of 1881 is recorded in the catalogue of birds in the U. S. National Museum.

Seiurus noveboracensis notabilis (Ridgw.). Grinnell Water-Thrush.

We saw our first water-thrush at Painted Stone Portage. It was close to the edge of the water and was running through the undergrowth which fringed the foot of a cliff. On the afternoon of the same day, June 26, we noted another at Robinson Portage, and on June 30 a third at Oxford Lake. When we arrived at Oxford House we found the species rather common, and from there to York Factory, as we descended the rivers, its sprightly song was heard daily. A pair seen at a portage on Hill River July 7 were feeding young just from the nest. Three specimens were taken in the marshy woods about York Factory, where the species was common July 11 to 17, and one was taken August 8 by Alfred E. Preble on Churchill River about 15 miles above Fort Churchill. These prove to be intermediate between *noveboracensis* and *notabilis*, but nearer to *notabilis*.

Murray received the species from Severn House.[c]

Geothlypis philadelphia (Wils.). Mourning Warbler.

Nutting records a specimen from the lower Saskatchewan (exact locality not stated) that is apparently intermediate in characters between *tolmiei* and *philadelphia*, but seems nearer to *tolmiei*.[d] Thompson, on the authority of Macoun, records the mourning warbler from Waterhen River and Swan Lake,[e] Manitoba, which are not far to the southward of where Nutting collected. As the species has been recorded also from various points to the southeast of Keewatin, there is little question that it breeds in the southern part of the Province.

Wilsonia pusilla (Wils.). Wilson Warbler.

Rather common in the undergrowth bordering a swamp at Robinson Portage June 27. Several were seen at York Factory July 10 to 17, one being taken on July 14.

---

[a] Proc. U. S. Nat. Mus., VIII, p. 237, 1885.
[b] Review of American Birds, p. 208, April, 1865.
[c] Edin. New Phil. Journ. (new ser.), IX, p. 222, 1859.
[d] Nat. Hist. Bull. Univ. Iowa, II, p. 278, 1893.
[e] Proc. U. S. Nat. Mus., XIII, p. 622, 1890.

The British Museum Catalogue records a specimen from Hudson Bay collected by Captain Herd. A specimen taken at Moose Factory in the summer of 1881 by Walton Haydon is in the U. S. National Museum collection.

Wilsonia canadensis (Linn.). Canadian Warbler.

The catalogue of birds in the U. S. National Museum collection records a specimen taken at Moose Factory in the summer of 1881 by Walton Haydon. Nutting records the species from Grand Rapids, Saskatchewan.[a]

Setophaga ruticilla (Linn.). Redstart.

Baird recorded a specimen taken at Moose Factory by Drexler,[b] and Thompson quotes Hutchins's manuscript to the effect that one was shot at Fort Albany.[c]

Anthus pensilvanicus (Lath.). Pipit.

Rather common July 24 to 30 on the rocky hills at Fort Churchill, where a pair were taken. A large flock was seen on lower Hayes River as we were ascending it August 29.

Murray recorded the species from Hudson Bay.[d]

Olbiorchilus hiemalis (Vieill.). Winter Wren.

The catalogue of birds in the U. S. National Museum collection records a specimen taken at Moose Factory by Walton Haydon.

Cistothorus stellaris (Licht.). Short-billed Marsh Wren.

A male was taken in a wet meadow at Norway House June 20. From its actions it probably had a nest in the vicinity, but despite a careful search none was found.

Sitta carolinensis Lath. White-breasted Nuthatch.

An adult female taken at Fort Churchill many years ago is recorded by Clarke.[e]

Sitta canadensis Linn. Red-breasted Nuthatch.

Heard on the Echimamish June 25, and a few seen on an island in Knee Lake on the forenoon of July 5.

In his narrative Dr. Rae speaks of taking one above Broad River, between York Factory and Fort Churchill.[f]

Parus atricapillus Linn. Chickadee.

Baird recorded a specimen taken at Moose Factory.[g] A chickadee in the U. S. National Museum that was taken at the same place by

---

[a] Nat. Hist. Bull. Univ. Iowa, II, p. 279, 1893.

[b] Review of American Birds, p. 256, May, 1865.

[c] Proc. U. S. Nat. Mus., XIII, p. 624, 1890.

[d] Edin. New Phil. Journ. (new ser.), IX, p. 222, 1859.

[e] Auk, VII, p. 322, 1890.

[f] Narrative of An Expedition to the Shores of the Arctic Sea, p. 11, 1850.

[g] Review of American Birds, p. 81, July, 1864.

Walton Haydon is referable to the eastern form. Forster recorded *Parus atricapillus* from Fort Albany.[a]

Parus atricapillus septentrionalis (Harris). Long-tailed Chickadee.

One taken on the lower Echimamish June 24. Nutting reported a specimen from Grand Rapids, Saskatchewan.[b]

Parus hudsonicus Forst. Hudsonian Chickadee.

We first met with this species on the Echimamish June 24. We noted it again at Robinson Portage three days later, and found it common at Oxford House, where we secured a male July 3. We saw several on an island in Knee Lake July 5, and a number near York Factory July 13, collecting two on the latter date. On our return we saw several on Hill River September 3.

The species was first described by Forster from Severn River.[c] It is recorded from Fort Churchill by Clarke,[d] and from Moose Factory by Rhoads.[e]

Regulus satrapa Licht. Golden-crowned Kinglet.

One was seen on the lower Echimamish June 24.

Regulus calendula (Linn.). Ruby-crowned Kinglet.

I saw one at Norway House June 17 and took a specimen on the Echimamish June 24. It was common at Oxford House June 30 to July 4. One was observed as we were ascending Hill River September 1.

Forster recorded a specimen probably sent from Severn River;[a] Bell recorded the species from York Factory;[f] and Clarke an adult from Fort Churchill.[d] The catalogue of birds in the U. S. National Museum collection records a specimen taken at Moose Factory in the summer of 1881 by Walton Haydon.

Hylocichla fuscescens salicicola Ridgw. Willow Thrush.

The characteristic 'veery' call note and song were heard several times, and the singer was seen once as we floated down Red River between Winnipeg and West Selkirk June 14. None were taken, but specimens from the region seem referable to the western form.

Hylocichla aliciæ (Baird). Gray-cheeked Thrush.

Not met with until we reached York Factory, where a female and two young just from the nest were taken in a dense willow thicket

---

[a] Phil. Trans., LXII, p. 407, 1772.
[b] Nat. Hist. Bull. Univ. Iowa, II, p. 279, 1893.
[c] Phil. Trans., LXII, p. 408, 430, 1772.
[d] Auk, VII, p. 322, 1890.
[e] Auk, X, p. 328, 1893.
[f] Rept. Prog. Can. Geol. Surv., 1882–3–4, App. III, p. 54DD (1885).

July 13. The young birds may be described as follows: Back and head dark olive-brown, each feather tipped with dusky and with a longitudinal spot of brown; rump and upper tail-coverts brownish spotted with rusty; lower parts white, slightly tinged on breast and sides with buffy, each feather tipped with a dusky bar, those on breast heavily marked, the marking decreasing in size posteriorly; throat almost unmarked; cheeks grayish, spotted with dusky; wings and tail olive-brown, the wing quills lighter on outer edges.

I again met with the species July 24 at Fort Churchill, where I saw several in stunted spruce woods.

Hylocichla ustulata swainsoni (Cab.). Olive-backed Thrush.

The song of this species was heard at Bull Head Point, Lake Winnipeg, on the morning of June 16. The birds were rather common at Norway House, and were seen or heard daily between there and Oxford House. They were common at Oxford House, and a specimen was taken at that point. While descending the streams between Oxford House and York Factory we found them abundant; every wooded islet in the lakes seemed to be the home of a pair, and wherever we camped we heard their songs, which began soon after midnight. A nest found in a bush overhanging Jack River, between Knee and Swampy lakes, July 6, contained eggs on the point of hatching. At York Factory, where we took two specimens, the species was apparently less abundant, and beyond that point we did not meet with it.

Baird recorded a specimen collected at Moose Factory in July, 1860, by Drexler.[a]

Hylocichla guttata pallasi (Cab.). Hermit Thrush.

Nutting found the hermit thrush abundant at Grand Rapids, Saskatchewan, in the summer of 1891.[b] Though it doubtless occurs in southern Keewatin, I find no published records of such occurrence.

Merula migratoria (Linn.). Robin.

Found throughout the region, but seldom seen elsewhere than in the vicinity of the posts, where, however, they were rather common. Many old and young were seen at Fort Churchill during the latter part of July. On our return trip we noted the species on Hayes River August 30, Steel River August 31, Hill River September 4, and between Oxford and Windy lakes September 12.

Forster recorded it from Severn River,[c] and Murray from Severn House and Trout Lake.[d]

---

[a] Review of American Birds, p. 21, June, 1864.

[b] Nat. Hist. Bull. Univ. Iowa, II, p. 279, 1893.

[c] Phil. Trans., LXII, p. 399, 1772.

[d] Edin. New Phil. Journ. (new ser.), IX, p. 222, 1859.

Saxicola œnanthe leucorhoa (Gmel.). Greenland Wheatear.

James Clark Ross recorded a specimen of *Sylvia œnanthe* obtained at Felix Harbor, Gulf of Boothia,[a] which is probably referable to the race recently recognized by Stejneger.[b] The British Museum Catalogue records from Albany River an adult male *œnanthe*, collected by Barnston, which is also probably referable to the Greenland race. If, as is probable, this bird inhabits the country to the north of Hudson Bay, its most natural route of migration would seem to be along the borders of the Bay, and it is not unlikely that it is a regular breeder about the northern shores.

Sialia sialis (Linn.). Bluebird.

The U. S. National Museum collection contains a specimen taken at Moose Factory in the summer of 1881 by Walton Haydon.

---

[a] Appendix to Ross's Second Voyage, p. xxvi, 1835.
[b] Proc. U. S. Nat. Mus., XXIII, p. 476, 1901.

We made a small collection of the species of frogs noted during the trip, and extended their previously recorded ranges. A gradual shortening of the hind legs as the northern limit of the ranges of these species is approached seems to be the rule, and is evidently correlated with the shorter period of activity. In addition to this collection I include references to several species of frogs and salamanders recorded by Cope from the region about James Bay, and from the mouth of Nelson River.

Rana pipiens Gmel. Leopard Frog.

We noted this species at two points—in the meadows near Norway House, where it was fairly common, and at Sea Falls, about 20 miles farther north, where we saw several individuals while we were making a portage. We collected two specimens at each place.

Rana palustris Le Conte. Le Conte Leopard Frog.

Cope records specimens from James Bay, collected by C. Drexler.[a] These specimens, which are preserved in the U. S. National Museum, have been re-examined in connection with this report.

Rana cantabrigensis latiremis Cope. Northern Wood Frog.

We collected a series of wood frogs that includes specimens from Taft's Fishery (on Great Playgreen Lake), Norway House, York Factory, and Fort Churchill. According to measurements, this series is referable to *latiremis* as restricted by Howe.[b]

Eleven well-grown specimens from Norway House average: Length of body (nose to anus) 46.3; femur (measured from central line) 19.1; tibia 19.9. Seven specimens from York Factory average: Length of body 43.9; femur 18.7; tibia 17.7. A single specimen taken at Fort Churchill measures: Length of body 50; femur 19; tibia 19.

Cope records *cantabrigensis* from the mouth of Nelson River, and from James Bay.[c] The James Bay specimens have not been critically examined, and may be nearer to *cantabrigensis* than to *latiremis*.

This frog was by far the most abundant species throughout the region between Lake Winnipeg and York Factory, and many were

[a] Batrachia of North America (Bull. 34, U. S. Nat. Mus.), p. 409, 1889.
[b] Proc. Boston Soc. Nat. Hist., vol. 28, No. 14, p. 373, Feb., 1899.
[c] Batrachia of North America (Bull. 34, U. S. Nat. Mus.), p. 437, 1889.

taken in our traps as well as by hand. At Fort Churchill, though frogs were reported to be frequently heard in the marshes, we were able to obtain but one specimen. This is characterized by extremely short legs, and differs further from any others collected in being heavily vermiculated on the sides with black. Most of the specimens collected have a grayish median dorsal stripe. The color from life of a York Factory specimen is as follows: Back grayish green, sparingly spotted with black, and with a greenish-white median dorsal stripe; beneath greenish white, darkest on sides.

Rana septentrionalis Baird.

Specimens recorded by Cope under the name *Rana cantabrigensis evittata* from Moose River, Ontario, are referable to *Rana septentrionalis*, as stated by Howe.[a]

Chorophilus septentrionalis Boulenger. Northern Chorophilus.

Chorophilus septentrionalis Boulenger, Cat. Batrachia Salientia Brit. Mus., p. 335, 1882.

We found this species throughout the region between Lake Winnipeg and York Factory, and took specimens at the following localities: Taft's Fishery (Great Playgreen Lake), 1; Norway House, 3; Oxford House, 1; York Factory, 3.

These specimens appear to be referable to *septentrionalis*, the type locality of which is Great Bear Lake, as the tibio-tarsal joint does not reach the tympanum when the leg is stretched forward. The following brief description of the color was taken from a live specimen from Oxford House: Body light green above, greenish white beneath; body stripes bronzy lavender; tympanum brownish; hind legs light green above, flesh color beneath.

Amblystoma jeffersonianum platineum Cope.

Cope records a specimen collected by C. Drexler on Moose River.[b]

Chondrotus microstomus Cope.

Cope records a specimen from Hudson Bay, probably from James Bay, collected by F. W. Hayden[c] (probably mistake for W. Haydon).

Plethodon cinereus Green.

Cope records four specimens from 'Hudson's Bay Territory,' probably from the region about James Bay, collected by C. Drexler.[d]

---

[a] Proc. Boston Soc. Nat. Hist., vol. 28, No. 14, p. 374, Feb., 1899.

[b] Batrachia of North America (Bull. 34, U. S. Nat. Mus.), p. 94, 1889.

[c] Ibid., p. 103, 1889.

[d] Ibid., p. 135, 1889.

INDEX.

[Names of new species in bold-face type.]

O

www.ingramcontent.com/pod-product-compliance
Lightning Source LLC
Chambersburg PA
CBHW020851210326
41598CB00018B/1634

* 9 7 8 3 7 4 3 3 2 3 6 0 5 *